Biology: Theory and Applications

Biology: Theory and Applications

Editor: Nicola Norris

R CALLISTO REFERENCE

www.callistoreference.com

Callisto Reference,
118-35 Queens Blvd., Suite 400,
Forest Hills, NY 11375, USA

Visit us on the World Wide Web at:
www.callistoreference.com

ISBN: 978-1-64116-142-8 (Hardback)

Trademark Notice: Registered trademark of products or corporate names are used only for explanation and identification without intent to infringe.

Cataloging-in-Publication Data

Biology : theory and applications / edited by Nicola Norris.
    p. cm.
Includes bibliographical references and index.
ISBN 978-1-64116-142-8
1. Biology. I. Norris, Nicola.
QH307.2 .B56 2019
570--dc23

Table of Contents

Preface

Biology is the branch of science that is concerned with the study of life and living organisms. The fundamental units of cells, genes and evolution are central to the development of biology. All structural, physiological, evolutionary, systematic, ecological and environmental aspects of organisms are explored in this domain. The sub-domains of biology are classified into biochemistry, molecular biology, cellular biology, physiology and ecology. The evolutionary relationships of living organisms are analyzed through phylogenetics, cladistics and phenetics. This book attempts to understand the multiple branches that fall under the discipline of biology and how such concepts have practical applications. The various advancements in biology are glanced at and their applications as well as ramifications are looked at in detail. This book will prove to be immensely beneficial to students and researchers in this field.

The information contained in this book is the result of intensive hard work done by researchers in this field. All due efforts have been made to make this book serve as a complete guiding source for students and researchers. The topics in this book have been comprehensively explained to help readers understand the growing trends in the field.

I would like to thank the entire group of writers who made sincere efforts in this book and my family who supported me in my efforts of working on this book. I take this opportunity to thank all those who have been a guiding force throughout my life.

Editor

Protein abundance of AKT and ERK pathway components governs cell type-specific regulation of proliferation

Lorenz Adlung[1,†] (iD), Sandip Kar[2,3,4,†], Marie-Christine Wagner[1,†], Bin She[1,†], Sajib Chakraborty[1], Jie Bao[5], Susen Lattermann[1], Melanie Boerries[5,6,7], Hauke Busch[5,6,7] (iD), Patrick Wuchter[8,9], Anthony D Ho[8], Jens Timmer[10], Marcel Schilling[1] (iD), Thomas Höfer[2,3,*] & Ursula Klingmüller[1,11,**] (iD)

Abstract

Signaling through the AKT and ERK pathways controls cell proliferation. However, the integrated regulation of this multistep process, involving signal processing, cell growth and cell cycle progression, is poorly understood. Here, we study different hematopoietic cell types, in which AKT and ERK signaling is triggered by erythropoietin (Epo). Although these cell types share the molecular network topology for pro-proliferative Epo signaling, they exhibit distinct proliferative responses. Iterating quantitative experiments and mathematical modeling, we identify two molecular sources for cell type-specific proliferation. First, cell type-specific protein abundance patterns cause differential signal flow along the AKT and ERK pathways. Second, downstream regulators of both pathways have differential effects on proliferation, suggesting that protein synthesis is rate-limiting for faster cycling cells while slower cell cycles are controlled at the G1-S progression. The integrated mathematical model of Epo-driven proliferation explains cell type-specific effects of targeted AKT and ERK inhibitors and faithfully predicts, based on the protein abundance, anti-proliferative effects of inhibitors in primary human erythroid progenitor cells. Our findings suggest that the effectiveness of targeted cancer therapy might become predictable from protein abundance.

Keywords 32D-EpoR; BaF3-EpoR; CFU-E; MAPK; PI3K
Subject Categories Cell Cycle; Quantitative Biology & Dynamical Systems; Signal Transduction

Introduction

Eukaryotic cells use a limited number of signal transduction pathways to integrate information from extracellular stimuli and to regulate cellular decisions such as proliferation, differentiation, and apoptosis. In particular, the PI3K/AKT and Ras/MEK/ERK pathways have been implicated in the control of cell growth and proliferation in many different cell types (Saez-Rodriguez *et al*, 2015). The key components of these pathways are highly conserved, which suggests that they form generic proliferation-control modules that function in a specialized way in different cell types. Indeed, previous studies suggest that the regulation of the AKT and the ERK pathway crucially depends on the cellular context (McCubrey *et al*, 2011).

Both pathways are activated by multiple growth factors and cytokines such as the hormone erythropoietin (Epo), which is the prime regulator of erythropoiesis. Epo is essential for survival, proliferation, and differentiation of erythroid progenitor cells and thereby facilitates continuous renewal of mature erythrocytes (Koury & Bondurant, 1990). The cognate Epo receptor (EpoR) is present on the cell surface of erythroid progenitor cells as a preformed homodimer (Livnah *et al*, 1999). At the stage of

1 Division of Systems Biology of Signal Transduction, German Cancer Research Center (DKFZ), Heidelberg, Germany
2 Division of Theoretical Systems Biology, German Cancer Research Center (DKFZ), Heidelberg, Germany
3 BioQuant Center, University of Heidelberg, Heidelberg, Germany
4 Department of Chemistry, Indian Institute of Technology, Mumbai, India
5 Systems Biology of the Cellular Microenvironment Group, IMMZ, ALU, Freiburg, Germany
6 German Cancer Consortium (DKTK), Freiburg, Germany
7 German Cancer Research Center (DKFZ), Heidelberg, Germany
8 Department of Medicine V, University of Heidelberg, Heidelberg, Germany
9 Institute for Transfusion Medicine and Immunology, University of Heidelberg, Mannheim, Germany
10 Center for Biological Signaling Studies (BIOSS), Institute of Physics, University of Freiburg, Freiburg, Germany
11 Translational Lung Research Center (TLRC), Member of the German Center for Lung Research (DZL), Heidelberg, Germany
*Corresponding author. E-mail: t.hoefer@dkfz.de
**Corresponding author. E-mail: u.klingmueller@dkfz.de
†These authors contributed equally to this work

colony-forming unit-erythroid (CFU-E), primary cells exhibit highest EpoR levels and are thus most responsive to Epo (Wu *et al*, 1995).

Epo-induced signaling comprises besides the activation of STAT5 (Klingmüller *et al*, 1996) the induction of PI3K and MAPK signal transduction (Miura *et al*, 1994; Haseyama *et al*, 1999). The activation of MAPK pathways has been associated with erythroblast enucleation and stress-induced erythropoiesis (Tamura *et al*, 2000; Schultze *et al*, 2012) while PI3K signaling has been shown to control maturation of erythroid progenitor cells by promoting cell survival and proliferation (Myklebust *et al*, 2002; Bouscary *et al*, 2003). The open question remains how these pathways are integrated to control proliferation.

PI3K/AKT is connected to the initiation of translation by the mammalian target of rapamycin (mTOR) in proliferating erythroid progenitor cells (Grech *et al*, 2008). Phosphorylation of the ribosomal protein S6, a downstream target of mTOR, is a crucial step for protein synthesis, and thus cell growth. If cells are treated with the mTOR inhibitor rapamycin, they are considerably smaller than untreated cells (Fingar *et al*, 2002, 2004). Lately, an mTOR-independent S6 activation mechanism through ERK and RSK (Roux *et al*, 2007) was found. S6 activation, and thus protein synthesis, is required for cell growth and proliferation of reticulocytes (Knight *et al*, 2014). Factor-induced proliferation is a complex process that can be divided into two steps (Smith & Martin, 1973). First, cells integrate growth factor signals and grow in G1 phase of the cell cycle if enough nutrients are available (Pardee, 1974). Second, if a critical mass is reached and the restriction point is crossed, cells progress through the cell cycle, synthesize DNA, and finally undergo cytokinesis to double their number (Jones & Kazlauskas, 2000).

It has been shown that Epo regulates proliferation of erythroid progenitor cells and modulates cell cycle regulators (Dai *et al*, 2000; Bouscary *et al*, 2003; Sivertsen *et al*, 2006). Epo stimulation of erythroblasts results in a rapid upregulation of Cyclin-D2 as well as Nupr1, Gstpt1, Egr1, Nab2 and a downregulation of Cyclin-G2 and p27 (Fang *et al*, 2007). The regulation of cell cycle progression is rather complex, as it is regulated by multiple feedforward and feedback loops (Ferrell, 2013), and for example, the negative regulators Cyclin-G2 and p27 do not necessarily act in a coordinated manner (Le *et al*, 2007). Mathematical models of the cell cycle in mammalian cells have been developed that describe the change of cyclins and CDKs with time (Yao *et al*, 2008; Alfieri *et al*, 2009); however, only lately mechanistic and dynamic links from signaling to cell cycle progression were established (Mueller *et al*, 2015).

The ample knowledge on molecular mechanisms contributing to the regulation of erythropoiesis has been facilitated, on the one hand, by the availability of factor-dependent, immortalized hematopoietic cell lines from mice. For example, the interleukin (IL) 3-dependent cell lines BaF3 of lymphoid origin (Palacios & Steinmetz, 1985) and 32D of myeloid origin (Greenberger *et al*, 1983) have been utilized for decades to unravel structure–function relationship of cytokine receptors such as the EpoR (Wang *et al*, 1993; Klingmüller *et al*, 1996). Exogenous expression of the EpoR renders these cell lines responsive to Epo and enables proliferation in the presence of Epo (D'Andrea *et al*, 1989). Due to their growth properties, BaF3 cells are currently widely used in kinase drug discovery and represent a reliable cellular system to access kinase activity (Jiang *et al*, 2005; Moraga *et al*, 2015). On the other hand, primary erythroid progenitor cells from mice (mCFU-E) are readily available

from fetal liver or bone marrow, and methods for their cultivation have been established (Rich & Kubanek, 1976; Landschulz *et al*, 1989). For the human system, a protocol has been devised (Broudy *et al*, 1991; Miharada *et al*, 2006) to expand and differentiate human erythroid progenitor (hCFU-E) cells from CD34[+] cells mobilized into the peripheral blood of healthy donors. With this strategy, sufficient material of hCFU-E can be obtained to confirm in functional studies the clinical relevance of observations.

Here, we present a mathematical model that links Epo-induced activation of AKT, ERK, and S6 to cell cycle progression and proliferation in the context of murine erythroid progenitor cells and murine hematopoietic cell lines exogenously expressing the EpoR. We uncover that the cell type-specific protein abundance is sufficient to explain alterations in the dynamics of the signaling pathways. Further, we demonstrate how mathematical modeling can establish a mechanistic connection from signaling to cell growth, cell cycle progression, and proliferation upon Epo stimulation and inhibitor treatment. We show that in murine erythroid progenitor cells, proliferation is primarily controlled by the regulation of cell growth, whereas regulation of cell cycle progression is the major determinant of proliferation of the murine hematopoietic cell lines as well as in human erythroid progenitor cells.

Results

Cell type-specific regulation of proliferation and signaling by erythropoietin

To quantitatively assess Epo-induced proliferative responses in murine primary erythroid progenitor cells at the colony-forming unit-erythroid stage (mCFU-E) and in the immortalized murine cell line BaF3 exogenously expressing the EpoR (BaF3-EpoR), we incubated the cells in the presence of different Epo doses and measured as a readout for proliferation DNA synthesis by thymidine incorporation. mCFU-E cells showed a higher sensitivity toward Epo with an EC_{50} of 0.26 ± 0.02 U/ml Epo as compared to 0.55 ± 0.04 U/ml Epo for BaF3-EpoR cells (Fig 1A). At saturating doses of 50 U/ml Epo, mCFU-E cells doubled their number within 13.1 h and BaF3-EpoR cells within 18.7 h (Fig 1B). We determined the size of unstimulated cells by imaging flow cytometry and observed average diameters of mCFU-E cells and BaF3-EpoR cells of 11 and 13.8 μm, respectively (Fig 1C). mCFU-E cells were smaller and more heterogeneous in size. The smaller average size of mCFU-E cells correlated with a higher sensitivity toward Epo and a shorter doubling time compared with BaF3-EpoR cells.

To investigate the Epo-dependent regulation of cell growth and proliferation, we examined the Epo-induced activation of AKT and ERK pathways in mCFU-E and BaF3-EpoR cells by immunoblotting. In the first step, the protein abundance and dynamics of phosphorylation of EpoR, AKT, and ERK in response to Epo stimulation were qualitatively assessed. To study the expression and phosphorylation of the EpoR and AKT, mCFU-E cells were stimulated with 2.5 U/ml Epo, while BaF3-EpoR cells were stimulated with 5 U/ml Epo to account for the at least twofold difference in sensitivities toward Epo (Fig 1A) and the observation that responses in mCFU-E cells already saturated at 2.5 U/ml Epo, whereas BaF3-EpoR required more than 5 U/ml Epo (Appendix Fig S1). As shown in Fig 1D, top

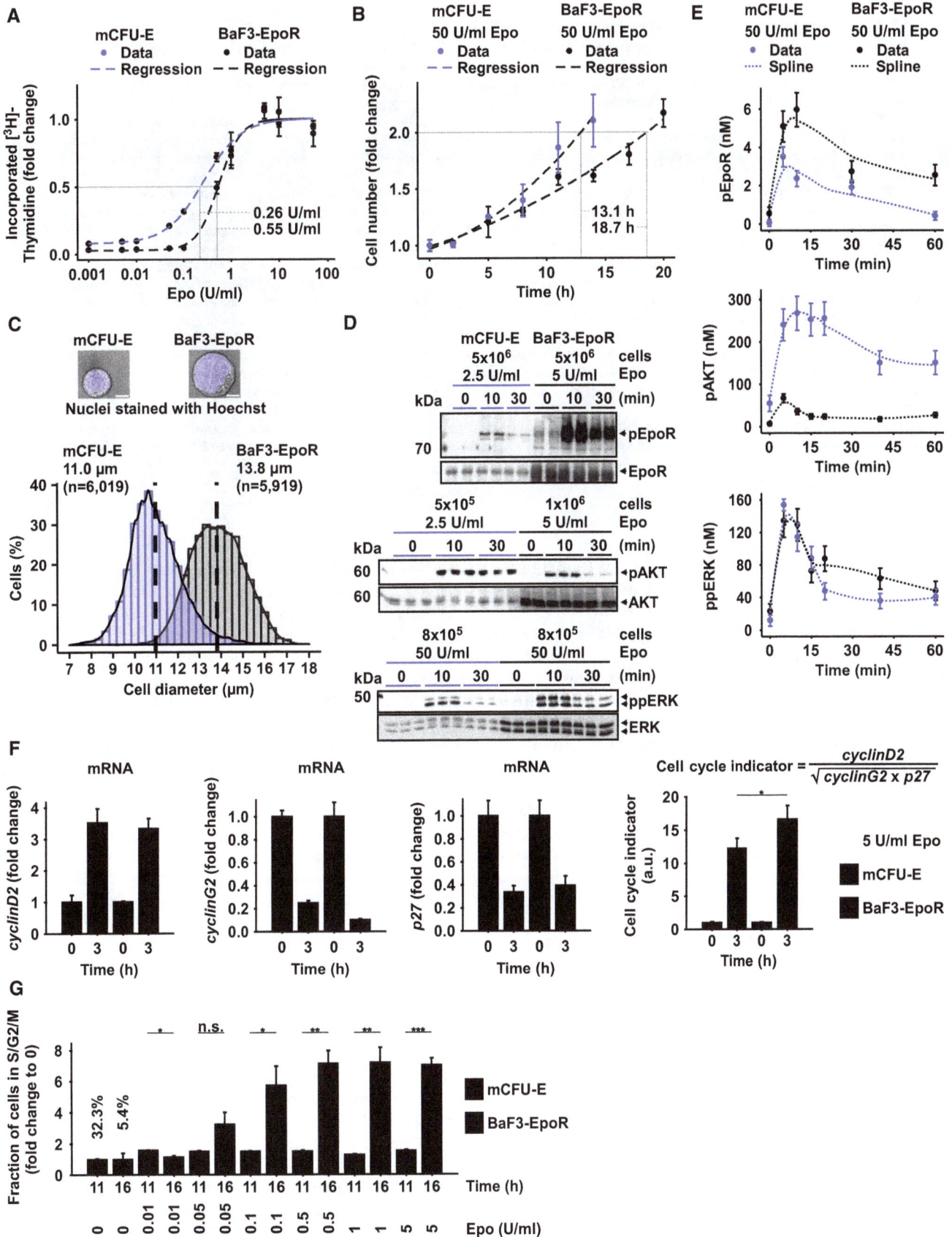

Figure 1.

◀

Figure 1. Characterization of Epo-induced proliferation and signaling.

A   DNA content of mCFU-E and BaF3-EpoR cells in response to different Epo concentrations. [³H]-thymidine incorporation was measured after 14 h (mCFU-E) or 38 h (BaF3-EpoR). Data represented as mean ± standard deviation, N = 3. Lines represent sigmoidal regression. EC_{50} values are given.

B   Cell doubling with time in response to 50 U/ml Epo. Cell numbers were determined by manual counting using trypan blue exclusion assay. Data represented as mean ± standard deviation, N = 3. Lines represent exponential regression. Doubling time is indicated.

C   Size determinations of mCFU-E and BaF3-EpoR cells. Exemplary fluorescence microscopy pictures upon Hoechst staining for nucleus visualization with 60× objective. The bar represents 10 μm distance (upper panel). Cell diameter was measured by imaging flow cytometry. Cytoplasm was stained with calcein, and nuclei were stained with DRAQ5. Probability density function of size distribution with indicated mean diameter of mCFU-E and BaF3-EpoR cells. All cells were growth factor-deprived and unstimulated.

D   Epo-induced phosphorylation of EpoR, AKT, and ERK of mCFU-E and BaF3-EpoR cells. Above each panel, the number of growth factor-deprived cells examined per time point is indicated as well as the concentration of Epo applied for stimulation. To the left, the position of the molecular weight marker is indicated in kDa and arrowheads indicate the position of the protein of interest. For the detection of the EpoR, 5×10^6 mCFU-E cells were stimulated with 2.5 U/ml Epo and 5×10^6 BaF3-EpoR cells were stimulated with 5 U/ml Epo. Cells were lysed, subjected to immunoprecipitation with anti-EpoR, and were analyzed by immunoblotting using either anti-pTyr (pEpoR) or anti-EpoR (EpoR) antibodies. For the detection of AKT, mCFU-E cells were stimulated with 2.5 U/ml Epo and BaF3-EpoR cells were stimulated with 5 U/ml Epo. Per time point, cellular lysates equivalent to 5×10^5 mCFU-E cells and 1×10^6 BaF3-EpoR cells were analyzed by immunoblotting using anti-pAKT and anti-AKT antibodies. For the detection of ERK, cells were stimulated with 50 U/ml Epo. Per time point, cellular lysates equivalent to 8×10^5 cells were analyzed by immunoblotting. Immunoblot detection was performed with chemiluminescence utilizing a CCD camera device (ImageQuant).

E   Absolute concentrations of pEpoR, pAKT, and ppERK in mCFU-E and BaF3-EpoR cells with time in response to 50 U/ml Epo. Experimental data of a representative experiment are depicted with filled circles, and dashed lines represent splines. Error bars indicate standard deviation estimated by an error model. N = 1.

F   Expression of cell cycle indicator genes of mCFU-E and BaF3-EpoR cells in response to 5 U/ml Epo. Genes were selected based on microarray analysis. Experimental data are shown as fold change to unstimulated cells with mean ± standard deviation, N = 3. Welch modified two-sample t-test, *P < 0.05.

G   Fold change of fractions of cells in S/G2/M phase of the cell cycle with respect to 0 U/ml Epo. Growth factor-deprived cells were stimulated with indicated Epo doses for given time. Fractions in sub-G1, G1, S, G2/M were determined by propidium iodide (PI) staining for DNA content. Data represented as mean ± standard deviation, N = 3. Welch modified two-sample t-test, n.s. = not significant, *P < 0.05, **P < 0.01, ***P < 0.005. The percentage of cells in S/G2/M phase in unstimulated cells is additionally indicated.

Source data are available online for this figure.

panel, the total expression level of the EpoR and the extent of EpoR phosphorylation were higher in BaF3-EpoR cells compared with mCFU-E cells. With regard to AKT, we consistently observed much lower levels of Epo-induced AKT phosphorylation in BaF3-EpoR cells. Therefore, to obtain reproducible results for Epo-induced phosphorylation of AKT in both cell types, we examined per time point 1×10^6 BaF3-EpoR cells and 5×10^5 mCFU-E cells. These studies showed that although the apparent abundance of total AKT was higher in BaF3-EpoR cells, the Epo-induced phosphorylation of AKT was higher and more sustained in mCFU-E cells compared with BaF3-EpoR cells (Fig 1D, middle panel). As an indicator for AKT activation, we focused on the analysis of Ser473 phosphorylation that is predictive for full kinase activation (Alessi *et al*, 1996; Scheid *et al*, 2002; Sarbassov *et al*, 2005) since we observed that it correlates with Thr308 phosphorylation in an Epo dose-dependent manner (Appendix Fig S2). We previously noted that cytokine receptors activated the MAP kinase pathway to a much lower extent compared with receptor tyrosine kinases (Iwamoto *et al*, 2016) and showed that 50 U/ml Epo is required to achieve an ERK phosphorylation degree of at least 10% in mCFU-E cells (Schilling *et al*, 2009). Therefore, we stimulated mCFU-E cells and BaF3-EpoR cells with 50 U/ml Epo to reliably examine ERK phosphorylation. As depicted in Fig 1D, bottom panel, higher ERK protein levels as well as elevated levels of ERK phosphorylation were observed in BaF3-EpoR cells, but overall both cell types exhibited a comparable transient ERK phosphorylation dynamics.

To quantitatively assess the cell type-specific differences, we determined the specific concentrations of key signaling molecules. We assumed spherical geometry of the cells and calculated the cytoplasmic volumes and cell surface areas by confocal microscopy (Table 1). For each protein of interest, calibrator proteins of known concentration were used to determine the absolute number of molecules per cell for mCFU-E, BaF3-EpoR, and 32D-EpoR cells, a cell

line that was used for validation experiments (Table 1). BaF3-EpoR cells exhibited a ten times higher density of EpoR molecules on their cell surface (26 molecules/μm^2) than mCFU-E cells (2.6 molecules/μm^2). The accurate quantification of total molecules per cell and the correction for the difference in cellular volume showed that indeed the concentration of total ERK was higher in BaF3-EpoR (2964 ± 166 nM) than in mCFU-E cells (1140 ± 64 nM), whereas the concentration of total AKT was comparable (510 ± 62 nM in BaF3-EpoR cells and 407 ± 16 nM in mCFU-E cells).

To quantitatively examine the dynamics of Epo-induced signal transduction in mCFU-E and BaF3-EpoR cells, we used randomized sample loading in combination with quantitative immunoblotting to determine in a time-resolved manner the phosphorylation of EpoR, AKT, and ERK in both cell types. In our previous studies (Becker *et al*, 2010), at maximum 75% of receptor dimers on the cell surface were bound to Epo. Therefore, we assumed the EpoR phosphorylation degree does not exceed 75% (Fig 1E). We experimentally determined the phosphorylated fraction of AKT by quantitative protein arrays that combine in-spot normalization and binding model-based calibration; 54% of AKT was phosphorylated in mCFU-E cells upon stimulation with 2.5 U/ml Epo for 10 min (Appendix Fig S3). For ppERK, we previously showed by quantitative mass spectrometry that at maximum 10% of ERK1/2 is double-phosphorylated in mCFU-E cells (Schilling *et al*, 2009). These numbers were used to derive the nanomolar concentrations of pEpoR, pAKT, and ppERK in mCFU-E cells. The concentrations of the respective abundance in BaF3-EpoR cells were scaled accordingly as mCFU-E and BaF3-EpoR cells were always analyzed on the same blot. In both cell types, the dynamics of the concentration of pEpoR was transient albeit with higher peak amplitude and steady state level in BaF3-EpoR cells (Fig 1E, top panel), which reflects their much larger total density of EpoR. Despite the higher EpoR activation in BaF3-EpoR cells, pAKT concentrations were higher in mCFU-E cells (Fig 1E, middle panel).

Table 1. Cytoplasmic concentration of pathway components of mCFU-E, BaF3-EpoR, and 32D-EpoR cells.

	mCFU-E	BaF3-EpoR	32D-EpoR
Cytoplasm (μm^3)	399	1,400	1,406
Cell surface area (μm^2)	378.5	600.3	607
EpoR surface (molecules/μm^2)	2.6	26.1	22.7
EpoR (nM)	4.16	18.62	16.76
PI3K/AKT			
AKT (nM)	407 ± 16.3	510 ± 62.1	607.7
PI3K (nM)	12.7 ± 2.5	12.4 ± 0.8	14.3
SHIP1 (nM)	15.4 ± 2.5	84.2 ± 10.4	127.8
PTEN (nM)	10.4 ± 0.5	107.4 ± 7.0	96.8
PDK1 (nM)	545 ± 80	763 ± 175	1,554.5
Gab1 (nM)	20.8 ± 2.3	–	–
Gab2 (nM)	–	30	1.1
Ras/ERK			
Ras (nM)	3,530 ± 249	9,531 ± 790	7,855.4
Raf (nM)	1,340 ± 298	3,886 ± 864	7,807.3
MEK (nM)	1,460	4,380	4,743.9
ERK (nM)	1,140 ± 64	2,964 ± 166	2,326.1
S6 (nM)	5,340 ± 694	2,590 ± 270	4,531.8
Ratio to mCFU-E:			
mTOR	1	6.18	1.8
Rictor	1	4.57	0.19
Raptor	1	3.08	2.3
RSK	1	7.4	2.8

Cytoplasmic volumes were estimated using imaging flow cytometry excluding the nucleus. 1,000 EpoR molecules on the surface of mCFU-E cells were reported (D'Andrea and Zon, 1990). 15,500 EpoR molecules on the surface of BaF3-EpoR cells were determined by a saturation-binding assay with [^{125}I]-labeled Epo (Becker et al, 2010). Number of EpoR molecules on the surface of 32D-EpoR cells was calculated in comparison to BaF3-EpoR cells using flow cytometry. All other concentrations were determined as fold differences to BaF3-EpoR expression level. "–" indicates the absence of protein. For details, see Appendix G.

This is not simply an effect of AKT expression, which was comparable in both cell types (Table 1). Moreover, the maximum concentration of ppERK was similar, despite the higher abundance of ERK in BaF3-EpoR cells (Table 1). The dynamics of ppERK in BaF3-EpoR and mCFU-E cells was slightly different at time points beyond 15 min of Epo stimulation (Fig 1E, bottom panel). These data indicate that EpoR activation is translated into cell type-specific patterns of activation of the AKT and ERK pathways.

To provide a link between signal transduction and cell cycle progression, transcriptome analysis was performed for up to 18.5 h after stimulation of BaF3-EpoR cells with 1 U/ml Epo (Appendix Fig S4A) and for up to 24 h after stimulation of mCFU-E cells with 0.5 U/ml Epo (Appendix Fig S4B). These analyses revealed that in both cell types several cell cycle regulator genes were differentially expressed upon Epo stimulation (Appendix Fig S4). Prominent among these cell cycle regulators affected by Epo were the activator Cyclin-D2, and the repressors Cyclin-G2 and p27, all of which jointly

control the progression from G1 phase to S phase—the key event for cell cycle entry (Fang et al, 2007). On the other hand, other genes involved in the regulation of the cell cycle such as cyclinE1 (CCNE1) and cyclinE2 (CCNE2) showed only little regulation in either cell types. To confirm the transcriptomics studies, we examined the selected Epo-responsive cell cycle-regulating genes by quantitative RT–PCR analysis (Fig 1F) and showed that after 3 h of stimulation with 5 U/ml Epo, a saturating Epo dose for proliferation in BaF3-EpoR and mCFU-E cells (Appendix Fig S1), mRNA induction of cyclinD2 (CCND2), and mRNA repression of cyclinG2 (CCNG2) and p27 (CDKN1B) exhibited comparable fold changes in BaF3-EpoR and mCFU-E cells. These results suggested that the quantification of the expression of cyclinD2, cyclinG2, and p27 might provide an early quantitative measure to compare Epo-induced cell cycle progression in BaF3-EpoR and mCFU-E cells. To summarize the contribution of the cell cycle activator and the two cell cycle repressors that counteract each other in controlling cell cycle progression, we defined a cell cycle indicator as follows:

$$\frac{[cyclinD2]}{\sqrt{[cyclinG2] \times [p27]}}$$

As evidenced in Fig 1F, after 3 h of Epo addition we observed in BaF3-EpoR and mCFU-E comparatively small changes in the expression of the individual components (e.g. only cyclinG2) but a strong increase in the cell cycle indicator, 16-fold for BaF3-EpoR cells and 13-fold for CFU-E cells, respectively. These results underscore that at this early time point the coefficient reflects the complex regulation of cell cycle progression in response to Epo stimulation better than any of its components alone.

Notably, the cell cycle indicator was significantly ($P = 0.04$) higher in BaF3-EpoR cells compared with CFU-E cells (Fig 1F, right panel). In line with this observation, we observed by propidium iodide staining after stimulation with 5 U/ml Epo for 16 h (BaF3-EpoR) or 11 h (mCFU-E) that the fold change of cells in the S/G2/M phase of the cell cycle in response to Epo stimulation was also significantly ($P = 0.002$) higher in BaF3-EpoR cells compared with mCFU-E cells (Fig 1G). This result supports our notion of the cell cycle indicator as an early measure for cell cycle progression and shows that, whereas mCFU-E cells are already committed to cell cycle progression, an increasing fraction of BaF3-EpoR cells enters S/G2/M phase in response to stimulation with increasing Epo doses. The dynamics of EpoR, AKT, and ERK phosphorylation were distinct between the two cell types, which also differed in their Epo sensitivity of the proliferative response and their proliferation rate.

Influence of cellular protein abundance on Epo-induced signaling dynamics

To understand how the cell type-specific signaling dynamics of AKT and ERK arise, we applied quantitative dynamical pathway modeling. Our mathematical model consists of coupled ordinary differential equations assuming mass action kinetics, and a Hill-type phenomenological term at the receptor level. As depicted in Fig 2A, the model describes preformed dimers of the EpoR to be phosphorylated (pEpoR) upon Epo stimulation. Dephosphorylation of the pEpoR is catalyzed by the pEpoR-activated phosphatase SHP1 and a constitutively active phosphatase. In the model, pEpoR forms

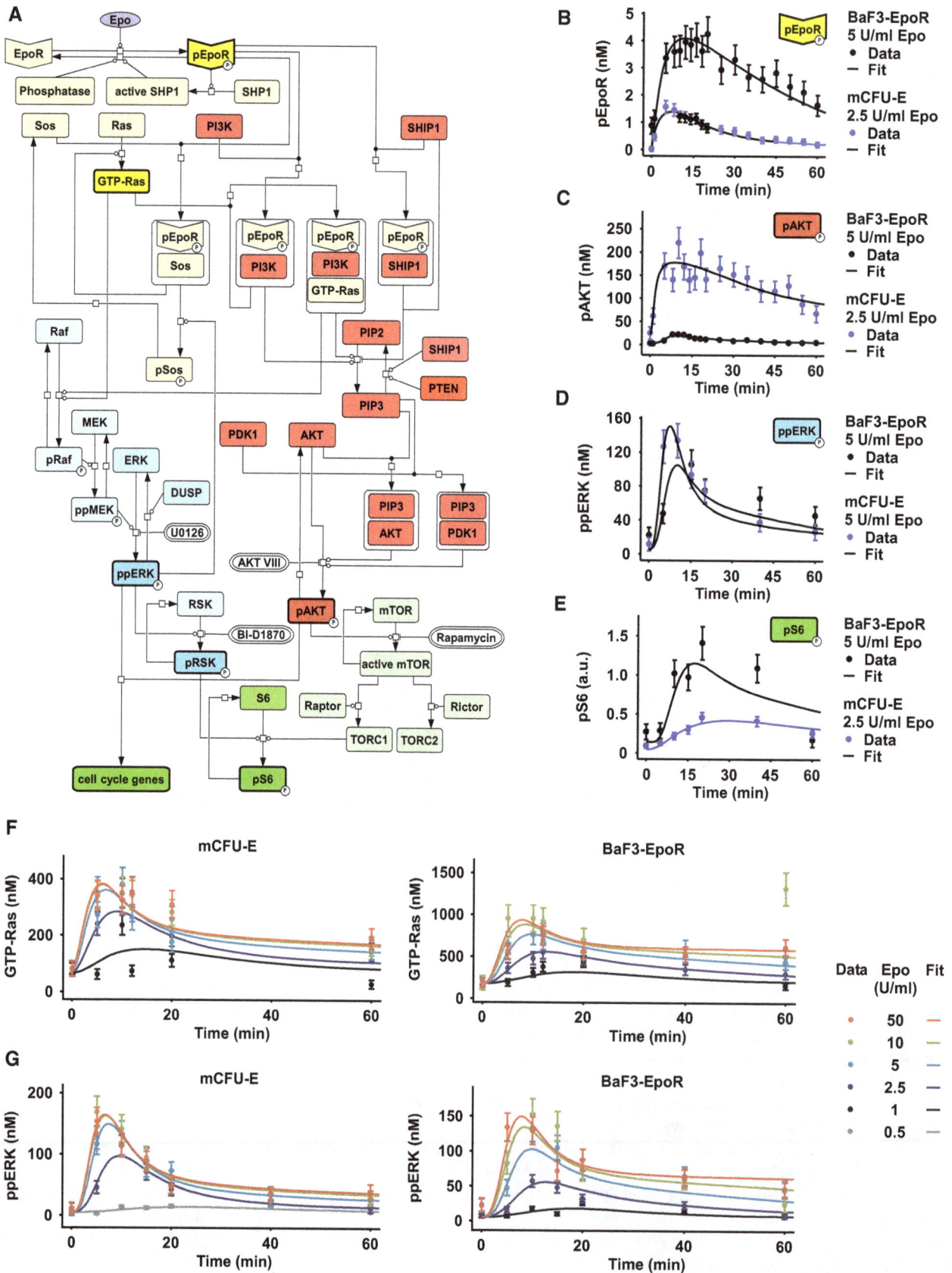

Figure 2.

Figure 2. Mathematical modeling of the Epo-induced AKT, ERK, and S6 activation.

A   The model is represented as a process diagram and reactions are modulated by enzyme catalysis (circle-headed lines) or inhibition (bar-headed lines). Prefix "p" represents phosphorylated species. Bold framed species were experimentally measured. White capsules represent inhibitors. Binding of the ligand Epo to its cognate receptor results in the phosphorylation of EpoR. Receptor and associated complexes are depicted in yellow, the AKT pathway is red, the ERK pathway is blue, and cell cycle genes and the S6 network are shown in green.

B–G   Model calibration with time-resolved quantitative immunoblot data of mCFU-E cells in blue and BaF3-EpoR cells in black. Growth factor-deprived mCFU-E cells (5×10^6 cells per condition) and BaF3-EpoR cells (1×10^7 cells per condition) were stimulated with different Epo doses, and absolute concentrations were determined for pEpoR (B), pAKT (C), ppERK (D). The scale for pS6 (E) was estimated in arbitrary units. GTP-Ras (F) and ppERK were determined upon stimulation with indicated, color-coded Epo doses. pEpoR was analyzed by immunoprecipitation followed by immunoblotting, GTP-Ras was analyzed after pulldown using a fusion protein harboring GST fused to the Ras binding domain of Raf-1 followed by detection by quantitative immunoblotting. For pAkt and ppERK, cellular lysates were subjected to quantitative immunoblotting. Calibrator proteins were used for EpoR, AKT, GTP-Ras, and ERK to facilitate the conversion to nM concentrations. Experimental data are represented by filled circles. Error bars represent standard deviation estimated by an error model. Solid lines represent model trajectories. $N = 1$.

Source data are available online for this figure.

complexes with Sos, PI3K, GTP-Ras, and SHIP1. PI3K generates PIP3 that recruits AKT and PDK1 to the plasma membrane, triggering phosphorylation of AKT. SHIP1 and PTEN dephosphorylate PIP3 and thus inhibit the activation of AKT. The pEpoR-Sos complex induces the formation of GTP-Ras that in turn activates the Raf/MEK/ERK cascade. ppERK is dephosphorylated by dual-specific phosphatases (DUSP). Both pAKT and ppERK regulate the expression of *cyclinD2*, *cyclinG2*, and *p27*. Phosphorylated S6 is linked to the regulation of cell growth and proliferation (Meyuhas & Dreazen, 2009). Phosphorylated AKT catalyzes the activation of mTOR that forms the complexes TORC1 and TORC2 with binding partners Raptor and Rictor, respectively. TORC1 facilitates S6 phosphorylation. ppERK can also trigger the phosphorylation of S6 via phosphorylation of RSK. Thus, pAKT and ppERK signals are integrated in the phosphorylation of S6 through mTOR and RSK (Appendix Fig S5) as well as in the expression of cell cycle regulators. To perturb the system, we used inhibitors acting at different processes (AKTVIII, phosphorylation of AKT; U0126, MEK activity; BID1870, phosphorylation of RSK; rapamycin, mTOR activity). By a systematic model reduction (Appendix F.2), we tested the binding rates of the adaptor proteins Gab1/2 (Sun *et al*, 2008) to the EpoR. We identified that the adapter proteins Gab1/2 may bind either very fast or slow and therefore play a negligible role in the fast equilibrium of receptor–adaptor complex formation. Additionally, we decomposed the enzymatic rate constants (e.g. for phosphatases and kinases) into the product of total enzyme concentration and a biochemical rate constant (also called catalytic efficiency, or turnover, k_{cat}). This decomposition enabled us to quantify the biochemical rate constant as a property of the enzyme, which therefore can be assumed to be independent of a given cell type, whereas the enzyme concentration is cell type-specific (Appendix F). For further details on the coupled ordinary differential equations, the dynamic variables, the parameter estimation as well as their annotation (Appendix Table S1), and their sensitivities toward inhibitors, see Appendix F. The full SBML model is available at FAIRDOMHub (https://fairdomhub.org/).

In the mathematical model, cell type-specific and global parameters were distinguished. The total concentrations of all proteins in the signaling network are specific for the particular cell type (Table 1; Appendix Fig S7). By contrast, the rate constants for reversible protein binding and enzymatic catalysis are global parameters independent of cell type. In principle, however, these kinetic parameters might still be affected by further regulatory proteins that have not been included in the model. To test this, the question was whether the different expression levels of the pathway components

we measured for mCFU-E and BaF3-EpoR cells could explain the observed cell type-specific signal processing through the ERK and AKT pathways. To this end, we used the measured protein concentrations as an input and otherwise assumed identical kinetic parameters for both cell types. The mathematical model was fitted first for receptor activation and deactivation to account for the different EpoR and JAK2 levels in the two cell types (Appendix Fig S8; Becker *et al*, 2010; Bachmann *et al*, 2011). In total, 432 data points (the raw data can be viewed as Source Data of the Figures and the Appendix) of Epo-induced pathway activation, measuring pEpoR, pAKT, GTP-Ras, ppERK, and pS6, were used to estimate the 82 global kinetic parameters of the model. The experimental conditions comprised different Epo doses and perturbation by inhibitor treatment or overexpression of negative regulators (Appendix Figs S9–11). We found that the distinct signaling dynamics and dose responses to Epo were captured by the mathematical model (Fig 2B–G; Appendix Fig S13).

In summary, our mathematical analysis indicates that differences in signal processing can be explained by different abundance in signaling proteins in mCFU-E and BaF3-EpoR cells, based on a mathematical model with global kinetic parameters.

Experimental validation of model predictions for negative regulators

Having established the mathematical model for the activation of the AKT and ERK pathways in CFU-E and BaF3-EpoR cells, the negative regulators of signaling came into focus.

First, the lipid phosphatases SHIP1 and PTEN were overexpressed, and the impact on AKT activation was monitored (Fig 3A). In mCFU-E cells, a strong effect of PTEN overexpression on Epo-induced AKT phosphorylation was experimentally observed, and a weaker effect of a similar overexpression of SHIP1, which were both captured by the model (Fig 3A, Appendix Fig S13). Further, we observed that the Epo-induced induction pAKT in wild-type BaF3-EpoR cells was even lower than in mCFU-E cells with overexpressed PTEN (Fig 3A), which is consistent with the high concentrations of SHIP1 and PTEN in BaF3-EpoR cells (Table 1). The mathematical model calibrated based on these data nevertheless predicted that overexpression of SHIP1 or PTEN would decrease AKT phosphorylation even further in these cells. Indeed in an independent experiment, the Epo-induced dynamics of pAKT in BaF3-EpoR cells overexpressing SHIP1 or PTEN was in agreement with the model trajectories (Fig 3B). Further, we predicted with the model and

A

mCFU-E, 2.5 U/ml Epo
- • Data — Fit

169 nM (12x) SHIP1
- • Data — Fit

125 nM (11x) PTEN
- • Data — Fit

BaF3-EpoR, 5 U/ml Epo
- • Data — Fit

B

BaF3-EpoR, 5 U/ml Epo

337 nM (4x) SHIP1
- — Model prediction
- • Experimental validation

322 nM (3x) PTEN
- — Model prediction
- • Experimental validation

C

D

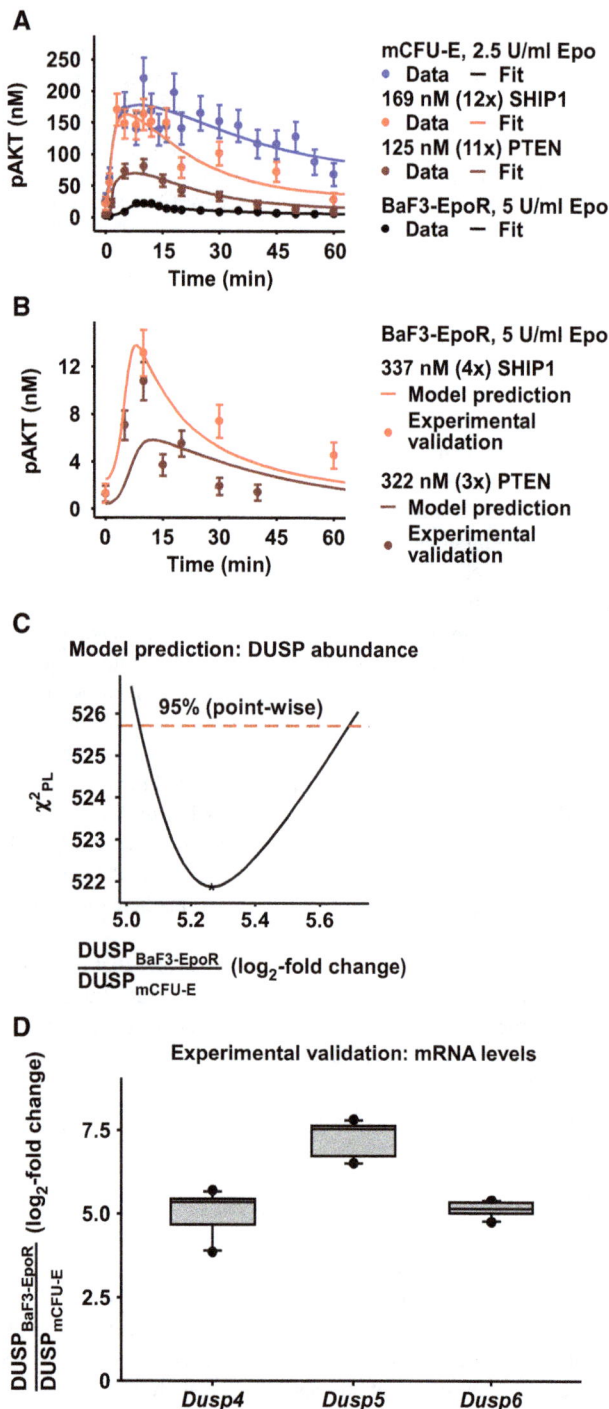

Figure 3. Model predictions and experimental validations for negative regulators of AKT and ERK signaling.

A Model calibration with mCFU-E wild-type cells, and mCFU-E cells overexpressing SHIP1 or PTEN, and BaF3-EpoR wild-type cells. Experimental data are represented by filled circles. Error bars represent standard deviation estimated by an error model. Solid lines represent model trajectories. $N = 1$.

B Model prediction of pAKT dynamics in BaF3-EpoR cells overexpressing PTEN or SHIP1. Model predictions are represented by solid lines. Experimental validation data obtained by quantitative immunoblotting are represented by filled circles. Error bars represent standard deviation estimated by an error model. A representative experiment is shown. $N = 1$.

C Model prediction for the basal expression level of the dual-specific phosphatase (DUSP). DUSP abundance ratio between BaF3-EpoR and mCFU-E cells was identified by the mathematical model. The solid line indicates the profile likelihood. The dashed red line indicates the threshold to assess point-wise 95% confidence interval. The asterisk indicates the optimal parameter value.

D Experimental validation of basal DUSP expression in mCFU-E and BaF3-EpoR. Quantitative RT–PCR was performed with all samples on the same plate, allowing a direct comparison of mRNA levels between mCFU-E and BaF3-EpoR cells. Data are normalized to the *Rpl32* gene. Ratios of expression in BaF3-EpoR cells compared with mCFU-E cells are shown as box plots. Boxes indicate the interquartile range and whiskers extend to 1.5 × interquartile range. $N = 10$. For details, see Appendix K.

Source data are available online for this figure.

mass spectrometry analysis of unstimulated mCFU-E, hCFU-E, and BaF3-EpoR cells, Epo-regulated DUSP family members were below the detection limit. Therefore, we used the mRNA expression levels as proxy, assuming at least some correlation with protein expression. The mathematical model predicted a $\log_2$-fold change of 5.27 higher basal expression of DUSP in BaF3-EpoR cells compared with mCFU-E cells (Fig 3C). To experimentally validate this model prediction, we first identified by microarray analysis of mCFU-E cells (Bachmann *et al*, 2011) and BaF3-EpoR cells (Appendix Fig S14) DUSP4, DUSP5, and DUSP6 as family members that are differentially expressed in response to Epo stimulation. The analysis of the basal mRNA expression of these DUSP by quantitative RT–PCR showed that the $\log_2$-fold difference in the basal expression of DUSP4, DUSP5, and DUSP6 in BaF3-EpoR cells compared with the expression in mCFU-E cells (Fig 3D) was in agreement with the prediction by the mathematical model.

In summary, the expression levels of negative regulators of the AKT and ERK pathways are critical for cell type-specific Epo signal processing.

Cell type-specific information flow through ERK and AKT pathways

To test our observation that the abundance of signal transduction proteins is a key determinant of the dynamics of cell type-specific signal processing, we examined another Epo-responsive hematopoietic cell line, 32D cells, which are derived from the myeloid branch, that exogenously express the EpoR, 32D-EpoR. We determined the abundance of pathway components in 32D-EpoR cells by quantitative immunoblotting (Table 1) and utilized these concentrations in our mathematical model as cell type-specific parameters. Without altering the previously determined global kinetic parameters, we simulated the putative response of pAKT, ppERK, and pS6 at 50 U/ml Epo in 32D-EpoR cells and observed good agreement with

validated experimentally that simultaneous downregulation of SHIP1 and PTEN to their respective concentrations in mCFU-E cells enhanced Epo-induced pAKT levels in BaF3-EpoR cells to the extent observed in mCFU-E cells (Appendix Fig S13).

Second, the DUSPs, a family of phosphatases that negatively regulate ERK signaling, were examined. The analysis of DUSP protein abundance is challenging because multiple isoforms with different functions exist and only very few antibodies, mostly with low specificity, are available. In our proteome-wide quantitative

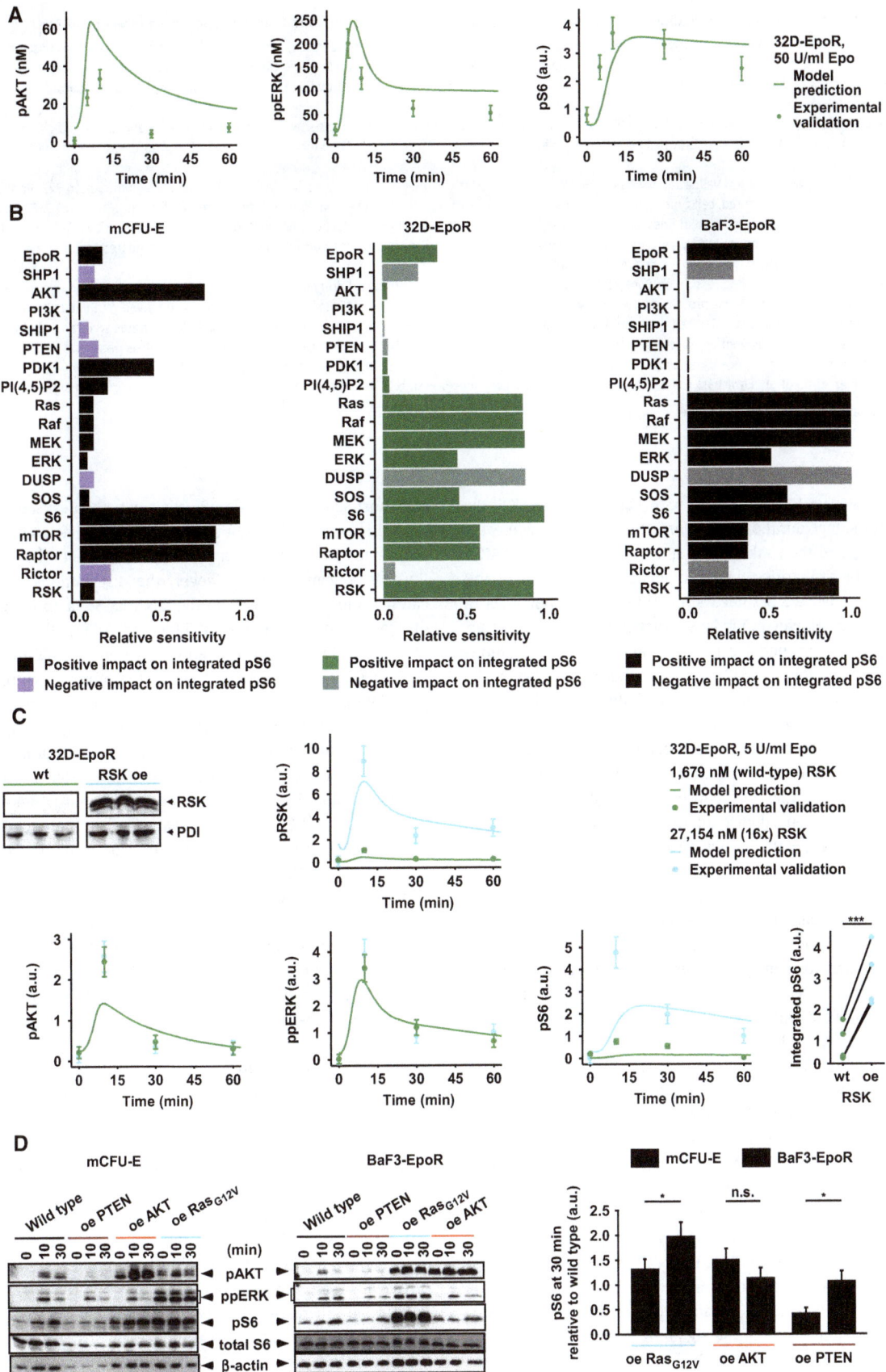

Figure 4.

◀ **Figure 4. Validation of Epo-induced signaling dynamics in 32D-EpoR cells under RSK wild-type conditions and upon overexpression of RSK.**

A   Model prediction and experimental validation for pAKT, ppERK, and pS6 dynamics of 32D-EpoR cells in response to 50 U/ml Epo. Simulations are based on measured, cell type-specific protein abundance of 32D-EpoR cells and global kinetic rates estimated from mCFU-E and BaF3-EpoR cells. Model predictions are represented by solid lines. Experimental validation data obtained by quantitative immunoblotting are represented by filled circles. Error bars represent standard deviation estimated by an error model, N = 1.

B   Sensitivity analysis for integrated pS6 in mCFU-E, 32D-EpoR, and BaF3-EpoR cells. Measured, cell type-specific protein abundance and estimated global kinetic rates were taken into account. Proteins were grouped according to the network modules. Integrated pS6 exhibited high sensitivity toward RSK in 32D-EpoR and BaF3-EpoR cells but not in mCFU-E cells.

C   Model prediction and experimental validation for RSK, AKT, ERK, and S6 activation upon RSK overexpression (oe) and 5 U/ml Epo stimulation in 32D-EpoR cells. Simulations are based on measured, cell type-specific protein abundance of 32D-EpoR cells and global kinetic rates estimated from mCFU-E and BaF3-EpoR cells. Model predictions are represented by solid lines. Experimental validation data obtained by quantitative immunoblotting are represented by filled circles. Error bars represent standard deviation estimated by an error model. Integrated pS6 was significantly higher upon RSK overexpression as compared to 32D-EpoR wild-type (wt) cells, N = 4 (lower right panel). Two-sample t-test, ***P < 0.005.

D   Impact of AKT, Ras, or PTEN overexpression on Epo-induced pS6 dynamics in mCFU-E and BaF3-EpoR cells. Quantitative immunoblotting upon overexpression of PTEN, AKT, a constitutively active Ras protein, or the empty vector control in mCFU-E cells or BaF3-EpoR cells. Growth factor-deprived mCFU-E cells (5×10^6 cells per condition) and BaF3-EpoR cells (1×10^7 cells per condition) were stimulated with 5 U/ml Epo for indicated time points. Cellular lysates were analyzed by immunoblotting employing sequential reprobing anti-pAKT, anti-ppERK, anti-pS6, anti-S6, and to ensure equal loading with anti-β-actin antibodies. Detection was performed with chemiluminescence using a CCD camera device (ImageQuant). Quantification of pS6 on the right is depicted as fold change to wild-type samples at 30 min after Epo stimulation. Error bars represent standard deviation. oe: overexpression. N = 3. Welch modified two-sample t-test, n.s. = not significant, *P < 0.05.

Source data are available online for this figure.

the experimental data for ppERK and pS6 (Fig 4A). For pAKT, the peak time and signal duration were correctly predicted, while the model overestimated the peak amplitude and steady state of pAKT. Further, given the similarities in the protein abundance of 32D-EpoR and BaF3-EpoR cells, we assumed similarities in the dynamics of pathway activation. However, model simulations in line with experimental data used for model calibration under those conditions (Fig 2F and G; Appendix Fig S12) indicated that differences in the peak amplitude, signal duration, and steady state existed between these two cell types (Appendix L). Further we showed that the goodness of fit of a mathematical model calibrated with data from mCFU-E and BaF3-Epor cells is superior to predict the dynamic activation of AKT, ERK, and S6 in response to 50 U/ml Epo stimulation in 32D-EpoR cells when adapted to the cell type-specific protein abundance compared with mathematical models calibrated with data obtained from mCFU-E cells or BaF3-EpoR cells alone (Appendix L).

To systematically characterize the signal flow through the underlying molecular network (see Fig 2A) in mCFU-E, BaF3-EpoR, and 32D-EpoR cells and to identify possible similarities or differences, the relative sensitivities (response coefficients) of the network output were computed with respect to the expression levels of these components. As output, the integrated response of pS6 was used as the most downstream molecular component that integrates the signals coming from the AKT and ERK pathway (Fig 4B). This analysis showed a remarkable difference between mCFU-E cells and BaF3-EpoR cells. The integrated response of pS6 in mCFU-E cells was primarily sensitive to the AKT pathway (Fig 4B, left column), whereas in BaF3-EpoR the influence of the Ras/MEK/ERK cascade dominated (Fig 4B, right column). The 32D-EpoR cells were similar to BaF3-EpoR cells but with a slightly higher impact of AKT pathway components AKT, PI3K, SHIP1, PTEN, PDK1, and PI(4,5)P2 on integrated pS6 (Fig 4B, middle column).

The high sensitivities in the specific cell types were overall associated with high abundance of the signaling proteins and pathway activities (Table 1; Figs 1E and 2B–D). For example, the AKT pathway was most active in mCFU-E cells and exhibited highest sensitivities there (Fig 4B, third row), while the Ras/MEK/ERK pathway

was most active in BaF3-EpoR cells having highest sensitivities there (Fig 4B). This result might seem counterintuitive, as high sensitivity is typically associated with network components that occur at low, limiting concentration. However, the distributions of sensitivities can be rationalized based on the fact that S6 is an integration node of signals from the Ras/MEK/ERK and AKT pathways. The more active a pathway is in a given cell type (e.g. AKT in CFU-E cells; Ras/MEK/ERK in BaF3-EpoR cells), the more it controls S6 phosphorylation and, hence, changes in such a pathway will have greater effects (higher sensitivities) than changes in quantitatively less important pathways if the system is not saturated. Therefore, the distribution of sensitivities for the integrated response of pS6 provides a quantitative measure for the differential signal flow along the Ras/MEK/ERK and AKT pathways in the different cell types.

The sensitivity analysis indicated, for example, that the RSK abundance (Fig 4B, bottom line) exhibits a high impact on the integrated pS6 response in BaF3-EpoR and 32D-EpoR cells but not in mCFU-E cells. These model-based insights are consistent with the high sensitivities obtained for the Ras/MEK/ERK pathway in the former two cell types, as RSK is downstream of ERK. Although wild-type 32D-EpoR cells already exhibited 2.8-fold higher levels of RSK than mCFU-E cells (Table 1), the sensitivity analysis taking this protein abundance into account suggested that RSK overexpression in 32D-EpoR cells would result in an increase in integrated pS6 in response to Epo stimulation. To test this counterintuitive model prediction, RSK was overexpressed in 32D-EpoR cells. Utilizing the amount of RSK experimentally detected in wild-type 32-EpoR cells as well as the amount of RSK present in the cells overexpressing RSK, the mathematical model predicted a major increase in pS6 in response to Epo stimulation, whereas pAKT and ppERK remain rather unaffected. In line with this model prediction, experimental overexpression of RSK had no effect on the Epo-induced dynamics of the upstream components pAKT and ppERK in 32D-EpoR cells but strongly increased the Epo-stimulated phosphorylation level of S6 (Fig 4C). The mathematical model correctly predicted the effect of RSK overexpression in 32D-EpoR cells on the Epo-induced dynamics of ppERK and pRSK, but the peak amplitude of pAKT and pS6 were underestimated. In four independent experiments, the

Figure 5. Evaluation of the effects of AKT inhibitor AKT VIII and MEK inhibitor U0126 on signaling and cell cycle.

A   Epo-induced signaling upon AKT or MEK inhibitor treatment. Growth factor-deprived cells were pretreated for half an hour with AKT VIII or U0126, respectively, and subsequently stimulated with 5 U/ml Epo. PDI served as loading control.

B   Model calibration with integrated pAKT, ppERK, and pS6 data upon inhibitor treatment and 5 U/ml Epo stimulation of mCFU-E, 32D-EpoR, and BaF3-EpoR cells. Area under curve from time-resolved, quantitative immunoblotting data was calculated and is represented by filled circles. $N = 3$. Error bars represent standard deviation estimated by an error model. Solid lines represent model trajectories.

C   Analysis of cell cycle indicator genes upon AKT VIII and U0126 treatment. Growth factor-deprived cells were pretreated for half an hour with indicated doses of a single inhibitor, followed by stimulation with 5 U/ml Epo for 0 and 3 h. The expression of *cyclinD2*, *cyclinG2*, and *p27* was measured by quantitative RT–PCR and normalized to the *Rpl32* gene. Genes were selected based on microarray analysis. Experimental data are shown as fold change to unstimulated cells with mean ± standard deviation, $N = 3$. Welch modified two-sample *t*-test, n.s. not significant, $*P < 0.05$, $**P < 0.01$, $***P < 0.005$.

Source data are available online for this figure.

integrated pS6 response was increased (Fig 4C, bottom right panel), validating the prediction of high RSK sensitivity of pS6 in this cell type. In further agreement with the sensitivity analysis, the observed experimental overexpression of constitutively active Ras resulted in comparison with wild-type cells in a significantly ($P = 0.04$) stronger elevation of Epo-induced S6 phosphorylation in BaF3-EpoR than in mCFU-E cells, whereas the overexpression of PTEN significantly ($P = 0.01$) diminished Epo-stimulated S6 phosphorylation more strongly in mCFU-E than in BaF3-EpoR cells (Fig 4D).

Taken together, our results show that the abundance of the network components directs the signal flow differentially through the AKT and Ras/MEK/ERK pathways. In mCFU-E cells, signaling to S6 occurs primarily through the AKT axis and in BaF3-EpoR cells primarily through the ERK pathway. In 32D-EpoR cells signaling to S6 is similar to BaF3-EpoR cells, but with slightly higher sensitivity toward AKT.

Effects of AKT and ERK inhibition on Epo signaling depend on cell type

Having established how the protein abundance of the network components controls the information flow through the AKT/ERK/S6 network from the EpoR, the question was how this network controls Epo-induced proliferation in a cell type-specific manner. To this end, inhibitors of specific network nodes in the AKT and ERK pathways were employed: AKT VIII, an inhibitor of AKT phosphorylation (Lindsley *et al*, 2005), and U0126, an inhibitor of ERK phosphorylation (Favata *et al*, 1998). The dynamics of AKT, ERK, and S6 phosphorylation upon inhibitor treatment and stimulation with 5 U/ml Epo were monitored in all three cell types by quantitative immunoblotting. Specifically, the levels of pAKT, ppERK, and pS6 were determined in the absence or presence of 0.05, 0.5 and 5 µM of each inhibitor after 0, 10, 30, and 60 min of Epo stimulation.

In mCFU-E cells, AKT VIII reduced the pAKT amplitude. In 32D-EpoR and BaF3-EpoR cells, higher AKT VIII doses reduced the steady state level of pAKT, which therefore became more transient (Fig 5A, upper panel). The cell type-specific impact of inhibitors on signaling dynamics was even more pronounced for U0126 treatment. The duration of the ppERK signal decreased with higher U0126 doses in mCFU-E cells, whereas the signal was overall reduced in BaF3-EpoR cells and to lesser extent also in 32D-EpoR cells (Fig 5A, lower panel). Surprisingly, none of the three cellular systems studied here exhibited a significant cross talk between the AKT and the ERK axes. Under wild-type conditions, AKT VIII reduced pAKT but not ppERK, and U0126 decreases ppERK but not

pAKT. S6 phosphorylation was primarily influenced by AKT VIII in mCFU-E cells, whereas U0126 more strongly diminished the pS6 response in BaF3-EpoR. 32D-EpoR cells occupied an intermediate position, showing both AKT VIII and U0126 effects on pAKT and ppERK, respectively. This might, however, depend on protein abundance because overexpression of PTEN, AKT, or Ras could shift the information flow (Fig 4D). The cell type-specific inhibitor effects are consistent with the signal flow in the network in the three cell types (cf. Fig 4B).

To account for the cell type-specific dynamics upon inhibitor treatment and for the fact that signaling components such as AKT, ERK, and S6 integrate information (Schneider *et al*, 2012), the integrated response within 1 h was calculated based on the experimentally determined data points. The values of these integrals were simulated with our mathematical model. Subsequently, the cell type-specific model parameters for the strength of the inhibitors were estimated. The experimentally observed effects of the two inhibitors on integrated pAKT, ppERK, and pS6 responses (Fig 5A) were reproduced by the mathematical model, except for a slight underestimation of the AKT VIII effect on pS6 in 32D-EpoR cells (Fig 5B).

To quantify the impact of the two inhibitors, AKTVIII and U0126, on the regulation of cell cycle progression, the expression levels of *cyclinD2*, *cyclinG2*, and *p27* in response to 5 U/ml Epo stimulation for 3 h and inhibitor treatment in all three cell types were determined by quantitative RT–PCR. The observed expression pattern of the individual genes *cyclinD2*, *cyclinG2*, and *p27* was complex (Appendix Fig S16). However, the cell cycle indicator, as a coefficient which summarizes the influence of the individual components, showed a graded alteration to the doses of the two inhibitors (Fig 5C). Specifically, the cell cycle indicator was significantly reduced already at low doses of AKT VIII in mCFU-E cells, at intermediate AKT VIII doses in 32D-EpoR cells, and only at high AKT VIII doses in BaF3-EpoR cells (Fig 5C). The effect of U0126 dose on the cell cycle indicator was graded in a similar manner for the three cell types (Fig 5C).

Taken together, these data show that the effect of inhibition of the AKT and ERK pathways depends on the cellular context, and the main determinant is protein abundance.

Linking Epo-induced signal processing to cell proliferation

Next, the molecular activity of the AKT-ERK signaling network was linked to cell proliferation. The integrated pS6 response and the cell cycle indicator quantify key cellular activities contributing to proliferation upon Epo stimulation and inhibitor treatment. On the one hand, pS6 serves as an indicator of the activity of the ribosomal

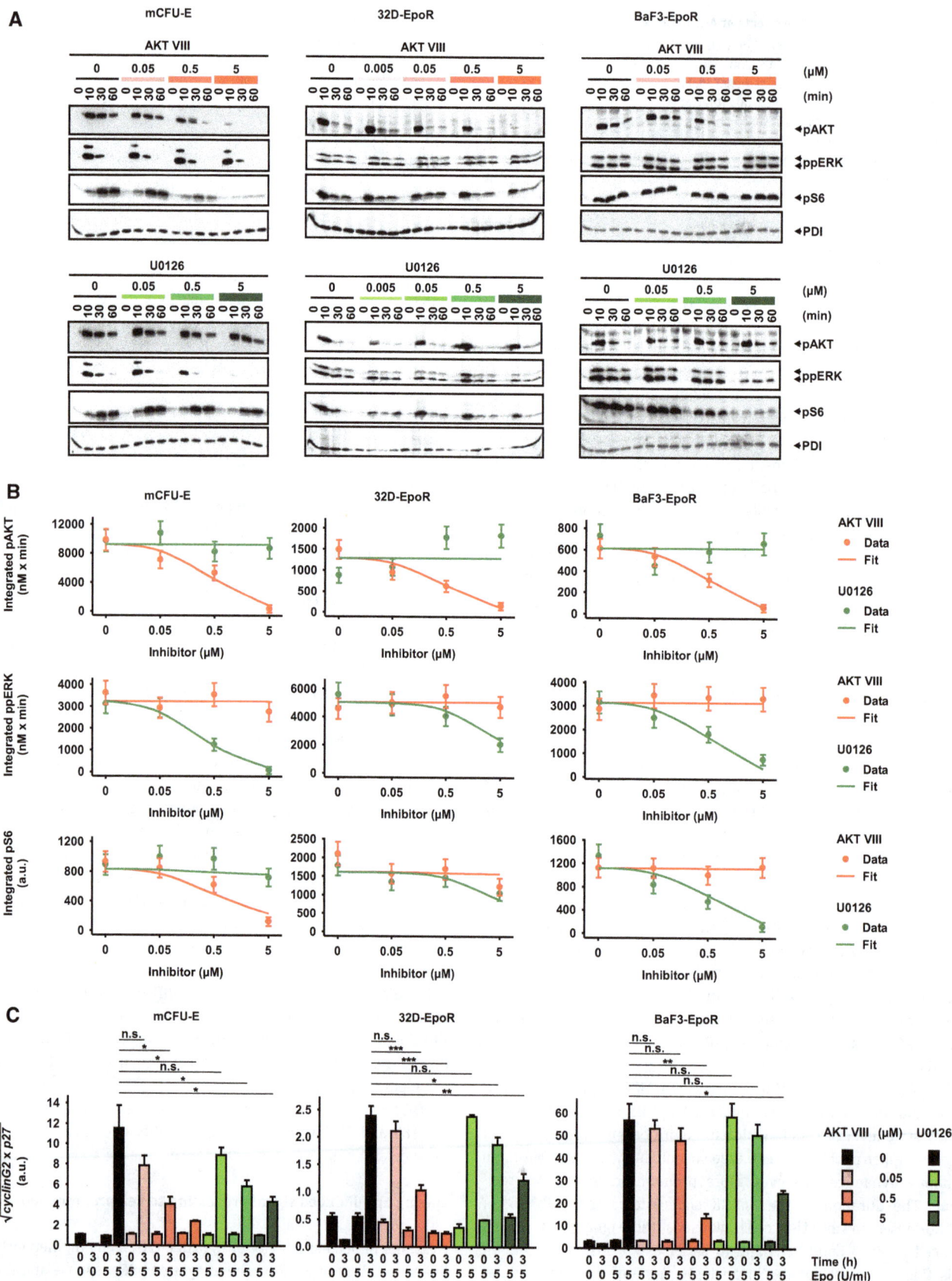

Figure 5.

protein S6 kinase, which is a pivotal regulator of protein synthesis and thus cell growth (Ruvinsky et al, 2006). On the other hand, the cell cycle regulator quantifies the balance of positive and negative regulators that control the entry of cells into the S phase of the cell cycle. Although the details of size regulation of mammalian cells remain poorly understood (Kafri et al, 2013), it is plausible that cellular context, such as protein abundance, will determine how protein synthesis and thus cell growth versus the G1-S transition rate control factor-induced proliferation. Cells with a long G1 phase will have sufficient time to grow, so that primarily the regulators of the G1-S transition, which are quantified by the cell cycle indicator, should control proliferation. Conversely, proliferation in cells with a short G1 phase should be more strongly controlled by growth as a necessary precondition for cell cycle progression.

Linear regression was applied to link the activity of AKT and Ras/MEK/ERK pathways with the cell cycle indicator (Fig 6A). The mathematical model of the signaling network was used to evaluate the integrated pAKT response and the integrated ppERK response in the absence or presence of 0.05, 0.5 or 5 µM of inhibitor upon 1-h stimulation with 5 U/ml Epo for mCFU-E, BaF3-EpoR, and 32D-EpoR cells. The measured values of the integrated pAKT response, the integrated ppERK response, and the cell cycle indicator for mCFU-E, BaF3-EpoR, and 32D-EpoR cells yielded high correlation, indicating that the effects of the inhibitor treatment on the cell cycle indicator are explained very well by changes in the integrated pAKT response and the integrated ppERK response with only slight deviation at the highest inhibitor doses (Fig 6B; Appendix N).

To quantitatively connect the integrated pS6 response and the cell cycle indicator to cell proliferation upon Epo stimulation and inhibitor treatment (Fig 6A), the proliferation of mCFU-E, BaF3-EpoR, and 32D-EpoR cells in response to 5 U/ml Epo and in the absence or presence of 0.005, 0.05, 0.5 and 5 µM of AKT VIII or U0126 was measured. The analysis shown in Appendix O (Fig 6C) revealed that both variables are not correlated and therefore are likely to be regulated independently. Multiple linear regression analysis was performed to link the integrated pS6 response and/or the cell cycle indicator with proliferation in mCFU-E, BaF3-EpoR, and 32D-EpoR cells upon Epo stimulation and inhibitor treatment, and the best model was selected based on Akaike's information criterion (Burnham & Anderson, 2002). For mCFU-E cells, Epo-induced proliferation under inhibitor treatment was described best as a function of the integrated pS6 response only ($R^2 = 0.89$), whereas for BaF3-EpoR cells and 32D-EpoR cells, the proliferation data were best described based on the cell cycle indicator ($R^2 = 0.86$ and 0.81, respectively; Fig 6D). Noteworthy, the models with contribution of both the integrated pS6 response and the cell cycle indicator to proliferation were not significantly more informative than the models of individual contributions (Appendix O) but predicting proliferation in mCFU-E, BaF3-EpoR, and 32D-EpoR cells upon Epo stimulation and inhibitor treatment similarly well (Appendix Fig S26).

In summary, the quantitative dynamical pathway model of Epo-induced signaling was linked to the phenotypic parameter proliferation rate by linear regression. The dependence of mCFU-E proliferation on pS6 indicates that for these rapidly proliferating cells, protein synthesis and cell growth primarily control proliferation. By contrast, for BaF3-EpoR and 32D-EpoR cells, the cell cycle indicator was best predictive for proliferation upon Epo stimulation and inhibitor treatment.

Combinatorial effects of AKT and ERK inhibitors predicted by the integrative model

To validate the quantitative link between Epo-induced signaling and proliferation, the integrative mathematical model (Fig 6A) was used, which was established for a single Epo dose, to predict proliferation in response to a broad range of Epo concentration and in response to overexpression of the negative regulators of AKT signaling, SHIP1, and PTEN (Fig 3A and B) for mCFU-E and BaF3-EpoR cells. Overall these phenotype predictions by the mathematical model (Fig 7A, upper panels) were in good agreement ($R^2 = 0.88$; Appendix Fig S19) with the experimental data (Fig 7A, lower panels). In agreement with the experimental data, the mathematical model predicted that there was no effect of overexpression of SHIP1 or PTEN on the EC_{50} of Epo-induced proliferation of BaF3-EpoR, whereas in mCFU-E cells a small effect was detectable and overexpression of PTEN consistently gave rise to the highest EC_{50} values. The EC_{50} values estimated for the experimentally measured proliferative responses of the wild-type mCFU-E (0.27 ± 0.05) and BaF3-EpoR (0.68 ± 0.46) cells were in line with our initial observations (see Fig 1A). At very low Epo concentrations, the mathematical model predicted an elevated baseline proliferation for wild-type and SHIP1-overexpressing mCFU-E cells that was not detected in the experiment. At these low Epo concentrations, residual phosphorylation of signaling components is detectable and this information was utilized for calibration of the mathematical model. However, in the experiments the activation of signal transduction below a certain threshold apparently was not sufficient to elicit proliferation and therefore baseline proliferation is absent. Yet, in line with the experimental observations the mathematical model predicted that overexpression of PTEN decreased the proliferative response of mCFU-E cells and BaF3-EpoR cells the most. Further, the mathematical model correctly predicted that BaF3-EpoR and 32D-EpoR cells showed comparable Epo dose-dependent proliferation, an observation that was experimentally validated (Appendix Fig S20).

The integrative mathematical model was used to predict the proliferation of mCFU-E, BaF3-EpoR, and 32D-EpoR cells upon stimulation with 5 U/ml Epo and cotreatment with AKT VIII and U0126. Note that these model predictions (Fig 7B, upper panels) were based on the results of the multiple linear regression analysis with the treatment of single inhibitors only. For mCFU-E and 32D-EpoR cells, AKT inhibition was predicted to control Epo-induced cell proliferation in a dose-dependent manner, without or negligible combined effect of MEK inhibition. Only for BaF3-EpoR cells, the model indicated that Epo-induced cell proliferation is strongly inhibited by increasing doses of both AKT VIII and U0126, resulting in a combined effect of both drugs together (Fig 7B, upper right panel). These model predictions were experimentally validated (Fig 7B, lower right panel) and imply that cellular context determines whether molecularly targeted inhibitors of proliferation have combined effect or not.

To highlight that the protein abundance governs cell type-specific regulation of Epo-induced proliferation and as a consequence the sensitivity toward inhibitors, we prepared human CFU-E cells from $CD34^+$ cells mobilized into the peripheral blood of three healthy donors. By means of mass spectrometry, we quantified 6,925 proteins, of which 5,912 proteins were shared among the hCFU-E cells from the three independent donors (Fig 8A). Next, we applied

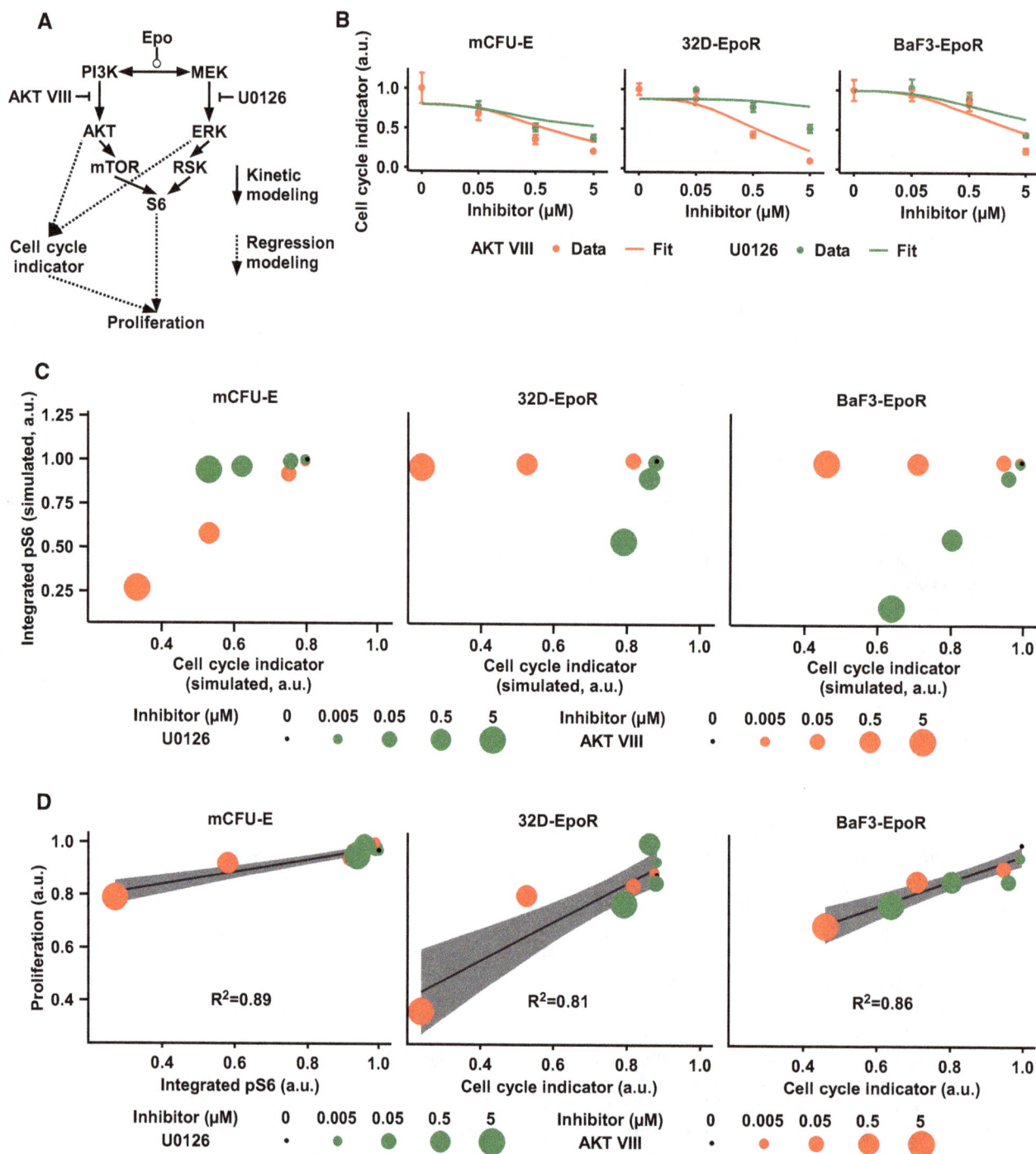

Figure 6. Linking cell cycle and integrated pS6 to proliferation.

A   Scheme of the mathematical model. Integrated pAKT and ppERK obtained from our kinetic model were linked by linear regression analysis to measured cell cycle indicator. Similarly, cell cycle indicator and integrated pS6 were linked to measured proliferation by linear regression analysis and model selection.

B   Linear regression fit for the respective cell cycle indicator of experimental data. Data points represent mean ± standard deviation, $N = 3$. Solid lines represent the linear regression fit.

C   Simulated integrated pS6 response versus simulated cell cycle indicator of mCFU-E, BaF3-EpoR, and 32D-EpoR cells upon 5 U/ml Epo stimulation and single inhibitor treatment at indicated doses.

D   Linear regression modeling and model selection revealed that while proliferation of mCFU-E cells is best described by integrated pS6 only, for 32D-EpoR and BaF3-EpoR cells proliferation correlates mainly with cell cycle indicator. Proliferation was measured, and integrated pS6 and cell cycle indicator were simulated with our mathematical model. The linear regression model selection is based on Akaike's information criterion. The respective correlation is given. For details, see Appendix N and O.

Source data are available online for this figure.

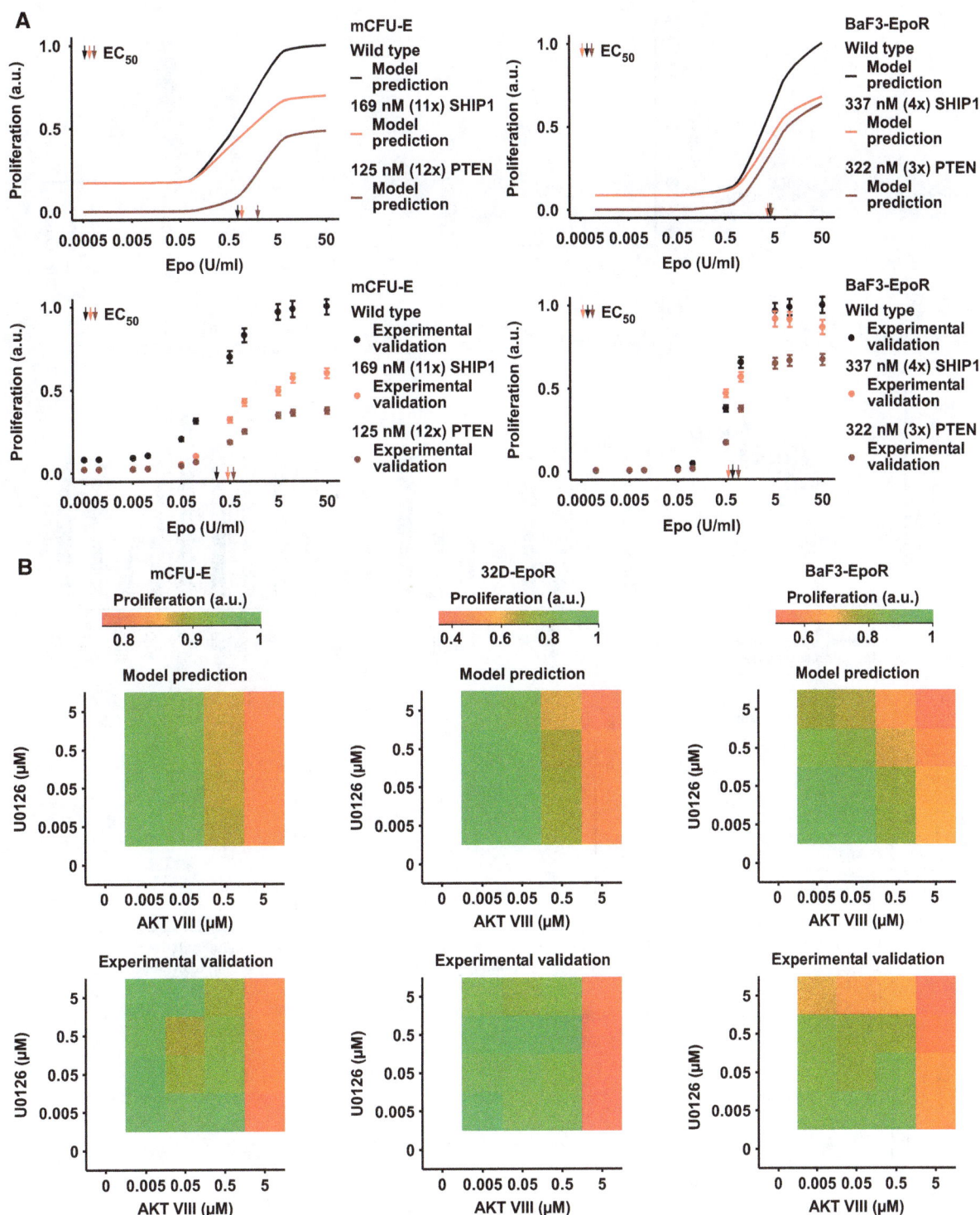

Figure 7. Prediction of cell type-specific proliferation.

A Upper panel: Model prediction of Epo-dependent proliferation in mCFU-E and BaF3-EpoR cells overexpressing SHIP1 or PTEN. Solid lines represent model trajectories. Lower panel: Experimental validation of PTEN and SHIP1 overexpression effects on Epo-dependent proliferation. Proliferation was assessed using [^{3}H]-thymidine incorporation 14 h (mCFU-E) or 38 h (BaF3-EpoR) after retroviral transduction with PTEN or SHIP1 construct for overexpression. Data are represented as mean ± standard deviation, N = 3. EC$_{50}$ values are given.

B Upper panel: Model prediction of proliferation upon combined inhibitor treatment and 5 U/ml Epo stimulation. Maximum proliferation was scaled to 1. Lower panel: Experimental validation of proliferation upon combined inhibitor treatment and 5 U/ml Epo stimulation. Proliferation was measured as cell numbers after 14 h (mCFU-E) or 38 h (BaF3-EpoR, 32D-EpoR) with Coulter Counter. Maximum was scaled to 1.

Source data are available online for this figure.

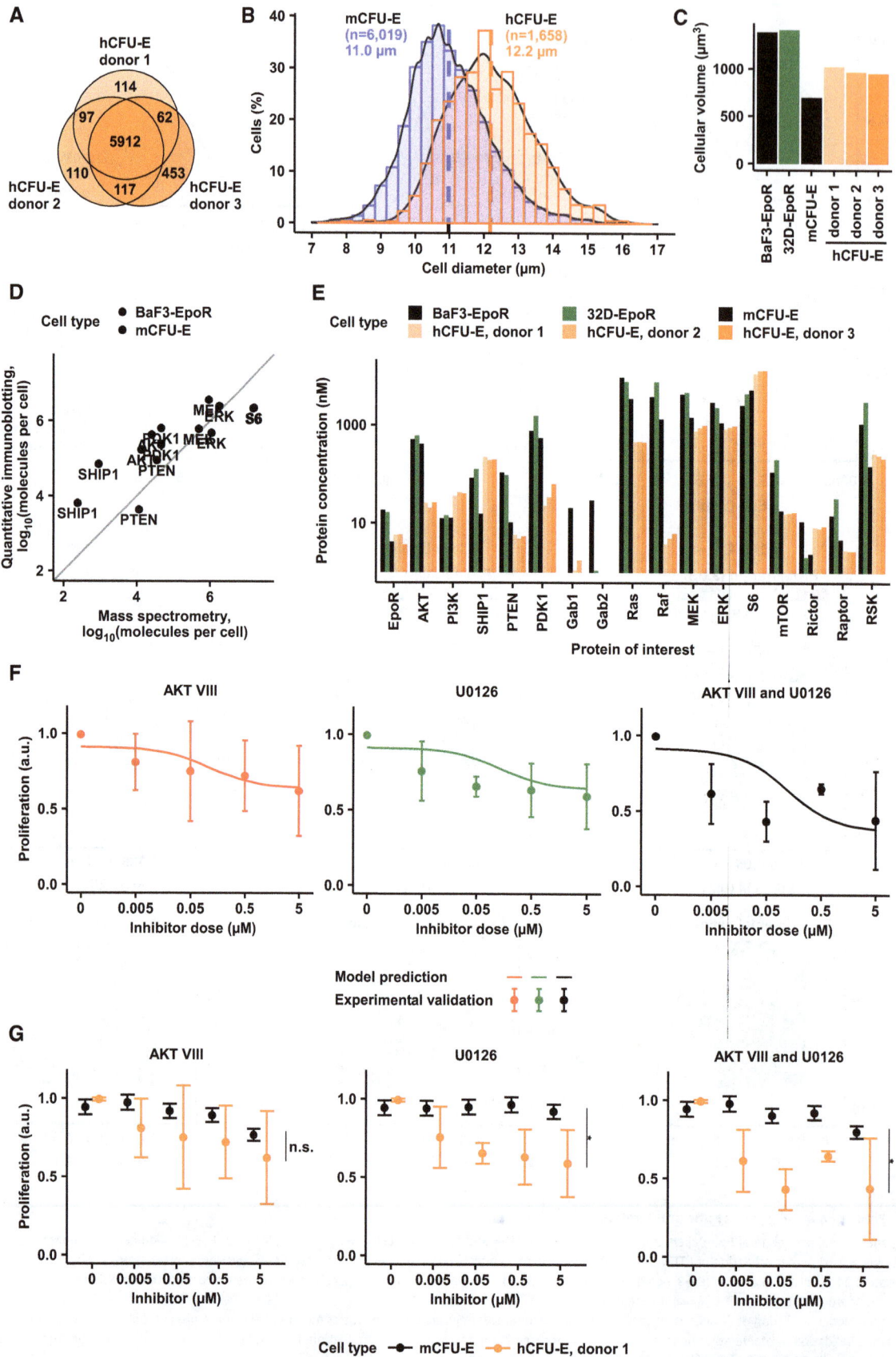

Figure 8.

Figure 8. Predicting combined effects of inhibitor treatment on Epo-induced proliferation of hCFU-E cells solely based on protein abundance.
Human CFU-E cells were prepared from CD34$^+$ cells mobilized into the peripheral blood of three independent stem cell donors.

A Venn diagram representing overlap of hCFU-E proteome from three independent healthy donors. In total, 6,925 proteins were quantified.

B Size determinations of hCFU-E compared to mCFU-E cells. Cell diameter was measured by imaging flow cytometry. Cytoplasm was stained with calcein, and nuclei were stained with DRAQ5. Probability density function of size distribution with indicated mean diameter of hCFU-E and mCFU-E cells. All cells were growth factor-deprived and unstimulated.

C Cytoplasmic volumes of BaF3-EpoR, 32D-EpoR, mCFU-E, and the hCFU-E cells. All volumes were determined using imaging flow cytometry as described in (B).

D Correlation between protein molecules per cell for key signaling components determined by quantitative immunoblotting or quantitative mass spectrometry and the "proteomic ruler" method (Wiśniewski *et al*, 2014) in BaF3-EpoR and mCFU-E cells. Diagonal line as guide to the eye. Spearman's rank-based coefficient of correlation ρ = 0.88.

E Protein abundance of key signaling components was determined by mass spectrometry of whole cell lysates of hCFU-E cells. Copy numbers of proteins per cell were obtained by the "proteomic ruler" method and converted to cytoplasmic concentrations with the volumes calculated from (B) and shown in (C).

F Model prediction and experimental validation of proliferation upon single or combined inhibitor treatment with AKT VIII and U0126 upon 5 U/ml Epo stimulation. Maximum proliferation was scaled to 1. Proliferation was measured as cell numbers, counted with a hemocytometer after 96 h by trypan blue exclusion assay. Solid lines represent model trajectories. Experimental data are represented as mean ± standard deviation. N = 3.

G Comparison of the effect of single or combinatorial inhibitor treatment on Epo-induced proliferation of murine and human CFU-E cells. Proliferation was measured upon single or combined inhibitor treatment with AKT VIII and U0126 upon 5 U/ml Epo stimulation. Numbers of mCFU-E cells were determined with the Coulter Counter after 14 h, whereas numbers of hCFU-E cells were determined with a hemocytometer after 96 h by the trypan blue exclusion assay. Maximum proliferation was scaled to 1. Data are represented as mean (N = 3), and error bars indicate standard deviation. Data of mCFU-E cells are presented in Figs 6D and 7B, and data of hCFU-E cells are taken from Fig 8F. Tukey multiple comparison of means, n.s. = not significant, *P < 0.05.

Source data are available online for this figure.

the "proteomic ruler" method (Wiśniewski *et al*, 2014) to calculate the copy numbers of individual proteins per cell. To convert these numbers into cytoplasmic concentrations, we measured the average cell size by imaging flow cytometry (Fig 8B) and calculated the cytoplasmic volume from these data. The average cytoplasmic volume of the hCFU-E of the three donors was comparable but considerably larger than the volume of the mCFU-E cells (Fig 8C). To relate the results obtained with the proteomic ruler method to the previous measurements in mCFU-E, BaF3-EpoR, and 32D-EpoR utilizing quantitative immunoblotting and recombinant protein standards, we performed additional mass spectrometric measurements for mCFU-E and BaF3-EpoR cells. As shown in Fig 8D, for the key signaling components in the two cell types, the number of molecules per cell determined with the different techniques showed a good correlation (Spearman's rank-based coefficient of correlation ρ = 0.88), validating our determinations by quantitative immunoblotting and confirming that the snapshot measurement by mass spectrometry yielded reliable results. Surprisingly, the protein abundance of the key signaling proteins determined for hCFU-E cells showed very low variance between the three independent donors but were highly distinct from the values obtained for mCFU-E cells as well as for the murine cells lines BaF3-EpoR and 32D-EpoR (Fig 8E). For example, Ras, Raf, and AKT were present at much lower levels in hCFU-E compared to mCFU-E, BaF3-EpoR, and 32-EpoR cells, whereas for PI3 kinase elevated levels were observed in hCFU-E cells. As expected, the levels of the EpoR were comparable in hCFU-E and mCFU-E and elevated to the same extent in Ba3-EpoR and 32D-EpoR cells. To link the signaling layer to proliferation, we applied our concept that larger (see Fig 8B) and more slowly dividing (Appendix Fig S28) cells regulate proliferation primarily by the control of G1-S progression (Appendix Q). Since the measured protein abundance of the key signaling proteins was highly comparable in the hCFU-E cells of the three donors, we used the average concentrations of these proteins to predict with our mathematical model the impact of the AKT inhibitor and MEK inhibitor on Epo-supported cell proliferation in hCFU-E cells. Without any further information (Appendix S), the mathematical model predicted the effect of the AKT inhibitor, the MEK inhibitor, and a combination

thereof on Epo-induced proliferation in hCFU-E cells (Fig 8F). To validate this model prediction experimentally, we quantified the numbers of hCFU-E cells after 96 h of stimulation with 5 U/ml Epo and individual or combined treatment with AKT VIII and U0126. Since hCFU-E cells from donor 3 responded most strongly to Epo, showing a significantly higher fold change in the numbers of cells compared with donor 1 (P = 0.038) and donor 2 (P = 0.009) after 3 days of subcultivation (Appendix Fig S31) and thus a advantageous signal-to-noise ratio, we focused the analyses of proliferation inhibition on hCFU-E cells from this donor in biological triplicates (Fig 8F). We observed that the Epo-induced proliferation of hCFU-E is impaired upon treatment with both inhibitors individually and their combination. Pearson's coefficient of correlation $R^2 = 0.88$ of the results obtained in two independent experiments confirmed the reproducibility of our experimental observations (Appendix Fig S32). When the impact of AKT VIII and U0126 and the combination thereof on the Epo-induced proliferation in hCFU-E and in mCFU-E was compared, we observed that the experimental data on proliferation (Fig 8G) were in line with the model predictions for hCFU-E (Fig 8F) and mCFU-E cells (Figs 6D and 7B), respectively. The results depicted in Fig 8G show that the impact of AKT VIII on Epo-induced proliferation was comparable in hCFU-E and mCFU-E (P = 0.3), whereas MEK inhibition (P = 0.03) and the combinatorial inhibitor treatment (P = 0.03) exhibited significantly larger effects on Epo-promoted proliferation in hCFU-E cells (Fig 8G).

These findings showcase that protein abundance can be reliably measured from snapshot data of human material. Based on these data, our integrative mathematical model allows to evaluate the impact of inhibitors *in silico* and thus may serve to improve the treatment of proliferative disorders such as tumors driven by exacerbated growth factor signaling.

Discussion

By a combination of quantitative measurements with mathematical modeling, we show that proliferation upon Epo stimulation and inhibitor treatment of mCFU-E cells is well predicted by integrated pS6 as a proxy for cell growth, whereas integrated pAKT and ppERK regulating cell cycle progression described proliferation upon Epo stimulation and inhibitor treatment of hCFU-E, BaF3-EpoR, and 32D-EpoR cells best. Importantly, the experimentally observed differences in the dynamics of Epo-induced activation of AKT, ERK, and S6 in mCFU-E, BaF3-EpoR, and 32D-EpoR cells are primarily due to cell type-specific abundance of key signaling components.

In principle, the link from Epo-induced signaling to cell proliferation could be established through cell cycle progression or cell growth or a combination of both. To investigate the connection of Epo-induced AKT and ERK pathway activation to proliferation, we linearly connected the integrated responses of pAKT and ppERK to cell cycle progression and/or the integrated pS6 response reflecting cell growth.

Since early measurements can be indicative of the outcome of cell decisions (Shokhirev et al, 2015), we analyzed the expression of three cell cycle genes after 3 h of Epo stimulation and calculated the cell cycle indicator as a coefficient to quantify cell cycle progression. The three cell cycle genes cyclinD2, cyclinG2, and p27 considered were identified from microarray data as differentially regulated genes (Appendix Fig S4). At saturating Epo doses, the individual genes cyclinD2, cyclinG2, and p27 were expressed to similar extent in mCFU-E and BaF3-EpoR cells (Fig 1F). However, treatment with AKT inhibitor had only a slight impact on the expression of cyclinD2 in mCFU-E cells (Appendix Fig S18), but due to upregulation of the cell cycle repressors cyclinG2 and p27 resulted in a strong reduction of the cell cycle indicator (Fig 6B). Treatment with MEK inhibitor had only mild effects on the expression of cyclinG2 and p27 alone (Appendix Fig S16), but the alterations in the expression level of both genes together in the denominator of the cell cycle indicator explained the observed effects of inhibitor treatment on Epo-induced proliferation in BaF3-EpoR cells (Fig 7B). Therefore, the cell cycle indicator as a coefficient summarizing alterations in the expression of three genes involved in the control of cell cycle progression is more informative than alterations in the expression of individual genes alone. To quantify cell growth, we utilized the integrated pS6 response. It was shown that embryonic fibroblasts from mutant mice, which cannot phosphorylate S6, are reduced in cell size but accelerated in cell division (Ruvinsky et al, 2005). Work in chicken erythroblasts suggested that the length of the G1 phase of the cell cycle ensures proper balancing between growth and cell cycle progression rates (Dolznig et al, 2004). We showed that at saturating Epo concentrations the doubling time, which is considered as a function of cell growth controlled by integrated pS6, correlates with difference in size of mCFU-E and BaF3-EpoR as well as 32D-EpoR cells (Appendix Q). In line with this assumption, hCFU-E cells, which are considerably larger than mCFU-E cells (Fig 8B), also doubled their number more slowly than mCFU-E cells (Fig 1B; Appendix Fig S28).

We observed that the impact of AKT or MEK inhibitors on Epo-induced proliferation in BaF3-EpoR cells and 32D-EpoR cells is explained by the sensitivity of the cell cycle indicator, whereas for mCFU-E the impact on the integrated pS6 response is most informative. To ensure sufficient oxygen supply, the oxygen-carrying capacity of mature erythrocytes has to be tightly controlled (Hawkey et al, 1991). Therefore, the size of erythroid progenitor cells is connected to a physiological function and is decisive to maintain functionality. On the contrary, the hematopoietic cell lines BaF3-EpoR and 32D-EpoR can proliferate unlimited in the presence of Epo

without fulfilling additional tasks. Therefore, here rapid cell cycle progression is key and the step controlled by integrated pS6 is no longer rate-limiting. This is in line with reports proposing that cell lines evolve toward rapid cell growth (Pan et al, 2008), whereas in primary cell maintenance of a specific function that depends on cell size can be key. Therefore, the specific link from factor-induced cell signaling to proliferation can be cell type specific.

In general, linear models are a simple and robust quantification of the input–output function. However, it is more difficult to rationalize mechanistically how the different variables are related. Therefore, it is still under debate to which detail signaling mechanisms should be modeled and how they can be connected to phenotypic behavior (Saez-Rodriguez et al, 2009; Birtwistle et al, 2013). This study investigates three layers of the cellular response to growth factors, which operate on distinct time scales: signal transduction, gene expression, and cell proliferation (Appendix Fig S27). While it is recognized that signal processing occurs across these layers and time scales (Klamt et al, 2006), so far data-based models of growth factor signaling have focused—with few exceptions (Kirouac et al, 2013)—on the fast (< 1 h) signal transduction layer. This focus has, at least in part, been due to the fact that the molecular details of signal transduction are overall better understood than those of the downstream layers. Consequently, data-based mathematical models that connect signal transduction, gene expression, and cell cycle regulation present a critical challenge that as yet is largely unmet (Gonçalves et al, 2013). To address this challenge, we developed a modular modeling approach that links mechanism-based models of signal transduction with conceptually simple but effective, linear regression models for the downstream layers. A practical rationale for this approach is that many targeted drugs for cancer therapy address signal transduction (reviewed in Saez-Rodriguez et al, 2015). Hence, our approach describes the immediate action of the drugs on signal transduction in mechanistic detail and, in turn, infers from the signaling dynamics the proliferative behavior of the cells by a linear regression model. This approach succeeded in quantitatively predicting proliferation inhibition by combinations of AKT and ERK pathway inhibitors in hCFU-E, mCFU-E, BaF3-EpoR, and 32D-EpoR cells.

Signal transduction networks, such as the AKT and ERK pathways, have a common core topology irrespective of cell type. Dynamic pathway models have usually been developed for specific cells by estimating a set of kinetic parameters (mainly enzymatic rate constants) from measurements of model variables (e.g. phosphorylation states) in a given cell (Kim et al, 2010). The concept of protein abundance determining the utilization of connections and the dynamics of cellular signal transduction (Appendix F) is not limited to hematopoietic cell types and can be extended to other cells (Merkle et al, 2016). It was shown that the abundance of growth factor receptors correlates with growth factor responses and AKT/ERK bias in diverse breast cancer cell lines (Niepel et al, 2014). We show here that the mere abundance of a cytokine receptor such as the EpoR is not sufficient to explain proliferative responses. Whereas mCFU-E and hCFU-E cells harbor comparable levels of the EpoR, the abundance of key signaling molecules is very distinct and culminates in major difference in the sensitivity of their Epo-induced proliferative responses toward the AKT and MEK inhibitor. Further refinements might become necessary, as the signaling network topology implemented here involves simplifications and neglects presence of different isoforms such as Gab1 in hCFU-E and mCFU-E cells and Gab2 in BaF3-EpoR and 32D-EpoR cells (Table 1, Fig 8E; Appendix F), which might need to be included in a larger context as these low-abundant proteins can be the bottlenecks of signaling (van den Akker *et al*, 2004; Shi *et al*, 2016). However, even the cell type-specific wiring of feedback loops in signal transduction (Klinger *et al*, 2013; D'Alessandro *et al*, 2015; Stites *et al*, 2015) ultimately depends on protein expression and can thus be captured by the conceptual framework proposed here. By combining the analysis of protein abundance and structure data, Kiel *et al* (2013) showed that the abundance of signaling components determined the cell context-specific topology of the ErbB signaling network. Further, predicting signaling dynamics from (comparatively simple) static measurements of protein abundance may become of practical use for prognosis, as shown for the JNK network in neuroblastoma (Fey *et al*, 2015). However, for cancer cells, oncogenic mutations that affect enzymatic activities of specific proteins or their binding interactions might also require adjustment of selected biochemical parameters or network topologies (Kiel & Serrano, 2009; AlQuraishi *et al*, 2014).

To determine the abundance of signaling components, we used in our approach quantitative immunoblotting (Schilling *et al*, 2005) and quantitative mass spectrometry in combination with the "proteomic ruler" method that is based on the determination of total protein concentrations relative to the abundance of histones (Wiśniewski *et al*, 2014). The results shown in Fig 8C demonstrate that the abundance of the signaling components determined by quantitative immunoblotting is very comparable to the results obtained by quantitative proteome-wide mass spectrometric measurements (Kulak *et al*, 2014; Hein *et al*, 2015). Protein abundance in tumor material can be quantified using a label-free approach as described above or a super-SILAC approach employing a labeled reference cell line (Zhang *et al*, 2014). It has been suggested that the abundance of proteins is characteristic for different cell types (Wilhelm *et al*, 2014) and can even be used to separate subtypes of cancer cell lines (Deeb *et al*, 2012). Even if not all proteins can be quantified, mathematical modeling linked with sensitivity analysis provides a hierarchy of important network components as it is widely applied to investigate parametric dependence of model properties (Schilling *et al*, 2009; Maiwald *et al*, 2010; Perumal & Gunawan, 2011). In our approach, the relative impact of protein abundance on integrated pS6 was calculated and it was shown that in mCFU-E cells, integrated pS6 was controlled mainly by AKT, whereas in 32D-EpoR and BaF3-EpoR cells, integrated pS6 mainly depended on Ras, Raf, MEK, ERK, and RSK (Table 1; Appendix Fig S27). These findings suggest that at points of signal integration (such as pS6), the more highly expressed signaling pathway(s) dominates signal processing. This is in stark contrast to metabolic regulation where the least abundant enzymes usually control the metabolic flux (Heinrich & Schuster, 1996).

We propose that differences in the protein abundance of signaling components can also explain differential sensitivity to inhibitor treatment. While the information flow in 32D-EpoR cells is similar to BaF3-EpoR cells, their proliferation behavior under inhibitor treatment is akin to mCFU-E cells. Profile likelihoods for the inhibitor parameters suggested that 32D-EpoR cells are significantly less sensitive to U0126 than mCFU-E and BaF3-EpoR cells (Appendix Fig S22). 32D-EpoR cells exhibit higher abundance of MEK than mCFU-E and BaF3-EpoR cells (Table 1). MEK is the target of U0126 and therefore the elevated MEK levels probably buffer the inhibitor's effect (Appendix Fig S27). This is in line with the observation that overexpression of activated MEK1 conferred U0126 resistance in HepG2 cells (Huynh *et al*, 2003).

In summary, the integrative mathematical model provides new insights into cell type-specific mechanisms regulating Epo-induced proliferation in primary erythroid progenitor cells and hematopoietic cell lines. We dissect the cell type-specific contribution of pAKT, ppERK, and pS6 on cell growth and cell cycle progression and thereby establish an important basis for rational interference of cellular information processing and the effects of inhibitor treatment on Epo-induced proliferation. Our study demonstrates that the determination of the abundance of signaling components is sufficient to adapt the integrative mathematical model and predict sensitivities for individual inhibitors or combinations thereof and thereby opens new possibility to test and verify therapeutic interventions.

Materials and Methods

Primary cell and cell line cultures

All animal experiments were approved by the governmental review committee on animal care of the state Baden-Württemberg, Germany (reference number DKFZ215). To obtain primary human erythroid progenitor CFU-E cells, CD34$^+$ cells, mobilized into the peripheral blood of healthy donors after written consent, were sorted by autoMACS (CD34-Multisort Kit, Miltenyi Biotech). Sample collection and data analysis were approved by the Ethics Committee of the Medical Faculty of Heidelberg. CD34$^+$ cells were expanded using Stem Span SFEM II (StemCell Technology) supplemented with Stem Span CC100 (StemCell Technology). After 7 days of expansion, cells were differentiated. For differentiation, cells were cultivated in Stem Span SFEM II supplemented with 10 ng/ml mIL-3 (R&D Systems), 50 ng/ml mSCF (R&D Systems), and 6 U/ml Epo alfa (Cilag-Jansen) as reported previously (Miharada *et al*, 2006). After 3 days of cultivation, human CFU-E cells were employed to perform experiments.

Primary murine erythroid progenitor CFU-E cells were prepared from fetal livers of E13.5 Balb/c mice. Fetal liver cells (FLC) were treated with Red Blood Cell Lysing Buffer (Sigma-Aldrich) to remove erythrocytes. For negative depletion, FLC of 40 livers were incubated with rat antibodies against the following surface markers: GR1, CD41, CD11b, CD14, CD45, CD45R/B220, CD4, CD8 (all BD Pharmingen), Ter119 (eBioscience), and with YBM/42 (gift from Suzanne M. Watt, Oxford, UK) for 60 min at 4°C. After washing, cells were incubated for 30 min at 4°C with anti-rat antibody-coupled magnetic beads (Miltenyi Biotech) and negative sorted with MACS columns according to manufacturer's instructions. Sorted mCFU-E cells were cultured for 14 h in Panserin 401 (Pan Biotech) and 50 μM β-mercaptoethanol supplemented with 0.5 U/ml Epo (Cilag-Jansen).

BaF3 and 32D cells were cultured in RPMI 1640 (Invitrogen) including 10% WEHI conditioned medium as a source of IL3 and supplemented with 10% FCS (Invitrogen), penicillin (100 U/ml) and streptomycin (100 mg/ml).

Plasmids and retroviral transduction

Retroviral expression vectors were pMOWS-puro-MCS/M2 (Ketteler *et al*, 2002). For stable transfection of BaF3 and 32D cells with murine EpoR, pMOWS-Kz-HA-EpoR was generated in our laboratory (Becker *et al*, 2008). The murine SHIP1, AKT1, RasV12G, RSK1, or human PTEN cDNAs were cloned into pMOWSnr-MCS/M2, in which the puromycin resistance gene was replaced by the LNGFR cDNA (Miltenyi Biotech) that allows magnetic bead selection of transduced cells. Transient transfection of Phoenix-eco cells with retroviral expression vectors was performed by calcium phosphate precipitation. To ensure an efficient uptake of DNA, Phoenix cells were incubated for 6 h in DMEM medium (Invitrogen) supplemented with 25 µM chloroquine and 10% FCS. Subsequently, the medium was replaced by IMDM (Invitrogen) containing 50 µM β-mercaptoethanol and 30% FCS and incubated for another 18 h. Each 250 µl of the filtered retroviral supernatant of pMOWS-Kz-HA-EpoR supplemented with 8 µg/ml polybrene was then used to transduce 1×10^5 BaF3 or 32D cells, which were centrifuged for 2 h at $340 \times g$ and 37°C. Selection with 1.5 µg/ml puromycin (Sigma-Aldrich) started 48 h after transduction resulting in BaF3-EpoR and 32D-EpoR cells. Surface expression of EpoR in BaF3 and 32D cells was verified by flow cytometry using an antibody against HA (Roche) and a Cy-5-labeled anti-rat antibody (Jackson Immuno Research).

For overexpression experiments, 5×10^6 cells were transduced using 4.5 ml retroviral supernatant supplemented with 8 µg/ml polybrene in a six-well plate and centrifuged for 3 h at $340 \times g$ and 37°C. Following spin infection, cells were cultured for 14–16 h in the standard media. Positively transduced cells were selected using MACSelect LNGFR selection kit (Miltenyi Biotech) according to manufacturer's instructions. Cells were either used immediately for experiments or further cultivated. Level of overexpression was always verified at the day of experiment using immunoblotting.

Time course experiments, cell lysis, quantitative immunoblotting, and mass spectrometry

Murine CFU-E cells were washed three times with Panserin 401 and were growth factor-deprived in the medium supplemented with 50 µM β-mercaptoethanol for 1–2 h while BaF3-EpoR cells were washed with RPMI 1640 and deprived for 4–5 h and 32D-EpoR cells for 3 h, in the medium supplemented with 1 mg/ml BSA (Sigma-Aldrich) at 37°C depending on type of experiment. Subsequently, cells were stimulated with 0.5–50 U/ml Epo (Cilag-Jansen) or cells were first pretreated with Akt inhibitor VIII (Millipore) and MEK1/2 inhibitor U0126 (Cell Signaling) for 30 min before Epo stimulation. For each time point, $0.4–1 \times 10^7$ cells were taken from the pool of cells and lysed by adding $2 \times$ 1% Nonidet P-40 lysis buffer as described elsewhere (Becker *et al*, 2010). Immunoprecipitation was performed by adding the respective antibody, protein A or G sepharose (GE Healthcare) to the lysates or the calibrator protein. Sample loading on SDS–PAGE was randomized and corrected with a spline-based normalization strategy to avoid correlated blotting errors (Schilling *et al*, 2005). Blots were

developed using ECL Western Blotting Reagents (GE Healthcare) and subsequently detected on a Lumi-Imager F1™ (Roche Diagnostics). Quantification of immunoblots was performed using Image Quant Software (GE). Antibodies were removed by treating the blots with stripping buffer as described previously (Klingmüller *et al*, 1995).

The following antibodies were used: anti-pTyr (4G10; Upstate/Millipore), anti-Ras (Calbiochem), anti-PDI (Stressgen), anti-β-actin (Sigma), and anti-PI3K p85 (N-SH2), anti-Gab1 (all from Upstate), and anti-EpoR (M20), anti-SHIP1, anti-Gab2, anti-Raf (all from Santa Cruz), and anti-Akt, anti-pAkt Ser473, anti-pAkt Thr308, anti-phospho-p44/p42 MAPK (Thr201/Tyr204), anti-p44/p42 MAPK, anti-PTEN, anti-S6, anti-pS6 (Ser235/236), anti-pS6 (Ser240/244), anti pRSK (Thr359/Ser363), anti-RSK, anti-mTOR, anti-Rictor, anti-Raptor, anti-PDK1 (all from Cell Signaling) as well as secondary horseradish peroxidase-coupled antibodies (Amersham Biosciences/Dianova).

For determination of protein concentrations of lysates of respective cells and calibrators for the respective protein, they were subjected to quantitative immunoblotting. Calibrators were either commercially available [Ras, Raf, PDK1, S6 (Abnova), ERK (Invitrogen)], or self-made.

Human CFU-E cells were washed three times with IMDM GlutaMAX and growth factor-deprived in IMDM GlutaMAX supplemented with 1 mg/ml BSA at 37°C for an hour. 2.5×10^6 cells were lysed by adding $2 \times$ RIPA buffer (100 mM Tris pH 7.4, 300 mM NaCl, 2 mM EDTA, 2 mg/ml deoxycholate, 1 mM Na_3VO_4, 5 mM NaF). Similarly, 1×10^7 growth factor-deprived BaF3-EpoR and mCFU-E cells were lysed. Whole cell lysates were sonicated and protein yield was determined by BCA assay (Thermo).

Each 75 µg of protein lysates of hCFU-E cells, 100 µg of BaF3-EpoR protein lysates, or 30 µg of mCFU-E protein lysates was fractionated by 10% 1D SDS–PAGE for 105 min. Gels were then stained with Coomassie (Invitrogen), each lane was divided into five segments, and each segment was cut into smaller pieces. In-gel digestion was performed as previously described (Boehm *et al*, 2014). Samples were analyzed by liquid chromatography, nanoelectrospray ionization, and tandem mass spectrometry with a Q-Exactive Plus (Thermo). Raw files obtained were then analyzed by MaxQuant, version 1.5.3.30, as described elsewhere (Cox & Mann, 2008) by MaxQuant (version: 1.5.0.12). MaxLFQ algorithm (Cox *et al*, 2014) was employed for the quantification purpose. Protein copy numbers per cell were obtained by applying the "proteomic ruler" method (Wiśniewski *et al*, 2014).

Cytoplasmic volumes of mCFU-E and BaF3-EpoR cells were determined by confocal microscopy with a Zeiss LSM 710. The cytoplasmic volume of 32D-EpoR cells was determined relatively to BaF3-EpoR cells by imaging flow cytometry. To determine cellular volumes of hCFU-E cells, the cytoplasm was stained with calcein (eBioscience) and DNA was stained with DRAQ5 (Cell Signaling). Imaging flow cytometry was performed on an Amnis ImageStream^X (Merck Millipore), and data were analyzed with the IDEAS Application v5.0 (Merck Millipore).

The raw data of all qualitative and quantitative immunoblots of this work can be accessed as Source Data for the respective figure. Data of quantified immunoblots have also been uploaded through Excemplify (Shi *et al*, 2013) to the SEEK platform (http://seek.sbe po.de/; Wolstencroft *et al*, 2015) and FAIRDOMHub (https://faird omhub.org/).

The mass spectrometry proteomics data have been deposited to the ProteomeXchange Consortium via the PRIDE (Vizcaíno et al, 2016) partner repository with the dataset identifier PXD004816.

Microarray analysis

After washing with serum-free medium, BaF3-EpoR cells were growth factor-deprived for 5 h and resuspended in RPMI supplemented with 1 mg/ml BSA. RNA samples were taken at 0, 1, 2, 3, 4, 5, 7, 18 h after 1 U/ml Epo stimulation.

Per time point, total RNA from 4×10^6 BaF3-EpoR cells was isolated using the RNeasy Mini Plus Kit (Qiagen). Gene expression analysis was conducted using Affymetrix Mouse Genome 2.0 Gene-Chip Arrays (Affymetrix).

Normalization was performed in the R environment together with the Bioconductor toolbox (http://www.bioconductor.org). Arrays were normalized via the Robust Multichip Analysis (Gautier et al, 2004). Subsequent probe annotation was handled with the Affymetrix mouse4302 annotation package (R package version 3.1.3). If multiple probes mapped to the same Gene ID, the one with the largest test interquartile range among all time points was selected. The expression data were deposited in the GEO database under accession number http://tinyurl.com/GSE72317.

Quantitative RT–PCR

To generate cDNA of mCFU-E, BaF3-EpoR, 32D-EpoR cells, 1–2 μg of total RNA was transcribed with the QuantiTect Reverse Transcription Kit (Qiagen). Quantitative RT–PCR was performed using a LightCycler 480 in combination with the hydrolysis-based Universal Probe Library (UPL) platform (Roche Diagnostics). Crossing point (CP) values were calculated using the Second Derivative Maximum method of the LightCycler 480 Basic Software (Roche Diagnostics). PCR efficiency correction was performed for each PCR setup individually based on a dilution series of template cDNA. Relative concentrations were normalized using *HPRT* or *RPL32* as reference genes. UPL probes and primer sequences were selected with the Universal ProbeLibrary Assay Design Center (Roche Diagnostics).

Gene name	Primer Forward	Reverse	UPL No
CCND2	CTGTGCATTTACACCGACAAC	CACTACCAGTTCCCACTCCAG	45
CCNG2	CCACGCGATTGTATTTTGTC	AGCTGCGCTTCGAGTTTATC	15
CDKN1B	GAGCAGTGTCCAGGGATGAG	TCTGTTCTGTTGGCCCTTTT	62
DUSP4	GTACCTCCCAGCACCAATGA	GAGGAAAGGGAGGATTTCCA	17
DUSP5	GATCGAAGGCGAGAGAAGC	GGAAGGGAAGGATTTCAACC	102
DUSP6	TGGTGGAGAGTCGGTCCT	TGGAACTTACTGAAGCCACCT	66
RPL32	GCTGCCATCTGTTTTACGG	TGACTGGTGCCTGATGAACT	12
HPRT	TCCTCCTCAGACCGCTTTT	CCTGGTTCATCATCGCTAATC	95

Cell proliferation assays

For proliferation assays, growth factor-deprived murine CFU-E cells were plated at a density of 20×10^4 cells/well growing for 14–20 h, while BaF3-EpoR and 32D-EpoR cells were plated at a density of 5×10^4 cells/well growing for 62 h, or 10×10^4 cells/well growing for 38 h, respectively. Growth factor-deprived human CFU-E cells were plated at a density of 8.75×10^4 cells/well growing for 96 h. Cells were cultured in their individual mediums supplemented with different doses of Epo (Cilag-Jansen), AKT inhibitor VIII (EMD Millipore), and MEK1/2 inhibitor U0126 (Cell Signaling). Cells were pre-incubated with the inhibitors for 30 min and subsequently stimulated with Epo. Human CFU-E cells were counted by trypan blue exclusion assay using a hemocytometer.

For Coulter Counter assay, cells were plated with appropriate densities in 24-well plates. After respective days, cell numbers were determined using a Coulter Counter Z2 (Beckman, particle size 5.00–12.00 μm for mCFU-E cells and 4.00–17.35 μm for BaF3-EpoR and 32D-EpoR cells).

[³H]-thymidine incorporation assay was performed as follows: murine CFU-E or BaF3-EpoR cells were plated in 96-well plates. After 4-h incubation, 1 μCi/well ³H-thymidine was added and cells were cultivated for respective hours. Cells were collected and the incorporated radioactivity was measured using a scintillation counter. To quantify the proliferation assay, regression lines were calculated with a four-parameter Hill regression ($y = y_0 + ax^b/(c^b + x^b)$). As the logarithmic transformation is a monotonic transformation, the sigmoidality of the curve is also true for a linear axis (Schilling et al, 2009).

For the propidium iodide (PI) staining, 2×10^6 cells were permeabilized with 70% ethanol at $-20°C$. Cells were washed with 0.3% BSA/PBS and incubated with ribonuclease reaction mixture at 23°C. After an additional washing step, fluorescence was measured by flow cytometry using a FACSCalibur (Becton Dickinson). Data analysis was performed with the MultiCycle (Phoenix Flow Systems) software.

Mathematical modeling

Quantitative dynamic modeling was performed in MATLAB (Mathworks) using the D2D software package from http://www.data2dynamics.org (Raue et al, 2015). For parameter estimation, a deterministic derivative-based optimization with a multi-start strategy based on Latin hypercube sampling was applied (Raue et al, 2013). For further details on mathematical modeling, see Appendix F.

As an error model for experimental data, 10% relative error plus 5% absolute error of the highest data point under this condition were assumed.

The relative sensitivity S_p^X shows the change in variable X with infinitesimal small changes in parameter p, scaled by the respective values:

$$S_p^X = \frac{\partial X}{\partial p}\frac{p}{X}$$

As variable, we used pS6 integrated for 1 h: $\int_{t=0}^{t=60\ min} pS6(t)dt$ Parameters were protein abundance of pathway components (Table 1).

Linear regression analysis and statistical testing was performed with R (http://www.r-project.org). Linear regression model selection was based on Akaike's information criterion (Burnham & Anderson, 2002). For further details on linear regression analyses, see Appendix N and O.

Acknowledgements

The authors thank Angela Lenze for autoMACS sorting of CD34[+] cells, Verena Lang for help with the imaging flow cytometry, Marvin Wäsch, Nora Schuhmacher, and Klara Zwadlow for excellent technical assistance, and Melania Barile, Helge Hass, and Bernhard Steiert for fruitful discussions about the mathematical model, as well as Aurelio Teleman and Kathrin Thedieck for their advice on the role of mTOR regulation. This work was supported by the SBCancer Network in the Helmholtz Alliance on Systems Biology as well as by the German Federal Ministry of Education and Research (BMBF)-funded CancerSys network LungSysII, the e:Bio network SBEpo, by the Helmholtz International Graduate School for Cancer Research at the German Cancer Research Center (DKFZ), the German Center for Lung Research (DZL) and the German Cancer Consortium (DKTK).

Author contributions

LA conducted all experiments with human CFU-E cells, flow cytometry, and cell doubling experiments. LA and SK developed the mathematical model together with JT, MS, and TH. M-CW conducted inhibitor experiments, proliferation dose–response experiments and cell cycle indicator and overexpression experiments. BS generated all the other experimental data for model calibration. SC performed mass spectrometric measurements of human CFU-E cells. SL assisted proliferation assays. JB, MB, and HB analyzed microarray data. PW and ADH provided the human material. LA, SK, M-CW, BS, MS, PW, ADH, JT, TH, and UK designed the project. LA, M-CW, TH, and UK wrote the manuscript with comments from SK, MS, and JT. All authors approved the paper.

References

van den Akker E, van Dijk T, Parren-van Amelsvoort M, Grossmann KS, Schaeper U, Toney-Earley K, Waltz SE, Löwenberg B, von Lindern M (2004) Tyrosine kinase receptor RON functions downstream of the erythropoietin receptor to induce expansion of erythroid progenitors. *Blood* 103: 4457–4465

Alessi DR, Andjelkovic M, Caudwell B, Cron P, Morrice N, Cohen P, Hemmings BA (1996) Mechanism of activation of protein kinase B by insulin and IGF-1. *EMBO J* 15: 6541–6551

Alfieri R, Barberis M, Chiaradonna F, Gaglio D, Milanesi L, Vanoni M, Klipp E, Alberghina L (2009) Towards a systems biology approach to mammalian cell cycle: modeling the entrance into S phase of quiescent fibroblasts after serum stimulation. *BMC Bioinformatics* 10(Suppl. 1): S16

AlQuraishi M, Koytiger G, Jenney A, MacBeath G, Sorger PK (2014) A multiscale statistical mechanical framework integrates biophysical and genomic data to assemble cancer networks. *Nat Genet* 46: 1363–1371

Bachmann J, Raue A, Schilling M, Böhm ME, Kreutz C, Kaschek D, Busch H, Gretz N, Lehmann WD, Timmer J, Klingmüller U (2011) Division of labor by dual feedback regulators controls JAK2/STAT5 signaling over broad ligand range. *Mol Syst Biol* 7: 516

Becker V, Schilling M, Bachmann J, Baumann U, Raue A, Maiwald T, Timmer J, Klingmüller U (2010) Covering a broad dynamic range: information processing at the erythropoietin receptor. *Science* 328: 1404 1408

Becker V, Sengupta D, Ketteler R, Ullmann GM, Smith JC, Klingmüller U (2008) Packing density of the erythropoietin receptor transmembrane domain correlates with amplification of biological responses. *Biochemistry* 47: 11771–11782

Birtwistle MR, Mager DE, Gallo JM (2013) Mechanistic versus Empirical network models of drug action. *CPT Pharmacometrics Syst Pharmacol* 2: e72

Boehm ME, Adlung L, Schilling M, Roth S, Klingmüller U, Lehmann WD (2014) Identification of isoform-specific dynamics in phosphorylation-dependent STAT5 dimerization by quantitative mass spectrometry and mathematical modeling. *J Proteome Res* 13: 5685–5694

Bouscary D, Pene F, Claessens Y-E, Muller O, Chrétien S, Fontenay-Roupie M, Gisselbrecht S, Mayeux P, Lacombe C (2003) Critical role for PI 3-kinase in the control of erythropoietin-induced erythroid progenitor proliferation. *Blood* 101: 3436–3443

Broudy VC, Lin N, Brice M, Nakamoto B, Papayannopoulou T (1991) Erythropoietin receptor characteristics on primary human erythroid cells. *Blood* 77: 2583–2590

Burnham KP, Anderson DR (2002) *Model selection and multimodel inference: a practical information-theoretic approach*, 2nd edn. New York: Springer

Cox J, Mann M (2008) MaxQuant enables high peptide identification rates, individualized p.p.b.-range mass accuracies and proteome-wide protein quantification. *Nat Biotechnol* 26: 1367–1372

Cox J, Hein MY, Luber CA, Paron I, Nagaraj N, Mann M (2014) Accurate proteome-wide label-free quantification by delayed normalization and maximal peptide ratio extraction, termed MaxLFQ. *Mol Cell Proteomics* 13: 2513–2526

Dai MS, Mantel CR, Xia ZB, Broxmeyer HE, Lu L (2000) An expansion phase precedes terminal erythroid differentiation of hematopoietic progenitor cells from cord blood *in vitro* and is associated with up-regulation of cyclin E and cyclin-dependent kinase 2. *Blood* 96: 3985–3987

D'Alessandro LA, Samaga R, Maiwald T, Rho S-H, Bonefas S, Raue A, Iwamoto N, Kienast A, Waldow K, Meyer R, Schilling M, Timmer J, Klamt S, Klingmüller U (2015) Disentangling the complexity of HGF signaling by combining qualitative and quantitative modeling. *PLoS Comput Biol* 11: e1004192

D'Andrea AD, Lodish HF, Wong GG (1989) Expression cloning of the murine erythropoietin receptor. *Cell* 57: 277–285

D'Andrea AD, Zon LI (1990) Erythropoietin receptor. Subunit structure and activation. *J Clin Invest* 86: 681–687

Deeb SJ, D'Souza RCJ, Cox J, Schmidt-Supprian M, Mann M (2012) Super-SILAC allows classification of diffuse large B-cell lymphoma subtypes by their protein expression profiles. *Mol Cell Proteomics* 11: 77–89

Dolznig H, Grebien F, Sauer T, Beug H, Müllner EW (2004) Evidence for a size-sensing mechanism in animal cells. *Nat Cell Biol* 6: 899–905

Fang J, Menon M, Kapelle W, Bogacheva O, Bogachev O, Houde E, Browne S, Sathyanarayana P, Wojchowski DM (2007) EPO modulation of cell-cycle regulatory genes, and cell division, in primary bone marrow erythroblasts. *Blood* 110: 2361–2370

Favata MF, Horiuchi KY, Manos EJ, Daulerio AJ, Stradley DA, Feeser WS, Van Dyk DE, Pitts WJ, Earl RA, Hobbs F, Copeland RA, Magolda RL, Scherle PA, Trzaskos JM (1998) Identification of a novel inhibitor of mitogen-activated protein kinase kinase. *J Biol Chem* 273: 18623–18632

Ferrell JE (2013) Feedback loops and reciprocal regulation: recurring motifs in the systems biology of the cell cycle. *Curr Opin Cell Biol* 25: 676–686

Fey D, Halasz M, Dreidax D, Kennedy SP, Hastings JF, Rauch N, Munoz AG, Pilkington R, Fischer M, Westermann F, Kolch W, Kholodenko BN, Croucher DR (2015) Signaling pathway models as biomarkers: patient-specific simulations of JNK activity predict the survival of neuroblastoma patients. *Sci Signal* 8: ra130

Fingar DC, Salama S, Tsou C, Harlow E, Blenis J (2002) Mammalian cell size is controlled by mTOR and its downstream targets S6K1 and 4EBP1/eIF4E. *Genes Dev* 16: 1472–1487

Fingar DC, Richardson CJ, Tee AR, Cheatham L, Tsou C, Blenis J (2004) mTOR controls cell cycle progression through its cell growth effectors S6K1 and 4E-BP1/eukaryotic translation initiation factor 4E. *Mol Cell Biol* 24: 200–216

Gautier L, Cope L, Bolstad BM, Irizarry RA (2004) Affy – analysis of affymetrix GeneChip data at the probe level. *Bioinformatics* 20: 307–315

Gonçalves E, Bucher J, Ryll A, Niklas J, Mauch K, Klamt S, Rocha M, Saez-Rodriguez J (2013) Bridging the layers: towards integration of signal transduction, regulation and metabolism into mathematical models. *Mol BioSyst* 9: 1576–1583

Grech G, Blázquez-Domingo M, Kolbus A, Bakker WJ, Müllner EW, Beug H, von Lindern M (2008) Igbp1 is part of a positive feedback loop in stem cell factor-dependent, selective mRNA translation initiation inhibiting erythroid differentiation. *Blood* 112: 2750–2760

Greenberger JS, Sakakeeny MA, Humphries RK, Eaves CJ, Eckner RJ (1983) Demonstration of permanent factor-dependent multipotential (erythroid/neutrophil/basophil) hematopoietic progenitor cell lines. *Proc Natl Acad Sci USA* 80: 2931–2935

Haseyama Y, Sawada KI, Oda A, Koizumi K, Takano H, Tarumi T, Nishio M, Handa M, Ikeda Y, Koike T (1999) Phosphatidylinositol 3-kinase is involved in the protection of primary cultured human erythroid precursor cells from apoptosis. *Blood* 94: 1568–1577

Hawkey CM, Bennett PM, Gascoyne SC, Hart MG, Kirkwood JK (1991) Erythrocyte size, number and haemoglobin content in vertebrates. *Br J Haematol* 77: 392–397

Hein MY, Hubner NC, Poser I, Cox J, Nagaraj N, Toyoda Y, Gak IA, Weisswange I, Mansfeld J, Buchholz F, Hyman AA, Mann M (2015) A human interactome in three quantitative dimensions organized by stoichiometries and abundances. *Cell* 163: 712–723

Heinrich R, Schuster S (1996) *The regulation of cellular systems.* Boston, MA: Springer Science & Business Media

Huynh H, Nguyen TTT, Chow K-HP, Tan PH, Soo KC, Tran E (2003) Over-expression of the mitogen-activated protein kinase (MAPK) kinase (MEK)-MAPK in hepatocellular carcinoma: its role in tumor progression and apoptosis. *BMC Gastroenterol* 3: 19

Iwamoto N, DAlessandro LA, Depner S, Hahn B, Kramer BA, Lucarelli P, Vlasov A, Stepath M, Bohm ME, Deharde D, Damm G, Seehofer D, Lehmann WD, Klingmuller U, Schilling M (2016) Context-specific flow through the MEK/ERK module produces cell- and ligand-specific patterns of ERK single and double phosphorylation. *Sci Signal* 9: ra13.

Jiang J, Greulich H, Jänne PA, Sellers WR, Meyerson M, Griffin JD (2005) Epidermal growth factor-independent transformation of Ba/F3 cells with cancer-derived epidermal growth factor receptor mutants induces gefitinib-sensitive cell cycle progression. *Cancer Res* 65: 8968–8974

Jones SM, Kazlauskas A (2000) Connecting signaling and cell cycle progression in growth factor-stimulated cells. *Oncogene* 19: 5558–5567

Kafri R, Levy J, Ginzberg MB, Oh S, Lahav G, Kirschner MW (2013) Dynamics extracted from fixed cells reveal feedback linking cell growth to cell cycle. *Nature* 494: 480–483

Ketteler R, Glaser S, Sandra O, Martens UM, Klingmüller U (2002) Enhanced transgene expression in primitive hematopoietic progenitor cells and embryonic stem cells efficiently transduced by optimized retroviral hybrid vectors. *Gene Ther* 9: 477–487

Kiel C, Serrano L (2009) Cell type-specific importance of ras-c-raf complex association rate constants for MAPK signaling. *Sci Signal* 2: ra38

Kiel C, Verschueren E, Yang J-S, Serrano L (2013) Integration of protein abundance and structure data reveals competition in the ErbB signaling network. *Sci Signal* 6: ra109

Kim KA, Spencer SL, Albeck JG, Burke JM, Sorger PK, Gaudet S, Kim DH (2010) Systematic calibration of a cell signaling network model. *BMC Bioinformatics* 11: 202

Kirouac DC, Du JY, Lahdenranta J, Overland R, Yarar D, Paragas V, Pace E, McDonagh CF, Nielsen UB, Onsum MD (2013) Computational modeling of ERBB2-amplified breast cancer identifies combined ErbB2/3 blockade as superior to the combination of MEK and AKT inhibitors. *Sci Signal* 6: ra68

Klamt S, Saez-Rodriguez J, Lindquist JA, Simeoni L, Gilles ED (2006) A methodology for the structural and functional analysis of signaling and regulatory networks. *BMC Bioinformatics* 7: 56

Klinger B, Sieber A, Fritsche-Guenther R, Witzel F, Berry L, Schumacher D, Yan Y, Durek P, Merchant M, Schäfer R, Sers C, Blüthgen N (2013) Network quantification of EGFR signaling unveils potential for targeted combination therapy. *Mol Syst Biol* 9: 673

Klingmüller U, Bergelson S, Hsiao JG, Lodish HF (1996) Multiple tyrosine residues in the cytosolic domain of the erythropoietin receptor promote activation of STAT5. *Proc Natl Acad Sci USA* 93: 8324–8328

Klingmüller U, Lorenz U, Cantley LC, Neel BG, Lodish HF (1995) Specific recruitment of SH-PTP1 to the erythropoietin receptor causes inactivation of JAK2 and termination of proliferative signals. *Cell* 80: 729–738

Knight ZA, Schmidt SF, Birsoy K, Tan K, Friedman JM (2014) A critical role for mTORC1 in erythropoiesis and anemia. *Elife* 3: e01913

Koury MJ, Bondurant MC (1990) Erythropoietin retards DNA breakdown and prevents programmed death in erythroid progenitor cells. *Science* 248: 378–381

Kulak NA, Pichler G, Paron I, Nagaraj N, Mann M (2014) Minimal, encapsulated proteomic-sample processing applied to copy-number estimation in eukaryotic cells. *Nat Methods* 11: 319–324

Landschulz KT, Noyes AN, Rogers O, Boyer SH (1989) Erythropoietin receptors on murine erythroid colony-forming units: natural history. *Blood* 73: 1476–1486

Le X-F, Arachchige-Don AS, Mao W, Horne MC, Bast RC (2007) Roles of human epidermal growth factor receptor 2, c-jun NH2-terminal kinase, phosphoinositide 3-kinase, and p70 S6 kinase pathways in regulation of cyclin G2 expression in human breast cancer cells. *Mol Cancer Ther* 6: 2843–2857

Lindsley CW, Zhao Z, Leister WH, Robinson RG, Barnett SF, Defeo-Jones D, Jones RE, Hartman GD, Huff JR, Huber HE, Duggan ME (2005) Allosteric Akt (PKB) inhibitors: discovery and SAR of isozyme selective inhibitors. *Bioorg Med Chem Lett* 15: 761–764

Livnah O, Stura EA, Middleton SA, Johnson DL, Jolliffe LK, Wilson IA (1999) Crystallographic evidence for preformed dimers of erythropoietin receptor before ligand activation. *Science* 283: 987–990

Maiwald T, Schneider A, Busch H, Sahle S, Gretz N, Weiss TS, Kummer U, Klingmüller U (2010) Combining theoretical analysis and experimental data generation reveals IRF9 as a crucial factor for accelerating interferon α-induced early antiviral signalling. *FEBS J* 277: 4741–4754

McCubrey JA, Steelman LS, Kempf CR, Chappell WH, Abrams SL, Stivala F, Malaponte G, Nicoletti F, Libra M, Bäsecke J, Maksimovic-Ivanic D, Mijatovic S, Montalto G, Cervello M, Cocco L, Martelli AM (2011) Therapeutic resistance resulting from mutations in Raf/MEK/ERK and PI3K/PTEN/Akt/mTOR signaling pathways. *J Cell Physiol* 226: 2762–2781

Merkle R, Steiert B, Salopiata F, Depner S, Raue A, Iwamoto N, Schelker M, Hass H, Wäsch M, Böhm ME, Mücke O, Lipka DB, Plass C, Lehmann WD, Kreutz C, Timmer J, Schilling M, Klingmüller U (2016) Identification of cell type-specific differences in erythropoietin receptor signaling in primary erythroid and lung cancer cells. PLoS Comput Biol 12: e1005049

Meyuhas O, Dreazen A (2009) Ribosomal protein S6 kinase from TOP mRNAs to cell size. Prog Mol Biol Transl Sci 90: 109−153

Miharada K, Hiroyama T, Sudo K, Nagasawa T, Nakamura Y (2006) Efficient enucleation of erythroblasts differentiated in vitro from hematopoietic stem and progenitor cells. Nat Biotechnol 24: 1255−1256

Miura Y, Miura O, Ihle JN, Aoki N (1994) Activation of the mitogen-activated protein kinase pathway by the erythropoietin receptor. J Biol Chem 269: 29962−29969

Moraga I, Wernig G, Wilmes S, Gryshkova V, Richter CP, Hong W-J, Sinha R, Guo F, Fabionar H, Wehrman TS, Krutzik P, Demharter S, Plo I, Weissman IL, Minary P, Majeti R, Constantinescu SN, Piehler J, Garcia KC (2015) Tuning cytokine receptor signaling by re-orienting dimer geometry with surrogate ligands. Cell 160: 1196−1208

Mueller S, Huard J, Waldow K, Huang X, D'Alessandro LA, Bohl S, Borner K, Grimm D, Klamt S, Klingmuller U, Schilling M (2015) T160-phosphorylated CDK2 defines threshold for HGF-dependent proliferation in primary hepatocytes. Mol Syst Biol 11: 795

Myklebust JH, Blomhoff HK, Rusten LS, Stokke T, Smeland EB (2002) Activation of phosphatidylinositol 3-kinase is important for erythropoietin-induced erythropoiesis from CD34(+) hematopoietic progenitor cells. Exp Hematol 30: 990−1000

Niepel M, Hafner M, Pace EA, Chung M, Chai DH, Zhou L, Muhlich JL, Schoeberl B, Sorger PK (2014) Analysis of growth factor signaling in genetically diverse breast cancer lines. BMC Biol 12: 20

Palacios R, Steinmetz M (1985) IL3-dependent mouse clones that express B-220 surface antigen, contain ig genes in germ-line configuration, and generate B lymphocytes in vivo. Cell 41: 727−734

Pan C, Kumar C, Bohl S, Klingmueller U, Mann M (2008) Comparative proteomic phenotyping of cell lines and primary cells to assess preservation of cell type-specific functions. Mol Cell Proteomics 8: 443−450

Pardee AB (1974) A restriction point for control of normal animal cell proliferation. Proc Natl Acad Sci USA 71: 1286−1290

Perumal TM, Gunawan R (2011) Understanding dynamics using sensitivity analysis: caveat and solution. BMC Syst Biol 5: 41

Raue A, Schilling M, Bachmann J, Matteson A, Schelker M, Kaschek D, Hug S, Kreutz C, Harms BD, Theis FJ, Klingmüller U, Timmer J (2013) Lessons learned from quantitative dynamical modeling in systems biology. PLoS ONE 8: e74335

Raue A, Steiert B, Schelker M, Kreutz C, Maiwald T, Hass H, Vanlier J, Tönsing C, Adlung L, Engesser R, Mader W, Heinemann T, Hasenauer J, Schilling M, Höfer T, Klipp E, Theis F, Klingmüller U, Schöberl B, Timmer J (2015) Data2Dynamics: a modeling environment tailored to parameter estimation in dynamical systems. Bioinformatics 31: 3558−3560

Rich IN, Kubanek B (1976) Erythroid colony formation (CFUe) in fetal liver and adult bone marrow and spleen from the mouse. Blut 33: 171−180

Roux PP, Shahbazian D, Vu H, Holz MK, Cohen MS, Taunton J, Sonenberg N, Blenis J (2007) RAS/ERK signaling promotes site-specific ribosomal protein S6 phosphorylation via RSK and stimulates cap-dependent translation. J Biol Chem 282: 14056−14064

Ruvinsky I, Sharon N, Lerer T, Cohen H, Stolovich-Rain M, Nir T, Dor Y, Zisman P, Meyuhas O (2005) Ribosomal protein S6 phosphorylation is a determinant of cell size and glucose homeostasis. Genes Dev 19: 2199−2211

Ruvinsky I, Ruvinsky I, Meyuhas O, Meyuhas O (2006) Ribosomal protein S6 phosphorylation: from protein synthesis to cell size. Trends Biochem Sci 31: 342−348

Saez-Rodriguez J, Alexopoulos LG, Epperlein J, Samaga R, Lauffenburger DA, Klamt S, Sorger PK (2009) Discrete logic modelling as a means to link protein signalling networks with functional analysis of mammalian signal transduction. Mol Syst Biol 5: 331

Saez-Rodriguez J, MacNamara A, Cook S (2015) Modeling signaling networks to advance new cancer therapies. Annu Rev Biomed Eng 17: 143−163

Sarbassov DD, Guertin DA, Ali SM, Sabatini DM (2005) Phosphorylation and regulation of Akt/PKB by the rictor-mTOR complex. Science 307: 1098−1101

Scheid MP, Marignani PA, Woodgett JR (2002) Multiple phosphoinositide 3-kinase-dependent steps in activation of protein kinase B. Mol Cell Biol 22: 6247−6260

Schilling M, Maiwald T, Bohl S, Kollmann M, Kreutz C, Timmer J, Klingmuller U (2005) Quantitative data generation for systems biology: the impact of randomisation, calibrators and normalisers. Syst Biol 152: 193−200

Schilling M, Maiwald T, Hengl S, Winter D, Kreutz C, Kolch W, Lehmann WD, Timmer J, Klingmuller U (2009) Theoretical and experimental analysis links isoform-specific ERK signalling to cell fate decisions. Mol Syst Biol 5: 334

Schneider A, Klingmüller U, Schilling M (2012) Short-term information processing, long-term responses: Insights by mathematical modeling of signal transduction: early activation dynamics of key signaling mediators can be predictive for cell fate decisions. BioEssays 34: 542−550

Schultze SM, Mairhofer A, Li D, Cen J, Beug H, Wagner EF, Hui L (2012) p38α controls erythroblast enucleation and Rb signaling in stress erythropoiesis. Cell Res 22: 539−550

Shi L, Jong L, Wittig U, Lucarelli P, Stepath M, Mueller S, D'Alessandro LA, Klingmüller U, Müller W (2013) Excemplify: a flexible template based solution, parsing and managing data in spreadsheets for experimentalists. J Integr Bioinform 10: 220

Shi T, Niepel M, McDermott JE, Gao Y, Nicora CD, Chrisler WB, Markillie LM, Petyuk VA, Smith RD, Rodland KD, Sorger PK, Qian W-J, Wiley HS (2016) Conservation of protein abundance patterns reveals the regulatory architecture of the EGFR-MAPK pathway. Sci Signal 9: rs6

Shokhirev MN, Almaden J, Davis-Turak J, Birnbaum HA, Russell TM, Vargas JAD, Hoffmann A (2015) A multi-scale approach reveals that NF-κB cRel enforces a B-cell decision to divide. Mol Syst Biol 11: 783

Sivertsen EA, Hystad ME, Gutzkow KB, Døsen G, Smeland EB, Blomhoff HK, Myklebust JH (2006) PI3K/Akt-dependent Epo-induced signalling and target genes in human early erythroid progenitor cells. Br J Haematol 135: 117−128

Smith JA, Martin L (1973) Do cells cycle? Proc Natl Acad Sci USA 70: 1263−1267

Stites EC, Aziz M, Creamer MS, Von Hoff DD, Posner RG, Hlavacek WS (2015) Use of mechanistic models to integrate and analyze multiple proteomic datasets. Biophys J 108: 1819−1829

Sun J, Pedersen M, Rönnstrand L (2008) Gab2 is involved in differential phosphoinositide 3-kinase signaling by two splice forms of c-Kit. J Biol Chem 283: 27444−27451

Tamura K, Sudo T, Senftleben U, Dadak AM, Johnson R, Karin M (2000) Requirement for p38alpha in erythropoietin expression: a role for stress kinases in erythropoiesis. Cell 102: 221−231

Vizcaíno JA, Csordas A, del-Toro N, Dianes JA, Griss J, Lavidas I, Mayer G, Perez-Riverol Y, Reisinger F, Ternent T, Xu Q-W, Wang R, Hermjakob H (2016) 2016 update of the PRIDE database and its related tools.

Wang Y, Kayman SC, Li JP, Pinter A (1993) Erythropoietin receptor (EpoR)-dependent mitogenicity of spleen focus-forming virus correlates with viral pathogenicity and processing of env protein but not with formation of gp52-EpoR complexes in the endoplasmic reticulum. *J Virol* 67: 1322−1327

Wilhelm M, Schlegl J, Hahne H, Moghaddas Gholami A, Lieberenz M, Savitski MM, Ziegler E, Butzmann L, Gessulat S, Marx H, Mathieson T, Lemeer S, Schnatbaum K, Reimer U, Wenschuh H, Mollenhauer M, Slotta-Huspenina J, Boese J-H, Bantscheff M, Gerstmair A *et al* (2014) Mass-spectrometry-based draft of the human proteome. *Nature* 509: 582−587

Wiśniewski JR, Hein MY, Cox J, Mann M (2014) A "proteomic ruler" for protein copy number and concentration estimation without spike-in standards. *Mol Cell Proteomics* 13: 3497−3506

Wolstencroft K, Owen S, Krebs O, Nguyen Q, Stanford NJ, Golebiewski M, Weidemann A, Bittkowski M, An L, Shockley D, Snoep JL, Mueller W, Goble C (2015) SEEK: a systems biology data and model management platform. *BMC Syst Biol* 9: 33

Wu H, Liu X, Jaenisch R, Lodish HF (1995) Generation of committed erythroid BFU-E and CFU-E progenitors does not require erythropoietin or the erythropoietin receptor. *Cell* 83: 59−67

Yao G, Lee TJ, Mori S, Nevins JR, You L (2008) A bistable Rb-E2F switch underlies the restriction point. *Nat Cell Biol* 10: 476−482

Zhang W, Wei Y, Ignatchenko V, Li L, Sakashita S, Pham N-A, Taylor P, Tsao MS, Kislinger T, Moran MF (2014) Proteomic profiles of human lung adeno and squamous cell carcinoma using super-SILAC and label-free quantification approaches. *Proteomics* 14: 795−803

2

Biphasic response as a mechanism against mutant takeover in tissue homeostasis circuits

Omer Karin [iD] & Uri Alon[*] [iD]

Abstract

Tissues use feedback circuits in which cells send signals to each other to control their growth and survival. We show that such feedback circuits are inherently unstable to mutants that misread the signal level: Mutants have a growth advantage to take over the tissue, and cannot be eliminated by known cell-intrinsic mechanisms. To resolve this, we propose that tissues have biphasic responses in which the signal is toxic at both high and low levels, such as glucotoxicity of beta cells, excitotoxicity in neurons, and toxicity of growth factors to T cells. This gives most of these mutants a frequency-dependent selective disadvantage, which leads to their elimination. However, the biphasic mechanisms create a new unstable fixed point in the feedback circuit beyond which runaway processes can occur, leading to risk of diseases such as diabetes and neurodegenerative disease. Hence, glucotoxicity, which is a dangerous cause of diabetes, may have a protective anti-mutant effect. Biphasic responses in tissues may provide an evolutionary stable strategy that avoids invasion by commonly occurring mutants, but at the same time cause vulnerability to disease.

Keywords calcium homeostasis; design principles; evolutionary dynamics; mathematical models of disease; stem-cell homeostasis; tissue homeostasis
Subject Categories Quantitative Biology & Dynamical Systems; Signal Transduction

Introduction

Maintaining proper tissue size is a fundamental problem for multicellular organisms. To do so, cells must precisely coordinate their proliferation and death rates, because an imbalance in these rates leads to either excessive growth or degeneration. Moreover, cells must coordinate their growth in the face of fluctuations, such as injury, or changes in the target size of the tissue, such as during development. This coordination requires feedback control.

Feedback circuits for controlling tissue size regulate cell growth by a signal that is affected by the size of the tissue. Thus, when the tissue is too small, the growth rate is positive, and when it is too large, the growth rate is negative. Only when the tissue reaches a desired size are the proliferation and death rates equal and the system reaches steady state. An example of such a feedback is the control of the concentration of T cells by IL-2 produced by the T cells (Hart *et al*, 2014). Another example occurs in endocrine tissues whose growth is regulated by the physiological variable that they control (Karin *et al*, 2016), such as in the control of beta-cell mass by blood glucose. These physiological feedback circuits can show dynamical compensation mechanisms that make them robust with respect to variation in parameters, such as insulin resistance in the case of glucose control (Karin *et al*, 2016). Feedback is also found in tissues that are renewed by proliferating stem cells (Lander *et al*, 2009), such as skeletal muscle and olfactory epithelium.

For such feedback circuits to function properly, each cell must respond precisely to the input signal. These responses depend on the activity and expression of receptors, signaling pathways, and regulatory proteins and are thus susceptible to mutations. When a mutant cell with a dysregulated proliferative or apoptotic response arises, it may invade the population and thus break the homeostatic control. Mutant takeover leads to aberrant tissue size and function. Thus, mechanisms must be in place to prevent such takeover.

One mechanism for protection from mutant invasion is cell intrinsic and concerns the paradoxical activation of apoptosis by c-myc (Lowe *et al*, 2004). C-myc is a transcription factor that drives proliferation in many cell types (Bouchard *et al*, 1998), yet it paradoxically induces apoptosis when overexpressed (Evan *et al*, 1992). This paradoxical induction of apoptosis plays an important role in tumor suppression because it eliminates transformed cells (Harrington *et al*, 1994; Lowe *et al*, 2004).

Here, we extend the idea of cell-intrinsic elimination of mutants to the level of circuits of communicating cells. We show that tissue feedback circuits are inherently sensitive to takeover by common types of mutants that misread the feedback signal, such as receptor loss-of-function or receptor locked-on mutations. The feedback loop gives these mutants a growth advantage relative to wild-type cells. We propose that mutant invasion can be prevented by a biphasic response mechanism, in which the signal is toxic to the cells at both low and high levels. Biphasic control gives the mutants a selective disadvantage compared to wild-type cells, and the mutants are hence eliminated.

Biphasic control of growth is prevalent in physiological systems. Examples include the control of beta-cell mass by glucose

Department of Molecular Cell Biology, Weizmann Institute of Science, Rehovot, Israel
*Corresponding author. E-mail: uri.alon@weizmann.ac.il

(Robertson *et al*, 2003), the control of mammary gland mass by estrogen (Lewis-Wambi & Jordan, 2009), the control of neuronal survival by glutamate (Hardingham & Bading, 2003), epidermal growth factor signaling (Högnason *et al*, 2001), and the control of T-cell concentration by IL2 and by antigen level (Critchfield *et al*, 1994; Hart *et al*, 2014). Biphasic control was also demonstrated for mechanical signaling—the control of epithelial cell proliferation by mechanical stretch through Piezo1 (Gudipaty *et al*, 2017). In all of these cases, signal is toxic at both high and low levels.

As mentioned above, we find that biphasic control can protect the tissue from invasion by commonly occurring mutants that missense the feedback signal. However, we show that this protective mechanism comes at a cost. The biphasic response introduces an unstable fixed point. If the input signal fluctuates beyond this unstable fixed point, a runaway phenomenon occurs in which cells are eliminated and signal diverges, potentially leading to disease. We discuss this tradeoff between stability to mutant invasion and risk of disease in several systems, including beta-cell control of glucose, parathyroid control of calcium, stem-cell differentiation, and excitotoxicity in neurons.

Results

Tissue homeostasis circuits are inherently vulnerable to invasion by sensing mutants

Feedback circuits that control tissue size act to balance the proliferation and removal rates of the cells. The cells adjust their growth rate (proliferation minus death/removal) as function of an input y, which in turn is affected by the size of the tissue (Figs 1A and B, and EV1A and B).

There are two possible cases. In the first case, the signal y increases with tissue size Z (i.e., Z activates y), and y inhibits the growth rate of the cells (Figs 1A and EV1A). If there are too many cells, y is large and growth rate is negative leading to reduction in tissue size. If there are too few cells, the opposite occurs and the tissue grows. This feedback loop guides the tissue to steady state at the point where growth rate is zero, at $y = y_{ST}$.

In the second case, the signal y decreases with tissue size (Z inhibits y) and y increases the growth rate of the cells (Figs 1B and EV1B). The same considerations show that the tissue stably settles at $y = y_{ST}$. These feedback circuits thus provide a stable tissue size, because different initial cell populations and different initial concentrations of y all converge on the same final population Z_{ST} (Fig 1C). At the same time, the circuits also provide a stable signal level $y = y_{ST}$.

We propose that such generic feedback circuits are susceptible to invasion by mutants that misread the signal. When such a mutant arises at steady state, it senses the actual signal level y_{ST} as a larger or smaller value y_{MUT}. In the case of Fig 1A, if the mutation causes a misreading of the signal as too low, as for example in a receptor-inactivating mutation, the feedback loop provides the mutant with a growth advantage and the mutant will take over the population. As a result, the tissue will show aberrant growth and, when the mutant is at high enough frequency, will show a level of y that is too high, $y > y_{ST}$.

In the case of Fig 1B, if the mutation causes a misreading of the signal as too high, as for example in a locked-on receptor mutation, the feedback loop provides the mutant with a growth advantage (Fig EV2A). The mutant will take over the population (Fig 1D). As a result, the tissue will show aberrant growth and, when the mutant is at high enough frequency, will show a level of y that is too low, $y < y_{ST}$ (because in this case the tissue acts to reduce y). In both cases, sensing mutants can take over the population and only reach equilibrium again at an aberrant tissue size, leading to a breakdown of homeostatic control. Importantly, the same conclusion holds whether the growth of the cells is modeled as logistic or exponential (see Appendix Section S1), and when y acts in delay (see Appendix Section S2).

Biphasic response can protect against mutant invasion but can cause vulnerability to disease

To overcome the problem of mutant invasion, the sensing mutants need to have a selective disadvantage. One way to do this is an alternative implementation of the feedback circuit, in which y affects the growth rate of Z in a *biphasic* manner (Figs 1E and F, and EV1C and D). The word biphasic means that the growth rate curve has an inverse-U shape, with a rising and a falling phase—y stimulates the growth of Z at low concentrations and inhibits the growth of Z at high concentrations, so the signal is toxic (negative growth rate) at both low and high levels of y.

As with the monophasic circuits, here there are also two possible cases. In the first case, the signal y increases with tissue size Z (i.e., Z activates y). This circuit has a stable fixed point at $y = y_{ST}$ and an unstable fixed point at $y = y_{UST}$ where $y_{UST} < y_{ST}$ (Figs 1E and EV1C). In the second case, the signal y decreases with tissue size (Z inhibits y). This circuit also has a stable fixed point at $y = y_{ST}$ and an unstable fixed point at $y = y_{UST}$, but here $y_{UST} > y_{ST}$ (Figs 1F and EV1D).

In comparison with the monophasic circuits depicted in Fig 1A and B, the biphasic circuits have *fewer* types of sensing mutations with a fitness advantage. In particular, they are protected from invasion by loss-of-sensing mutants and locked-on sensing mutants. Whereas in the monophasic circuit of Fig 1A, loss-of-sensing mutations invade the population, in Fig 1E, the biphasic response gives these mutants a negative growth rate. They are thus eliminated. Similarly, whereas locked-on sensing mutants invade the monophasic circuit of Fig 1B, in the biphasic case of Fig 1F, they are eliminated (Fig EV2B). Thus, mutants with strong inactivation (or strong activation) in the response to y have a fitness disadvantage (Fig 1H). This robustness to mutants is very important since such mutations may be common. For example, many constitutively active mutations of diverse G-coupled protein receptors have been observed (Seifert & Wenzel-Seifert, 2002), and it is common for mutations to lead to loss of function (Eyre-Walker & Keightley, 2007; Sarkisyan *et al*, 2016). Mutations with intermediate effects may be rarer; for example, a study in yeast (González *et al*, 2015) showed that mutations that destroy protein function are much more common than those that reduce its activity to an intermediate level.

The elimination of sensing mutants by the biphasic mechanism is frequency dependent: Mutants are eliminated if they have low frequency compared with wild-type cells. The reason for this is that when mutants are rare, the tissues maintain a proper signal y_{ST} which the mutants mis-sense as y_{MUT} and therefore have a fitness

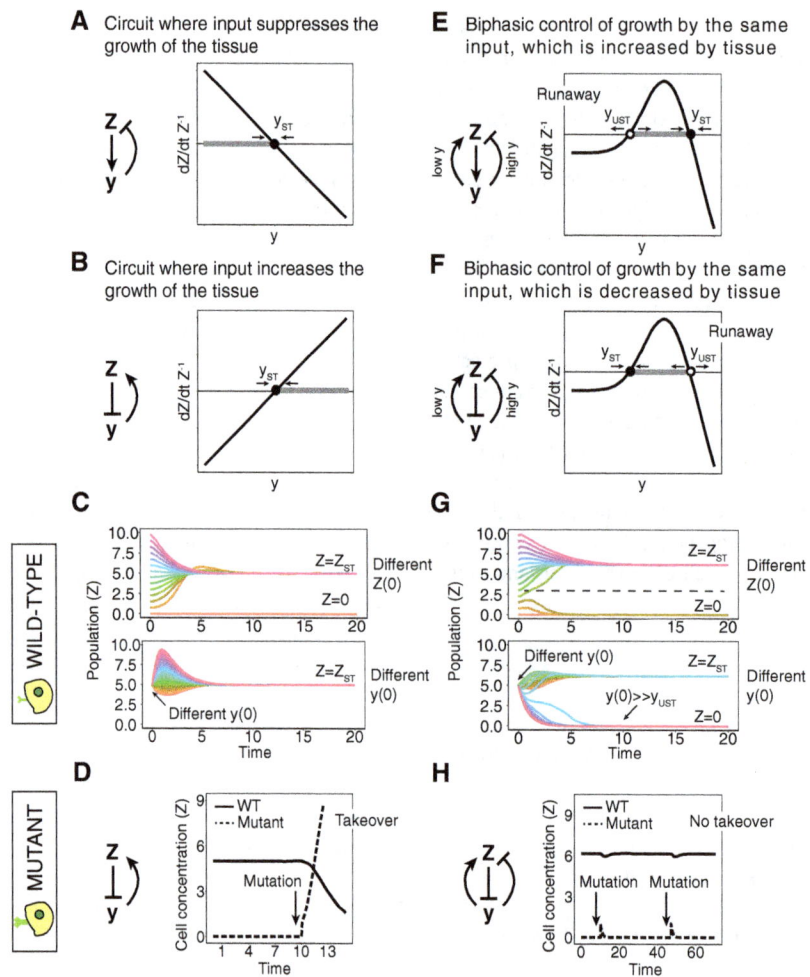

Figure 1.   Biphasic control can resist mutant invasion of feedback circuits.

A   A monophasic feedback circuit in which cells Z generate an input y that inhibits their growth rate. The population is at steady state $Z = Z_{ST}$ when $y = y_{ST}$.

B   A monophasic feedback circuit where cells Z decrease an input y, which increases their growth rate. The population is at steady state $Z = Z_{ST}$ when $y = y_{ST}$.

C   Trajectories of Z from different initial concentrations of cells (Z) (i) or y (ii) for the circuit of (B). The healthy concentration $Z = Z_{ST}$ is reached regardless of initial concentration of Z, as long as it is nonzero, and regardless of the initial concentration of y.

D   An arrow marks the time when a mutant with a strong activation of the sensing of y arises (for the circuit depicted in B). This mutant has a selective advantage and takes over the population.

E   A biphasic feedback circuit where Z generates a signal y, which, in turn, decreases the growth rate of Z at high concentrations and increases the growth rate of Z at low concentrations. The population is at steady state $Z = Z_{ST}$ when $y = y_{ST}$, and there is also an unstable fixed point at $y_{UST} < y_{ST}$.

F   A biphasic feedback circuit where cells Z inhibit y, which, in turn, decreases the growth rate of Z at high concentrations and increases the growth rate of Z at low concentrations. The population is at steady state $Z = Z_{ST}$ when $y = y_{ST}$, and there is also an unstable fixed point at $y_{UST} > y_{ST}$.

G   Trajectories of Z from different initial concentrations of Z (i) or y (ii) for the circuit depicted in (F). The healthy concentration $Z = Z_{ST}$ is not reached for small values of Z ($Z << Z_{ST}$) or large values of y ($y >> y_{UST}$).

H   The arrows mark the times when a mutant with a strong activation of the sensing of y arises (for the biphasic circuit depicted in F). This mutant has a selective disadvantage and is thus eliminated.

disadvantage. On the other hand, if the mis-sensing mutant appears at high enough frequency, it is prevalent enough to change the level of y and force it to reach an improper level that it mis-reads as y_{ST}. In this case, the population of mis-sensing mutants will be at steady state and will not be eliminated.

Biphasic circuits still have a range of mild-effect mutants with a growth advantage. These mutants mis-interpret the normal signal y_{ST} as a different value lying in the gray-shaded regions of Fig 1E and F, namely y_{MUT} lies between y_{ST} and y_{UST}. Later, we discuss

mechanisms that can reduce the growth advantage of these mild mutants.

The biphasic mechanism of resistance to mutants, however, comes at a cost in terms of the robustness of the circuit to perturbations in the input y. The biphasic response curve crosses zero twice and therefore introduces a new *unstable* fixed point, denoted by a white circle in Fig 1E and F. The stable fixed point (full black circle) still exists, and the circuit can maintain the cell concentration constant in the face of small fluctuations around this fixed point.

However, large fluctuations in signal y that exceed the unstable fixed point, or large fluctuations in Z, may lead to negative growth rate and to the risk of the elimination of the cell population (Fig 1G). Beyond the unstable fixed point, a runaway phenomenon occurs in which the cell population shrinks, leading to change in y that pushes it deeper into the unstable region, leading to faster shrinkage and so on. This runaway phenomenon has the hallmarks of certain disease as described below.

To summarize so far, circuits with biphasic control avoid invasion by mutants with strong activation or inactivation of sensing. This robustness is useful because such mutations have a severe effect if they take over the population. This mechanism has two vulnerabilities: Mutations with mild effect on sensing may still invade, and an unstable fixed point introduced by the biphasic control provides risk of runaway behavior if signal fluctuates too widely. We next provide several examples of biphasic control.

Glucotoxicity can protect from mutant beta cells, but can cause diabetes

The first example occurs in the endocrine circuit that regulates blood glucose by pancreatic beta cells (Fig 2). Fasting blood glucose (y) is maintained within a tight range around approximately $y_{ST} = 5$ mM, and blood glucose dynamics are precise in response to perturbations (Allard et al, 2003; Ferrannini et al, 1985). To achieve this tight regulation, beta cells (Z) secrete insulin, which reduces glucose by increasing its uptake by peripheral tissues and decreasing its endogenous production. Thus, in this case, Z inhibits y (the case of Fig 1B and E).

The response of beta-cell growth to glucose is biphasic (Figs 2A and EV3A). Both low and high levels of glucose are toxic to beta cells. The response curve therefore has a stable fixed point at $y = 5$ mM (Karin et al, 2016), and it also has an unstable fixed point at a higher glucose concentration. The toxicity at high levels of glucose is known as glucotoxicity (Efanova et al, 1998; Del Prato, 2009; Bensellam et al, 2012).

The unstable fixed point caused by glucotoxicity is puzzling, because, as described by Topp et al (2000), it provides a potential susceptibility to the system. If glucose levels exceed the unstable fixed point for extended periods (e.g., due to insulin resistance), beta cells will have negative growth rate and be removed, leading to an increase in glucose and a vicious cycle which can eliminate the beta-cell population. This process has been suggested to lead to type II diabetes (Topp et al, 2000; De Gaetano et al, 2008; Ha et al, 2016).

Glucotoxicity of beta cells is therefore detrimental, because it adds instability to perturbations in blood glucose. Since glucotoxicity is mediated by reactive oxygen species, it could have been mitigated by antioxidants, but antioxidants and oxidative stress protective genes are expressed at an exceptionally low level in beta cells (Robertson et al, 2003). This raises the question of why beta cells have not evolved to resist glucotoxicity. Here, we suggest that glucotoxicity may have a biological function: It eliminates beta-cell mutants with impaired glucose sensing.

Mutants that affect glucose sensing can occur at many steps in the glucose sensing and insulin secretion process. These steps include glucose import, phosphorylation and metabolism, closure of K_{ATP} channels, and opening of voltage-dependent calcium channels

(MacDonald et al, 2005). Many mutations are known to affect glucose sensing by beta cells (Fajans et al, 2001; James et al, 2009), among them autosomal dominant mutations in the enzyme glucokinase (GCK) (Froguel et al, 1993; Glaser et al, 1998; Fajans et al, 2001; Matschinsky, 2002; James et al, 2009).

We focus on GCK because it performs the rate-limiting step in glucose sensing. GCK is a hexokinase isozyme expressed in beta cells that phosphorylates the glucose that is transported into the cells. It has a half-maximal activation at $K = 8.4$ mM glucose and a Hill coefficient of $n = 1.8$ (Matschinsky, 2002). Hence, its activity level at the homeostatic set point of 5 mM glucose is ~30% of maximal activation. Glucose sensing is also affected by the expression level of GCK (Wang & Iynedjian, 1997). Germ line mutations in GCK result in low and high blood glucose levels for activating and inactivating mutations, respectively (Matschinsky, 2002). GCK mutations also affect the rates of beta-cell death and proliferation (Porat et al, 2011; Tornovsky-Babeay et al, 2014). This means that somatic mutations in GCK cause impaired glucose sensing that may alter circuit function.

Our theory makes a specific prediction on the fate of GCK mutations in beta cells, namely that their survival will be dependent on their frequency in the tissue (Fig 2B). We predict that a strong activating mutation in GCK will be lost when the majority of the population is wild type, because the mutants will be eliminated by the biphasic mechanism: Glucose level is set by the wild-type tissue to be 5 mM, but the mutants mis-sense a much higher level and succumb to glucotoxicity. In contrast, when the mutation is transmitted through the germ line, it is expected to survive and instead pull the blood glucose level to a low point, which the mutants interpret incorrectly as 5 mM.

This frequency-dependent survival was observed in an experiment by Tornovsky-Babeay et al (2014), which studied a strong (~6-fold) activating mutation of GCK (Y214C). This mutation, when transmitted via germ line, and thus to all beta cells, results in large and hyperfunctional islets and severe hypoglycemia (Cuesta-Munoz et al, 2004) as occurs in rare human patients. The experimenters conditionally expressed this mutation in beta cells of adult mice, such that only a subset of the beta cells expressed the transgene (~25%: 3 days after conditional expression). Both the proliferation and apoptosis rates increased in the cells expressing the transgene, but the increase in apoptosis rate was higher and after 22 days only 5% of the cells that expressed the transgene were left. Thus, the mutated cells were eliminated, whereas the wild-type cells remained.

We simulated this experiment using the biphasic control circuit (Fig 2B–D, see Appendix Section S3 for simulation details). The results of the simulation are consistent with the experimental observations—despite having a higher proliferation rate, the population of induced mutants was eliminated by their even higher apoptosis rate (Fig 2C). This elimination restores blood glucose levels to baseline after an initial hypoglycemia (Fig 2D), in agreement with the experimental blood glucose measurements (Fig 2C and D).

Resistance to mutant invasion is enhanced by low proliferation, low cell number, and spatial compartments

We have thus seen that strong hypersensing mutants are eliminated from the beta-cell population. We now address the question of sensitivity to *mild* sensing mutants. These mutants misread the

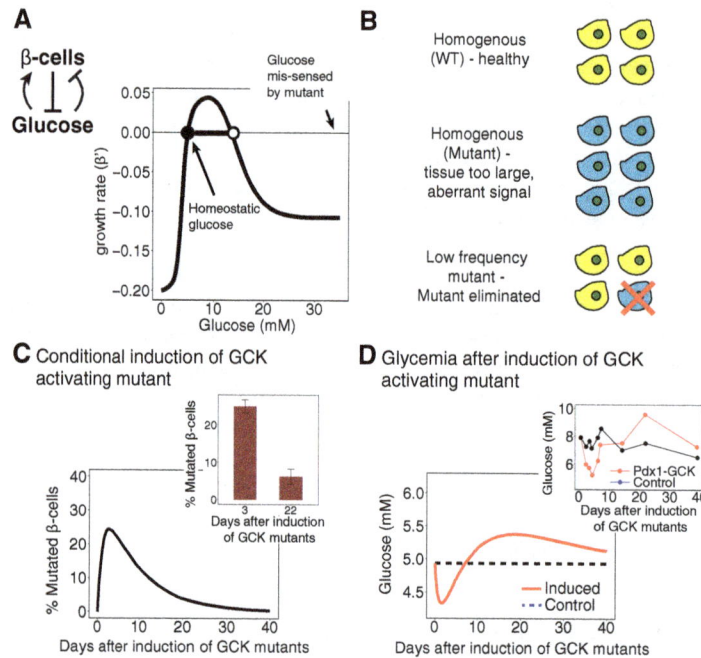

Figure 2.   Frequency-dependent selection of mutant pancreatic beta cells.

A   Feedback circuit in which beta cells secrete insulin, which lowers blood glucose levels. Blood glucose levels, in turn, affect beta-cell growth rate in a biphasic manner, with beta-cell growth being negative at both low and high glucose concentrations. The system is stable at the homeostatic glucose concentration at 5 mM. It also has an unstable fixed point at a higher glucose concentration. A mutant with a sixfold increase in glucokinase affinity senses the glucose level y as $y_{MUT} = 6y$.

B   Biphasic control leads to frequency-dependent selection of the sensing mutant. The cell population in the tissue reaches a stable steady state only when it is homogenous with respect to the sensing of y. When a mutant with low frequency arises somatically, it is eliminated from the tissue; in contrast, if it is transmitted in the germ line, it will spawn a tissue with aberrant size.

C, D   Mathematical simulation of a tamoxifen-induced conditional knock-in of a sixfold activating GCK mutant in beta cells. (C) The percentage of beta cells with mutated GCK increases to ~25% after 3 days, but then decreases and is eliminated after a few weeks. (D) Glucose levels initially decrease after the tamoxifen injection, but return to normal after a few weeks. Insets: Experimental results of Tornovsky-Babeay *et al* (2014).

signal at a level that lies between the stable and unstable fixed points, $y_{ST} < y < y_{UST}$, leading to a growth advantage (shaded area of Fig 2A).

Using evolutionary dynamics theory (Nowak, 2006), we quantify the probability that such sensing mutants will invade the population of beta cells during a normal life span. The analysis results in several design principles to reduce the probability of such invasion.

To make simple approximation, we approximate the growth rate of beta cells when $y_{ST} < y < y_{UST}$ as constant, where λ_+ is the proliferation rate and λ_- is the death rate in this range ($\lambda_+ > \lambda_-$). We also approximate by a constant the probability that a k-fold activating mutant will arise after a cell division, $\mu(k) = \mu_0$. The population of beta cells is sub-divided into compartments—pancreatic islets of Langerhans—each consisting of about $N \approx 3 \times 10^3 - 4 \times 10^3$ beta cells (Leslie & Robbins, 1995). We define the evolutionary stability of the circuit as the probability that no mutant will invade a single pancreatic islet by time t. This probability is given by a Moran process term (Appendix Sections S4 and S5):

$$\zeta(t) = e^{-N\tau^{-1}\delta\mu_0\left(1-\frac{1}{v}\right)t} \tag{1}$$

where $\delta = \frac{y_{UST} - y_{ST}}{y_{ST}}$ is the range of inputs to which the circuit is stable given by the relative distance between the stable and

unstable fixed points (dynamic stability of the circuit), N is the number of beta cells in each islet, τ^{-1} is beta-cell turnover rate, and $v = \lambda_+/\lambda_-$ is the ratio of proliferation to removal for the mutants.

We can now approximate the evolutionary stability of the glucose homeostasis circuit using (equation 1). Typical parameters are $\delta \approx 1$ which corresponds to glucotoxicity around 10 mM glucose (Maedler *et al*, 2006) and $v \approx 3$ (Stolovich-Rain *et al*, 2012). For an average beta-cell turnover of $\tau^{-1} \approx 0.001$/day (Meier *et al*, 2008; Saisho *et al*, 2013) and mutation probability with a target size of 100 bp, $\mu_0 = 10^{-7}$, the evolutionary stability of the circuit is $\zeta(t) \approx e^{-2 \times 10^{-6}t}$. For a 70-year life span, the stability is $\zeta \approx 0.994$, so that ~0.6% of the pancreatic islets have an invading mutant by 70 years for these parameters. Note that we have only analyzed here mutations that are due to cell division and excluded other possible sources of somatic mutation, which may increase the overall number of islets with invading mutants. Increasing the mutation rate by 10-fold leads to ~6–7% of islets being taken over by an invading mutant.

The analysis shows that evolutionary stability is in a tradeoff with dynamic stability. The circuit can be made more dynamically stable by pushing the unstable fixed point to higher glucose (increasing δ). This, however, increases the range of mild mutants that can invade (Fig 3A–C). At the extreme, one can push the unstable fixed point to infinity and end up with a monostable circuit, which is susceptible to all activating mutations in sensing.

Similarly, evolutionary stability is in tradeoff with the response time of the circuit. The response time to a glucose perturbation depends on the growth rate of the cells, which is, for $\tau^{-1} = \lambda_+$: $\lambda_+ - \lambda_- = \tau^{-1} \cdot (1 - \frac{1}{\nu})$. To make the response time more rapid, either τ^{-1} or ν should increase, but this will make the circuit less evolutionarily stable [decrease ζ according to (equation 1)]. The intuitive reason for this tradeoff is that fast response requires faster proliferation, but this gives mutants a bigger growth advantage. We conclude that if mutations that activate glucose sensing are sufficiently likely, glucotoxicity may be selected for despite its harmful potential for diabetes.

Evolutionary stability of the parathyroid gland

Another prediction of (equation 1) is that changes in the parameters, such as a large increase in cell proliferation rate, can lead to invasion by mild sensing mutants. We hypothesize that this occurs in the circuit that controls calcium homeostasis, leading to the disease known as tertiary hyperparathyroidism.

The parathyroid (PT) gland (Z) controls plasma calcium (y) by secreting parathyroid hormone (PTH) which increases calcium production. This circuit is analogous to the glucose homeostasis circuit discussed previously—calcium controls both the secretion of

PTH and the mass dynamics of the PT gland (Naveh-Many et al, 1995; Wada et al, 1997; Mizobuchi et al, 2007). The signs of the circuit are opposite to the glucose circuit, because Z acts to increase y. It is unclear whether this circuit has biphasic control, so the circuit is similar to either one of the circuits in Fig 1A or Fig 1E (Fig EV3B).

This circuit is sensitive to invasion by deactivating mutants in calcium sensing (mis-sensing calcium as lower than it actually is). Under normal conditions, however, the PT gland has a very low turnover (very small τ^{-1}) (Bilezikian et al, 2001). Therefore, these mutants have a low probability to arise and invade due to the rarity of cell divisions.

However, in cases of increased demand for PTH, which occurs in hypocalcaemia such as that caused by renal failure, excessive proliferation of the parathyroid cells takes place (τ^{-1} increases). Such conditions are termed secondary hyperparathyroidism (SHPT). In such conditions, we expect, from (equation 1), that mutants will arise and have a large probability to invade the PT gland.

Indeed, the invasion of mutants with calcium-sensing inactivation often occurs in secondary hyperparathyroidism (Gogusev et al, 1997; Yano et al, 2000; Fraser, 2009). The invasion of such mutants alters the calcium homeostatic set point (Malberti, 1999) and leads to tertiary hyperparathyroidism. The new set point is mildly higher calcium, which is due to the mis-sensing of the mutants. The common mutations that lead to tertiary hyperparathyroidism are known to cause only an intermediate reduction in the expression of

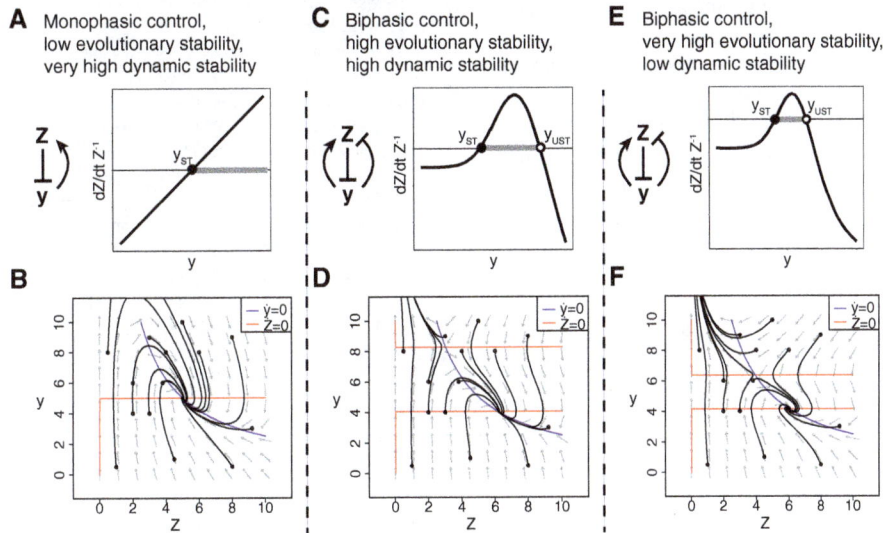

Figure 3. Tradeoff between evolutionary stability and dynamical stability.

A A monophasic feedback circuit where cells Z inhibit y which increases their growth rate. This circuit has low evolutionary stability—any mutant that mis-senses y to a higher value may invade the population

B For every initial value of $Z > 0$ or y, the circuit converges to the homeostatic set point. Nullclines are indicated by red and blue lines in the phase plots.

C A biphasic feedback circuit where Z inhibits y, which, in turn, decreases the growth rate of Z at high concentrations and increases the growth rate of Z at low concentrations. This circuit has high dynamical stability (large y_{UST}-y_{ST}) and high evolutionary stability, but mild activating mutants may invade the population.

D Large perturbations in either Z or y may result in the elimination of the cell population Z due to a runaway process. Nullclines are indicated by red and blue lines in the phase plots.

E A biphasic feedback circuit with lower dynamical stability (small y_{UST}-y_{ST}) and higher evolutionary stability, since only few mild activating mutants may invade the population (gray region).

F Small perturbations in either Z or y may result in the elimination of the cell population Z due to a runaway process. Nullclines are indicated by red and blue lines in the phase plots.

the calcium-sensing receptor (Yano *et al*, 2000) and not a strong inactivation. This is what we expect if calcium controls PT gland growth in a biphasic manner, eliminating the strong deactivating mutations.

Biphasic control in secrete-and-sense circuits in T cells and bacteria

As an additional example, we consider the evolutionary stability of a motif suggested to control cell populations known as secrete and sense (You *et al*, 2004). An experimental characterization of such a circuit *in vitro* employed the control of T-cell population size by IL2, a cytokine secreted by the T cells (Hart *et al*, 2014). In this circuit, $y = IL2$ increases both the death rate and proliferation rate of the cells at different rates, similar to the circuit depicted in Fig 1E. The resulting overall growth rate is biphasic, with negative growth rate at low ($y < y_{UST}$) and high ($y > y_{ST}$) concentrations of $IL2$ (Fig EV3C). This causes the population to have a stable fixed point at $y = y_{ST}$ and an unstable fixed point at $y = y_{UST}$. Initial seeding of T cells in plates across 4 decades of concentration led to convergence after 7 days to the same steady-state population to within a factor of 2 (Hart *et al*, 2014). This steady-state population was much lower than the carrying capacity of the system and resulted from vigorous balance of cell proliferation and death. Seeding with too few T cells, or experimental reduction in IL2, led to the elimination of the T cells. The present analysis suggests that the biphasic effects of IL2 can protect against loss-of-sensing mutants in IL2 signaling. This suggests an experiment in which such mutants are predicted to be eliminated if present at low concentrations within a wild-type population, but to take over if present at high numbers (frequency-dependent selection).

A secrete-and-sense circuit has also been synthetically engineered in bacteria by You *et al* (2004), by placing a death gene under control of a quorum sensing signal so the gene is activated when quorum signal is strong, similar to the circuit depicted in Fig 1A (Fig EV3D). This circuit maintains cell concentration constant. However, homeostatic control is rapidly lost (Balagadde, 2005) since selection favors mutants which inactivate the synthetic signaling pathway, in accord with the present predictions.

Evolutionary stable strategies in tissues with stem cells

The cases discussed so far have a population of dividing cells that is under size control. Many tissues, however, are made of nondividing differentiated cells that originate from a pool of stem cells. We consider the case of tissues in which the differentiated cells are constantly removed and must be replenished, such as blood cells and the epithelia of lungs and skin. In these cases, tissue size control requires feedback from the differentiated cells back to the stem cells (Bullough, 1975). The dividing stem-cell population is sensitive to takeover by mutants. Here, we suggest that a biphasic control mechanism can provide protection from invasion of mutants also in this case (Fig 4).

We demonstrate the effect of biphasic control by considering a monophasic circuit presented by Buzi *et al* (2015). In this circuit, differentiated cells secrete a molecule y that affects the differentiation probability of the stem cells (Fig 4A and B). The molecule y increases the differentiation rate of the stem cells and thus limits their expansion rate. This type of feedback has been demonstrated

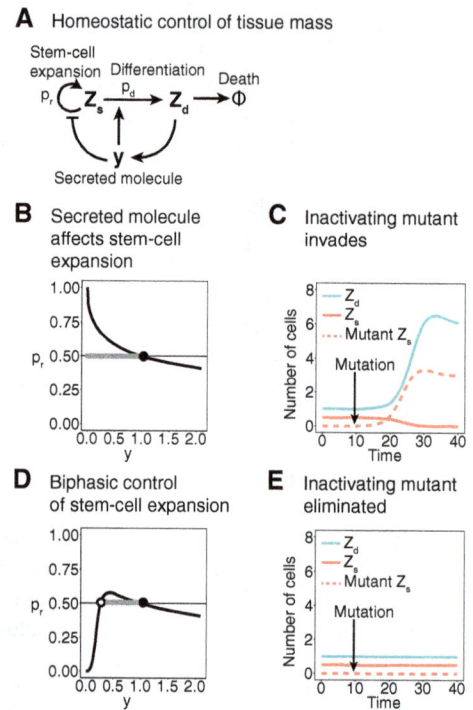

Figure 4. Biphasic control can provide mutant resistance to stem-cell homeostatic circuits.

A   Homeostatic control of a population of cells Z_d, which differentiate from a population of dividing stem cells Z_s. Differentiated cells secrete a factor y which increases the differentiation rate p_d of Z_s and therefore decreases the rate of stem-cell expansion $p_r = 1 - p_d$.

B   In a monophasic model, stem-cell expansion rate decreases with y. The system has a stable fixed point at the concentration of y where $p_r = 0.5$.

C   A mutated stem cell with a strong inactivation of the sensing of y has a growth advantage (differentiates less), and therefore, it invades the stem-cell population. As a result, both the stem-cell pool and the number of terminally differentiated cells increase.

D   Biphasic control of stem-cell expansion, where stem-cell expansion is low both at high and low concentrations of y. The system has a stable fixed point at the concentration of y where $p_r = 0.5$ and an unstable fixed point at some lower concentration of y.

E   A mutated stem cell with a strong inactivation on y sensing now has a growth disadvantage and is therefore eliminated from the stem-cell population.

for many tissues, such as blood, skin, skeletal muscle, olfactory epithelium, bone, hair, and more (Lander *et al*, 2009; Buzi *et al*, 2015). In many of these tissues, the secreted molecule belongs to the TGF-β family (Lander *et al*, 2009). The dynamic equations for the stem cells Z_s and differentiated cells Z_d are as follows:

$$\dot{Z}_s = (2p_r(y) - 1)\lambda_+ Z_s \qquad (2)$$

$$\dot{Z}_d = 2p_d(y)\lambda_+ Z_s - \lambda_- Z_d \qquad (3)$$

$$y \propto Z_d \qquad (4)$$

where λ_+ is the stem-cell division rate, λ_- is the differentiated cell removal rate, p_r is the probability that a stem cell that divided will not differentiate, and p_d is the probability that it will ($p_r = 1 - p_d$). The differentiated cells secrete molecule y that increases differentiation rate $p_d(y)$—forming a negative feedback loop (because differentiation is akin to loss of stem cells, leading to less differentiated tissue in the long term). Too many differentiated cells Z_d lead to a high level of y and to a decrease in the stem-cell population, leading to a reduction back to tissue size set point Z_{dST}. This monophasic circuit thus maintains a stable, constant population of differentiated cells (Buzi *et al*, 2015).

As above, this monophasic circuit is susceptible to invasion by loss-of-sensing mutations: stem cells that cannot sense y or that mis-sense y as too low. Within a single compartment, such a sensing mutant cannot coexist with wild-type stem cells. As this mutant stem cell differentiates less than the wild-type stem cells, it self-renews more often and has an evolutionary advantage over other stem cells. It is likely to invade the compartment and disrupt tissue homeostasis (Fig 4C). Invasion of the mutant means an exponential growth in both (mutant) stem cell and differentiated cell populations.

We find that adding a biphasic response curve can increase the evolutionary stability of this circuit (Fig 4D and E). In such biphasic control, y stimulates the growth of the stem cells at low concentrations and also stimulates differentiation at high concentrations (and thus inhibits renewal at high concentrations). Therefore, stem cells with a strong inactivating mutants that mis-sense y as too low grow less than wild type and thus have a selective disadvantage relative to other stem cells and are eliminated from the stem-cell population. The TGF-β feedback has indeed been demonstrated to have biphasic control in several cell types (Battegay *et al*, 1990; McAnulty *et al*, 1997; Cordeiro *et al*, 2000; Fosslien, 2009).

Neuronal excitotoxicity as an additional putative case for evolutionary stability/disease tradeoff

Several other diseases are associated with a biphasic response of cells to their input. Glutamate, a common neurotransmitter, has a biphasic effect on neurons—it increases neuronal survival at intermediate concentrations and causes neuronal death at low and high levels (Lipton & Nakanishi, 1999; Fig EV3E). The latter effect is called neuronal excitotoxicity. Excitotoxicity is associated with neurodegenerative diseases such as Alzheimer's, Parkinson's, and Huntington's (Coyle & Puttfarcken, 1993; Dong *et al*, 2009).

We speculate that the biphasic effect of glutamate on neuronal survival may be beneficial for the elimination of neurons with improper sensing. Such defective neurons may arise due to somatic mutation in the brain, either in mature neurons (Lodato *et al*, 2015) or in neuronal progenitors (Poduri *et al*, 2013). In order to evaluate the role of the biphasic effect of glutamate on neuronal evolutionary stability, it is necessary to better characterize the homeostatic feedback circuits that control neuronal mass dynamics.

Discussion

In this study, we raise the question of the stability of circuits that control tissue size with respect to invasion by mutants. We consider feedback circuits that provide size control by regulating cell growth according to an input signal proportional to the number of cells. We show that such feedback mechanisms can be invaded by commonly occurring mutants, which have loss-of-sensing or locked-on sensing of the input signal. Invasion leads to aberrant tissue size and function.

We find that these common mutants can be eliminated by a biphasic control mechanism. In biphasic control, the signal is toxic at both high and low levels, giving the mutants a selective disadvantage. The biphasic protection mechanism comes at a cost: It introduces an unstable fixed point that can cause a runaway phenomenon under strong fluctuations in the input signal, potentially leading to disease. This study thus provides an explanation for several well-studied toxicity phenomena associated with diseases, by suggesting that they have a beneficial function of protecting tissues from invasion by common mutants.

The biphasic control mechanism protects against strong sensing mutations, such as loss-of-function or locked-on receptors. These strong mutations presumably have a large mutational target size and are thus the most commonly arising mutations in a tissue. The control mechanism is sensitive, however, to a range of mild sensing mutations. These mutations cause the cell to mis-interpret the signal level, to a level that lies between the desired steady-state level y_{ST} and the unstable fixed point y_{UST} (Fig 1). It is likely that such mild mutations are more rare than loss-of-function or locked-on mutations. The vulnerability to these mild mutations might explain the recurrence of a few specific point mutations of mild effect in sensing pathways in cancer, presumably because mutations of larger effect are eliminated (Hanahan & Weinberg, 2011).

There is a tradeoff between evolutionary stability—the range of mild mutations that can invade, and dynamic stability—the position of the unstable fixed point. The closer the y_{ST} is to y_{UST}, the higher the evolutionary stability and the lower the dynamical stability. As y_{UST} approaches y_{ST}, we expect to see critical slowing down of the dynamics of the system and a general loss of resilience to perturbations (Scheffer *et al*, 2009). Such critical slowing down was shown to occur in populations of yeast in response to dilution (Dai *et al*, 2012) as well as in genetic circuits (Axelrod *et al*, 2015).

This study maps the concept of evolutionary stable strategies (ESS) from evolutionary ecology (Smith & Price, 1973) to the level of cell circuits in tissues. In ecology, an ESS is defined when a population of organisms with that strategy cannot be invaded by any other strategy. In the present study, a mis-sensing mutant is analogous to the invading strategy. In this sense, the biphasic mechanism is evolutionarily stable with respect to strong mutations. It is unstable to a range of mild mutations. The evolutionary instability to mild mutations can be reduced using compartments with small cell numbers, low turnover rates, and proximity of the stable and unstable fixed points, as described by equation (1).

As in ESS in ecology, selection of sensing mutants is frequency dependent: If the entire population is mutant, it can survive. But a single-mutant cell on a background of wild-type cells is eliminated by the biphasic mechanism. Experiments show the predicted frequency-dependent selection of a strong glucokinase mutant in beta cells: When present at low frequency, the mutants are eliminated; when present in the germ line, they survive and cause hypoglycemia.

Many of the biphasic toxicity phenomena considered here are mediated by excess production of reactive oxygen species (ROS) which leads to apoptosis (Coyle & Puttfarcken, 1993; Hildeman *et al*, 1999; Schulz *et al*, 2002; Robertson, 2004). ROS are implicated in both beta-cell glucotoxicity and neuronal excitotoxicity. Such toxicity can be mitigated by antioxidants, which reduce ROS levels (Skulachev, 1998). Thus, the level of antioxidants may, in principle, tune the tradeoff between evolutionary stability and dynamic stability described here. High antioxidants can reduce the toxicity of high signal level and thus push the unstable fixed point farther from the stable fixed point. This can reduce the risk of disease, but increase susceptibility to invasion by mild mutants. This tradeoff may provide a viewpoint to understand the conflicting effects of antioxidants on health (Bjelakovic *et al*, 2012; Sayin *et al*, 2014; Le Gal *et al*, 2015).

In this study, we discussed circuits where a tissue regulates its own size. Some tissues, however, regulate the size of other tissues. For example, the ovaries regulate mammary epithelial mass by secreting estrogen, and the pituitary gland regulates the mass of the thyroid and adrenal glands by secreting TSH and ACTH, respectively. Depending on the feedback loops at play, such circuits may be susceptible to mutant invasion both in the regulating and regulated tissue. The considerations of this study indicate that biphasic control reduces the susceptibility to invading mutants in these cases as well. We therefore predict biphasic responses also when tissues regulate each other. For example, estrogen controls mammary growth in a biphasic manner (Lewis-Wambi & Jordan, 2009), therefore reducing the target range of mutants with a fitness advantage in the mammary epithelium.

Finally, biphasic control raises the question of how tissues can start growing. Consider the tissue in Fig 1E, in which Z produces y. If initially $y = 0$, $Z = \varepsilon$, then the tissue has negative growth rate and cannot grow to reach $Z = Z_{ST}$. This can be resolved if y is determined externally during tissue development. For example, during gestation, metabolites and factors are supplied to the fetus externally by the mother at levels close to y_{ST}. Another possibility is that tissue development is determined by a different program that is later suppressed.

In summary, we show that physiological feedback circuits are inherently vulnerable to takeover by mutants that mis-sense the feedback signal. Biphasic mechanisms, in which the signal is toxic at both high and low levels to the relevant tissue, can protect against such mutant invasion. We therefore hypothesize that phenomena such as glucotoxicity and excitotoxicity may reflect the bad side of a good anti-mutant strategy (Stearns & Medzhitov, 2016). Characterizing physiological homeostatic circuits and the tradeoffs they face in quantitative detail may thus lead to a better understanding of diabetes (Topp *et al*, 2000), neurodegenerative diseases (Doble, 1999; Lipton & Nakanishi, 1999), and possibly other pathologies associated with biphasic control.

Materials and Methods

Circuits with monophasic and biphasic control

To simulate the circuits of Fig 1 in the main text, we used a circuit where a cell mass Z either increases the level of its input y (Fig 1A and E) or decreases the level of y (Fig 1B and F). The equation used for Z is follows:

$$\dot{Z} = Z \cdot (\lambda_+(y) - \lambda_-(y)) = Z \cdot \lambda(y) \tag{5}$$

where λ_+ is the y-dependent proliferation rate of Z, and λ_- is the y-dependent removal rate of Z.

In Fig 1C, D, G and H, we simulated two cases—a monophasic circuit, where y increases the growth rate of Z, and a biphasic circuit, where y increases the growth rate of Z at low concentrations and decreases the growth rate of Z at high concentrations. The monophasic circuit was simulated using the growth rate equations:

$$\lambda_+(y) = \frac{y}{10} \tag{6}$$

$$\lambda_-(y) = 0.5 \tag{7}$$

and the biphasic circuit was simulated by using the growth rate equations:

$$\lambda_+(y) = \frac{4.8}{1 + \left(\frac{7}{y}\right)^5} \tag{8}$$

$$\lambda_-(y) = \frac{6}{1 + \left(\frac{8}{y}\right)^5} + 0.1 \tag{9}$$

These circuits were also used to simulate the phase plots in Fig 3A and B. For Fig 3C, we used the following circuit:

$$\lambda_+(y) = \frac{4.8}{1 + \left(\frac{5.5}{y}\right)^6} \tag{10}$$

$$\lambda_-(y) = \frac{6}{1 + \left(\frac{6.3}{y}\right)^6} + 0.3 \tag{11}$$

We used the following equation for the dependence of y on Z:

$$\dot{y} = \mu \cdot (M - Zy) \tag{12}$$

This equation means that Z increases the degradation rate of y, and at steady state, we get $Z_{st}y_{st} = M$. We chose the parameters $M = 25, \mu = 0.25$.

Mutant invasion simulation

We simulated the effect of a mutation by adding a term Z_{mut} such that:

$$\dot{y} = \mu \cdot (M - (Z + Z_{mut})y) \tag{13}$$

Z_{mut} represents the mass of cells with a (given) k-fold sensing mutation on y, so the growth rate of Z_{mut} is given as follows:

$$\dot{Z}_{mut} = Z_{mut}\lambda(ky) = Z_{mut}(\lambda_+(ky) - \lambda_-(ky)) \tag{14}$$

Note that for the monophasic circuit simulated in Fig 1, the removal rate λ_- does not depend on y, and therefore, it is not affected by the sensing mutation (only λ_+ is affected). We simulated the invasion of a fourfold sensing mutant in Fig 1D and H by setting $Z_{mut} \leftarrow 1$ at specific time intervals in the simulation ($t = 10$ for the monophasic circuit and $t = 10$, $t = 47$ for the biphasic circuit). The initial values for the simulations were $Z_{mut0} \leftarrow 0$, $Z_0 \leftarrow 5$, $y_0 \leftarrow 4$ for the monophasic circuit in Fig 1D, and $Z_{mut0} \leftarrow 0$, $Z_0 \leftarrow 6.16$, $y_0 \leftarrow 4.06$ for the biphasic circuit in Fig 1H.

Circuits of communicating stem cells

In this study, we presented two circuits that regulate the functional mass of differentiated cells, based on the model that is presented in Buzi *et al* (2015). For the monophasic circuit, the equations are as follows:

$$\dot{Z}_s = (2p_r(y) - 1)\lambda_+ Z_s \tag{15}$$

$$\dot{Z}_{s_{mut}} = (2p_r(ky) - 1)\lambda_+ Z_{s_{mut}} \tag{16}$$

$$\dot{Z}_d = 2(1 - p_r(y))\lambda_+ Z_s + 2(1 - p_r(ky))\lambda_+ Z_{s_{mut}} - \lambda_- Z_d \tag{17}$$

$$y \propto Z_d \tag{18}$$

where λ_+ is the stem-cell division rate, λ_- is the differentiated cell removal rate, p_r is the probability that a stem cell that divided will not differentiate, and $1 - p_r$ is the probability that it will differentiate. The population $Z_{s_{mut}}$ is the population of stem cells with a k-fold sensing mutation. The monophasic replication rate $p_r(y)$, which is depicted in Fig 3B, was set as follows:

$$p_r(y) = \frac{1}{1 + \sqrt{y}} \tag{19}$$

The exact function used is not important, since as long as it is monotonically decreasing, an invading mutant will take over. In the biphasic case, the replication rate used is as follows:

$$p_r(y) = \frac{1}{1 + \sqrt{y}} \cdot \frac{1}{1 + \left(\frac{1}{5y}\right)^4} \tag{20}$$

The simulation of invading mutants is the same as for Fig 1 (which is explained in the mutant invasion simulation section). For the simulations, we set $\lambda_+ \leftarrow 1, \lambda_- \leftarrow 0.5$, $k = \frac{1}{6}$, and with the initial conditions $Z_{s0} \leftarrow 0.5$, $Z_{s_{mut}0} \leftarrow 0$, $Z_{d0} \leftarrow 1$. A mutation event was set such that $Z_{s_{mut}} \leftarrow 0.01$ at $t = 10$.

Acknowledgements

This work was supported by the Israel Science Foundation (1349/15) and the Minerva Foundation. UA is the incumbent of the Abisch-Frenkel Professorial Chair. OK is supported by the Azrieli Center for Systems Biology grant.

Author contributions

OK and UA conceived and performed the research. OK and UA wrote the manuscript.

References

Allard P, Delvin EE, Paradis G, Hanley JA, O'Loughlin J, Lavallée C, Levy E, Lambert M (2003) Distribution of fasting plasma insulin, free fatty acids,

and glucose concentrations and of homeostasis model assessment of insulin resistance in a representative sample of Quebec children and adolescents. *Clin Chem* 49: 644–649

Axelrod K, Sanchez A, Gore J (2015) Phenotypic states become increasingly sensitive to perturbations near a bifurcation in a synthetic gene network. *eLife* 4: e07935

Balagadde FK (2005) Long-term monitoring of bacteria undergoing programmed population control in a microchemostat. *Science* 309: 137–140

Battegay EJ, Raines EW, Seifert RA, Bowen-Pope DF, Ross R (1990) TGF-beta induces bimodal proliferation of connective tissue cells via complex control of an autocrine PDGF loop. *Cell* 63: 515–524

Bensellam M, Laybutt DR, Jonas J-C (2012) The molecular mechanisms of pancreatic β-cell glucotoxicity: recent findings and future research directions. *Mol Cell Endocrinol* 364: 1–27

Bilezikian JP, Marcus R, Levine MA (eds) (2001) *The parathyroids: basic and clinical concepts*, 2nd edn. San Diego, CA: Academic Press

Bjelakovic G, Nikolova D, Gluud LL, Simonetti RG, Gluud C (2012) Antioxidant supplements for prevention of mortality in healthy participants and patients with various diseases. *Cochrane Database Syst Rev* CD007176

Bouchard C, Staller P, Eilers M (1998) Control of cell proliferation by Myc. *Trends Cell Biol* 8: 202–206

Bullough WS (1975) Mitotic control in adult mammalian tissues. *Biol Rev* 50: 99–127

Buzi G, Lander AD, Khammash M (2015) Cell lineage branching as a strategy for proliferative control. *BMC Biol* 13: 13–28

Cordeiro MF, Bhattacharya SS, Schultz GS, Khaw PT (2000) TGF-beta1, -beta2, and -beta3 *in vitro*: biphasic effects on Tenon's fibroblast contraction, proliferation, and migration. *Invest Ophthalmol Vis Sci* 41: 756–763

Coyle J, Puttfarcken P (1993) Oxidative stress, glutamate, and neurodegenerative disorders. *Science* 262: 689–695

Critchfield JM, Racke MK, Zúñiga-Pflücker JC, Cannella B, Raine CS, Goverman J, Lenardo MJ (1994) T cell deletion in high antigen dose therapy of autoimmune encephalomyelitis. *Science* 263: 1139–1143

Cuesta-Munoz AL, Huopio H, Otonkoski T, Gomez-Zumaquero JM, Nanto-Salonen K, Rahier J, Lopez-Enriquez S, Garcia-Gimeno MA, Sanz P, Soriguer FC, Laakso M (2004) Severe persistent hyperinsulinemic hypoglycemia due to a *de novo* glucokinase mutation. *Diabetes* 53: 2164–2168

Dai L, Vorselen D, Korolev KS, Gore J (2012) Generic indicators for loss of resilience before a tipping point leading to population collapse. *Science* 336: 1175–1177

De Gaetano A, Hardy T, Beck B, Abu-Raddad E, Palumbo P, Bue-Valleskey J, Porksen N (2008) Mathematical models of diabetes progression. *Am J Physiol Endocrinol Metab* 295: E1462–E1479

Del Prato S (2009) Role of glucotoxicity and lipotoxicity in the pathophysiology of Type 2 diabetes mellitus and emerging treatment strategies. *Diabet Med* 26: 1185–1192

Doble A (1999) The role of excitotoxicity in neurodegenerative disease: implications for therapy. *Pharmacol Ther* 81: 163–221

Dong X, Wang Y, Qin Z (2009) Molecular mechanisms of excitotoxicity and their relevance to pathogenesis of neurodegenerative diseases. *Acta Pharmacol Sin* 30: 379–387

Efanova IB, Zaitsev SV, Zhivotovsky B, Köhler M, Efendić S, Orrenius S, Berggren PO (1998) Glucose and tolbutamide induce apoptosis in pancreatic beta-cells. A process dependent on intracellular Ca^{2+} concentration. *J Biol Chem* 273: 33501–33507

Evan GI, Wyllie AH, Gilbert CS, Littlewood TD, Land H, Brooks M, Waters CM, Penn LZ, Hancock DC (1992) Induction of apoptosis in fibroblasts by c-myc protein. *Cell* 69: 119–128

Eyre-Walker A, Keightley PD (2007) The distribution of fitness effects of new mutations. *Nat Rev Genet* 8: 610–618

Fajans SS, Bell GI, Polonsky KS (2001) Molecular mechanisms and clinical pathophysiology of maturity-onset diabetes of the young. *N Engl J Med* 345: 971–980

Ferrannini E, Bjorkman O, Reichard GA, Pilo A, Olsson M, Wahren J, DeFronzo RA (1985) The disposal of an oral glucose load in healthy subjects: a quantitative study. *Diabetes* 34: 580–588

Fosslien E (2009) The hormetic morphogen theory of curvature and the morphogenesis and pathology of tubular and other curved structures. *Dose Response* 7: 307–331

Fraser WD (2009) Hyperparathyroidism. *Lancet* 374: 145–158

Froguel P, Zouali H, Vionnet N, Velho G, Vaxillaire M, Sun F, Lesage S, Stoffel M, Takeda J, Passa P (1993) Familial hyperglycemia due to mutations in glucokinase. Definition of a subtype of diabetes mellitus. *N Engl J Med* 328: 697–702

Glaser B, Kesavan P, Heyman M, Davis E, Cuesta A, Buchs A, Stanley CA, Thornton PS, Permutt MA, Matschinsky FM, Herold KC (1998) Familial hyperinsulinism caused by an activating glucokinase mutation. *N Engl J Med* 338: 226–230

Gogusev J, Duchambon P, Hory B, Giovannini M, Goureau Y, Sarfati E, Drüeke TB (1997) Depressed expression of calcium receptor in parathyroid gland tissue of patients with hyperparathyroidism. *Kidney Int* 51: 328–336

González C, Ray JCJ, Manhart M, Adams RM, Nevozhay D, Morozov AV, Balázsi G (2015) Stress-response balance drives the evolution of a network module and its host genome. *Mol Syst Biol* 11: 827

Gudipaty SA, Lindblom J, Loftus PD, Redd MJ, Edes K, Davey CF, Krishnegowda V, Rosenblatt J (2017) Mechanical stretch triggers rapid epithelial cell division through Piezo1. *Nature* 543: 118–121

Ha J, Satin LS, Sherman AS (2016) A mathematical model of the pathogenesis, prevention, and reversal of type 2 diabetes. *Endocrinology* 157: 624–635

Hanahan D, Weinberg RA (2011) Hallmarks of cancer: the next generation. *Cell* 144: 646–674

Hardingham GE, Bading H (2003) The yin and yang of NMDA receptor signalling. *Trends Neurosci* 26: 81–89

Harrington EA, Fanidi A, Evan GI (1994) Oncogenes and cell death. *Curr Opin Genet Dev* 4: 120–129

Hart Y, Reich-Zeliger S, Antebi YE, Zaretsky I, Mayo AE, Alon U, Friedman N (2014) Paradoxical signaling by a secreted molecule leads to homeostasis of cell levels. *Cell* 158: 1022–1032

Hildeman DA, Mitchell T, Teague TK, Henson P, Day BJ, Kappler J, Marrack PC (1999) Reactive oxygen species regulate activation-induced T cell apoptosis. *Immunity* 10: 735–744

Högnason T, Chatterjee S, Vartanian T, Ratan RR, Ernewein KM, Habib AA (2001) Epidermal growth factor receptor induced apoptosis: potentiation by inhibition of Ras signaling. *FEBS Lett* 491: 9–15

James C, Kapoor RR, Ismail D, Hussain K (2009) The genetic basis of congenital hyperinsulinism. *J Med Genet* 46: 289–299

Karin O, Swisa A, Glaser B, Dor Y, Alon U (2016) Dynamical compensation in physiological circuits. *Mol Syst Biol* 12: 886

Lander AD, Gokoffski KK, Wan FYM, Nie Q, Calof AL (2009) Cell lineages and the logic of proliferative control. *PLoS Biol* 7: e1000015

Le Gal K, Ibrahim MX, Wiel C, Sayin VI, Akula MK, Karlsson C, Dalin MG, Akyurek LM, Lindahl P, Nilsson J, Bergo MO (2015) Antioxidants can increase melanoma metastasis in mice. *Sci Transl Med* 7: 308re8.

Leslie RDG, Robbins DC (1995) *Diabetes: clinical science in practice Cambridge.* New York, NY: Cambridge University Press

Lewis-Wambi JS, Jordan VC (2009) Estrogen regulation of apoptosis: how can one hormone stimulate and inhibit? *Breast Cancer Res* 11: 206

Lipton SA, Nakanishi N (1999) Shakespeare in love—with NMDA receptors? *Nat Med* 5: 270–271

Lodato MA, Woodworth MB, Lee S, Evrony GD, Mehta BK, Karger A, Lee S, Chittenden TW, D'Gama AM, Cai X, Luquette LJ, Lee E, Park PJ, Walsh CA (2015) Somatic mutation in single human neurons tracks developmental and transcriptional history. *Science* 350: 94–98

Lowe SW, Cepero E, Evan G (2004) Intrinsic tumour suppression. *Nature* 432: 307–315

MacDonald PE, Joseph JW, Rorsman P (2005) Glucose-sensing mechanisms in pancreatic -cells. *Philos Trans R Soc B Biol Sci* 360: 2211–2225

Maedler K, Schumann DM, Schulthess F, Oberholzer J, Bosco D, Berney T, Donath MY (2006) Aging correlates with decreased beta-cell proliferative capacity and enhanced sensitivity to apoptosis: a potential role for Fas and pancreatic duodenal homeobox-1. *Diabetes* 55: 2455–2462

Malberti F (1999) The PTH-calcium curve and the set point of calcium in primary and secondary hyperparathyroidism. *Nephrol Dial Transplant* 14: 2398–2406

Matschinsky FM (2002) Regulation of pancreatic beta-cell glucokinase: from basics to therapeutics. *Diabetes* 51(Suppl 3): S394–S404

McAnulty RJ, Hernández-Rodriguez NA, Mutsaers SE, Coker RK, Laurent GJ (1997) Indomethacin suppresses the anti-proliferative effects of transforming growth factor-beta isoforms on fibroblast cell cultures. *Biochem J* 321(Pt 3): 639–643

Meier JJ, Butler AE, Saisho Y, Monchamp T, Galasso R, Bhushan A, Rizza RA, Butler PC (2008) Cell replication is the primary mechanism subserving the postnatal expansion of -cell mass in humans. *Diabetes* 57: 1584–1594

Mizobuchi M, Ogata H, Hatamura I, Saji F, Koiwa F, Kinugasa E, Koshikawa S, Akizawa T (2007) Activation of calcium-sensing receptor accelerates apoptosis in hyperplastic parathyroid cells. *Biochem Biophys Res Commun* 362: 11–16

Naveh-Many T, Rahamimov R, Livni N, Silver J (1995) Parathyroid cell proliferation in normal and chronic renal failure rats. The effects of calcium, phosphate, and vitamin D. *J Clin Invest* 96: 1786–1793

Nowak MA (2006) *Evolutionary dynamics: exploring the equations of life Cambridge.* Mass: Belknap Press of Harvard University Press

Poduri A, Evrony GD, Cai X, Walsh CA (2013) Somatic mutation, genomic variation, and neurological disease. *Science* 341: 1237758–1237758

Porat S, Weinberg-Corem N, Tornovsky-Babaey S, Schyr-Ben-Haroush R, Hija A, Stolovich-Rain M, Dadon D, Granot Z, Ben-Hur V, White P, Girard CA, Karni R, Kaestner KH, Ashcroft FM, Magnuson MA, Saada A, Grimsby J, Glaser B, Dor Y (2011) Control of pancreatic β cell regeneration by glucose metabolism. *Cell Metab* 13: 440–449

Robertson RP, Harmon J, Tran PO, Tanaka Y, Takahashi H (2003) Glucose toxicity in beta-cells: type 2 diabetes, good radicals gone bad, and the glutathione connection. *Diabetes* 52: 581–587

Robertson RP (2004) Chronic oxidative stress as a central mechanism for glucose toxicity in pancreatic islet beta cells in diabetes. *J Biol Chem* 279: 42351–42354

Saisho Y, Butler AE, Manesso E, Elashoff D, Rizza RA, Butler PC (2013) Cell mass and turnover in humans: effects of obesity and aging. *Diabetes Care* 36: 111–117

Sarkisyan KS, Bolotin DA, Meer MV, Usmanova DR, Mishin AS, Sharonov GV, Ivankov DN, Bozhanova NG, Baranov MS, Soylemez O, Bogatyreva NS, Vlasov PK, Egorov ES, Logacheva MD, Kondrashov AS, Chudakov DM, Putintseva EV, Mamedov IZ, Tawfik DS, Lukyanov KA *et al* (2016) Local fitness landscape of the green fluorescent protein. *Nature* 533: 397–401

Sayin VI, Ibrahim MX, Larsson E, Nilsson JA, Lindahl P, Bergo MO (2014) Antioxidants accelerate lung cancer progression in mice. *Sci Transl Med* 6: 221ra15

Scheffer M, Bascompte J, Brock WA, Brovkin V, Carpenter SR, Dakos V, Held H, van Nes EH, Rietkerk M, Sugihara G (2009) Early-warning signals for critical transitions. *Nature* 461: 53–59

Schulz JB, Henshaw DR, Siwek D, Jenkins BG, Ferrante RJ, Cipolloni PB, Kowall NW, Rosen BR, Beal MF (2002) Involvement of free radicals in excitotoxicity *in vivo*. *J Neurochem* 64: 2239–2247

Seifert R, Wenzel-Seifert K (2002) Constitutive activity of G-protein-coupled receptors: cause of disease and common property of wild-type receptors. *Naunyn Schmiedebergs Arch Pharmacol* 366: 381–416

Skulachev VP (1998) Cytochrome c in the apoptotic and antioxidant cascades. *FEBS Lett* 423: 275–280

Smith JM, Price GR (1973) The logic of animal conflict. *Nature* 246: 15–18

Stearns SC, Medzhitov R (2016) *Evolutionary medicine* Sunderland, MA: Sinauer Associates, Inc., Publishers

Stolovich-Rain M, Hija A, Grimsby J, Glaser B, Dor Y (2012) Pancreatic beta cells in very old mice retain capacity for compensatory proliferation. *J Biol Chem* 287: 27407–27414

Topp B, Promislow K, deVries G, Miura RM, Finegood DT (2000) A model of beta-cell mass, insulin, and glucose kinetics: pathways to diabetes. *J Theor Biol* 206: 605–619

Tornovsky-Babeay S, Dadon D, Ziv O, Tzipilevich E, Kadosh T, Schyr-Ben Haroush R, Hija A, Stolovich-Rain M, Furth-Lavi J, Granot Z, Porat S,

Philipson LH, Herold KC, Bhatti TR, Stanley C, Ashcroft FM, In't Veld P, Saada A, Magnuson MA, Glaser B et al (2014) Type 2 diabetes and congenital hyperinsulinism cause DNA double-strand breaks and p53 activity in β cells. *Cell Metab* 19: 109–121

Wada M, Furuya Y, Sakiyama J, Kobayashi N, Miyata S, Ishii H, Nagano N (1997) The calcimimetic compound NPS R-568 suppresses parathyroid cell proliferation in rats with renal insufficiency. Control of parathyroid cell growth via a calcium receptor. *J Clin Invest* 100: 2977–2983

Wang H, Iynedjian PB (1997) Modulation of glucose responsiveness of insulinoma beta-cells by graded overexpression of glucokinase. *Proc Natl Acad Sci USA* 94: 4372–4377

Yano S, Sugimoto T, Tsukamoto T, Chihara K, Kobayashi A, Kitazawa S, Maeda S, Kitazawa R (2000) Association of decreased calcium-sensing receptor expression with proliferation of parathyroid cells in secondary hyperparathyroidism. *Kidney Int* 58: 1980–1986

You L, Cox RS, Weiss R, Arnold FH (2004) Programmed population control by cell–cell communication and regulated killing. *Nature* 428: 868–871

3

RNA polymerase II primes Polycomb-repressed developmental genes throughout terminal neuronal differentiation

Carmelo Ferrai[1,2,3,*,†] (iD), Elena Torlai Triglia[1,†] (iD), Jessica R Risner-Janiczek[3,4,5], Tiago Rito[1], Owen JL Rackham[6], Inês de Santiago[2,3,§], Alexander Kukalev[1], Mario Nicodemi[7], Altuna Akalin[8], Meng Li[3,4,¶], Mark A Ungless[3,5,**] (iD) & Ana Pombo[1,2,3,9,***] (iD)

Abstract

Polycomb repression in mouse embryonic stem cells (ESCs) is tightly associated with promoter co-occupancy of RNA polymerase II (RNAPII) which is thought to prime genes for activation during early development. However, it is unknown whether RNAPII poising is a general feature of Polycomb repression, or is lost during differentiation. Here, we map the genome-wide occupancy of RNAPII and Polycomb from pluripotent ESCs to non-dividing functional dopaminergic neurons. We find that poised RNAPII complexes are ubiquitously present at Polycomb-repressed genes at all stages of neuronal differentiation. We observe both loss and acquisition of RNAPII and Polycomb at specific groups of genes reflecting their silencing or activation. Strikingly, RNAPII remains poised at transcription factor genes which are silenced in neurons through Polycomb repression, and have major roles in specifying other, non-neuronal lineages. We conclude that RNAPII poising is intrinsically associated with Polycomb repression throughout differentiation. Our work suggests that the tight interplay between RNAPII poising and Polycomb repression not only instructs promoter state transitions, but also may enable promoter plasticity in differentiated cells.

Keywords cell plasticity; chromatin bivalency; gene regulation; RNA polymerase II; transcriptional poising
Subject Categories Chromatin, Epigenetics, Genomics & Functional Genomics; Genome-Scale & Integrative Biology; Transcription

Introduction

Embryonic differentiation starts from a totipotent cell and culminates with the production of highly specialized cells. In ESCs, many genes important for early development are repressed in a state that is poised for subsequent activation (Azuara et al, 2006; Bernstein et al, 2006; Stock et al, 2007; Brookes et al, 2012). These genes are mostly GC-rich (Deaton & Bird, 2011), and their silencing in pluripotent cells is mediated by Polycomb repressive complexes (PRCs). Genes with more specialized cell type-specific functions are neither active nor Polycomb repressed in ESCs, have AT-rich promoter sequences, and their activation is associated with specific transcription factors (Sandelin et al, 2007; Brookes et al, 2012).

Polycomb repressive complex proteins have major roles in modulating gene expression during differentiation and in disease (Prezioso & Orlando, 2011; Richly et al, 2011). They assemble in two major complexes, PRC1 and PRC2, which catalyze H2AK119 monoubiquitination and H3K27 methylation, respectively (Simon & Kingston, 2013). Both PRC-mediated histone marks are important for chromatin repression, and synergize in a tight feedback loop to recruit each other's modifying enzymes (Blackledge et al, 2015).

1 Epigenetic Regulation and Chromatin Architecture, Max Delbrück Center for Molecular Medicine, Berlin, Germany
2 Genome Function, MRC London Institute of Medical Sciences (previously MRC Clinical Sciences Centre), London, UK
3 Institute of Clinical Sciences (ICS), Faculty of Medicine, Imperial College London, London, UK
4 Stem Cell Neurogenesis, MRC London Institute of Medical Sciences (previously MRC Clinical Sciences Centre), London, UK
5 Neurophysiology Group, MRC London Institute of Medical Sciences (previously MRC Clinical Sciences Centre), London, UK
6 Duke-NUS Medical School, Singapore, Singapore
7 Dipartimento di Fisica, Università di Napoli Federico II and INFN Napoli, Complesso Universitario di Monte Sant'Angelo, Naples, Italy
8 Scientific Bioinformatics Platform, Berlin Institute for Medical Systems Biology, Max Delbrück Center for Molecular Medicine, Berlin, Germany
9 Institute for Biology, Humboldt-Universität zu Berlin, Berlin, Germany
*Corresponding author. E-mail: carmelo.ferrai@mdc-berlin.de
**Corresponding author. E-mail: mark.ungless@imperial.ac.uk
***Corresponding author (lead contact). E-mail: ana.pombo@mdc-berlin.de
†These authors contributed equally to this work
§Present address: Seven Bridges Genomics UK Ltd, London, UK
¶Present address: Neuroscience and Mental Health Research Institute, School of Medicine and School of Biosciences, Cardiff, UK

Although imaging studies suggest that PRC-repressed chromatin has a compact conformation (Francis *et al*, 2004; Eskeland *et al*, 2010; Boettiger *et al*, 2016), molecular and cell biology approaches show that PRC repression in ESCs coincides with the occupancy of poised RNAPII complexes and active histone marks in the vast majority of PRC-repressed promoters (Azuara *et al*, 2006; Bernstein *et al*, 2006; Stock *et al*, 2007; Brookes *et al*, 2012; Tee *et al*, 2014; Kinkley *et al*, 2016; Weiner *et al*, 2016). The co-occurrence of RNAPII and PRC enzymatic activities, RING1B and EZH2, on chromatin has been confirmed in ESCs by sequential ChIP (Brookes *et al*, 2012) and is mirrored by the simultaneous presence of H3K4me3 and H3K27me3 marks, called chromatin bivalency (Azuara *et al*, 2006; Bernstein *et al*, 2006; Voigt *et al*, 2012; Kinkley *et al*, 2016; Sen *et al*, 2016; Weiner *et al*, 2016).

RNAPII function is regulated through complex post-translational modifications at the C-terminal domain (CTD) of its largest subunit, RPB1, which coordinate the co-transcriptional recruitment of chromatin modifiers and RNA processing machinery to chromatin, leading to productive transcription events and mRNA expression (Brookes & Pombo, 2009; Zaborowska *et al*, 2016). In mammals, the CTD comprises 52 repeats of the heptapeptide sequence N-Tyr1-Ser2-Pro3-Thr4-Ser5-Pro6-Ser7-C. At active genes, RNAPII is phosphorylated on Ser5 (S5p) to mark transcription initiation, on Ser7 (S7p) during the transition to productive transcription, and on Ser2 (S2p) during elongation. S5p and S7p are mediated by CDK7, while S2p is mediated by CDK9. RNAPII also exists in a paused state of activation characterized by short transcription events at promoter regions, followed by promoter-proximal termination and re-initiation events (Adelman & Lis, 2012). RNAPII pausing is identified at genes that produce mRNA at lower levels and is often measured as the amount of RNAPII at gene promoters relative to its occupancy throughout coding regions. Paused states of RNAPII are therefore a feature of genes that are active to a lower extent, are characterized by the presence of S5p and S7p at gene promoters, low abundance of S2p throughout the coding regions, and they are recognized by 8WG16, an antibody which has a preference for unphosphorylated Ser2 residues. The paused RNAPII complex is also characterized by methylation and acetylation of the non-canonical Lys7 residues at the CTD (Schröder *et al*, 2013; Dias *et al*, 2015; Voss *et al*, 2015).

The RNAPII complex that primes PRC-repressed genes in mouse ESCs has a unique configuration of post-translational modifications of the CTD, which is different from the paused RNAPII, and was originally referred to as *poised* RNAPII (Stock *et al*, 2007; Brookes & Pombo, 2009). The poised RNAPII is characterized by exclusive phosphorylation of S5p in the absence of S7p, S2p, K7me1/2, K7ac, or recognition by 8WG16 (Brookes *et al*, 2012; Dias *et al*, 2015). Poised RNAPII-S5p, in the absence of 8WG16, has not been described in *Drosophila* (Gaertner *et al*, 2012), consistent with lack of chromatin bivalency (Vastenhouw & Schier, 2012; Voigt *et al*, 2013). Importantly, Ser5 phosphorylation of poised RNAPII complexes at Polycomb-repressed genes in ESCs is mediated by different kinases, ERK1/2 (Tee *et al*, 2014; Ma *et al*, 2016), instead of CDK7 which phosphorylates both S5p and S7p at active genes, irrespectively of pausing ratio (Akhtar *et al*, 2009; Glover-Cutter *et al*, 2009). Loss of ERK1/2 activity in ESCs results in the loss of poised RNAPII-S5p and decreased occupancy of PRC2 at

Polycomb-repressed developmental genes (Tee *et al*, 2014), suggesting a tight functional link between the presence of poised RNAPII-S5p at Polycomb-target genes and the recruitment of Polycomb.

While histone bivalency has been studied to some extent during mammalian cell differentiation and found present at smaller proportion of genes (Mohn *et al*, 2008; Lien *et al*, 2011; Wamstad *et al*, 2012; Xie *et al*, 2013), it remains unexplored whether the co-occupancy of poised RNAPII-S5p at PRC targets is a property of ESCs or extends beyond pluripotency. The poised RNAPII-S5p state was observed at Polycomb-repressed genes in ESCs grown in the presence of serum and leukemia inhibitor factor (LIF; Stock *et al*, 2007; Brookes *et al*, 2012; Tee *et al*, 2014; Ma *et al*, 2016). Other studies grow ESCs in 2i conditions to simulate a more naïve pluripotent state, through inhibition of GSK3 and MEK signaling, which in turn inhibits ERK signaling. In these conditions, the occupancy of poised RNAPII complexes is reduced at Polycomb-target genes (Marks *et al*, 2012; Williams *et al*, 2015), consistent with the effects of ERK1-2 inhibition (Tee *et al*, 2014). Interestingly, the decreased occupancy of poised RNAPII-S5p at PRC-repressed genes in 2i conditions is accompanied by reduced occupancy of PRC2 catalytic subunit EZH2 and H3K27me3 modification, suggesting a tight interplay between the presence of poised RNAPII-S5p and Polycomb occupancy at Polycomb-repressed genes in ESCs, which is interfered upon in 2i conditions. Interestingly, prolonged 2i treatment was shown to impair ESC developmental potential and cause widespread loss of DNA methylation (Choi *et al*, 2017; Yagi *et al*, 2017), leading to renewed interest in understanding the regulation of developmental genes, and in particular whether poised RNAPII complexes are a more general feature of Polycomb repression mechanisms in the early and late stages of differentiation. Recent studies of DNA methylation in differentiated tissues show that many silent developmental regulator genes remain hypomethylated in wide genomic regions (also called DNA methylation valleys; or DMVs) in differentiated tissues (Xie *et al*, 2013), raising the question of whether poised RNAPII complexes might prime developmental regulator genes in cell lineages irrespectively of future activation. In this scenario, Polycomb repression may represent the universal mode of repression of this group of CpG-rich genes that recruit poised RNAPII-S5p and which are not targeted by DNA methylation. Silencing of developmental regulator genes through Polycomb repression mechanisms in fully differentiated cells, especially in the presence of poised RNAPII complexes, may nevertheless have roles in the remodeling of cell function, for example in response to specific stimuli such as tissue injury, or in disease such as in cancers associated with Polycomb dysfunction.

To investigate RNAPII poising at Polycomb-repressed genes, from pluripotency to terminal differentiation, we mapped H3K27me3 (a marker of Polycomb repression), RNAPII-S5p (present at both active and poised RNAPII complexes), and RNAPII-S7p (a marker of productive gene expression) and produced matched mRNA-seq datasets in ESCs and in four stages of differentiation of functionally mature dopaminergic neurons. We show compelling evidence that the presence of poised RNAPII at H3K27me3-marked chromatin is not a specific feature of ESCs, but a general property common to differentiating and post-mitotic cells. We also observe *de novo* Polycomb repression during neuronal cell commitment and neuronal maturation that promotes waves of transient downregulation of gene expression. We discover a group of genes that maintain poised RNAPII-S5p and Polycomb silencing throughout neuronal differentiation, and which are developmental transcription factors important for cell specification toward non-neuronal lineages. Although these genes are unlikely to be subsequently reactivated in the neuronal lineage, their silencing in neuronal precursors and mature neurons is sensitive to Polycomb inhibition or knockout. We also show that the presence of poised RNAPII-S5p at specific subsets of Polycomb-repressed genes in terminally differentiated neurons coincides with their wide hypomethylation in mouse brain. Our study reveals the interplay between RNAPII poising and Polycomb repression in the control of regulatory networks and cell plasticity throughout cell differentiation.

Results

Capturing distinct stages of differentiation from ESCs to dopaminergic neurons

To study the dynamic changes in Polycomb and RNAPII occupancy at gene promoters during differentiation, we optimized neuronal differentiation protocols to obtain large quantities of pure cell populations required for mapping chromatin-associated histone marks and RNAPII at five states of neuronal differentiation that leads to the production of functional dopaminergic neurons (ESC, days 1, 3, 16, and 30; Fig 1A). To capture the early exit from pluripotency, we adopted an approach that starts from mouse ESCs grown in serum-free and 2i-free conditions and which within 3 days achieves synchronous exit from pluripotency toward the production of neuronal progenitors (Abranches *et al*, 2009; Fig EV1A, top row). To obtain terminally differentiated dopaminergic neurons, we used an approach that commits ESCs to a midbrain neuron phenotype (Jaeger *et al*, 2011; Fig EV1A, bottom row).

The expression of pluripotency markers *Nanog* and *Oct4* decreases dramatically at days 1 and 3 of differentiation, respectively (Fig 1B). The early differentiation marker *Fgf5* is transiently expressed in days 1–3, whereas neuronal markers *Blbp*, *Hes5*, and *Mash1/Ascl1* are increasingly expressed from day 2 (Figs 1B and EV1B). The expression of *Sox2*, which encodes for a transcription factor expressed in ESCs and by most central nervous system progenitors (Graham *et al*, 2003), is detected from ESC to day 4, as expected (Fig EV1B; Abranches *et al*, 2009). After sixteen days, we obtained neurons that no longer express OCT4 protein, are positive for the neuronal marker TUBB3 (detected by Tuj1 antibody), and no longer divide, as assessed by lack of BrdU incorporation into newly replicated DNA (Fig 1C).

To confirm the dopaminergic phenotype, we performed immunofluorescence for tyrosine hydroxylase (TH), the rate-limiting enzyme in dopamine synthesis (Fig 1D). The neuronal populations obtained expressed TH from day 16, reaching close to ubiquitous expression at day 30 (Fig 1D). Moreover, 70% of cells co-express LMX1A and FOXA2 at day 16 (Fig EV1C), two markers specific to the dopaminergic ventral midbrain, confirming that the neuronal populations produced are highly enriched for the dopaminergic lineage (Hegarty *et al*, 2013). Taken together, these results indicate that day 16 neurons represent an immature stage of differentiation committed to the ventral midbrain lineage, which further mature until day 30.

To characterize the changes in gene expression that accompany neuronal differentiation, we produced mRNA-seq datasets for ESCs, day 1, day 3, day 16, and day 30. We identified 4656 genes whose expression levels peaked at one specific time point (Fig 1E). Genes peaking in ESCs are enriched in Gene Ontology (GO) terms typical of pluripotency, such as *"stem cell maintenance"*, *"regulation of gene silencing"*, and *"sugar utilization"*, and include *Nanog*, *Tet1*, and *Hk2*. Genes with highest expression on day 1 have roles in the exit from pluripotency, such as *Wt1*, *Foxd3*, and *Dnmt3b*, and are enriched in GO terms *"cell morphogenesis"*, *"pattern specification process"*, and *"gene silencing"*. On day 3, the expression of *Fgf8*, *Gli3* and *HoxA1* peaked, reflecting an early stage of neuronal commitment, highlighted by enrichment in GO terms such as *"cellular developmental process"*, *"multicellular organismal process"*, and *"neuronal nucleus development"*. Day 16 coincided with highest expression of genes associated with GO terms such as *"nervous system development"*, *"axon guidance"*, and *"neuron migration"* (including *Nova1*, *Sema3f*, *Ascl1*, *Neurog2*), and day 30 with genes important for dopaminergic synaptic transmission, for the *"G-protein coupled receptor protein signaling pathway"* and *"response to alkaloid"* (such as *Lpar3*, *Th*, *Park2*, *Chrnb4*). The complete list of enriched GO terms is presented in Dataset EV1. These expression profiles show that each time point captures a specific stage of neuronal development, and suggest that days 16 and 30 reflect early and late stages, respectively, of maturation of dopaminergic neurons.

To further confirm the quality of our samples, we also explored the expression profiles of specific single genes (Fig EV2). In addition to confirming the expression of the differentiation markers studied by quantitative PCR and immunofluorescence (Fig EV2A), we also found that the proneural gene *Ngn2* (expressed in immature, but not in mature, dopaminergic neurons) peaks at day 16 and drops by day 30, while *Nurr1* (required for maintenance of dopaminergic neurons) is upregulated at day 16 but remains expressed at day 30 (Fig EV2B; Ang, 2006). Other markers of dopaminergic neurons,

---

Figure 1. Model of differentiation from pluripotent stem cells to terminal dopaminergic neurons.

A Schematic representation of the differentiation system used and the temporal expression dynamics of differentiation stage markers.

B RNA levels of differentiation stage markers were measured by qRT–PCR. Relative levels are normalized to *Actb* internal control, and values are plotted relative to the highest expressed time point. Mean and standard deviation (SD) are from three biological replicates.

C Indirect immunofluorescence confirms expression of stage-specific markers at the single cell level. OCT4 is a marker of pluripotent ESCs. Tuj1 is an antibody that detects neuronal marker TUBB3 at day 16 and day 30 neurons. The cycling activity of ESCs, day 16, and day 30 neurons was assessed by BrdU incorporation (24 h) into replicating BrDNA. Nuclei are counterstained with DAPI. Scale bar, 100 μm.

D Tyrosine hydroxylase (TH; in red) is a marker of dopaminergic neurons. It is not expressed in ESCs and detected weakly in day 16 and broadly in day 30 neurons. Nuclei are counterstained with DAPI. Scale bar, 100 μm.

E Gene expression dynamics across the differentiation time line for genes whose expression peaks in a single time point (z-score > 1.75; from mRNA-seq). Representative enriched GO terms were calculated using as background all genes expressed (> 1 TPM) in at least one time point. *n*, number of genes per group. Permute *P*-value (GO-Elite) is shown.

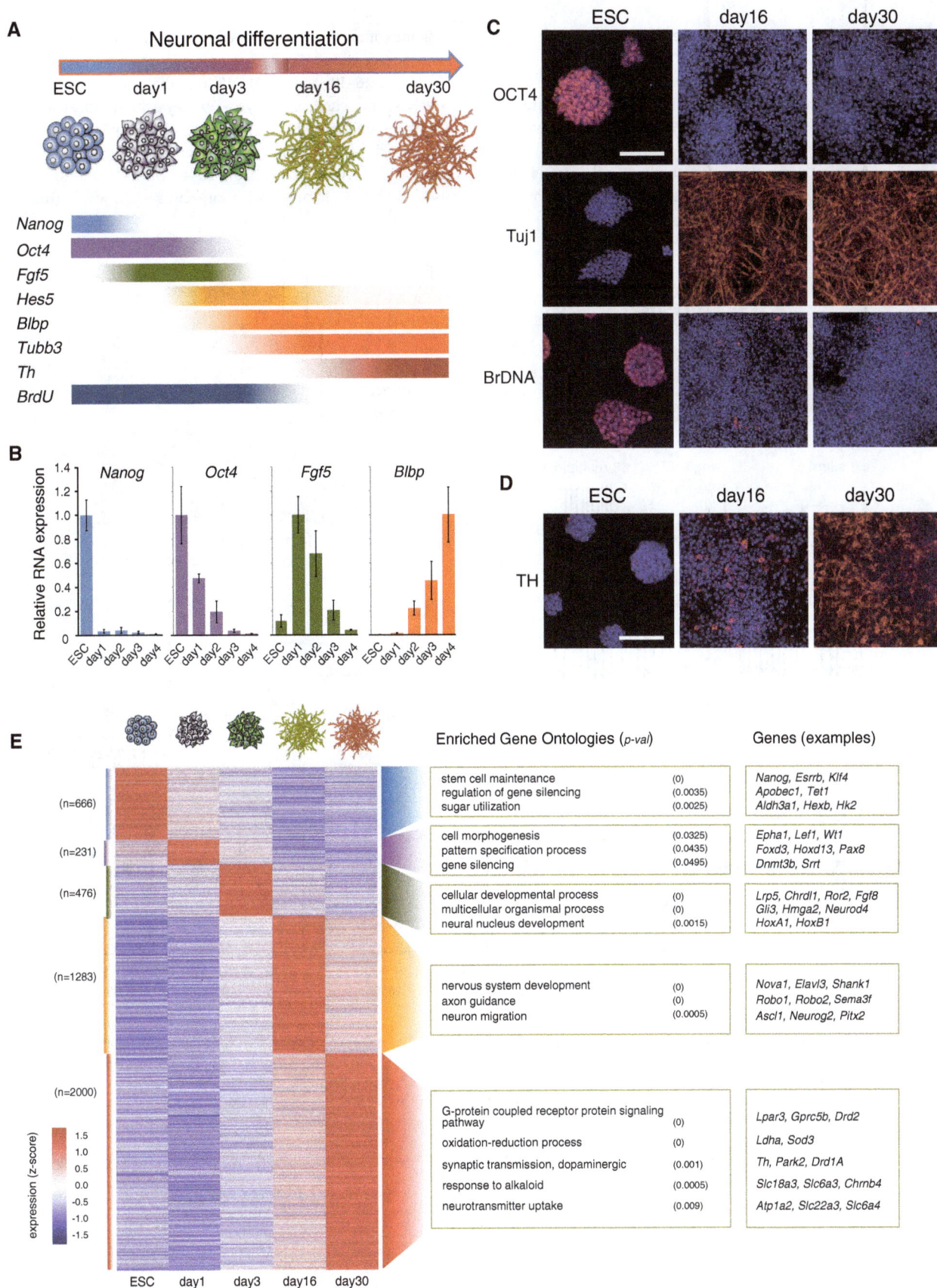

Figure 1.

such as *Pitx3*, *Aadc*, *Vmat,* and *Dat*, are most highly expressed at day 30 (Fig EV2B).

Electrophysiological measurements demonstrate distinct stages of neuronal maturation on days 16 and 30 of differentiation

To directly investigate the state of maturation of neurons upon prolonged culture, we measured their action potential activity and

synaptic connectivity by conducting targeted whole-cell electrophysiological recordings (Fig 2A) at four different time points (days 14–16, 20–25, 26–30, and > 30). We find that days 14–16 neurons are largely silent, whereas at days > 30 they exhibit robust spontaneous action potential activity (Figs 2B and EV3A), similar to the activity of midbrain dopaminergic neurons from *ex vivo* slice preparations and *in vivo* (Marinelli & McCutcheon, 2014). During maturation, neurons also exhibit a hyperpolarization-activated inward (Ih) current

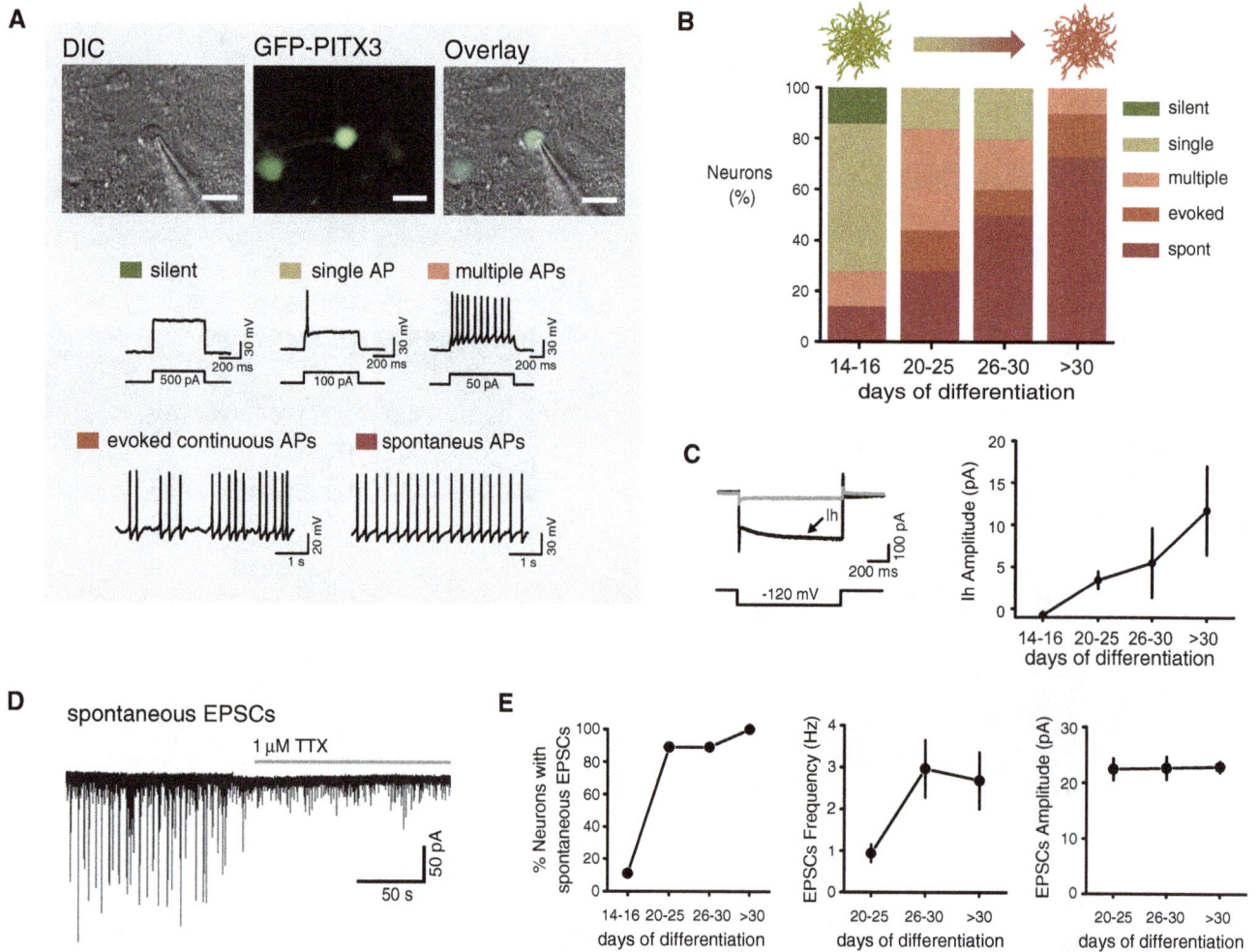

Figure 2. Whole-cell electrophysiological recordings from *in vitro* differentiated dopaminergic neurons show maturation of action potential firing and synaptic activity.

A   Differential Interference Contrast (DIC) and GFP images showing a glass micropipette recording from a Pitx3-GFP+ cell and examples of action potential (AP) measurements. Scale bars, 35 μm. Immature neurons were generally silent or fired only a single AP; they progressively started firing multiple APs in response to a depolarizing current pulse, evoked spontaneous APs generated in response to a constant depolarizing current injection, or spontaneous pacemaker-like APs when they were fully mature.

B   Graph shows progressive functional neuronal maturation toward spontaneous AP firing from days 14–16 to day 30 based on the % of their AP firing categories (n = 21, 25, 10, 30; Spontaneous firing; chi-squared test for trend = 20.58, P-value < 0.0001).

C   Two traces from different cells illustrating the presence (black, arrow) or absence (gray) of a hyperpolarization-activated inward current (Ih), evoked by a hyperpolarizing voltage pulse of −120 mV. Graph shows mean and standard error of the mean (SEM). Ih amplitude progressively increases across differentiation days (n = 22, 10, 12, 20; ANOVA = 2.565, P-value = 0.0483).

D   Example trace of spontaneous excitatory post-synaptic currents (sEPSCs) recorded at −70 mV and application of 1 μM tetrodotoxin (TTX), which revealed miniature EPSCs (mEPSCs).

E   Spontaneous synaptic activity emerged at days 20–25. Graph shows percentage of neurons with spontaneous synaptic activity across differentiation days (n = 10, 19, 8, 5; chi-squared test for trend = 20.76, P-value < 0.0001). Frequency, but not amplitude of mEPSCs, continues to increase once synaptic activity has emerged (n = 7, 7, 10; Frequency ANOVA = 6.555, P-value = 0.0377; Amplitude ANOVA = 0.04457, P-value = 0.978). Graph shows SEM.

(Fig 2C), an electrophysiological feature commonly used to identify dopaminergic neurons (Ungless & Grace, 2012). Strikingly, after prolonged culture many cells exhibit burst-like events (Fig EV3B and C), often seen *in vivo* in midbrain dopaminergic neurons (Grace & Bunney, 1984), which are thought to be driven in part by synaptic inputs (Paladini & Roeper, 2014). We also observed maturation of functional synaptic connectivity. Spontaneous synaptic events were rare at days 14–16, but were present in most neurons from days 20–25 onwards (Fig 2D and E). Further experiments in the presence of TTX (which blocked action potential activity; Fig EV3D) revealed the presence of glutamatergic AMPA receptor-mediated miniature excitatory post-synaptic currents (mEPSCs; Fig EV3E) and GABAA receptor-mediated miniature inhibitory post-synaptic currents (mIPSCs; Fig EV3F), similar to those seen in *ex vivo* dopaminergic neurons. Consistent with this, we also observed a small number of GABAergic and glutamatergic neurons in our cultures (Fig EV3G), as seen in the ventral tegmental area (VTA) of the midbrain *in vivo*, and expression of GABAergic and glutamatergic neuronal markers (Fig EV3H). Further evidence in support of the quality of our neuronal samples was the absence of glutamatergic autapses (Fig EV3I), which are seen in cultured *ex vivo* dissociated neurons, but not in *ex vivo* brain slices (Sulzer *et al*, 1998).

We conclude that the neuronal differentiation approach established here yields functional neurons that undergo progressive maturation, as shown by their action potential activity and synaptic connectivity. Taken together, gene expression and electrophysiological analyses confirm that the five time points used here represent distinct stages of neuronal fate commitment and maturation.

Polycomb repression dynamics during terminal differentiation of dopaminergic neurons

To identify the genes regulated by Polycomb during neuronal specification, we mapped the genome-wide occupancy of H3K27me3, the PRC2-mediated histone modification. In ESCs grown in serum-free conditions, 4,107 genes are marked by H3K27me3 (Fig 3A), most of which (78–86%) were also found targeted by PRC in ESCs in serum-containing media (Lienert *et al*, 2011; Young *et al*, 2011; Brookes *et al*, 2012), demonstrating the important roles of Polycomb repression mechanisms in ESCs independent from the use of serum.

Upon early exit from the pluripotent state, the number of H3K27me3[+] genes increases to ~4,600 in days 1 and 3, before decreasing to ~2,200 genes in days 16 and 30 (Fig 3A). The abundance of PRC-marked promoters in dopaminergic neurons suggests an important role of Polycomb repression in differentiated cells, which is consistent with previous reports in other specialized cell types: pyramidal glutamatergic neurons (2,178 genes; Mohn *et al*, 2008), striatal GABAergic neurons (2,057 genes; von

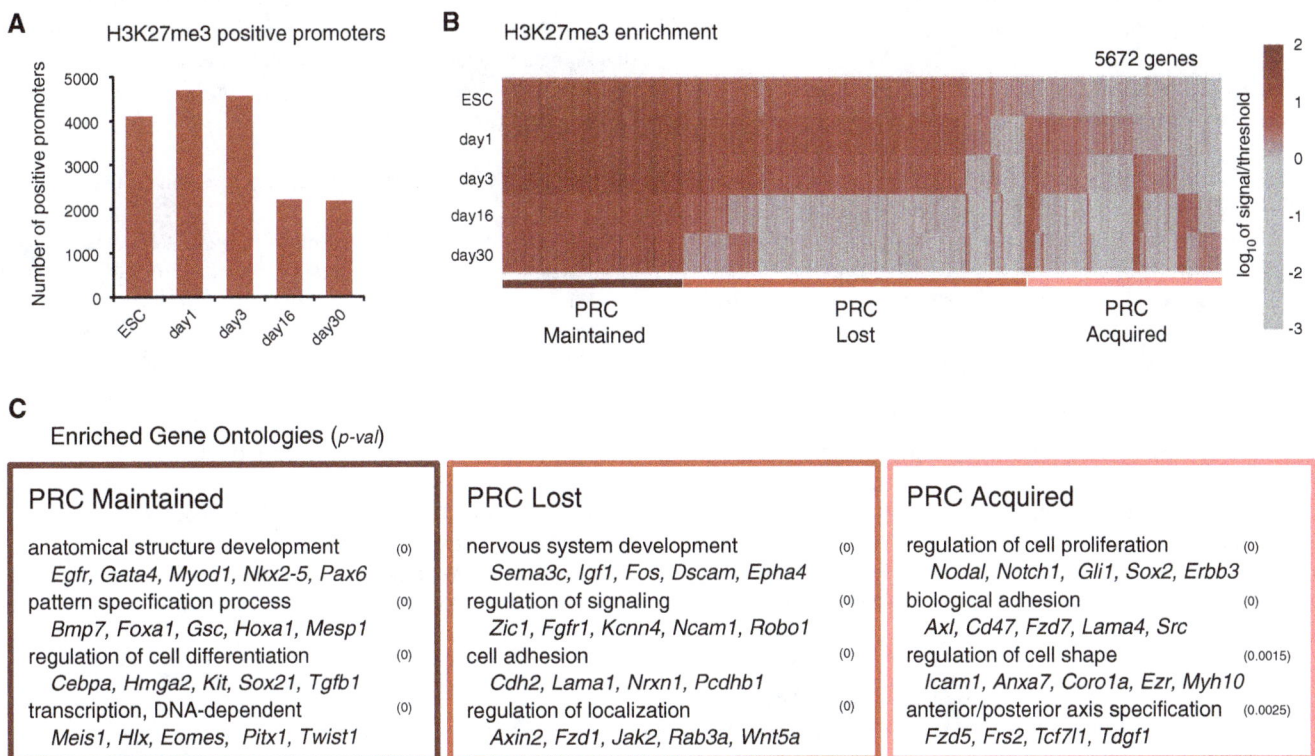

Figure 3. **PRC occupancy at gene promoters is highly dynamic throughout all stages of the neuronal differentiation.**

A Numbers of PRC[+] promoters from ESC to day 30. In terminally differentiated neurons, 2,178 genes are marked by H3K27me3.
B Dynamic changes in H3K27me3 presence at gene promoters. Vertical lines represent each of the promoters marked by H3K27me3[+] in at least one time point. Throughout differentiation, 1,408 genes are marked by H3K27me3 (PRC Maintained), 2,699 lose H3K27me3 (PRC Lost), and 1,565 gain H3K27me3 (PRC Acquired). Color represents number of reads in transcription start site (TSS) window, scaled for visualization.
C Examples of enriched GO terms and gene examples among the groups of genes classified as PRC Maintained, Lost, and Acquired, using as background all genes.

Schimmelmann *et al*, 2016), functional endocrine cells (1,273 genes; Xie *et al*, 2013), and in E12.5 heart ventricle apex (1,052 genes; He *et al*, 2012).

To study the dynamics of Polycomb repression across the different stages of differentiation, we grouped promoters according to H3K27me3 presence or absence (Fig 3B). H3K27me3 marks a total of 5,672 genes in at least one time point (Dataset EV2). Over half of the H3K27me3$^+$ genes in ESCs lose H3K27me3 between days 1 and 30 ("*PRC lost*" genes), while 1,408 genes remain H3K27me3$^+$ throughout the differentiation timeline ("*PRC maintained*"; Fig 3C). Genes that lose H3K27me3 include *Fos*, *Zic1*, *Ncam1*, and *Wnt5a*, and are enriched in GO terms such as "*nervous system development*", "*regulation of signaling*", and "*cell adhesion*", consistent with their activation upon acquisition of the neuronal phenotype. Genes associated with PRC in all time points include developmental transcription factors such as *Nkx2-5*, *Pax6*, and *Gata4*, and are enriched in functions such as "*anatomical structural development*", "*pattern specification process*", "*regulation of cell differentiation*", and "*DNA dependent transcription*". *De novo* acquisition of H3K27me3 is detected throughout differentiation ("*PRC acquired*" genes), with a transient appearance of H3K27me3 at 931 genes between days 1 and 3, and at 351 genes in post-mitotic neurons (Fig 3C). These genes are enriched in GO terms such as "*regulation of cell proliferation*" and include important signaling regulators as *Nodal*, *Notch1*, and *Fzd5*. Our results show that genes targeted by PRC during neuronal differentiation undergo dynamic changes at appropriate stages of differentiation.

Poised RNAPII-S5p is present at H3K27me3$^+$ promoters throughout differentiation

To investigate whether poised RNAPII-S5p is lost from Polycomb-repressed promoters upon exit from pluripotency, or associated with Polycomb throughout terminal differentiation, we produced ChIP-seq datasets for RNAPII-S5p and RNAPII-S7p in ESCs and at days 1, 3, 16, and 30 (Fig 4A). Inspection of single-gene profiles revealed different promoter states and dynamic changes (Fig 4B). For example, the housekeeping gene *Eif2b2* is always expressed at the mRNA level and occupied by RNAPII-S5p and RNAPII-S7p, whereas testis-specific *Brs3* is always inactive and lacks H3K27me3 and RNAPII-S5p and RNAPII-S7p throughout our differentiation timeline. In contrast, *Adcy5*, a gene implicated in axonal elongation and in addiction (del Puerto *et al*, 2012; Procopio *et al*, 2013), is marked by RNAPII-S5p and H3K27me3 in ESCs and becomes Active from day 16, acquiring S7p, losing H3K27me3, and maintaining S5p. *Pkp1*, which has roles in epidermal morphogenesis (South *et al*,

2003), is always marked by H3K27me3, but associated with poised RNAPII-S5p only in ESCs and in early differentiation. Notably, *Pitx1*, a transcription factor with roles in limb morphogenesis (Infante *et al*, 2013), is marked by RNAPII-S5p and H3K27me3 at all five time points, without S7p or mRNA expression. Examples of *de novo* PRC repression include *Ajuba*, a gene implicated in cell-cycle progression (Hirota *et al*, 2003) which is downregulated during differentiation while acquiring H3K27me3, and retaining S5p, S7p, and low levels of mRNA expression. Genes marked with H3K27me3, RNAPII-S5p, and RNAPII-S7p in bulk ChIP assays have been previously described in ESCs and shown to correspond to promoter fluctuations between active and PRC-repressed states in different cells or alleles (Brookes *et al*, 2012; Kar *et al*, 2017).

For an unbiased exploration of the transitions between states of Polycomb repression and productive transcription across differentiation, we identified the promoters that are associated with RNAPII or Polycomb in each time point. RNAPII-S5p marks both active and PRC-repressed promoters, RNAPII-S7p marks only at active promoters, and H3K27me3 marks PRC-repressed promoters. As previously reported, most promoters marked by H3K27me3 in ESCs are also bound by RNAPII-S5p (81%; Fig 4C; see also Brookes *et al*, 2012). Importantly, we find that many promoters are positively marked by RNAPII-S5p and H3K27me3 throughout differentiation, showing that RNAPII poising at PRC-marked promoters is a general property of Polycomb repression and not only specific to ESCs. During terminal differentiation, PRC/S5p genes become increasingly marked by RNAPII-S7p (Fig 4D) and expressed at the mRNA level (Fig 4E), suggesting a role for PRC repression not limited to strict gene silencing, but also in modulating gene expression levels, as seen previously in ESCs (Brookes *et al*, 2012; Kar *et al*, 2017).

Genome-wide transitions in Polycomb repression states

To understand the dynamic transitions in promoter states throughout differentiation, we classified genes according to the presence of RNAPII-S5p, RNAPII-S7p, and H3K27me3, in each time point (Fig 5A). PRC-positive promoters were classified as "PRC Only" (H3K27me3$^+$S5p$^-$S7p$^-$), as "PRC/S5p" (H3K27me3$^+$S5p$^+$ S7p$^-$), and as "PRC/Active" (H3K27me3$^+$S5p$^+$S7p$^+$). PRC-negative promoters were classified as "Active" (H3K27me3$^-$S5p$^+$S7p$^+$), or as "Inactive" (H3K27me3$^-$S5p$^-$S7p$^-$). The other three combinations of promoter marks (S7p Only, S5p Only, and PRC/S7p) were uncommon and not explored further in the present study. The numbers of genes in each state are summarized in Fig EV4A; gene promoter states are listed in Dataset EV2. We found that the decrease in the number of promoters marked by PRC is accompanied by an increase in Active

Figure 4. RNAPII-S5p overlaps with Polycomb at all stages of differentiation.

A Schematic summarizing the strategy used to define promoter states through differentiation based on the interplay between PRC and RNAPII.

B Examples of single-gene ChIP-seq profiles across differentiation. *Eif2b2* and *Brs3* are always Active and Inactive, respectively. *Adcy5* and *Pkp1* are PRC/S5p in ESC and days 1/3, but become Active or PRC Only, respectively. *Pitx1* maintains the poised PRC/S5p conformation from ESCs until day 30 neurons. *Ajuba* is Active in ESCs but acquires H3K27me3 at day 16 while becoming downregulated.

C Proportion of H3K27me3$^+$ promoters (dark red) marked by RNAPII-S5p (purple) throughout differentiation. The vast majority of H3K27me3$^+$ promoters are bound by RNAPII-S5p in all time points.

D Proportion of H3K27me3$^+$ and S5p$^+$ promoters (purple) marked by S7p (green) throughout differentiation. The number of promoters marked by H3K27me3$^+$S5p$^+$S7p$^+$ increases during differentiation.

E Proportion of H3K27me3 and RNAPII-S5p promoters (purple) that are also transcribed (gold; TPM > 1). The number of actively transcribed promoters increases during differentiation, in agreement with the increased presence of S7p.

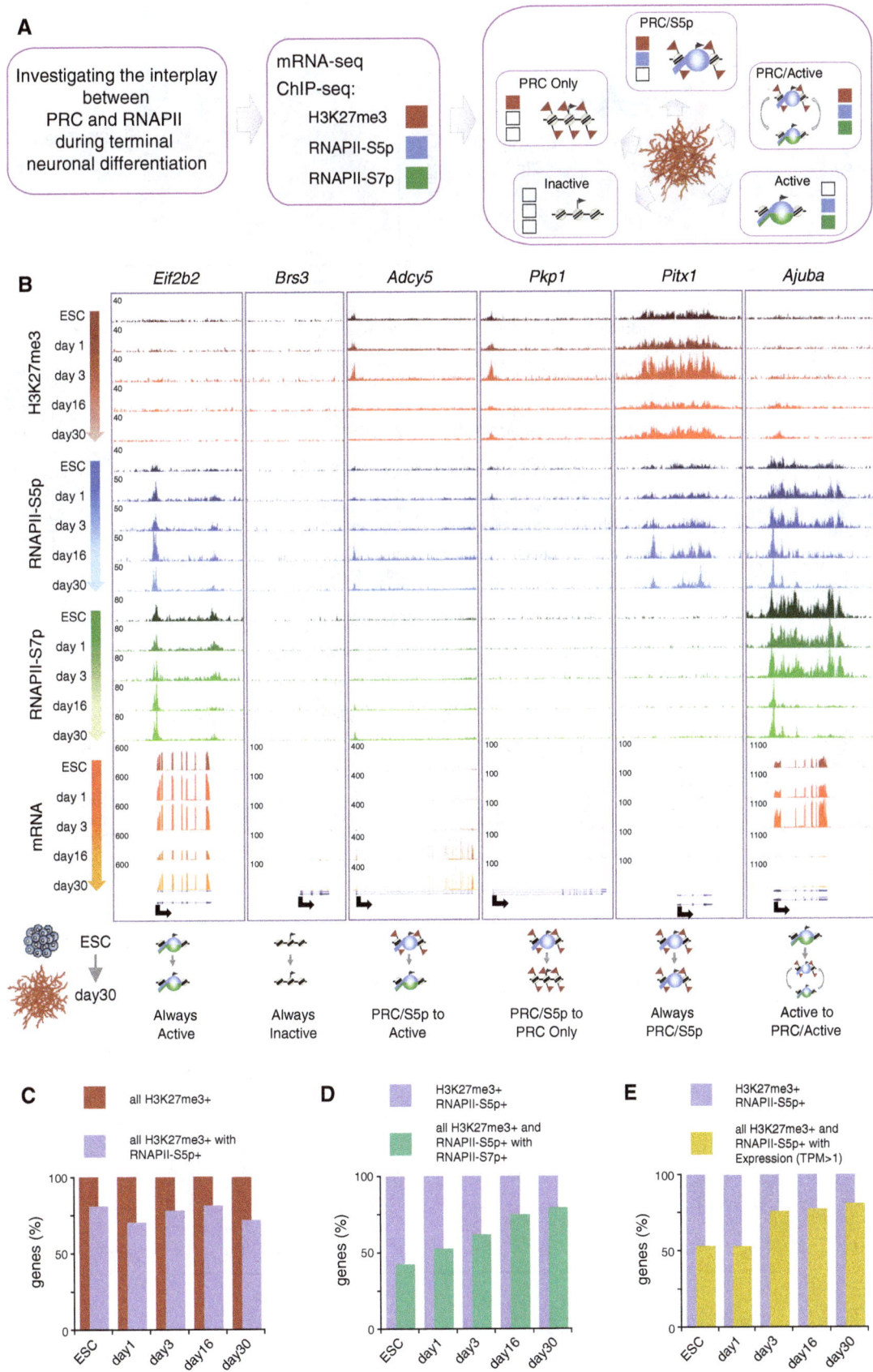

Figure 4.

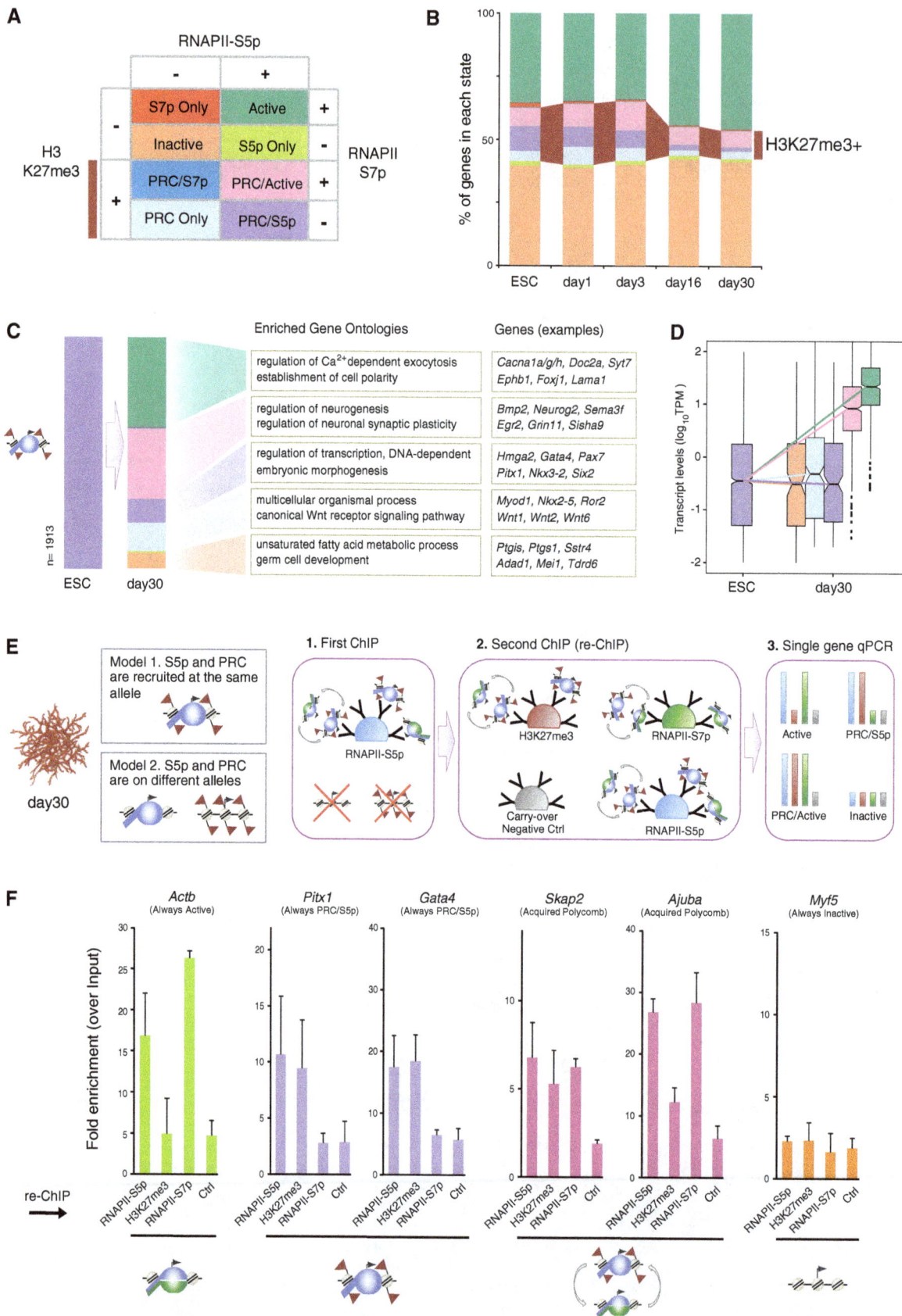

Figure 5.

promoters, while the number of Inactive promoters remains stable (Fig 5B). The decrease in PRC-positive promoters occurs across all PRC classes, except for a relative increase in the PRC/Active state (Fig 5B). By following the changes in promoter state between time points (Fig EV4B), we show that the increase in the Active promoter state in days 16 and 30 originates mostly from genes that were PRC repressed in ESCs to day 3; this transition is accompanied by increased mRNA expression, as expected from the acquisition of S7p and loss of H3K27me3 at this group of genes (Fig EV4C). These genes with newly Active promoter states have specific neuronal functions (Fig EV4D).

Next, we investigated the promoter state transitions of the genes associated with the poised PRC/S5p promoter state in ESCs (Fig 5C). Many of the PRC/S5p genes in ESCs become Active (743/1913) or PRC/Active (580/1913) in day 30, and have functions in cell polarity or synaptic activity. Acquisition of S7p is accompanied by increased mRNA expression, as expected (Fig 5D). Notably, a group of PRC/S5p genes in ESCs retain the poised conformation in day 30 neurons (200/1,913 genes; Fig 5C). These genes are important developmental transcription factors (TFs) which regulate non-neuronal cell lineages (e.g., *Gata4*, *Pax7*, *Nkx3-2*). We also find PRC/S5p genes that resolve to Inactive (129/1,913), with functions in "*germ cell development*", or that become PRC Only (236/1,913), with functions including "*Wnt signalling*".

Other interesting Polycomb-dependent transitions in promoter activation states are those that result in *de novo* acquisition of H3K27me3 in either days 1/3 or days 16/30 (Fig EV5A and B). Most genes that gain H3K27me3 in days 1 or 3 tend to be Active in ESCs, and go back to their Active promoter state in day 30, with the acquisition of H3K27me3 in days 1/3 being accompanied by decreased gene expression (Fig EV5A). The genes that acquire PRC at days 16 or 30 are downregulated at this stage of differentiation, and tend to be Active in ESCs (Fig EV5B).

Taken together, our results show a role for Polycomb, most often accompanied by poised RNAPII-S5p complexes, in the silencing of non-neuronal TFs in neurons and in fine-tuning gene expression states during differentiation.

H3K27me3 chromatin is co-occupied with RNAPII-S5p complexes in mature neurons

To test the simultaneous occupancy of RNAPII-S5p and H3K27me3 at Polycomb-repressed genes in terminally differentiated neurons, we performed sequential ChIP (or re-ChIP; Fig 5E). We immunoprecipitated chromatin from day 30 neurons with the antibody used to map RNAPII-S5p, before re-immunoprecipitation with different antibodies. We show that the promoter of the Active gene *Actb* is co-bound by RNAPII-S5p and RNAPII-S7p, but not by H3K27me3, as expected (Fig 5F). Notably, the promoters of non-neuronal transcription factors *Pitx1* and *Gata4*, which are classified as PRC/S5p in all time points, are co-occupied in neurons by H3K27me3 and RNAPII-S5p, but not RNAPII-S7p. These results show that RNAPII-S5p and H3K27me3 simultaneously occupy Polycomb-repressed promoters in mature dopaminergic neurons. We also tested the co-association of H3K27me3 with RNAPII-S5p at *Skap2* and *Ajuba*, which are PRC/Active genes that acquire H3K27me3 *de novo* in day 30, and found that H3K27me3 and RNAPII-S5p also co-localize at these promoters. In agreement with the PRC/Active classification, RNAPII-S5p co-immunoprecipitates with RNAPII-S7p, consistent with the presence of the active (S5p$^+$S7p$^+$) form of RNAPII at some alleles or in different cells (Brookes *et al*, 2012; Kar *et al*, 2017). None of the marks tested were enriched at the inactive gene (*Myf5*), confirming the specificity of the assay. Re-ChIP with an unrelated antibody, against plant steroid digoxigenin, was used as a negative control to confirm minimal antibody carryover from the first to the second immunoprecipitation. The sequential ChIP analyses performed here in terminally differentiated neurons show that RNAPII poising is present at PRC-repressed genes in post-mitotic neurons.

Poised PRC/S5p promoter states are associated with increased potential for derepression upon Polycomb knockout or chemical inhibition

To explore the biological relevance of the poised PRC/S5p promoter state in ESCs, neuronal progenitors, and neurons, we investigated the effect of Polycomb knockout at PRC/S5p genes compared to PRC Only genes. First, we took advantage of published transcriptome resources in ESCs which measured transcriptional derepression after stable knockout of PRC2 component *Eed* or conditional knockout of PRC1 component *Ring1B* in *Ring1A*$^{-/-}$ background (Endoh *et al*, 2008; Cruz-Molina *et al*, 2017). We find that a larger proportion of PRC/S5p genes are upregulated in both PRC knockout cell lines, compared to PRC Only genes (Fig 6A). Only a small fraction of Active and Inactive genes responded to PRC knockout, which may arise through more indirect regulatory effects associated with adaptation in the knockout cell lines. Second, we performed similar analyses using a published microarray dataset from an *in vivo* PRC2

◄ **Figure 5. RNAPII-S5p co-localizes with H3K27me3 in terminally differentiated neurons.**

A Schematic of gene classification according to the occupancy of H3K27me3, S5p, and S7p at promoter regions (2-kb windows centered on Transcription Start Sites; TSSs).

B Proportion of genes with each promoter state during terminal neuronal differentiation. Dark red: H3K27me3$^+$ genes.

C PRC/S5p promoters in ESCs and their promoter state in day 30 neurons. Examples of enriched GO terms (determined using as background all PRC/S5p genes in ESCs) and genes in each category.

D Expression level of PRC/S5p promoters in ESCs and day 30 neurons. Acquisition of S7p correlates with increased expression level. Genes with PRC/S5p, PRC Only, or Inactive state remain repressed in day 30 neurons.

E Sequential ChIP (re-ChIP) was performed using day 30 neurons to test RNAPII and H3K27me3 co-occupancy, as described in the schematic. The first ChIP with anti-RNAPII-S5p antibody was followed by a second ChIP using antibodies against RNAPII-S5p (positive control), H3K27me3, RNAPII-S7p, or Digoxigenin (carryover negative control "Ctrl").

F Sequential ChIP shows RNAPII-S5p co-association with H3K27me3 at PRC/S5p (*Pitx1* and *Gata4*) and PRC/Active (*Skap2* and *Ajuba*) promoters but not at Active (*Actb*) or Inactive (*Myf5*) promoters. The mean and SD are from two biological replicates.

knockout in GABAergic striatal medium spiny neurons which results in neurodegeneration (von Schimmelmann et al, 2016). Although the neuronal subtypes are not fully matched between the promoter state classifications from our neuronal cultures, that simulate the ventral tegmental area, and the GABAergic neurons from the striatum, we find that most genes upregulated upon PRC2 loss in ex vivo striatal neurons have the PRC/S5p promoter state in our mature day 30 neurons (52 out of the total 150 upregulated genes; hypergeometric test for enrichment, $P = 3.1 \times 10^{-56}$; of note, 39 of those are PRC/S5p at all stages; Fig 6B). In contrast, only 17 genes upregulated in the striatal neurons are classified as PRC Only in the dopaminergic cultures ($P = 6.8 \times 10^{-7}$), and a smaller proportion of Active or Inactive genes become upregulated. These results suggest that the PRC/S5p state leads to increased likelihood for derepression upon Polycomb loss in both ESCs and differentiated neurons, which has implications for our understanding of Polycomb roles in maintaining cell differentiation states.

Lastly, we tested the response of PRC/S5p and PRC Only genes to PRC2 enzymatic inhibition at three different stages of differentiation (ESC, day 3, and day 30). We used UNC1999, a chemical drug that blocks the enzymatic activity of both EZH1 and EZH2 (Konze et al, 2013), in two concentrations (1 and 3 μM), and measured relative RNA levels by quantitative PCR. We tested six genes that maintain the PRC/S5p state throughout differentiation (Tal1, Gata4, Skor1, Pitx1, HoxA7, and Mesp1), three genes that are PRC/S5p in ESCs but become PRC Only from day 3 onwards (Pouf3, Pkp1, She) and three genes that maintain the PRC Only promoter state throughout differentiation (Fam150b, Fgf20, Chdrl2). We find that five out of the six PRC/S5p genes that maintain their promoter state are upregulated in ESCs, day 3 and day 30 upon enzymatic inhibition of PRC2, with increased derepression at the highest inhibitor concentration (Fig 6C). In contrast, none of the analyzed genes acquiring or maintaining the PRC Only state show detectable RNA levels or respond to EZH1/2 inhibition at either UNC1999 concentration in any of the three time points tested. Quality of PCR primers was verified by agarose gel electrophoresis using genomic DNA (Appendix Fig S1).

Taken together, these results suggest that the poised RNAPII-S5p is associated with increased potential for derepression upon loss of Polycomb repression in ESCs and in differentiated cells at all stages of maturation, including terminally differentiated neurons both in vitro and in vivo.

Major non-neuronal transcription factors maintain PRC/S5p occupancy upon terminal differentiation

To investigate further the significance of the poised PRC/S5p promoter state during neuronal cell commitment, we focused on 115 genes that maintain the PRC/S5p state during all the sampled stages of neuronal differentiation (Fig 7A). For comparison, we chose two other groups of genes which are also PRC/S5p in ESCs but remain silent in neurons through other promoter states; one group becomes PRC Only (losing S5p; 127 genes), and the other becomes Inactive (losing both H3K27me3 and S5p; 92 genes; Figs 7A and EV6A). Average profiles of H3K27me3 and RNAPII-S5p occupancy show that these three groups of genes are clearly marked by both H3K27me3 and RNAPII-S5p in ESCs (Fig 7B; Brookes et al, 2012). The genes that maintain the PRC/S5p promoter state throughout neuronal differentiation have similar S5p occupancy at their promoters and throughout the coding region in ESCs, but show higher average levels of H3K27me3 occupancy in comparison with genes that resolve to PRC Only or Inactive (Fig 7B). We noticed that the PRC/S5p genes that maintain their state during neuronal differentiation are associated with wider regions of H3K27me3 occupancy compared to genes that lose RNAPII-S5p or to all other neuronal H3K27me3^{+} promoters in both ESCs and day 30 neurons (Fig 7B–D).

To explore the biological relevance of the three groups of genes, we performed GO enrichment analyses (Fig EV6B). Interestingly, the group of 115 PRC/S5p genes that maintain their state is enriched in TFs important for non-neuronal lineages such as Gata4, Mesp1, Tbx2 (GO term: "heart morphogenesis"), Barx1, Gata6, Hoxb5 ("epithelial cell differentiation"), and Osr2, Pax1, Tcfap2a ("bone morphogenesis"), suggesting a role for the poised PRC/S5p state in terminally differentiated neurons in maintaining cell identity and plasticity by repressing non-neuronal genes.

Key non-neuronal transcription factors that maintain the PRC/S5p state in neurons have high CpG content

Polycomb occupancy in ESCs is highly correlated to CpG content (Ku et al, 2008; Deaton & Bird, 2011) and unmethylated CpGs (Sen et al, 2016). To test whether the retention of the PRC/S5p state throughout neuronal differentiation is related to high CpG content, we measured the proportion of the promoter regions covered by CpG islands, the GC content of TSSs, and the extent of CpG island coverage of gene bodies. We observe that the promoter of PRC/S5p genes that maintain the poised state in neurons is more extensively covered by CpG islands than the genes that become PRC Only or Inactive (Fig 7E). Their promoter GC content is also higher (Fig EV6C), and the CpG islands cover longer portions of gene body (Fig EV6D) than at genes that become Inactive. These results suggest that Polycomb repression in co-association with poised RNAPII-S5p correlates with specific sequence features of gene promoters and gene bodies, namely with high CpG content.

Figure 6.   Poised PRC/S5p promoter states have increased potential for reactivation upon loss of Polycomb repression.

A   Effect of PRC deletion in ESCs (PRC2 component EED and PRC1 components RING1B and RING1A). Barplots represent the percentage of genes which are upregulated more than twofold according to their promoter state in ESCs.

B   Effect of PRC2 deletion in striatal GABAergic medium spiny neurons in 6-month-old mice. Barplot represents the percentage of genes which are upregulated more than twofold according to their promoter state in day 30 neuron.

C   RNA levels of PRC/S5p genes were measured by qRT–PCR upon treatment with EZH1/2 inhibitor UNC1999 at 1 and 3 μM, for 24 h prior to RNA extraction. Relative levels are normalized to Actb internal control. Three or two biological replicates were performed in ESC and day 3 or in day 30 neurons, respectively. Black lines connect mean values.

▶

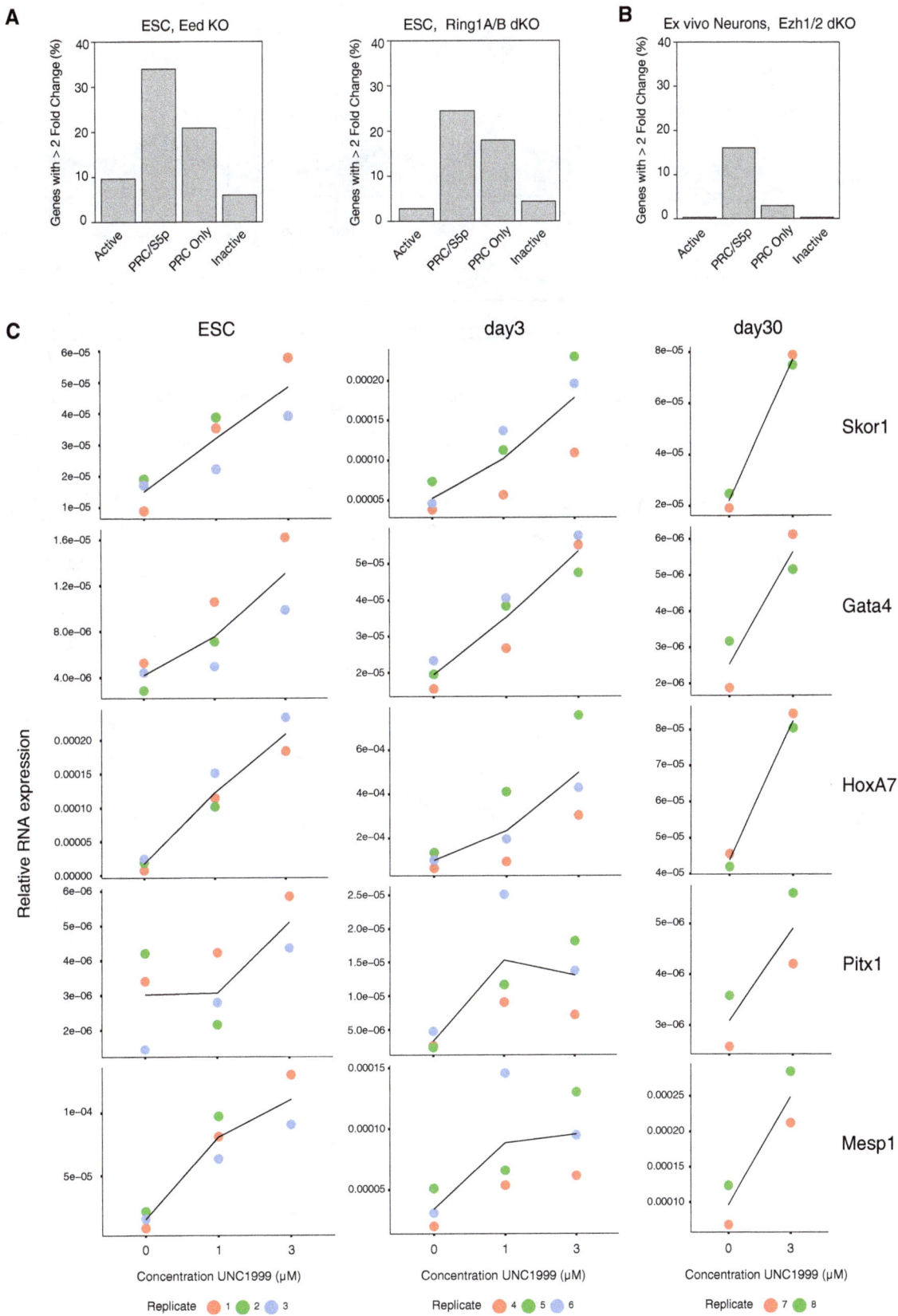

Figure 6.

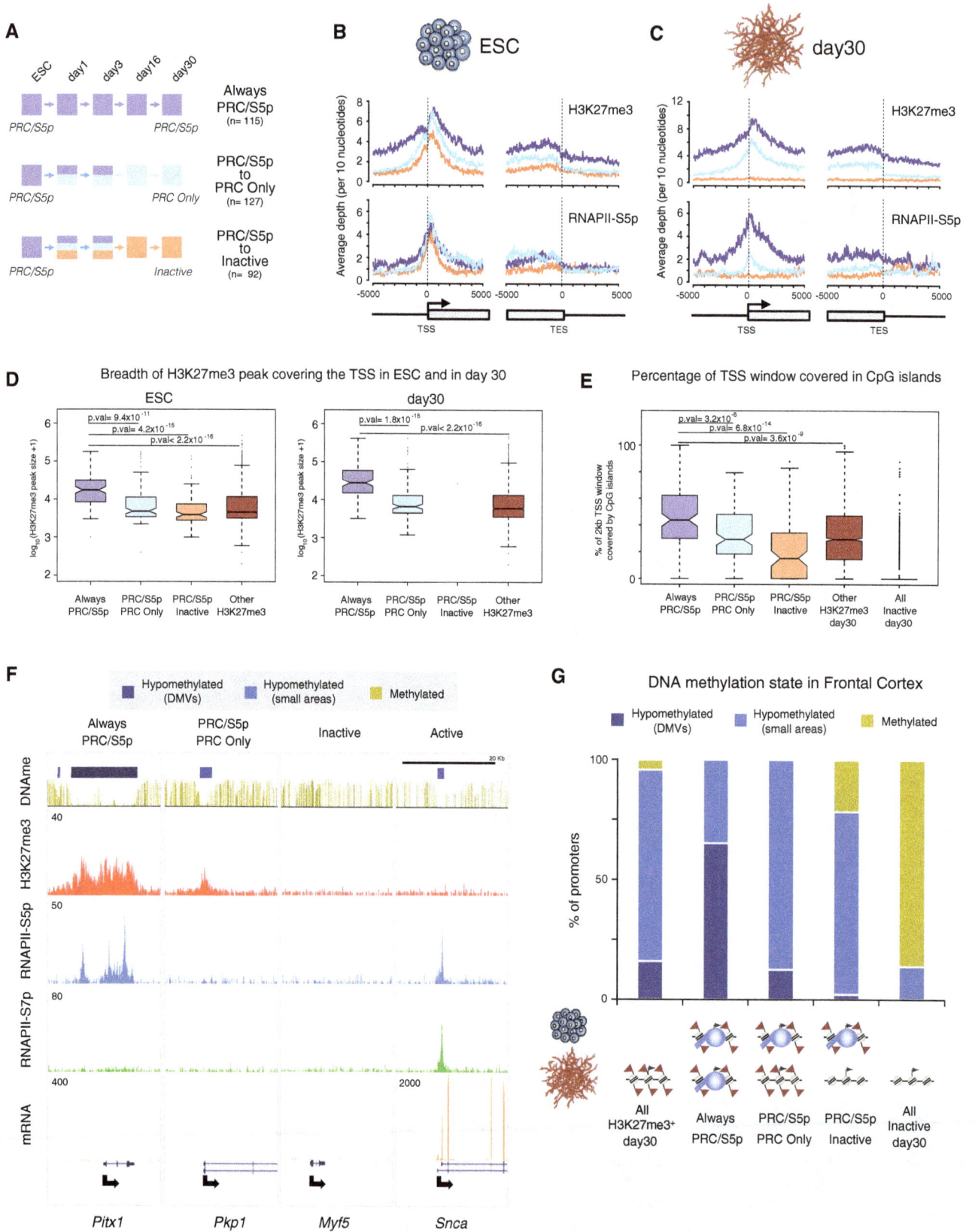

Figure 7.

Figure 7. PRC/S5p genes that maintain the poised state throughout differentiation exhibit broader distributions of H3K27me3 and RNAPII-S5p which coincide with DMVs.

A   PRC/S5p genes are kept silenced through different mechanisms during neuronal differentiation by maintaining PRC/S5p (Always PRC/S5p; purple), resolving to PRC Only (light blue), or resolving to Inactive (orange). Colors indicate promoter states in each time point.

B   Average ChIP-seq enrichment plots of H3K27me3 and S5p in 10-kb genomic windows centered on transcription start site (TSS) and transcription end sites (TES), in ESCs.

C   Average ChIP-seq enrichment plots of H3K27me3 and S5p in 10-kb genomic windows centered on TSS and TES, in day 30 neurons.

D   Boxplot of the length of the genomic regions enriched for H3K27me3 that overlap with promoters in ESC or in day 30 neurons. The length of H3K27me3-enriched regions covering the remaining H3K27me3+ genes is shown for comparison. P-values are calculated with Wilcoxon rank-sum test.

E   Percentage of TSS window covered by CpG islands of PRC/S5p that maintain their state is significantly larger compared to the other gene subgroups. P-values are calculated with Wilcoxon rank-sum test.

F   UCSC Browser tracks of genes classified in day 30 neurons as Always PRC/S5p, PRC Only, Inactive, and Active genes. First row shows methylated (gold) or hypomethylated DNA methylation valleys (DMVs, dark blue; and smaller regions of hypomethylated DNA, light blue) in mouse frontal cortex.

G   Proportion of promoters in the three gene groups which overlap with methylated and hypomethylated DNA. Always PRC/S5p genes are preferentially located at DMVs. All H3K27me3+ and Inactive genes in day 30 neurons are shown for comparison.

Key non-neuronal transcription factors that maintain the PRC/S5p state in neurons are located in DNA methylation valleys

Previous reports have identified DMVs (wide depletions of DNA methylation, longer than 5 kb) in different cell lineages, which coincide with CpG-rich regions often located at loci encoding for developmental transcription factors (Xie *et al*, 2013). To explore the association of genes that maintain the poised PRC/S5p conformation throughout differentiation with DMVs, we mined a published dataset for DNA methylation from mouse frontal cortex (Lister *et al*, 2013), and classified DMVs (Xie *et al*, 2013). Visual inspection of single-gene profiles shows that the regions occupied by H3K27me3 and RNAPII-S5p in neurons often coincide with large stretches of hypomethylated DNA, as shown for *Pitx1* (Fig 7F). Expanding these observations genome-wide, we find that all genes with PRC/S5p promoters that maintain their poised state throughout neuronal differentiation are hypomethylated in mouse cortex, with the majority (65%; 70/115) overlapping with DMVs (Fig 7G). In contrast, only 13% (16/127) of genes that become PRC Only or 2% (2/92) of genes that become Inactive reside in DMVs. Therefore, the maintenance of the PRC/S5p state is a property of important developmental transcription factors that are also rich in CpG and lack DNA methylation in terminally differentiated tissues.

Maintenance of PRC/S5p state occurs at the promoters of developmental transcription factors that are candidate drivers of trans-differentiation

The observation that key developmental TFs are repressed in post-mitotic neurons through Polycomb repression in the presence of poised RNAPII-S5p led us to hypothesize that this conformation contribute to a state that makes neurons permissive to losing their identity and potentially to trans-differentiation to other cell types. To explore this hypothesis, we took advantage of Mogrify, a predictive system of the TFs necessary to induce cell-type reprogramming, which combines gene expression data with regulatory network information (Rackham *et al*, 2016; Fig 8A). We overlapped the TFs that Mogrify predicts to be important for trans-differentiation to various cell fates with the promoter states that we have identified in day 30 neurons. We find that the Mogrify lists of candidate genes for trans-differentiation toward non-neuronal lineages, such as cardiomyocytes, skin fibroblasts, and skeletal muscle, are significantly enriched in genes marked by H3K27me3 in day 30 neurons, and especially in genes with the poised PRC/S5p promoter state (Fig 8B; for additional examples see Fig EV7A). In contrast, the Mogrify lists of reprogramming TFs are depleted of genes that have Inactive promoter states in day 30 neurons. We also investigated the lists of reprogramming TFs predicted to be important for trans-differentiation to neuronal lineages, including to hippocampus, neuron, or brain. These genes are more often Active or PRC/Active in day 30 neurons, unsurprisingly due to their expression in the neuronal lineage, or alternatively associated with poised PRC/S5p promoter state (Fig 8B).

Mogrify also ranks TFs according to their importance for trans-differentiation when they are predicted to regulate a larger number of the genes required to drive changes in phenotype; these genes are scored as having *high influence* on trans-differentiation. In contrast, the genes that are found to be less important for trans-differentiation (and scored as *low influence*) tend to be less connected or to regulate genes that are less specific to the target cell type (see schematic

Figure 8. Genes with the poised PRC/S5p promoter state in terminally differentiated neurons are enriched for TFs with trans-differentiation potential.

A   Cells can trans-differentiate into other cell types by forced expression of lineage-specific TFs. The Mogrify algorithm identifies candidate TFs that drive cell conversion as exemplified in the schematic (Rackham *et al*, 2016).

B   Promoter state of TFs in day 30 which are predicted by Mogrify to drive trans-differentiation toward non-neuronal or neuronal cell types. H3K27me3+ promoters are enriched in the Mogrify lists of trans-differentiation candidates. PRC/S5p and the PRC/Active genes are the mostly enriched promoter classes. Inactive genes are depleted. Enrichment and depletion were tested with the hypergeometric test.

C   Mogrify ranks all TFs by the likelihood to induce conversion from cell types A to B. By averaging these ranks, it is possible to identify those TFs most frequently required for a conversion into a certain cell type. The ranks are used to estimate the importance of a given TF to drive the conversion.

D   TFs important for driving trans-differentiation toward different cell types were grouped according to their promoter state and tested for enrichment in high or low importance factors (upwards/downwards arrows). Purple and white backgrounds correspond to enrichment tested for promoter states in day 30 or maintained along all differentiation, respectively. Enrichment in high or low importance was tested with Gene Set Enrichment Analysis, considered significant when false discovery rate (FDR) was < 25%.

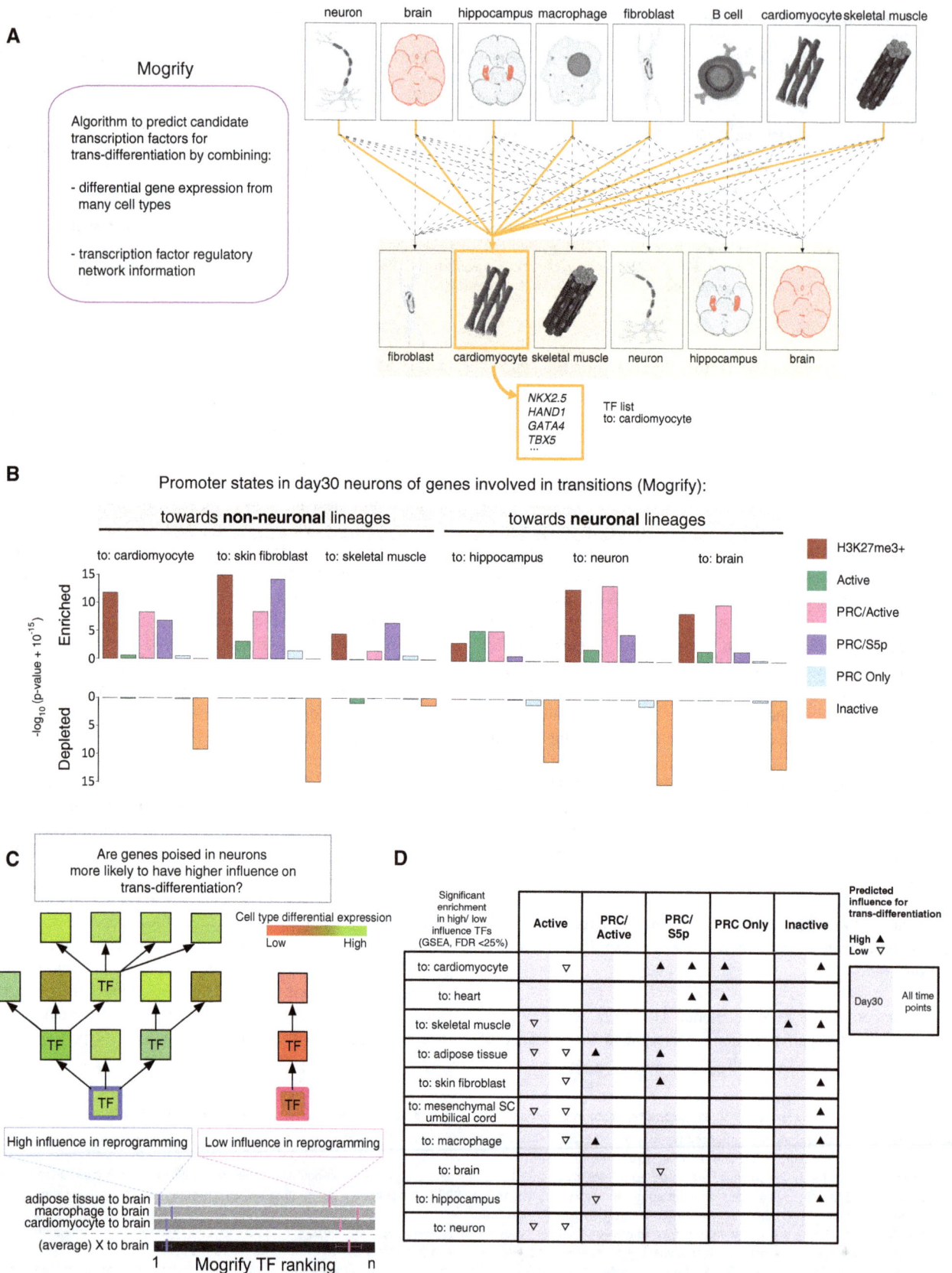

Figure 8.

in Fig 8C). To test whether genes with *high influence* are enriched in PRC-repressed promoter states, we used Gene Set Enrichment Analysis (GSEA; Subramanian *et al*, 2005). Interestingly, we find that genes that have Active promoter states in day 30 neurons are significantly enriched for *low-influence* TFs (false discovery rate < 25%), whereas poised PRC/S5p genes are significantly enriched in *high-influence* TFs which are candidates for promoting conversions toward non-neuronal cell fates (Fig 8D; see example in Fig EV7B). These results show that H3K27me3 and RNAPII-S5p mark silent genes in neurons which are predicted to be important to promote trans-differentiation to non-neuronal lineages, suggesting a role of the poised RNAPII-S5p state in allowing cell identity conversions and of Polycomb repression in stabilizing specific differentiated states.

Discussion

Most genes repressed by Polycomb in ESCs are kept in a poised state through their association with an unusual form of RNAPII characterized by the S5p modification (Stock *et al*, 2007; Brookes *et al*, 2012; Tee *et al*, 2014; Ma *et al*, 2016). The PRC/S5p promoter state is thought to play an important role in the regulation of cell plasticity, conferring pluripotent stem cells with the ability to rapidly activate specific subgroups of genes in response to developmental stimuli (Brookes & Pombo, 2009; Voigt *et al*, 2012; Di Croce & Helin, 2013). However, how the interplay between Polycomb and poised RNAPII evolves upon differentiation has remained unexplored. Here, we optimized an *in vitro* differentiation approach to obtain functional dopaminergic neurons at different stages of maturation, and produced a comprehensive collection of datasets that provides an in depth profile of the transcriptional and promoter state dynamics during differentiation. We observe that promoter state transitions during differentiation are instructed by a tight interplay between RNAPII poising and Polycomb repression. We show for the first time that poising of genes by RNAPII-S5p occurs at intermediate and terminal stages of differentiation, that it associates with increased potential for derepression upon loss of Polycomb repression, and that it may have roles in trans-differentiation events *in vivo* or *in vitro*.

We show *de novo* acquisition of PRC marks at many promoters during differentiation, which coincides with downregulation of gene expression. Notably, we identify a subgroup of genes that maintain the poised PRC/S5p conformation throughout all stages of differentiation studied, including many transcription factors important for specification of non-neuronal lineages. We directly demonstrate the simultaneous co-association of H3K27me3 and RNAPII-S5p at PRC/S5p and PRC/Active genes in differentiated neurons using sequential ChIP, both at genes that maintain the PRC/S5p state and that acquire PRC *de novo*.

We observe that the genes that maintain the PRC/S5p state throughout differentiation have a specific chromatin structure characterized by high CpG content, broad distributions of RNAPII-S5p and H3K27me3, and overlap with large hypomethylated regions or DMVs. This cohort of genes is enriched for developmental factors with key roles in non-neuronal lineages, such as heart, bone, or epithelium. Differentiated cells can be reprogrammed and their identities altered in a process called trans-differentiation (Ladewig *et al*, 2013). To explore the implication of maintaining genes that encode

for lineage-commitment TFs in a poised Polycomb-repressed configuration throughout differentiation, we considered whether they are candidate mediators of trans-differentiation. We discovered that the TFs that maintain the PRC/S5p promoter state during differentiation are important candidate drivers of trans-differentiation conversions. Future studies in other terminally differentiated cell types will help understand whether the poised chromatin state is an inherent state of repression for these important transcription factors, or alternatively whether distinct sets of poised PRC/S5p genes are specifically chosen in each specific cell type to restrict the potential to selectively reprogramming into another cell type, either during the normal physiology of a tissue or in disease.

In conclusion, our study identifies key roles of Polycomb repression and RNAPII-S5p poising in modulating the dynamic changes in gene expression that occur during ESC terminal differentiation into dopaminergic neurons, and provides comprehensive maps of gene expression and promoter states of repression which are a valuable resource for regenerative and developmental biology. Studying gene regulation in terminally differentiated neurons with a dopaminergic phenotype is of particular interest for understanding the biology behind Parkinson's disease and for the generation of cells for replacement therapy as possible treatment.

Materials and Methods

Cell culture and differentiation

Cells were grown at 37°C in a 5% (v/v) CO_2 incubator. Mouse embryonic stem cells (ESC-46C; ES cell line E14tg2a that expresses GFP under Sox1 control; Ying *et al*, 2003), kindly provided by Domingos Henrique, were grown in GMEM medium (Invitrogen, Cat# 21710025), supplemented with 10% (v/v) Fetal Calf Serum (FCS; BioScience LifeSciences, Cat# 7.01 batch number 110006), 2,000 U/ml LIF (Millipore, Cat# ESG1107), 0.1 mM β-mercaptoethanol (Invitrogen, Cat# 31350-010), 2 mM L-glutamine (Invitrogen, Cat# 25030-024), 1 mM sodium pyruvate (Invitrogen, Cat# 11360039), 1% penicillin-streptomycin (Invitrogen, Cat# 15140122), 1% MEM Non-Essential Amino Acids (Invitrogen, Cat# 11140035) on gelatin-coated (0.1% v/v) Nunc T25 flasks. The medium was changed every day, and cells were split every other day. Before sample collection for ChIP-seq and mRNA-seq, ESCs were plated on gelatin-coated (0.1% v/v) Nunc 10-cm dishes in serum-free ESGRO Complete Clonal Grade Medium (Millipore, Cat# SF001-B) to which was added 1,000 U/ml LIF. Cells were grown for 48 h, with a medium change at 24 h. ESC batches were tested for mycoplasma infection.

Early neuronal differentiation was optimized based on the method described in Abranches *et al* (2009). ESCs were plated with high density (1.5×10^5 cells/cm^2) in serum-free ESGRO Complete Clonal Grade Medium (Millipore, Cat# SF001-B) to which 1,000 U/ml LIF was added. After 24 h, ESCs were washed 3 times with PBS without magnesium and calcium, incubated in PBS for 3 min at room temperature, and then dissociated by incubating in 0.05% (v/v) Trypsin (Gibco, Cat# 25300-054) for 2 min at 37°C. ESCs were plated onto 0.1% (v/v) gelatin-coated 10-cm dishes (Nunc) at 1.6×10^6 cells/dish in RHB-A (Takara-Clontech, Cat# Y40001), changing media every day.

Mouse EpiSCs were established from ESC-46C after growth (4 weeks) in N2B27 basal medium containing 20 ng/ml of Activin (R&D, Cat# 338-AC-050) and 12 ng/ml FGF2 (Peprotech, Cat# 100-18B). The composition of the N2B27 basal medium was as follows: half of DMEM/F12 (Invitrogen, Cat# 21331-020), half of Neurobasal Medium (Invitrogen, Cat# 21103-049), 0.5× N2 (Invitrogen, Cat# 17502-048), 0.5× B27 (Invitrogen, Cat# 12587-010), 0.05 M β-mercaptoethanol (Invitrogen, Cat# 31350-010), and 2 mM L-glutamine (Invitrogen, Cat# 25030-024). EpiSCs were grown on Nunc plates coated with FCS (BioScience LifeSciences, Cat# 7.01 batch number 110006). Culture medium was changed every day and cells were split every other day, by washing 3 times with PBS without magnesium and calcium, incubating in PBS for 3 min at room temperature, gently scraping them from the plate pipetting up and down 3 times, before transferring to a new FCS-coated plate. The mouse EpiSC-Pitx3-GFP line was previously established in the Li laboratory (Jaeger et al, 2011), from ESC-Pitx3-GFP (Zhao et al, 2004). Frozen cell batches were tested for mycoplasma infection.

EpiSCs were differentiated into day 16 and 30 neurons with midbrain properties using a protocol optimized based on the method developed in Jaeger et al (2011).

The day before starting the differentiation protocol (day "−1"), growing EpiSCs were plated on Nunc plates coated with 15 μg/ml human plasma fibronectin (Millipore, Cat# FC010) and cultured in N2B27 basal medium containing Activin and FGF2, to reach 70–80% confluency after 24 h. Differentiation was started by rinsing cells twice with PBS, and culturing in N2B27 basal medium plus 1 μM PD 0325901 (Axon, Cat# 1408) for 2 days. Medium was refreshed every day. On day 2, cells were washed with PBS, scraped, replated on Nunc 10-cm dishes coated with 15 μg/ml human plasma fibronectin (Millipore, Cat# FC010), and cultured in N2B27 basal medium for 3 days. At this stage, medium was refreshed every day by removing half of the volume and adding half of the volume of freshly prepared medium.

After 72 h, the medium was replaced with N2B27 basal medium plus 100 ng/ml FGF8 (Peprotech, Cat# 100-25-25) and 200 ng/ml Shh (R&D, Cat# 464-sh-025). Medium was refreshed every day by removing half of the volume and adding half of the volume of freshly prepared medium.

After 96 h, cells were washed with PBS, and the medium replaced with N2B27 basal medium plus 10 ng/ml BDNF (R&D, Cat# 450-02-10), 10 ng/ml GDNF (R&D, Cat# 450-10-10), and 200 μM L-ascorbic acid (Sigma, Cat# A4544). Medium was refreshed every day by removing half of the volume and adding half of the volume of freshly prepared medium until days 16 and 30.

EZH1 and EZH2 enzymatic inhibition on ESCs, day 3, and day 30 differentiated cells was carried out adding 1 or 3 μM of UNC1999 (Konze et al, 2013; Sigma, Cat# SML0778) dissolved in DMSO, which were added directly to the medium. Cells treated with UNC1999 and control cells treated with DMSO were incubated for 24 h with drug or DMSO alone before RNA collection.

mRNA expression

Total RNA was prepared from cells using TRIzol (Invitrogen, Cat# 15596-018) extraction following the manufacturer's instructions. Total RNA was treated with TURBO DNase I (Ambion, Cat#

AM1907) according to the manufacturer's instruction. Extracted RNA (1 μg) was reverse transcribed with 50 ng random primers and 10 U reverse transcriptase (Superscript II kit, Invitrogen, Cat# 11904-018) in a 20-μl reaction. The synthesized cDNA was diluted 1:10, and 2.5 μl used for qRT–PCR. Results were normalized to Actb. mRNA libraries were made using TruSeq RNA Sample

Table 1. List of primers for qRT–PCR (5′–3′).

Actb	Fw	TCTTTGCAGCTCCTTCGTTG
	Rev	ACGATGGAGGGGAATACAGC
Nanog	Fw	ATGAAGTGCAAGCGGTGGCAGAAA
	Rev	CCTGGTGGAGTCACAGAGTAGTTC
Oct4	Fw	CTGAGGGCCAGGCAGGAGCACGAG
	Rev	CTGTAGGGAGGGCTTCGGGCACTT
Fgf5	Fw	CTTCTGCCTCCTCACCAGTC
	Rev	CACTCTCGGCCTGTCTTTTC
Blbp	Fw	GGGTAAGACCCGAGTTCCTC
	Rev	ATCACCACTTTGCCACCTTC
Hes5	Fw	AAGTACCGTGGCGGTGGAGAT
	Rev	CGCTGGAAGTGGTAAAGCAGC
Mash1	Fw	TGGAGACGCTGCGCTCGGC
	Rev	CGTTGCTTCAATGGAGGCAAATG
Sox2	Fw	CATGTGAGGGCTGGACTGCG
	Rev	GCTGTCGTTTCGCTGCGG
Skor1	Fw	GCTAAGTGCATCAAGTGCGG
	Rev	CACGAGTTAAAGTTGGCGGC
Gata4	Fw	GAGGCTCAGCCGCAGTTGCAG
	Rev	CGGCTAAAGAAGCCTAGTCCTTGCTT
HoxA7	Fw	AAGCCAGTTTCCGCATCTACC
	Rev	GTAGCGGTTGAAATGGAATTCC
Pitx1	Fw	CAGATCAGCGTCGGACGATT
	Rev	CACCACCACCGCACGAC
Mesp1	Fw	CGCAGTCGCTCGGTCC
	Rev	GGATGCTGTTTCTGCGTACAG
Pou2f3	Fw	GATCCTCTTTAGAGCCCCACCTG
	Rev	TCAGTGGGCTCATCATTCCCTC
Pkp1	Fw	TACGAATGCTTCCAGGACCA
	Rev	ACTTCTGCCGTTTGACGGT
She	Fw	CCGTAAGAACTCCCGGGTCA
	Rev	TCCCATTCTTCTCCTGCGTG
Fam150b	Fw	CGCTGCTGAGGTTGCTAGTT
	Rev	CTCTGCTCCTCTTGGTCTGC
Fgf20	Fw	GAGGCAGCAAGAAGTGCGA
	Rev	ATTTCAGTCATCCTGTCCGGC
Chrdl2	Fw	GAAAAGATATATACCCCCGGCCA
	Rev	TCAGAGCAGGTACAGCGCAC
Tal1	Fw	GCACCTAGTCCTGCTCAACG
	Rev	GGCCCTTTAAGTCCCTCGC

Preparation Kits v2 setA (Illumina, Cat# RS-122-2001) and sequenced paired-end using Illumina Sequencing Technology by an Illumina HiSeq2000 following the manufacturer's instructions. The primers used are reported in Table 1.

Incorporation of BrdU in replicating cells

To measure replication index, cells were grown in their normal medium in the presence of 50 μM BrdU (Sigma, Cat# 59-14-3) for 24 h. Cells were fixed (10 min) with 4% freshly depolymerized paraformaldehyde in 125 mM HEPES (pH 7.8). After washing (2×) in PBS, cells were permeabilized (20 min) with 0.2% Triton X-100 in PBS and blocked (30 min) with blocking solution (0.5% BSA, 0.2% fish skin gelatin in PBS). Cells were incubated in blocking solution (1 h, humid dark chamber) to which were added 0.5 U/μl DNase I (Sigma, Cat# D4263), 3 mM MgCl$_2$, and anti-BrdU antibody (1:150; raised in mouse which detected BrDNA after in vivo incorporation of BrdU into replicating DNA; Caltag Laboratories, Cat# MD5110). Cells were washed (3×, 3 min each) with blocking solution and incubated (40–60 min) with blocking solution containing the secondary antibody anti-Mouse IgG conjugated with Alexa Fluor 555 (1:1,000; raised in donkey; Invitrogen, Cat# A31570), in a humid dark chamber. Cells were rinsed (2×) with blocking solution and (2×) with PBS. Cells were stained (5 min) in 0.1 mg/ml DAPI in PBS, washed (1×) with PBS, and mounted using Dako Fluorescence Mounting Medium (Dako Agilent Technologies, Cat# S3023).

Protein immunofluorescence

Cells were fixed (10 min) with 4% freshly depolymerized paraformaldehyde in 125 mM HEPES, washed twice with PBS, and permeabilized (3×, 5 min each) with 0.3% Triton X-100 in PBS (PBST), by gently rocking. Cells were blocked (30 min) with 3% donkey serum in PBST and then incubated (at 4°C, overnight), with primary antibody in PBST+3% donkey serum. Primary antibodies used were as follows: mouse Tuj1 (1:500; Sigma, Cat# T8660); mouse anti-OCT4 (1:50; BD Biosciences, Cat# 611202); sheep anti-TH (1:500; Pel-Freez, Rogers-Arkansas, Cat# P60101-0). After washing (3×, 10 min each) with PBST, cells were incubated (1 h, in a humid dark chamber) with secondary antibodies Alexa Fluor 555 anti-mouse IgG (1:1,000; raised in donkey; Invitrogen, Cat# A31570) or Alexa Fluor 555 anti-sheep IgG (1:500; raised in donkey; Invitrogen, Cat# A21436) in PBST+3% donkey serum. After three washes in PBS (10 min each), cells were incubated (5 min) in 0.1 mg/ml DAPI in PBS, washed with PBS (3×, 10 min each), before mounting in Dako Fluorescence Mounting Medium (Dako, Cat# S3023).

For double immunostaining of LMX1A and FOXA2, cells were permeabilized (3×, 5 min each) with 0.3% Triton X-100 in PBS (PBST), by gently rocking. Cells were blocked (30 min) with 3% donkey serum in PBST and then incubated (at 4°C, overnight), with anti-FOXA2 antibodies (1:200; raised in goat; Santa Cruz, Cat# sc-6554) in PBST+3% donkey serum. After washing (3×, 10 min each) with PBST, cells were incubated (1 h, in a humid dark chamber) with secondary antibodies Alexa Fluor 555 anti-goat IgG (1:500; raised in donkey; Invitrogen, Cat# A21432) in PBST+3% donkey serum. Cells were washed (3×, 10 min each) with PBST+3% donkey serum cells, stored (5 h) in 3% donkey serum in PBST, before incubation (at 4°C, overnight), with guinea pig LMX1A (1:500; clone GP6; custom-made antibody generated in the Li laboratory) in PBST+3% donkey serum. After washing (3×, 10 min each) with PBST, cells were incubated (1 h, in a humid dark chamber) with secondary antibodies anti-guinea pig IgG conjugated with Alexa Fluor 488 (1:500; raised in goat; Invitrogen, Cat# A11073) in PBST+3% donkey serum.

After three washes in PBS (10 min each), cells were incubated (5 min) in 0.1 mg/ml DAPI in PBS, washed with PBS (3×, 10 min each), before mounting in Dako Fluorescence Mounting Medium (Dako, Cat# S3023).

For triple immunostaining of dopaminergic, GABAergic, and glutamatergic markers (Fig EV3G), fixed cells were washed (3×) in PBST (PBS containing 0.1% Triton X-100) and incubated 2 h in a blocking solution that contained 5% donkey serum and 3% BSA in PBST. Primary antibodies used were as follows: sheep anti-TH (1:500, Pel-Freez, Rogers-Arkansas, Cat# P60101-0), mouse anti-VGluT2 (vesicular glutamate transporter 2, 1:1,000, Millipore, Cat# MAB5504), and rabbit anti-GABA (gamma-aminobutyric acid, 1:1,000, Sigma, Cat# A2052). The primary antibodies were diluted in blocking solution and incubated overnight at 4°C. The next day, cells were washed (3×) with PBST and bathed (1 h) in secondary antibodies Alexa Fluor 488 anti-sheep IgG (1:200, raised in donkey, Molecular Probes, Cat# A-11015), Alexa Fluor 555 anti-rabbit IgG (1:200, raised in donkey, Molecular Probes, Cat# A-31572), and Alexa Fluor 647 anti-mouse IgG (1:200, raised in donkey, Molecular Probes, Cat# A-31571) diluted in blocking solution, and after three washes in PBS, mounted with Dako Fluorescence Mounting Medium (Dako, Cat# S3023).

Microscopy

Single immunofluorescence images (Fig 1C and D) were collected sequentially on a confocal laser-scanning microscope (Leica TCS SP8 STED; HC PLAPOCS2 20×/0.75 IMM objective), equipped with a 405 diode and a WLL (Supercontinuum visible) laser, and pinhole equivalent to 1 Airy disk. For comparison of the different staining during neuronal differentiation, images were collected on the same day using the same settings, and without saturation of the intensity signal.

Double and triple immunofluorescence images (Figs EV1C and EV3G) were acquired on a confocal laser-scanning microscope (Leica TCS SP5; HCX PL APO CS 40.0× 1.25 OIL objective) equipped with a 405 nm diode, and Argon (488 nm), HeNe (543 nm), and HeNe (633 nm) lasers.

Electrophysiology

Coverslips of mature neuronal cultures were placed onto a recording chamber and viewed using an Olympus BX51WI microscope with a 40× water immersion lens and DIC (differential interference contrast) optics. Cells were bathed at room temperature in a solution containing (in mM): 140 NaCl, 3.5 KCl, 1.25 NaH$_2$PO$_4$, 2 CaCl$_2$, 1 MgCl$_2$, 10 glucose, and 10 HEPES (pH 7.4). For whole-cell electrophysiological recordings, low resistance recording pipettes (8–14 MΩ) were pulled from capillary glass (Harvard Apparatus Ltd.) and coated with ski wax (Toko) to reduce pipette capacitance. Recording pipettes were filled with a solution containing: 140 mM K-gluconate, 5 mM NaCl, 2 mM ATP, 0.5 mM GTP, 0.1 mM CaCl$_2$,

1 mM MgCl$_2$, 1 mM ethylene glycol-bis (b-aminoethyl ether)-N,N, N',N'-tetraacetic acid (EGTA), and 10 mM HEPES (pH 7.4). For recordings of mIPSCs, a high-chloride solution was used: 135 mM CsCl, 4 mM NaCl, 2 mM ATP, 0.3 mM GTP, 0.5 mM MgCl$_2$, 2 mM EGTA, and 10 mM HEPES (pH 7.4). Pitx3-GFP positive neurons were identified and targeted using fluorescence via a GFP selective filter (X-Cite series 120, EXFO). For analysis of synaptic physiology, cells were held in voltage-clamp mode at −70 mV and events were detected over 3 min of recording. The first 30 s of recording was used for analysis of action potentials (i.e., frequency and CV-ISI). Cells were excluded from analysis if they exhibited a holding current > −150 pA, a series resistance > 30 MΩ, or if the holding current did not remain stable over the course of recording. Antagonists were dissolved in the extracellular solution and applied to the cultures via bath perfusion at 30–33°C. The sodium channel blocker tetrodotoxin citrate (TTX) was used at 1 µM (Tocris Bioscience, Cat# 1069). In order to block glutamatergic synaptic activity, 2,3-dioxo-6-nitro-1,2,3,4-tetrahydrobenzo[f]quinoxaline-7-sulfonamide (NBQX; Sigma, Cat# N171), an AMPA receptor antagonist, was used at 5 µM, while Picrotoxin (100 µM, Sigma, Cat# P1675), a GABA$_A$ receptor antagonist, was used to identify GABAergic synaptic events. Data were acquired using an Axon Multiclamp 700B amplifier and a Digidata 1440a acquisition system, with pClamp 10 software (Molecular Devices). Data analysis was carried out using Clampfit 10.2 software (Axon Instruments), OriginPro 8.1 (OriginLab Corporation), Spike2v5 software (Cambridge Electronic Design), and MiniAnalysis software v6.0.7 (Synaptosoft). Data are presented as either percentage of cells or the mean ± standard error of the mean (SEM). Statistical analyses were performed using either chi-squared test for trend or a one-way ANOVA (Kruskal–Wallis; GraphPad Prism, GraphPad Software, San Diego, CA). Significance was noted when $P < 0.05$.

Chromatin immunoprecipitation (ChIP), ChIP-seq, and re-ChIP

The chromatin from ESCs or differentiated cultures was prepared by cross-linking in 1% formaldehyde for 10 min as previously described (Stock et al, 2007; Brookes et al, 2012). Arbitrary "chromatin concentration" was obtained by measuring absorbance (280 nm) of alkaline-lysed chromatin, and using the conversion 50 mg/ml for 1 absorbance unit. For mouse anti-RNAPII-S5p (clone CTD4H8, BioLegend, Cat# 904001), mouse anti-digoxigenin (Jackson Immunoresearch, Cat# 200-002-156), and rat anti-RNAPII-S7p (4E12; Chapman et al, 2007; kindly provided by Dirk Eick) antibodies, protein-G-magnetic beads (Active Motif, Cat# 53014) were incubated with rabbit anti-mouse (IgG + IgM) bridging antibodies (Jackson Immunoresearch, Cat# 315-005-048; 10 µg per 50-µl beads) for 1h at 4°C and washed with sonication buffer. For rabbit anti-H3K27me3 (Millipore, Cat# 07-449) antibodies, magnetic beads were just washed with sonication buffer. Fixed (700 µg) chromatin was immunoprecipitated (4°C, overnight) with the specific antibody and 50-µl beads (with/without bridging antibody), as previously described (Stock et al, 2007) and (Brookes et al, 2012). After immunoprecipitation using IgG antibodies, beads were washed as described previously (Stock et al, 2007). Immune complexes were eluted from beads (65°C, 5 min then room temperature, 15 min) with 50 mM Tris–HCl pH 8.0, 1 mM EDTA, and 1% SDS. Elution was repeated once and eluates pooled. Reverse cross-linking was

carried out (8 h, 65°C) after addition of NaCl and RNase A. EDTA was then increased to 5 mM, and samples were incubated (50°C, 2 h) with 200 µg/ml proteinase K (Roche, Cat# 3115836001). DNA was recovered by phenol-chloroform extraction and ethanol precipitation. The final DNA concentration was determined by Quant-iT PicoGreen dsDNA Assay Kit fluorimetry (Thermo Fisher Scientific, Cat# P7589). The quality of immunoprecipitated DNA used for high-throughput sequencing was confirmed prior to library preparation by quantitative real-time PCR (qPCR) analyses of Active, PRC/S5p, and Inactive genes with previously characterized chromatin states (Brookes et al, 2012); these quality control tests are not shown. Samples were diluted to the same concentration (0.2 ng/µl). The same amount of immunoprecipitated and input DNA (0.5 ng) was analyzed by qPCR using primers with the previously published sequences (Brookes et al, 2012). Amplifications (40 cycles) were performed using SensiMix No-ROX Kit (Bioline, Cat# QT650-05) with DNA Engine Opticon 1/2 Real-Time PCR system (Bio-Rad, Hemel Hempstead, Hertfordshire, UK).

ChIP-seq libraries were prepared from 10 ng DNA (quantified by PicoGreen and Qubit) using the Next ChIP-Seq Library Prep Master Mix Set for Illumina (NEB, Cat# E6240) following the NEB protocol, with some modifications. The intermediate products from the different steps of the NEB protocol were purified using MinElute PCR purification kit (Qiagen, Cat# 28004). Adaptors, PCR amplification primers, and indexing primers were from the Multiplexing Sample Preparation Oligonucleotide Kit (Illumina, Cat# PE-400-1001). Samples were PCR amplified prior to size selection on an agarose gel (250–600 bp including adapters). The size range was selected due to the broad distribution of RNAPII and H3K27me3 at promoters, where there is evidence that protection of DNA may occur (Ferrai et al, 2007, 2010). After purification by QIAquick Gel Extraction kit (Qiagen, Cat# 28704), libraries were quantified by Qubit (Invitrogen) and by qPCR using KAPA Library Quantification Universal Kit (KapaBiosystems, Cat# KK4824). Library size was assessed before high-throughput sequencing by Bioanalyzer (Agilent) using the High Sensitivity DNA analysis kit (Agilent, Cat# 5067-4626). Fragment sizes were within the selected size distribution. ChIP-seq libraries were sequenced single-end using Illumina Sequencing Technology using an Illumina HiSeq2000, according to the manufacturer's instructions.

Re-ChIP experiments were performed as previously described (Brookes et al, 2012). The first ChIP was performed as described above with the exception that after the final wash in TE buffer, immune complexes were eluted from the beads twice (65°C, 5 min, and room temperature, 15 min) using elution buffer (50 mM Tris pH 8.0, 1 mM EDTA, and 1% SDS). "Re-ChIP dilution" buffer (55 mM HEPES pH 7.9, 154 mM NaCl, 1.0 mM EDTA, 1.1% Triton X-100, 0.11% Na-deoxycholate) was added to dilute the eluate 10-fold to reach a final SDS concentration of 0.1%. The second round of immunoprecipitation was set up following the standard ChIP protocol. Enrichment relative to the input chromatin material was obtained using qPCR (see above) using the same amount of DNA in each PCR. To test carryover from the first immunoprecipitation, an unrelated antibody anti-digoxigenin (raised in mouse; Jackson Immunoresearch, Cat# 200-002-156) was used in the second round of immunoprecipitation. As a positive control, RNAPII-S5p ChIP was repeated in the second immunoprecipitation. The primers used for Re-ChIP are reported in Table 2.

Table 2. List of Re-ChIP primers (5'–3').

Actb	Fw	GCAGGCCTAGTAACCGAGACA
	Rev	AGTTTTGGCGATGGGTGCT
Pitx1	Fw	CAGATCAGCGTCGGACGATT
	Rev	CACCACCACCGCACGAC
Gata4	Fw	TTCCCAGAAAACCGGCGCGA
	Rev	CCCTTAGGCCAGTCAGCGCA
Skap2	Fw	TCCGGCTGTCCAGGGAGGAT
	Rev	CACCCTATGCGGACGGTGGG
Ajuba	Fw	GCCCTGCCTCTGCCTCTGTC
	Rev	CGAGTGACGGCTCTCCTGGC
Myf5	Fw	GGAGATCCGTGCGTTAAGAATCC
	Rev	CGGTAGCAAGACATTAAAGTTCCGTA

Bioinformatics

Read mapping and visualization

Sequenced single-end (50–51 bp) reads originating from ChIP-seq libraries were aligned to the Mouse reference genome mm10 (Dec. 2011, GRCm38/mm10) using Bowtie2 v2.0.5 (Langmead & Salzberg, 2012). The reference genome was indexed, and the alignments were performed with default parameters. Replicated reads (i.e., identical reads, aligned to the same genomic location), occurring more often than the 95th percentile of the frequency distribution of each dataset, were removed. When reads originated from multiple runs (technical replicates of sequencing), datasets were merged after mapping and removal of replicated reads.

Sequenced paired-end (2 × 100bp) reads from mRNA-seq libraries were aligned for visualization using TopHat v2.0.8 (Trapnell et al, 2009). The gene references provided to TopHat were the gtf annotation from UCSC Known Genes (mm10, version 6) and the associated isoform-gene relationship information from the Known Isoforms table (UCSC). Tables were downloaded from the UCSC Table browser (http://genome.ucsc.edu/cgi-bin/hgTables). UCSC Genome Browser (http://genome.ucsc.edu) was used for Figs 4B and 6E, and EV2A, and EV3G using default view settings, except: Windowing function—Mean, Smoothing Window—2 pixels and inclusion of zero on y-axis. Number of mapped reads for the datasets produced for this study is shown in Table 3.

Non-redundant gene list selection

To investigate the most representative isoform for each gene in each time point, we followed the same strategy as in Brookes et al (2012), applied to mouse genome assembly mm10. To define isoforms and link them to gene clusters, we used UCSC Known Gene table (mm10, version 6) and the associated UCSC Known Isoforms table. Genes belonging to the mitochondrial genome (chrM) or "random chromosomes" (chr_random) were discarded. We also discarded clusters not linked to any RefSeq annotation (from the associated UCSC kgXref table). After filtering, we obtained a dataset of 23,177 gene clusters, associated with one or more RefSeq genes.

Around half of the gene clusters, 11,331, had a single isoform. When more than one isoform (UCSC Known Gene identifier) was present in a cluster, we selected one isoform per time point using

Table 3. List of genome-wide ChIP-seq and mRNA-seq datasets produced in the present study.

	Cell type	Read type	Mapped reads (millions)
ChIP-seq dataset			
RNAPII-S5p	46C- ESC	1 × 50 bp	46
RNAPII-S5p	46C- day1	1 × 50 bp	82
RNAPII-S5p	46C- day3	1 × 50 bp	75
RNAPII-S5p	46C- day16	1 × 50 bp	95
RNAPII-S5p	46C- day30	1 × 50 bp	58
RNAPII-S7p	46C- ESC	1 × 50 bp	68
RNAPII-S7p	46C- day1	1 × 50 bp	65
RNAPII-S7p	46C- day3	1 × 50 bp	87
RNAPII-S7p	46C- day16	1 × 50 bp	50
RNAPII-S7p	46C- day30	1 × 50 bp	81
H3K27me3	46C- ESC	1 × 50 bp	40
H3K27me3	46C- day1	1 × 50 bp	42
H3K27me3	46C- day3	1 × 50 bp	62
H3K27me3	46C- day16	1 × 50 bp	48
H3K27me3	46C- day30	1 × 50 bp	32
Control (Dig)	46C- ESC	1 × 51 bp	67
Control (Dig)	46C- day16	1 × 50 bp	21
polyA mRNA-seq dataset			
ESC	46C- ESC	2 × 100 bp	48
day1	46C- day1	2 × 100 bp	46
day3	46C- day3	2 × 100 bp	47
day16	46C- day16	2 × 100 bp	40
day30	46C- day30	2 × 100 bp	51

the following criteria: (i) We selected the gene isoform with the highest amount of reads for RNAPII-S5p in a 2-kb window centered on the TSS (Transcription Start Site; in 15.7–17.1% of gene clusters the isoform was selected at this step); (ii) if ambiguity was still present, we selected the isoform with the highest RNAPII-S2p in a 4-kb window centered on the TES (Transcription End Site; in 13.3–13.9% of gene clusters the isoform was selected at this step); (iii) if ambiguity was still present, we selected the canonical isoform annotated in the associated UCSC mm10 known Canonical table (the gene with the highest number of coding bases; in 20.5–21.3% of gene clusters, the isoform was selected at this step); and (iv) if ambiguity was still present, one isoform was selected randomly (in 0.1–0.2% of gene clusters, the isoform was selected at this step).

Promoter classification

To define genome-wide enriched regions for RNAPII-S5p, RNAPII-S7p, and H3K27me3, for each time point, we used Bayesian Changepoint Model (BCP) peak-finder (Xing et al, 2012). BCP performs well in detecting enriched regions for broad chromatin features such as (broad) histone marks or RNAPII (Harmanci et al, 2014; Thomas et al, 2016). BCP was run in Histone Mark (HM) mode using as control dataset: (i) control (digoxigenin) ChIP-seq from ESCs, for

ESC, day 1, and day 3 datasets or (ii) control (digoxigenin) ChIP-seq from day 16 neurons, for day 16 and 30 datasets. Gene promoters were considered positive for RNAPII-S5p, RNAPII-S7p, or H3K27me3 when: (i) the 2-kb windows centered on the TSS overlapped with a region enriched for the mark, and (ii) the amount of reads in the TSS window was above a threshold. The threshold was defined as the 5th percentile of the distribution of reads in the TSS window of positive genes (5% tail cut).

To remove overlapping genes, we excluded genes whose (positive) TSS window overlapped positive windows for the same mark of other genes for more than 10% (200 bp). Positive genes that were inside other positive genes for the same mark were also excluded (internal gene removed). Excluded genes are marked as NA in Dataset EV2. Genes classified as NA for at least one of the three marks used for the classification (RNAPII-S5p, RNAPII-S7p, or H3K27me3) in one time point were excluded from further analyses. The genes that passed these filters (19,352) were classified, in each time point, according to the combination of RNAPII and H3K27me3 as belonging to one of eight different promoter states using the logics presented in main text (Fig 5A). The complete promoter classification is shown in Dataset EV2.

mRNA-seq analyses

To calculate expression estimates, mRNA-seq reads were mapped with STAR (Spliced Transcripts Alignment to a Reference, v2.4.2a; Dobin et al, 2013) and processed with RSEM (RNA-Seq by Expectation-Maximization, v1.2.25; Li & Dewey, 2011). The references provided to STAR and RSEM were the gtf annotation from UCSC Known Genes (mm10, version 6) and the associated isoform-gene relationship information from the Known Isoforms table (UCSC). Both tables were downloaded from the UCSC Table browser (http://genome.ucsc.edu/cgi-bin/hgTables). Gene-level expression estimates in Transcript Per Million (TPM) were used for all the analyses and are reported in Dataset EV2.

To select genes that peak in one specific time point (Fig 1E), the expression levels from genes in the non-redundant list were standardized across time points using z-scores (calculated as the TPM value in each time point minus the mean value across the five time points, divided by the standard deviation). Then, genes whose standardized expression was higher than 1.75 in one time point were selected. Genes needed to be expressed (> 1 TPM) in at least one time point to be included in the analysis.

Statistical tests were performed using R, and the name of the test is specified in main text.

Gene ontology enrichment

Gene ontology (GO) enrichment analysis was performed using GO-Elite version 1.2.5 (Gladstone Institutes; http://genmapp.org/go_e lite). Pruned results (to decrease term redundancy) are reported. Default parameters were used as filters: z-score threshold more than 1.96, P-value < 0.05, number of genes changed more than 2. Over-representation analysis was performed with "permute P-value" option, 2,000 permutations. Permute P-value is reported in figures. UCSC Known Gene IDs were converted into the correspondent Ensembl Gene IDs (using the UCSC KnownToEnsembl table, downloaded from the UCSC Table browser http://genome.ucsc.edu/cgi-bin/hgTables) before performing the GO enrichment analyses. The group of genes used as background for each analysis is specified in

the associated figure legend. The complete list of enriched pruned GO terms is shown in Dataset EV1.

Comparisons with published datasets

To compare PRC-positive genes in this study with previously published work, the list of genes whose promoter is classified as positive for H3K27me3 in ESC-46C was overlapped with available resources as follows: (i) List of genes positive for H3K27me3 in Young et al, 2011 defined using the classification provided in Supplementary Table 2 of the corresponding paper. In particular, genes classified as "Marked", "Broad", "Promoter", or "TSS" in the "ES H3K27me3" column were considered. Genes were matched with our classification using Ensembl Gene ID identifiers and converted from UCSC IDs as previously described. (ii) List of genes positive for H3K27me3 in Lienert et al, 2011 or in Brookes et al, 2012 were obtained from Table S2 in Brookes et al, 2012; that is, the genes classified as "1" in columns "H3K27me3 (TSS) (Lienert et al, 2011)" or "H3K27me3 (TSS) (Mikkelsen et al, 2007)", respectively. Genes were matched with our classification using Gene Symbols. Percentages were calculated as the number of genes positive in both classifications divided for the number of ESC-46C H3K27me3-positive genes present in the published lists.

To compare PRC/S5p genes in this study with bivalent genes, the list of genes whose promoter is classified as positive for H3K27me3 and RNAPII-S5p in ESC-46C was overlapped with list of bivalent genes ("K4+K27", "ESC") in Supplementary Table 3 in Mikkelsen et al, 2007. Genes were matched with our classification using Gene Symbols. We find 69% of bivalent genes in Mikkelsen et al, 2007 are positive for H3K27me3 and RNAPII-S5p in our ESC datasets.

Upregulated genes in published Polycomb-knockout datasets

To explore the effect of PRC1/2 knockout in ESCs on the genes which are marked by H3K27me3 in presence or absence of RNAPII-S5p, we mined published resources. Expression values of wild-type and Eed-knockout ESCs (Cruz-Molina et al, 2017) were obtained from GEO Gene Expression Omnibus repository (GSE89210). Identifiers were then matched via Ensembl Gene IDs. Fold change was calculated as ratio between the average FPKM value in Eed-knockout ESCs relative to the average FPKM value in wild-type ESCs. Genes were classified as upregulated if the Fold change was > 2 and the expression in Eed-knockout ESC was > 1 FPKM. Genes classified as PRC/S5p, PRC Only or Inactive that were expressed > 1 FPKM in the published datasets for wild-type ESCs were excluded from the analysis, as these are unlikely to have matched repressed promoter states. The percentage of upregulated genes in each promoter state was calculated relative to the number of genes in that promoter state found in the published dataset.

Expression values of control cells and tamoxifen-induced conditional Ring1B knockout in Ring1A$^{-/-}$ knockout cells measured by microarray (Endoh et al, 2008) were obtained from GEO (GSM265042 and GSM265043) and analyzed as in (Brookes et al, 2012). Identifiers were matched via Gene Symbols. The percentage of genes upregulated in each promoter state was calculated in relation to the number of genes with the same promoter state that are present in the published datasets.

To explore the effect of PRC2 knockout in neurons on genes marked by H3K27me3 in the presence or absence of RNAPII-S5p, we mined a published list of upregulated genes from ex vivo medium spiny neurons from 6 month old control and $Ezh1^{-/-}$; $Ezh2^{fl/fl}$; $Camk2a$-cre knockout mice, measured by microarray (from the Supplementary Table 3 in von Schimmelmann et al, 2016). Genes in the published list were matched with our day 30 list of promoter states using Gene Symbols. Microarray probes matching to more than one Gene Symbol were not considered. In the absence of the total list of detected genes, the percentage of genes upregulated in each promoter state was calculated relative to the total number of genes in that group.

Plot generation

Heatmaps in Figs 1E and 3B, and EV4B were produced using the CRAN package *pheatmap* (Kolde, 2015) in R.

Bar plots, dot plots, and boxplots were produced using Excel or R. For mRNA-seq, a pseudo-count of 10^{-4} was added to TPM prior to logarithmic transformation and plotting. In Fig 7D, a pseudo-count of 1 was added to peak size H3K27me3 prior to logarithmic transformation and plotting.

To facilitate comparison between days, in Fig 3B, the number of reads mapping in 2-kb windows around the TSS (transformed in log scale) for each gene was divided by the threshold of minimal number of reads in positive genes (5% Threshold, see section on Promoter Classification). To help visualization, these values were then scaled to have the same maximum value, and the minimum and maximum values of the color scale were set to 1 and 99% of the distribution, respectively.

Average ChIP-seq profiles (Fig 7B and C) were generated as previously (Brookes et al, 2012), by plotting the average coverage in non-overlapping windows of 10 bp, across genomic windows centered on the TSS and the TES using custom scripts.

Peak size, GC content, and CpG coverage calculation

To calculate the breath of H3K27me3 windows covering each gene (Fig 7D), we calculated the size of the BCP-detected enriched region that overlaps the 2-kb window centered on the TSS of each gene. If more than one peak was present, the longest was selected. To calculate the overlap between TSS window or gene body (TSS to TES) with CpG islands, we used the CpG island definition from the mm10 cpgIslandExt table, downloaded from UCSC Table Browser. For each gene, we used the isoform selected in day 30 neurons. The overlap was computed with IntersectBed, part of Bedtools v2.17.0 (Quinlan & Hall, 2010). To calculate the proportion of TSS window composed of G or C nucleotides, the sequence of the 2-kb window centered on the TSS used in day 30 was extracted using bedtools 2.17.0 getfast and analyzed using a custom script. Statistical tests were performed using R, and the name of the test is specified in the text.

DNA hypomethylated region and DNA methylation valley definition

Hypomethylated regions (HMRs) from mouse brain, Frontal Cortex (Lister et al, 2013); dataset name: MouseFrontCortexMale22Mo), were downloaded from Methbase (Song et al, 2013); http://smithlabresearch.org/software/methbase), a resource from the Smith laboratory, accessed through the UCSC Table Browser. HMRs were downloaded as a bed file, already mapped to mm10. DNA methylation valleys (DMVs) were defined starting from HMR using custom scripts for proximity clustering and 5-kb size cut-off, following a previously described approach (Xie et al, 2013).

Mogrify

To identify candidate TFs required to convert a given cell type into another regardless of their original identity, we extracted the influence ranks provided by the Mogrify algorithm (Rackham et al, 2016). These ranks are calculated for each TF for a conversion between any of 173 cell types or 134 tissue types based on the TFs estimated ability to regulate the differential expression of its immediate network neighborhood. For any given cell type, we take the average rank of the TF over every possible conversion; for instance, the TF Sox2 is often ranked highly for cell conversions into human neurons and as such has an average rank of 34.86 (the maximum possible rank is 303). The full list of genes and extracted average ranks for the transitions investigated here can be found in Dataset EV3.

Each human gene was then matched with the mouse homolog using homology table from Ensembl, release 84 (downloaded through BioMart http://www.ensembl.org/biomart/). Matching was performed on Gene Names. Enrichment or depletion in particular promoter states from day 30 neurons was tested in R with hypergeometric test.

Promoter state enrichment in genes with high or low influence on trans-differentiation (estimated with their rank) was tested using Gene Set Enrichment Analysis (Subramanian et al, 2005), in "GSEAPreranked" mode. GSEA tests whether a ranked list (here, the TFs involved in trans-differentiation) is enriched in certain Gene Sets (here, different promoter states in day 30 or all days). In "GSEAPreranked" mode, GSEA is run against a user-supplied, ranked list of genes (the list of TFs involved in trans-differentiation, using the influence ranks calculated by Mogrify) and tests for statistically significant enrichment at either end of the ranking. Enrichment was considered significant at a false discovery rate < 25%.

Acknowledgements

We thank Laurence Game and Adam Giess (MRC LMS Sequencing Facility) for sequencing and raw data processing, Elsa Abranches, Domingos Henrique and Ines Jaeger for neuronal differentiation advice, and Dirk Eick for the RNAPII-S7p antibody. We thank the Pombo laboratory for helpful discussions and comments on the manuscript. CF and AP thank the BBSRC (UK); CF, ETT, TR, AK, AA, and AP thank the Helmholtz Association (Germany) for support. CF was recipient of a Wellcome Trust VIP award. This work was supported by the UK Medical Research Council (MRC; MC_U120061476) to AP; by the MRC (G117/560, U120005004) and EU framework program 7 (NeuroStemcell, no. 222943) to ML; by the MRC (U120085816) and a Royal Society University Research Fellowship to MAU. ML was an MRC Senior Non-Clinical Research Fellow. MN thanks CINECA ISCRA HP10CYFPS5 and HP10CRTY8P, computer resources at INFN and Scope at the University of Naples, and the Einstein BIH Fellowship Award.

Author contributions

CF and AP conceived and designed the project; CF, JRR-J, and AK conducted the experiments; ETT, TR, OJLR, IS, and AA designed and conducted bioinformatic analysis; MN helped with data analysis; ML supported the development of the neuronal differentiation protocol; MAU supervised the electrophysiological experiments; CF, ETT, JRR-J, TR, OJLR, MAU, and AP analyzed and interpreted the results; CF, ETT, MAU, and AP prepared the manuscript with the help of the other authors.

References

Abranches E, Silva M, Pradier L, Schulz H, Hummel O, Henrique D, Bekman E (2009) Neural differentiation of embryonic stem cells *in vitro*: a road map to neurogenesis in the embryo. *PLoS One* 4: e6286

Adelman K, Lis JT (2012) Promoter-proximal pausing of RNA polymerase II: emerging roles in metazoans. *Nat Rev Genet* 13: 720−731

Akhtar MS, Heidemann M, Tietjen JR, Zhang DW, Chapman RD, Eick D, Ansari AZ (2009) TFIIH kinase places bivalent marks on the carboxy-terminal domain of RNA polymerase II. *Mol Cell* 34: 387−393

Ang SL (2006) Transcriptional control of midbrain dopaminergic neuron development. *Development* 133: 3499−3506

Azuara V, Perry P, Sauer S, Spivakov M, Jorgensen HF, John RM, Gouti M, Casanova M, Warnes G, Merkenschlager M, Fisher AG (2006) Chromatin signatures of pluripotent cell lines. *Nat Cell Biol* 8: 532−538

Bernstein BE, Mikkelsen TS, Xie X, Kamal M, Huebert DJ, Cuff J, Fry B, Meissner A, Wernig M, Plath K, Jaenisch R, Wagschal A, Feil R, Schreiber SL, Lander ES (2006) A bivalent chromatin structure marks key developmental genes in embryonic stem cells. *Cell* 125: 315−326

Blackledge NP, Rose NR, Klose RJ (2015) Targeting Polycomb systems to regulate gene expression: modifications to a complex story. *Nat Rev Mol Cell Biol* 16: 643−649

Boettiger AN, Bintu B, Moffitt JR, Wang SY, Beliveau BJ, Fudenberg G, Imakaev M, Mirny LA, Wu CT, Zhuang XW (2016) Super-resolution imaging reveals distinct chromatin folding for different epigenetic states. *Nature* 529: 418−422

Brookes E, Pombo A (2009) Modifications of RNA polymerase II are pivotal in regulating gene expression states. *EMBO Rep* 10: 1213−1219

Brookes E, de Santiago I, Hebenstreit D, Morris KJ, Carroll T, Xie SQ, Stock JK, Heidemann M, Eick D, Nozaki N, Kimura H, Ragoussis J, Teichmann SA, Pombo A (2012) Polycomb associates genome-wide with a specific RNA polymerase II variant, and regulates metabolic genes in ESCs. *Cell Stem Cell* 10: 157−170

Chapman RD, Heidemann M, Albert TK, Mailhammer R, Flatley A, Meisterernst M, Kremmer E, Eick D (2007) Transcribing RNA polymerase II is phosphorylated at CTD residue serine-7. *Science* 318: 1780−1782

Choi J, Huebner AJ, Clement K, Walsh RM, Savol A, Lin K, Gu H, Di Stefano B, Brumbaugh J, Kim SY, Sharif J, Rose CM, Mohammad A, Odajima J, Charron J, Shioda T, Gnirke A, Gygi S, Koseki H, Sadreyev RI et al (2017) Prolonged Mek1/2 suppression impairs the developmental potential of embryonic stem cells. *Nature* 548: 219−223

Cruz-Molina S, Respuela P, Tebartz C, Kolovos P, Nikolic M, Fueyo R, van Ijcken WFJ, Grosveld F, Frommolt P, Bazzi H, Rada-Iglesias A (2017) PRC2 facilitates the regulatory topology required for poised enhancer function during pluripotent stem cell differentiation. *Cell Stem Cell* 20: 689−705

Deaton AM, Bird A (2011) CpG islands and the regulation of transcription. *Gene Dev* 25: 1010−1022

Di Croce L, Helin K (2013) Transcriptional regulation by Polycomb group proteins. *Nat Struct Mol Biol* 20: 1147−1155

Dias JD, Rito T, Torlai Triglia E, Kukalev A, Ferrai C, Chotalia M, Brookes E, Kimura H, Pombo A (2015) Methylation of RNA polymerase II non-consensus Lysine residues marks early transcription in mammalian cells. *Elife* 4: e11215

Dobin A, Davis CA, Schlesinger F, Drenkow J, Zaleski C, Jha S, Batut P, Chaisson M, Gingeras TR (2013) STAR: ultrafast universal RNA-seq aligner. *Bioinformatics* 29: 15−21

Endoh M, Endo TA, Endoh T, Fujimura Y, Ohara O, Toyoda T, Otte AP, Okano M, Brockdorff N, Vidal M, Koseki H (2008) Polycomb group proteins Ring1A/B are functionally linked to the core transcriptional regulatory circuitry to maintain ES cell identity. *Development* 135: 1513−1524

Eskeland R, Leeb M, Grimes GR, Kress C, Boyle S, Sproul D, Gilbert N, Fan YH, Skoultchi AI, Wutz A, Bickmore WA (2010) Ring1B compacts chromatin structure and represses gene expression independent of histone ubiquitination. *Mol Cell* 38: 452−464

Ferrai C, Munari D, Luraghi P, Pecciarini L, Cangi MG, Doglioni C, Blasi F, Crippa MP (2007) A transcription-dependent micrococcal nuclease-resistant fragment of the urokinase-type plasminogen activator promoter interacts with the enhancer. *J Biol Chem* 282: 12537−12546

Ferrai C, Xie SQ, Luraghi P, Munari D, Ramirez F, Branco MR, Pombo A, Crippa MP (2010) Poised transcription factories prime silent uPA gene prior to activation. *PLoS Biol* 8: e1000270

Francis NJ, Kingston RE, Woodcock CL (2004) Chromatin compaction by a polycomb group protein complex. *Science* 306: 1574−1577

Gaertner B, Johnston J, Chen K, Wallaschek N, Paulson A, Garruss AS, Gaudenz K, De Kumar B, Krumlauf R, Zeitlinger J (2012) Poised RNA polymerase II changes over developmental time and prepares genes for future expression. *Cell Rep* 2: 1670−1683

Glover-Cutter K, Larochelle S, Erickson B, Zhang C, Shokat K, Fisher RP, Bentley DL (2009) TFIIH-associated Cdk7 kinase functions in phosphorylation of C-terminal domain Ser7 residues, promoter-proximal pausing, and termination by RNA polymerase II. *Mol Cell Biol* 29: 5455−5464

Grace AA, Bunney BS (1984) The control of firing pattern in nigral dopamine neurons − burst firing. *J Neurosci* 4: 2877−2890

Graham V, Khudyakov J, Ellis P, Pevny L (2003) SOX2 functions to maintain neural progenitor identity. *Neuron* 39: 749−765

Harmanci A, Rozowsky J, Gerstein M (2014) MUSIC: identification of enriched regions in ChIP-Seq experiments using a mappability-corrected multiscale signal processing framework. *Genome Biol* 15: 474

He A, Ma Q, Cao J, von Gise A, Zhou P, Xie H, Zhang B, Hsing M, Christodoulou DC, Cahan P, Daley GQ, Kong SW, Orkin SH, Seidman CE, Seidman JG, Pu WT (2012) Polycomb repressive complex 2 regulates normal development of the mouse heart. *Circ Res* 110: 406−415

Hegarty SV, Sullivan AM, O'Keeffe GW (2013) Midbrain dopaminergic neurons: a review of the molecular circuitry that regulates their development. *Dev Biol* 379: 123−138

Hirota T, Kunitoku N, Sasayama T, Marumoto T, Zhang DW, Nitta M, Hatakeyama K, Saya H (2003) Aurora-A and an interacting activator, the LIM protein Ajuba, are required for mitotic commitment in human cells. *Cell* 114: 585−598

Infante CR, Park S, Mihala AG, Kingsley DM, Menke DB (2013) Pitx1 broadly associates with limb enhancers and is enriched on hindlimb cis-regulatory elements. *Dev Biol* 374: 234−244

Jaeger I, Arber C, Risner-Janiczek JR, Kuechler J, Pritzsche D, Chen IC, Naveenan T, Ungless MA, Li M (2011) Temporally controlled modulation of FGF/ERK signaling directs midbrain dopaminergic neural progenitor fate in mouse and human pluripotent stem cells. *Development* 138: 4363−4374

Kar G, Kim JK, Kolodziejczyk AA, Natarajan KN, Torlai Triglia E, Mifsud B, Elderkin S, Marioni JC, Pombo A, Teichmann SA (2017) Flipping between Polycomb repressed and active transcriptional states introduces noise in gene expression. *Nat Commun* 8: 36

Kinkley S, Helmuth J, Polansky JK, Dunkel I, Gasparoni G, Frohler S, Chen W, Walter J, Hamann A, Chung HR (2016) reChIP-seq reveals widespread bivalency of H3K4me3 and H3K27me3 in CD4(+) memory T cells.

Kolde R (2015) pheatmap: Pretty Heatmaps. http://cran.r-project.org/web/packages/pheatmap/.

Konze KD, Ma A, Li F, Barsyte-Lovejoy D, Parton T, Macnevin CJ, Liu F, Gao C, Huang XP, Kuznetsova E, Rougie M, Jiang A, Pattenden SG, Norris JL, James LI, Roth BL, Brown PJ, Frye SV, Arrowsmith CH, Hahn KM et al (2013) An orally bioavailable chemical probe of the Lysine Methyltransferases EZH2 and EZH1. ACS Chem Biol 8: 1324–1334

Ku M, Koche RP, Rheinbay E, Mendenhall EM, Endoh M, Mikkelsen TS, Presser A, Nusbaum C, Xie XH, Chi AS, Adli M, Kasif S, Ptaszek LM, Cowan CA, Lander ES, Koseki H, Bernstein BE (2008) Genomewide analysis of PRC1 and PRC2 occupancy identifies two classes of bivalent domains. PLoS Genet 4: e1000242

Ladewig J, Koch P, Brustle O (2013) Leveling Waddington: the emergence of direct programming and the loss of cell fate hierarchies. Nat Rev Mol Cell Biol 14: 225–236

Langmead B, Salzberg SL (2012) Fast gapped-read alignment with Bowtie 2. Nat Methods 9: 357–359

Li B, Dewey CN (2011) RSEM: accurate transcript quantification from RNA-Seq data with or without a reference genome. BMC Bioinformatics 12: 323

Lien WH, Guo XY, Polak L, Lawton LN, Young RA, Zheng DY, Fuchs E (2011) Genome-wide maps of histone modifications unwind in vivo chromatin states of the hair follicle lineage. Cell Stem Cell 9: 219–232

Lienert F, Mohn F, Tiwari VK, Baubec T, Roloff TC, Gaidatzis D, Stadler MB, Schübeler D (2011) Genomic prevalence of heterochromatic H3K9me2 and transcription do not discriminate pluripotent from terminally differentiated cells. PLoS Genet 7: e1002090

Lister R, Mukamel EA, Nery JR, Urich M, Puddifoot CA, Johnson ND, Lucero J, Huang Y, Dwork AJ, Schultz MD, Yu M, Tonti-Filippini J, Heyn H, Hu S, Wu JC, Rao A, Esteller M, He C, Haghighi FG, Sejnowski TJ et al (2013) Global epigenomic reconfiguration during mammalian brain development. Science 341: 1237905

Ma C, Karwacki-Neisius V, Tang H, Li W, Shi Z, Hu H, Xu W, Wang Z, Kong L, Lv R, Fan Z, Zhou W, Yang P, Wu F, Diao J, Tan L, Shi YG, Lan F, Shi Y (2016) Nono, a bivalent domain factor, regulates Erk signaling and mouse embryonic stem cell pluripotency. Cell Rep 17: 997–1007

Marinelli M, McCutcheon JE (2014) Heterogeneity of dopamine neuron activity across traits and states. Neuroscience 282: 176–197

Marks H, Kalkan T, Menafra R, Denissov S, Jones K, Hofemeister H, Nichols J, Kranz A, Stewart AF, Smith A, Stunnenberg HG (2012) The transcriptional and epigenomic foundations of ground state pluripotency. Cell 149: 590–604

Mikkelsen TS, Ku M, Jaffe DB, Issac B, Lieberman E, Giannoukos G, Alvarez P, Brockman W, Kim TK, Koche RP, Lee W, Mendenhall E, O'Donovan A, Presser A, Russ C, Xie X, Meissner A, Wernig M, Jaenisch R, Nusbaum C et al (2007) Genome-wide maps of chromatin state in pluripotent and lineage-committed cells. Nature 448: 553–560

Mohn F, Weber M, Rebhan M, Roloff TC, Richter J, Stadler MB, Bibel M, Schübeler D (2008) Lineage-specific polycomb targets and de novo DNA methylation define restriction and potential of neuronal progenitors. Mol Cell 30: 755–766

Paladini CA, Roeper J (2014) Generating bursts (and Pauses) in the dopamine midbrain neurons. Neuroscience 282: 109–121

Prezioso C, Orlando V (2011) Polycomb proteins in mammalian cell differentiation and plasticity. FEBS Lett 585: 2067–2077

Procopio DO, Saba LM, Walter H, Lesch O, Skala K, Schlaff G, Vanderlinden L, Clapp P, Hoffman PL, Tabakoff B (2013) Genetic markers of comorbid depression and alcoholism in women. Alcohol Clin Exp Res 37: 896–904

del Puerto A, Diaz-Hernandez JI, Tapia M, Gomez-Villafuertes R, Benitez MJ, Zhang J, Miras-Portugal MT, Wandosell F, Diaz-Hernandez M, Garrido JJ (2012) Adenylate cyclase 5 coordinates the action of ADP, P2Y1, P2Y13 and ATP-gated P2X7 receptors on axonal elongation. J Cell Sci 125: 176–188

Quinlan AR, Hall IM (2010) BEDTools: a flexible suite of utilities for comparing genomic features. Bioinformatics 26: 841–842

Rackham OJL, Firas J, Fang H, Oates ME, Holmes ML, Knaupp AS, Suzuki H, Nefzger CM, Daub CO, Shin JW, Petretto E, Forrest ARR, Hayashizaki Y, Polo JM, Gough J, Consortium F (2016) A predictive computational framework for direct reprogramming between human cell types. Nat Genet 48: 331–335

Richly H, Aloia L, Di Croce L (2011) Roles of the Polycomb group proteins in stem cells and cancer. Cell Death Dis 2: e204

Sandelin A, Carninci P, Lenhard B, Ponjavic J, Hayashizaki Y, Hume DA (2007) Mammalian RNA polymerase II core promoters: insights from genome-wide studies. Nat Rev Genet 8: 424–436

von Schimmelmann M, Feinberg PA, Sullivan JM, Ku SM, Badimon A, Duff MK, Wang Z, Lachmann A, Dewell S, Ma'ayan A, Han MH, Tarakhovsky A, Schaefer A (2016) Polycomb repressive complex 2 (PRC2) silences genes responsible for neurodegeneration. Nat Neurosci 19: 1321–1330

Schröder S, Herker E, Itzen F, He D, Thomas S, Gilchrist DA, Kaehlcke K, Cho S, Pollard KS, Capra JA, Schnolzer M, Cole PA, Geyer M, Bruneau BG, Adelman K, Ott M (2013) Acetylation of RNA polymerase II regulates growth-factor-induced gene transcription in mammalian cells. Mol Cell 52: 314–324

Sen S, Block KF, Pasini A, Baylin SB, Easwaran H (2016) Genome-wide positioning of bivalent mononucleosomes. BMC Med Genomics 9: 60

Simon JA, Kingston RE (2013) Occupying chromatin: polycomb mechanisms for getting to genomic targets, stopping transcriptional traffic, and staying put. Mol Cell 49: 808–824

Song Q, Decato B, Hong EE, Zhou M, Fang F, Qu J, Garvin T, Kessler M, Zhou J, Smith AD (2013) A reference methylome database and analysis pipeline to facilitate integrative and comparative epigenomics. PLoS One 8: e81148

South AP, Wan H, Stone MG, Dopping-Hepenstal PJC, Purkis PE, Marshall JF, Leigh IM, Eady RAJ, Hart IR, McGrath JA (2003) Lack of plakophilin 1 increases keratinocyte migration and reduces desmosome stability. J Cell Sci 116: 3303–3314

Stock JK, Giadrossi S, Casanova M, Brookes E, Vidal M, Koseki H, Brockdorff N, Fisher AG, Pombo A (2007) Ring1-mediated ubiquitination of H2A restrains poised RNA polymerase II at bivalent genes in mouse ES cells. Nat Cell Biol 9: 1428–1435

Subramanian A, Tamayo P, Mootha VK, Mukherjee S, Ebert BL, Gillette MA, Paulovich A, Pomeroy SL, Golub TR, Lander ES, Mesirov JP (2005) Gene set enrichment analysis: a knowledge-based approach for interpreting genome-wide expression profiles. Proc Natl Acad Sci USA 102: 15545–15550

Sulzer D, Joyce MP, Lin L, Geldwert D, Haber SN, Hattori T, Rayport S (1998) Dopamine neurons make glutamatergic synapses in vitro. J Neurosci 18: 4588–4602

Tee WW, Shen SS, Oksuz O, Narendra V, Reinberg D (2014) Erk1/2 activity promotes chromatin features and RNAPII phosphorylation at developmental promoters in mouse ESCs. Cell 156: 678–690

Thomas R, Thomas S, Holloway AK, Pollard KS (2016) Features that define the best ChIP-seq peak calling algorithms. Brief Bioinform 18: 441–450

Trapnell C, Pachter L, Salzberg SL (2009) TopHat: discovering splice junctions with RNA-Seq. Bioinformatics 25: 1105–1111

Ungless MA, Grace AA (2012) Are you or aren't you? Challenges associated with physiologically identifying dopamine neurons. *Trends Neurosci* 35: 422–430

Vastenhouw NL, Schier AF (2012) Bivalent histone modifications in early embryogenesis. *Curr Opin Cell Biol* 24: 374–386

Voigt P, LeRoy G, Drury WJ, Zee BM, Son J, Beck DB, Young NL, Garcia BA, Reinberg D (2012) Asymmetrically modified nucleosomes. *Cell* 151: 181–193

Voigt P, Tee WW, Reinberg D (2013) A double take on bivalent promoters. *Genes Dev* 27: 1318–1338

Voss K, Forne I, Descostes N, Hintermair C, Schuller R, Maqbool MA, Heidemann M, Flatley A, Imhof A, Gut M, Gut I, Kremmer E, Andrau JC, Eick D (2015) Site-specific methylation and acetylation of lysine residues in the C-terminal domain (CTD) of RNA polymerase II. *Transcription* 6: 91–101

Wamstad JA, Alexander JM, Truty RM, Shrikumar A, Li F, Eilertson KE, Ding H, Wylie JN, Pico AR, Capra JA, Erwin G, Kattman SJ, Keller GM, Srivastava D, Levine SS, Pollard KS, Holloway AK, Boyer LA, Bruneau BG (2012) Dynamic and coordinated epigenetic regulation of developmental transitions in the cardiac lineage. *Cell* 151: 206–220

Weiner A, Lara-Astiaso D, Krupalnik V, Gafni O, David E, Winter DR, Hanna JH, Amit I (2016) Co-ChIP enables genome-wide mapping of histone mark co-occurrence at single-molecule resolution. *Nat Biotechnol* 34: 953–961

Williams LH, Fromm G, Gokey NG, Henriques T, Muse GW, Burkholder A, Fargo DC, Hu G, Adelman K (2015) Pausing of RNA polymerase II regulates mammalian developmental potential through control of signaling networks. *Mol Cell* 58: 311–322

Xie W, Schultz MD, Lister R, Hou Z, Rajagopal N, Ray P, Whitaker JW, Tian S, Hawkins RD, Leung D, Yang H, Wang T, Lee AY, Swanson SA, Zhang J, Zhu Y, Kim A, Nery JR, Urich MA, Kuan S et al (2013) Epigenomic analysis of multilineage differentiation of human embryonic stem cells. *Cell* 153: 1134–1148

Xing H, Mo Y, Liao W, Zhang MQ (2012) Genome-wide localization of protein-DNA binding and histone modification by a Bayesian change-point method with ChIP-seq data. *PLoS Comput Biol* 8: e1002613

Yagi M, Kishigami S, Tanaka A, Semi K, Mizutani E, Wakayama S, Wakayama T, Yamamoto T, Yamada Y (2017) Derivation of ground-state female ES cells maintaining gamete-derived DNA methylation. *Nature* 548: 224–227

Ying QL, Stavridis M, Griffiths D, Li M, Smith A (2003) Conversion of embryonic stem cells into neuroectodermal precursors in adherent monoculture. *Nat Biotechnol* 21: 183–186

Young MD, Willson TA, Wakefield MJ, Trounson E, Hilton DJ, Blewitt ME, Oshlack A, Majewski IJ (2011) ChIP-seq analysis reveals distinct H3K27me3 profiles that correlate with transcriptional activity. *Nucleic Acids Res* 39: 7415–7427

Zaborowska J, Egloff S, Murphy S (2016) The pol II CTD: new twists in the tail. *Nat Struct Mol Biol* 23: 771–777

Zhao S, Maxwell S, Jimenez-Beristain A, Vives J, Kuehner E, Zhao J, O'Brien C, de Felipe C, Semina E, Li M (2004) Generation of embryonic stem cells and transgenic mice expressing green fluorescence protein in midbrain dopaminergic neurons. *Eur J Neurosci* 19: 1133–1140

4

Lysine acetylome profiling uncovers novel histone deacetylase substrate proteins in *Arabidopsis*

Markus Hartl[1,2,3,§] (iD), Magdalena Füßl[1,2,4,§], Paul J Boersema[5,†], Jan-Oliver Jost[6,‡], Katharina Kramer[1], Ahmet Bakirbas[1,4,¶], Julia Sindlinger[6] (iD), Magdalena Plöchinger[2], Dario Leister[2], Glen Uhrig[7], Greg BG Moorhead[7] (iD), Jürgen Cox[5], Michael E Salvucci[8], Dirk Schwarzer[6] (iD), Matthias Mann[5] (iD) & Iris Finkemeier[1,2,4,*] (iD)

Abstract

Histone deacetylases have central functions in regulating stress defenses and development in plants. However, the knowledge about the deacetylase functions is largely limited to histones, although these enzymes were found in diverse subcellular compartments. In this study, we determined the proteome-wide signatures of the RPD3/HDA1 class of histone deacetylases in *Arabidopsis*. Relative quantification of the changes in the lysine acetylation levels was determined on a proteome-wide scale after treatment of *Arabidopsis* leaves with deacetylase inhibitors apicidin and trichostatin A. We identified 91 new acetylated candidate proteins other than histones, which are potential substrates of the RPD3/HDA1-like histone deacetylases in *Arabidopsis*, of which at least 30 of these proteins function in nucleic acid binding. Furthermore, our analysis revealed that histone deacetylase 14 (HDA14) is the first organellar-localized RPD3/HDA1 class protein found to reside in the chloroplasts and that the majority of its protein targets have functions in photosynthesis. Finally, the analysis of HDA14 loss-of-function mutants revealed that the activation state of RuBisCO is controlled by lysine acetylation of RuBisCO activase under low-light conditions.

Keywords *Arabidopsis*; histone deacetylases; lysine acetylation; photosynthesis; RuBisCO activase
Subject Categories Methods & Resources; Plant Biology; Post-translational Modifications, Proteolysis & Proteomics

Introduction

Optimal plant growth and development are dependent on fine-regulation of the cellular metabolism in response to environmental conditions (Nunes-Nesi *et al*, 2010). During a day or a season, plants often face rapidly changing environmental conditions such as changes in temperature, light intensity, and water and nutrient availability (Calfapietra *et al*, 2015). Due to their sessile life style, plants cannot escape from environmental perturbations. Instead, plants activate a variety of cellular response mechanisms that allow them to acclimate their metabolism to the environment. Cellular signaling networks are activated within seconds when metabolic homeostasis is perturbed, and these networks regulate the plant's physiology (Dietz, 2015; Mignolet-Spruyt *et al*, 2016). Such signaling networks regulate gene expression, translation, protein activity, and turnover. Post-translational modifications (PTMs) of proteins like phosphorylation, ubiquitination, methylation, and acetylation play a pivotal role in all of these regulatory processes (Hartl & Finkemeier, 2012; Johnova *et al*, 2016). Except for phosphorylation, most of the cellular protein targets and the regulating enzymes of these PTMs are largely unexplored in plants (Huber, 2007). Here, we study the regulation of lysine acetylation.

Lysine acetylation is a post-translational modification (PTM), which was first discovered on histone tails where it is now known to regulate chromatin structure and gene expression (Allfrey *et al*, 1964). The transfer of the acetyl group to lysine neutralizes the positive charge of the amino group, which can affect the biological function of proteins such as enzyme activities, protein–protein, and protein–DNA interactions (Yang & Seto, 2008). Acetyl-CoA serves as

1 Plant Proteomics, Max Planck Institute for Plant Breeding Research, Cologne, Germany
2 Plant Molecular Biology, Department Biology I, Ludwig-Maximilians-University Munich, Martinsried, Germany
3 Mass Spectrometry Facility, Max F. Perutz Laboratories (MFPL), Vienna Biocenter (VBC), University of Vienna, Vienna, Austria
4 Plant Physiology, Institute of Plant Biology and Biotechnology, University of Muenster, Muenster, Germany
5 Proteomics and Signal Transduction, Max-Planck Institute of Biochemistry, Martinsried, Germany
6 Interfaculty Institute of Biochemistry, University of Tübingen, Tübingen, Germany
7 Department of Biological Sciences, University of Calgary, Calgary, AB, Canada
8 US Department of Agriculture, Agricultural Research Service, Arid-Land Agricultural Research Center, Maricopa, AZ, USA
*Corresponding author. E-mail: iris.finkemeier@uni-muenster.de
§These authors contributed equally to this work
†Present address: Department of Biology, Institute of Biochemistry, ETH Zurich, Zurich, Switzerland
‡Present address: Leibniz-Forschungsinstitut für Molekulare Pharmakologie im Forschungsverbund Berlin e.V. (FMP), Berlin, Germany
¶Present address: Plant Biology Graduate Program, University of Massachusetts Amherst, Amherst, USA
[Correction added on 7 November 2017 after first online publication: current address affiliation has been added.]

substrate for lysine acetylation in an enzymatic process catalyzed by different types of lysine acetyltransferases (KATs) (Kleff *et al*, 1995; Parthun *et al*, 1996; Yuan & Marmorstein, 2013; Drazic *et al*, 2016). However, lysine acetylation can also occur non-enzymatically especially at a cellular pH higher than eight (Wagner & Payne, 2013; König *et al*, 2014a); a level that can be reached during active respiration in the mitochondrial matrix, as well as in the chloroplast stroma during photosynthesis (Hosp *et al*, 2016). Non-enzymatic acetylation is of particular abundance in bacteria, which additionally contain the highly reactive acetyl-phosphate as metabolite (Weinert *et al*, 2013). In plants, many organellar proteins from mitochondria and chloroplasts were previously identified as lysine-acetylated (Finkemeier *et al*, 2011; Wu *et al*, 2011; König *et al*, 2014a; Nallamilli *et al*, 2014; Smith-Hammond *et al*, 2014; Fang *et al*, 2015; He *et al*, 2016; Hosp *et al*, 2016; Xiong *et al*, 2016; Zhang *et al*, 2016; Zhen *et al*, 2016).

Lysine acetylation can be reversed by lysine deacetylases (KDACs), which were named histone deacetylases (HDA/HDAC) before the more recent discovery of non-histone protein acetylation. KDACs can be grouped into three different families: (i) reduced potassium dependency 3/histone deacetylase 1 (RPD3/HDA1)-like, (ii) HD-tuins (HDT), and (iii) silent information regulator 2 (Sir2) (Pandey *et al*, 2002; Alinsug *et al*, 2009; Shen *et al*, 2015). While the RPD3/HDA1 family has primarily been found in eukaryotes, the Sir2-type deacetylases also occur in bacteria, and the HDT-type deacetylases only occur in plants. The *Arabidopsis* genome encodes 18 KDACs from the three different families. The largest family comprises the RPD3/HDA1-like with 12 genes, four genes belong to the HDTs and two to the Sir2 family. The RPD3/HDA1 family can be further subdivided into class I (RPD3-like), class II (HDA1-like), and class IV KDACs, of which *Arabidopsis* possesses 6, 5, and 1 putative members, respectively (Pandey et al, 2002; Alinsug *et al*, 2009; Shen *et al*, 2015). Numerous studies have characterized different genes from the *Arabidopsis* KDAC families over the last two decades (Shen *et al*, 2015). In particular, HDA6, HDA9, and HDA19 from class I are the most well studied *Arabidopsis* KDACs and they have been implicated in many important developmental processes such as seed germination, flowering time, as well as plant hormone-related stress responses (Zhou *et al*, 2005; Benhamed *et al*, 2006; Chen *et al*, 2010; Choi *et al*, 2012; Cigliano *et al*, 2013; Zheng *et al*, 2016; Mengel *et al*, 2017). In terms of protein targets for deacetylation, very little is known about the preferences and targets of the different plant KDACs. While *Arabidopsis* sirtuin 2 deacetylates selected mitochondrial proteins such as the ATP/ADP carrier (König *et al*, 2014b), mainly histone H3 and H4 deacetylation has been studied for the other two families of KDACs (Shen *et al*, 2015).

Here, we report the first comprehensive profiling of putative *Arabidopsis* KDAC targets by using two different inhibitors of the RPD3/HDA1 family. By this approach, we identify several heretofore-unknown potential targets of the *Arabidopsis* KDACs in the nucleus and other subcellular localizations including plastids. Additionally, by the use of a peptide-based KDAC-probe, we were able to identify the first KDAC of the RPD3/HDA1 family, which is active in organelles and regulates the activity and activation state of ribulose-1,5-bisphosphate-carboxylase/oxygenase, the key enzyme in photosynthetic CO_2 fixation, and the most abundant protein on earth.

Results

The *Arabidopsis* leaf lysine acetylome 2.0

The first two lysine acetylomes of *Arabidopsis* leaves were reported in 2011, with only around 100 lysine acetylation sites identified (Finkemeier *et al*, 2011; Wu *et al*, 2011). Tremendous advances in mass spectrometry, improvements in antibody reagents, and the optimization of the overall protocol now allows a more in-depth profiling of the *Arabidopsis* lysine acetylome. To be able to quantify acetylome changes upon KDAC inhibitor treatment, we applied an isotopic dimethyl-labeling approach to differentially label two different protein samples (e.g., treatment and control), combined with an enrichment strategy for lysine-acetylated peptides (Fig 1A). For this procedure, proteins extracted from leaves were processed and trypsin-digested via filter-assisted sample preparation. Peptides were isotopically dimethyl-labeled, and samples for comparison were pooled. For proteome quantifications, samples were collected at this step and the rest of the sample was further processed by hydrophilic interaction liquid chromatography fractionation to reduce the peptide complexity. Six to seven fractions were collected and used for immuno-affinity enrichment using anti-acetyllysine agarose beads. Peptides were further processed for high-resolution mass spectrometry, and MaxQuant was used for the data analysis.

Altogether the datasets presented here comprise 2,152 lysine acetylation sites (localization probability > 0.75) on 1,022 protein groups (6,672 identified protein groups in total, Table 1, Datasets EV1–EV5)—this corresponds to 959 novel acetylated proteins and 2,057 novel acetylation sites when compared to the previously published datasets for *Arabidopsis* (Finkemeier *et al*, 2011; Wu *et al*, 2011; König *et al*, 2014a). A MapMan functional annotation analysis (Thimm *et al*, 2004) was used for the classification of the lysine-acetylated proteins, applying the TAIR mapping and selecting all identified proteins of the proteome analysis as background population. From the different cellular processes, the functional categories photosynthesis, tetrapyrrole synthesis, gluconeogenesis, redox, TCA cycle, as well as DNA and RNA regulation of transcription were identified as overrepresented as determined by a Fisher's exact test, while processes such as hormone metabolism, cell wall, and secondary metabolism were underrepresented (Fig 1B, at 5% FDR and a 1.5-fold enrichment/depletion cut-off). Based on the classification of localization of proteins using SUBA consensus (Heazlewood *et al*, 2007), proteins from plastids and nucleus were clearly overrepresented, while proteins from endoplasmic reticulum, vacuole, mitochondrion, plasma membrane, and extracellular space were significantly underrepresented in our dataset (Fig 1C).

Additionally, we analyzed the local sequence context around the acetylation sites using iceLogo (Maddelein *et al*, 2015) in combination with the *Arabidopsis* TAIR10 database with all identified proteins as background reference (Fig 1D). Overall, negatively charged amino acids, such as glutamate and aspartate, were significantly enriched in the −1, −2, −3 as well as +1 positions surrounding the lysine acetylation site. In more distant positions, lysine residues were the most strongly enriched on either side of the lysine acetylation site. The sequence motif surrounding the lysine acetylation site appeared different depending on the subcellular localization of the respective proteins. For example, the negatively charged amino acids were more prominent on cytosolic and plastidial

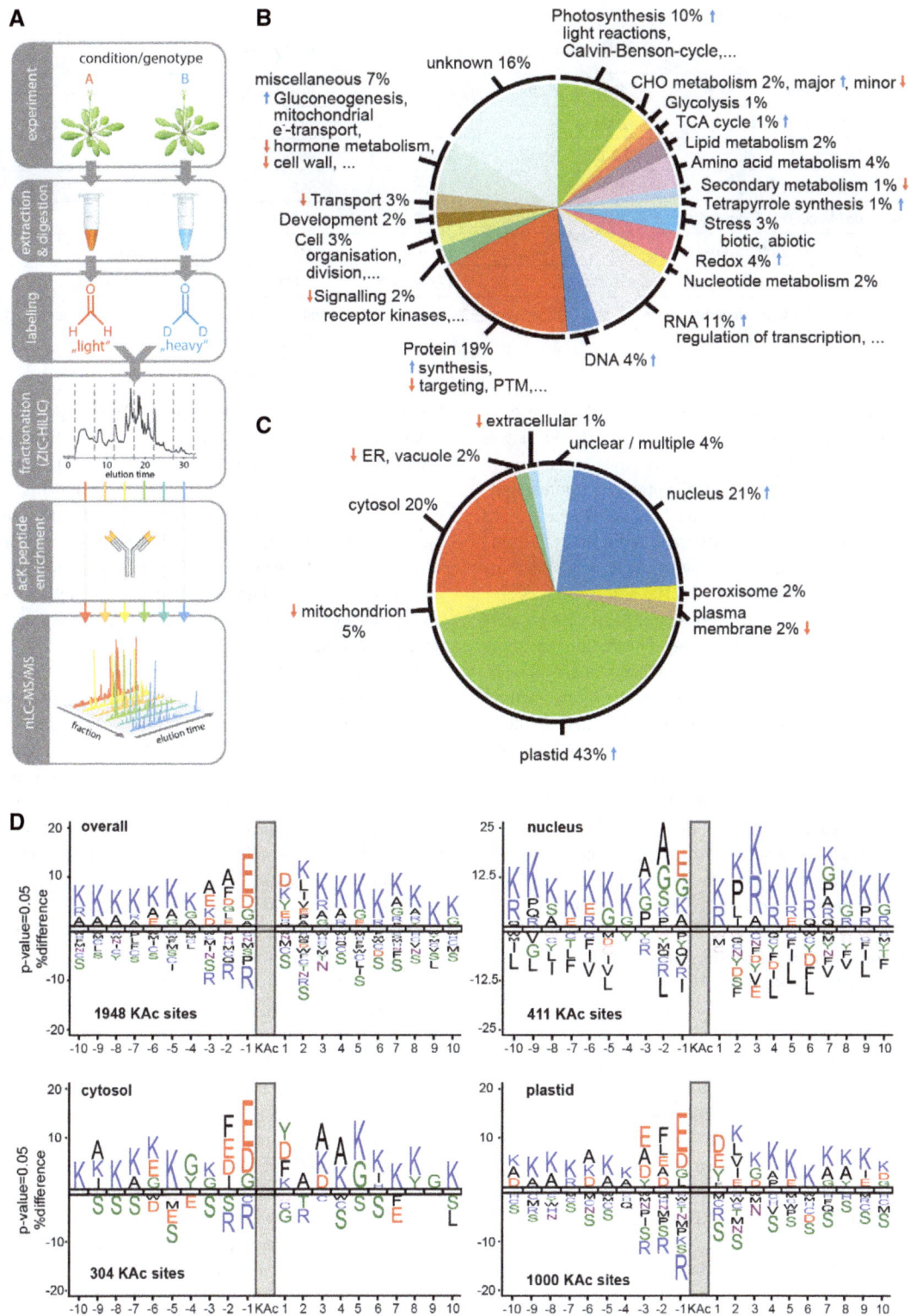

Figure 1. **Proteome-wide identification and classification of the *Arabidopsis thaliana* lysine acetylome.**

A   Experimental overview.

B, C   Functional classification and subcellular localization of identified lysine-acetylated proteins. Lysine-acetylated proteins identified over all experiments were classified according to MapMan categories and SUBA4 localization information, respectively. Over- or underrepresentation of categories was determined using a Fisher's exact test with all proteins identified at 1% FDR as background population. Blue and red arrows mark categories significantly enriched at 5% FDR (Benjamini–Hochberg) and a 1.5-fold-change cut-off.

D   Sequence logos for all lysine acetylation sites with all proteins identified as background population (sequence logos were generated using iceLogo, Maddelein *et al*, 2015).

Table 1. Summary of identified features.

Experiment	Description	Whole proteome analysis		Acetyllysine-containing		
		Protein groups	Peptides	Protein groups	Peptides	Sites
1	Apicidin versus Ctrl	2,384	11,188	538	1,064	1,041
2	TSA versus Ctrl	5,107	32,809	493	1,002	930
3	*hda14* versus WT	2,889	13,755	545	1,133	920
4	*hda14* versus WT low-light	4,138	27,835	367	756	700
5	*hda14* versus WT thylakoids	2,904	15,064	237	592	546
Total		6,672	47,338	1,022	2,405	2,152

Filters applied: 1% FDR at PSM and protein level, score for modified peptides $\geq$ 35, delta score for modified peptides $\geq$ 6, acetyllysine site localization probability $\geq$ 0.75; contaminants removed.

proteins in comparison with nuclear proteins, as well as the presence of a phenylalanine at position −2. Tyrosine at position +1 was found on cytosolic and plastid proteins, while phenylalanine at position +1 was only found enriched on cytosolic proteins. Interestingly, on the nuclear sequence motifs only positively charged amino acids were found at position +1 as well as generally more neutral amino acids such as glycine and alanine at positions −1 to −3, which are dominating on histone sequence motifs (Fig 1D).

Since 43% of all identified lysine-acetylated proteins are putative plastid proteins, we further analyzed the distribution of those proteins and number of acetylation sites in photosynthesis (Fig 2). About 24% of the proteins from the photosynthetic light reactions were acetylated on at least four lysine residues (Fig 2A). Proteins from the light harvesting complexes (LHC) of both PSII and PSI were heavily acetylated with 29 lysine acetylation sites on LHCII and 16 on LHCI proteins (Fig 2A). All enzyme complexes involved in the carbon fixing reactions (Calvin–Benson cycle) as well as RuBisCO activase (RCA) contained four or more lysine acetylation sites. With 18 lysine acetylation sites, the large subunit of RuBisCO was the most heavily modified of all the proteins (Fig 2B).

Identification of novel lysine acetylation sites targeted by *Arabidopsis* RPD3/HDA1-type KDACs

Different types of lysine deacetylase inhibitors have been developed in the past decade, which are widely used to modulate the activities of human KDACs in diseases (Newkirk *et al*, 2009). Here, we selected two commonly used KDAC inhibitors, apicidin and trichostatin A (TSA), to target the RPD3/HDA1-type family of KDACs and to profile their potential protein substrates. While apicidin was shown to specifically inhibit class I KDACs, TSA was described as a general inhibitor of class I and class II KDACs in HeLa cells (Scholz *et al*, 2015). For inhibitor treatment, *Arabidopsis* leaf strips were infiltrated either with a mock control or with 5 μM apicidin and 5 μM trichostatin A (TSA), respectively. Experiments were performed in three independent biological replicates, and the leaf strips were incubated for 4 h in the light before harvest. The protein intensities of the biological replicates had a Pearson correlation coefficient of > 0.87–0.98 (Appendix Fig S1), which indicates the robustness of the approach. Site-specific acetylation changes were quantified (Fig 3A and B) in addition to changes on total proteome level as control (Fig 3C and D). No significant changes in the regulation of protein abundances were observed after the inhibitor

treatments, which covered about 67–88% of proteins carrying the identified acetylated sites (Appendix Fig S2). However, the whole proteome analysis did not cover very low abundant proteins without enrichment. Therefore, we cannot exclude that the other sites, for which we were not able to quantify protein ratios, were not regulated due to bona fide stoichiometry differences from inhibited KDAC activity. However, we restricted inhibitor treatment to 4-h incubation time in order to minimize potential changes in protein abundances that might result from KDAC-dependent alterations in gene expression.

For apicidin treatment, 832 lysine acetylation sites were quantified, of which 148 were significantly regulated according to a LIMMA statistical analysis with a FDR cut-off < 5% (Fig 3A; Dataset EV1). As expected for a KDAC inhibitor treatment, most of the lysine acetylation sites (136 in total) were up-regulated (log2-FC 0.4–7.4) after apicidin treatment. The 12 down-regulated lysine acetylation sites comprise mainly multiply acetylated peptides for which peptide variants of lower acetylation status show a down-regulation of particular sites, whereas the corresponding peptide with higher acetylation status shows up-regulation in comparison. Interestingly, while the overall 832 lysine acetylation sites were detected on proteins from various subcellular compartments, 139 of the regulated lysine acetylation sites were found on proteins exclusively localized to the nucleus, such as histones, HATs, proteins involved in the regulation of transcription and signaling (G-protein and light signaling), DNA-repair and cell cycle, as revealed from a SUBAcon analysis (Dataset EV1). Three up-regulated lysine acetylation sites were detected on plastidial proteins including proteins involved in the light reactions (K99, PSAH-1; PSAH-2 log2-FC 0.43) as well as in the Calvin–Benson cycle (K305, SBPase, log2-FC 0.61). Looking at a less stringent *P*-value cut-off < 0.05 (Fig 1B), 182 lysine acetylation sites were found up-regulated of which 29 were found on organellar proteins. While most of these 29 lysine acetylation sites occur on proteins from the plastids, they only show a rather small increase in acetylation level (log2-FC 0.2–0.8) (Dataset EV1).

After TSA treatment, only 37 sites of the 385 quantified lysine acetylation sites were significantly up-regulated with an FDR < 5% (log2-FC 1.4–6.1) (Fig 3B, Dataset EV2). This low number of regulated sites, compared to apicidin treatment, was mainly due to a higher variability in the biological replicates of the TSA treatment. Of the 37 up-regulated lysine acetylation sites, only one was detected on a protein with a plastidial localization (K165,

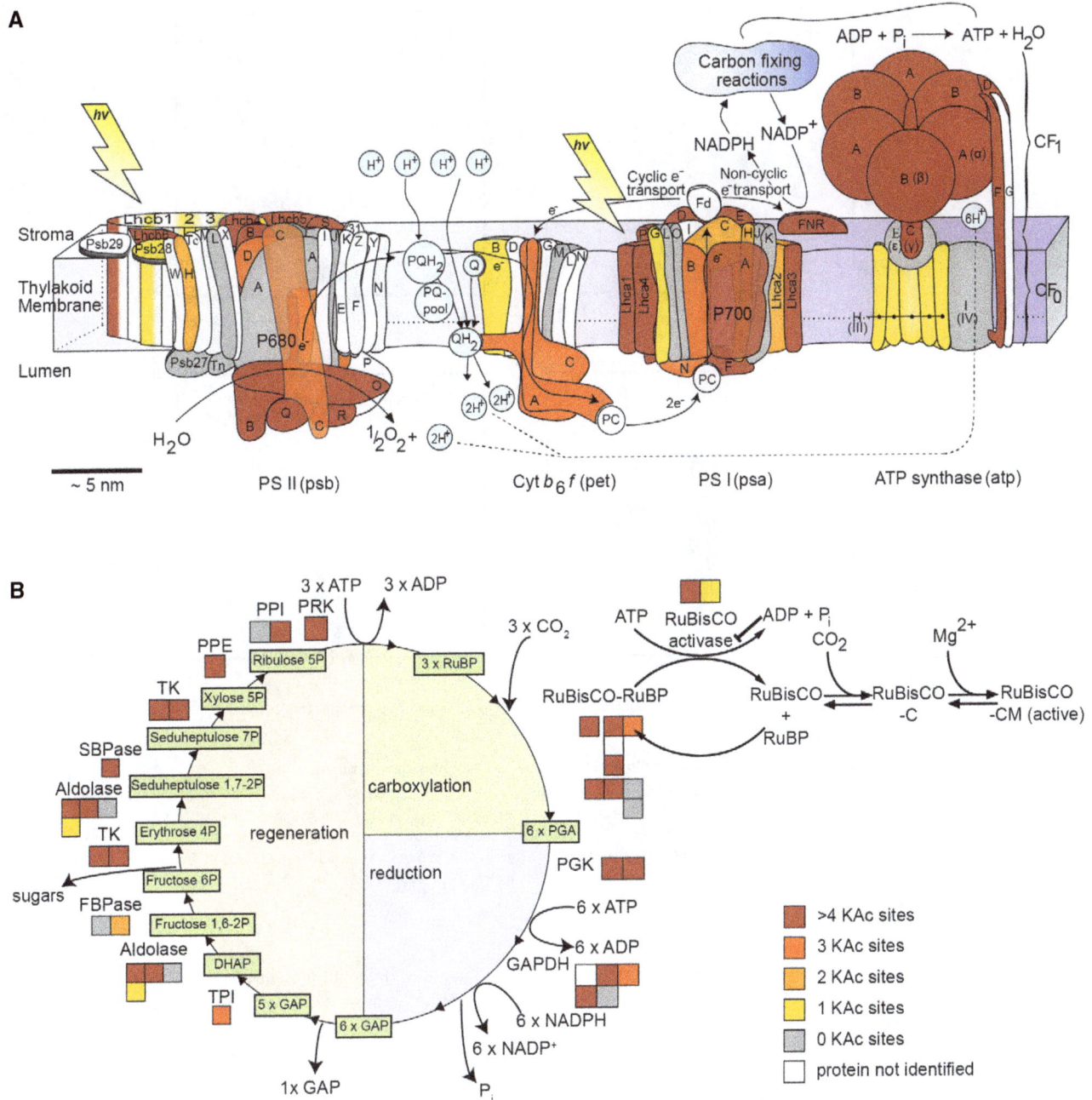

Figure 2. Overview of lysine-acetylated proteins in the light reactions (A) and the Calvin–Benson cycle (B) identified in this study in *Arabidopsis*.

A, B The classification of proteins into functional bins was performed using MapMan (Thimm *et al*, 2004). Color code: proteins not identified in the LC-MS/MS analyses (white), proteins without identified lysine-acetylated sites (gray), and proteins with one (yellow), two (orange), three (dark orange), or four or more acetylation sites (red). For the Calvin–Benson cycle, each box indicates a separate *Arabidopsis* AGI identifier as indicated in Dataset EV6. Cytb$_6$f, cytochrome b$_6$f; FBPase, fructose-1,6-bisphosphatase; GAPDH, glyceraldehyde-3-phosphate dehydrogenase; PGK, phosphoglycerate kinase; PPE, phosphopentose epimerase; PPI, phosphopentose isomerase; PRK, phosphoribulokinase; PSII, photosystem II; PSI, photosystem I; RuBisCO, ribulose-1,5-bisphosphate-carboxylase/oxygenase; SBPAse, seduheptulose-1,7-bisphosphatase; TPI, triose phosphate isomerase; TK, transketolase. A template of the light-reaction schematic was kindly provided by Jon Nield and modified.

At3g17930, log2-FC 3.6). Analyzing the data with a less stringent *P*-value cut-off ($P < 0.05$) resulted in 72 up-regulated lysine acetylation sites with an average log2-FC of 0.8–6. Among those were two more plastidial proteins, the RCA β1-isoform (K438, log2-FC 1.37) and PSAD-1/2 (K187/K191, log2-FC 1.77). Among the common nuclear targets of apicidin and TSA were several histone proteins. In total, 19 regulated lysine acetylation sites were found on histone 2A (H2A) and histone 2B (H2B) proteins. On HTB1 two different lysine acetylation sites were specifically and strongly (log$_2$FC at least 2) up-regulated either upon apicidin (K39, K40) or TSA (K28, K33) inhibition (Appendix Fig S3, Dataset EV1A, Dataset EV2A). The same was true for other histones of the H2B family (HTB9, HTB2),

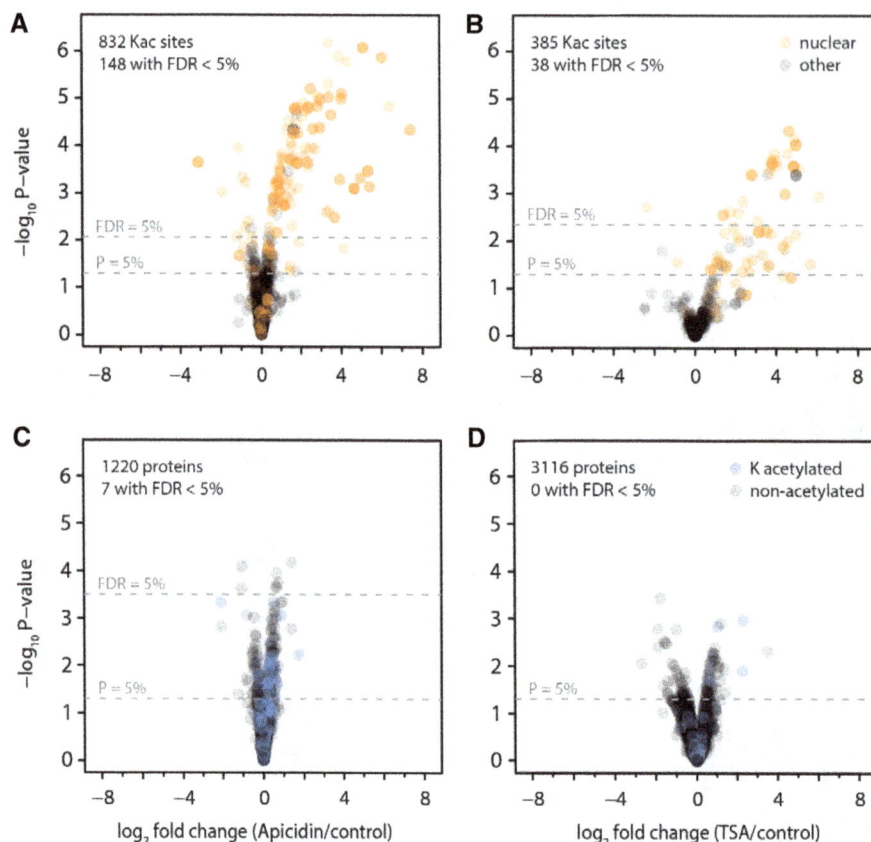

Figure 3. **Differential lysine acetylation and protein expression in *Arabidopsis* leaves after inhibitor treatment.**

A–D Vacuum infiltration of leaf strips with solutions containing either of the two deacetylase inhibitors apicidin (A, C) or trichostatin A (B, D) versus a buffer control for 4 h leads to differential accumulation of lysine acetylation sites. Volcano plots depict lysine acetylation site ratios (A, B) or protein ratios (C, D) for inhibitor treatment versus control, with *P*-values determined using the LIMMA package. Orange, protein with nuclear localization according to SUBA4 database. Blue, proteins with lysine acetylation sites identified. Dashed lines indicate significance thresholds of either uncorrected *P*-values < 5% or Benjamini–Hochberg corrected FDR < 5%. A missing line indicates that the significance threshold was not reached by any of the data points.

which also showed unique up-regulated sites depending on the inhibitor used. Interestingly, histones of the H2A family showed the same up-regulated lysine acetylation sites upon apicidin and TSA treatment. Overall, between apicidin and TSA treatment, there was an overlap of 25 protein groups (*P*-value < 0.05), which showed enhanced lysine acetylation sites after both treatments (Dataset EV1A, Dataset EV2A). The KDAC inhibitor study revealed that most of the RPD3/HDA1 classes KDACs of *Arabidopsis* have their potential substrate proteins in the nucleus, but that some members also seem to have their targets in other subcellular compartments, such as the plastids.

HDA14 is the first member of a RPD3/HDA1-family protein to be localized in organelles

Members of the RPD3/HDA1 family are usually localized in the nucleus and/or in the cytosol. Here, we had clear indications that proteins targeted to the chloroplast were found hyper-acetylated upon inhibitor treatment. However, it was not clear whether the hyper-acetylation already occurred due to KDAC inhibition in the cytosol during transit of the proteins to the plastid or whether there exists a plastid-localized member of the RPD3/HDA1-class. Since

KDACs are low abundant proteins, they are usually not detected in leaf proteomes and therefore need to be enriched before detection. Here, we used a recently developed peptide-based KDAC-probe, mini-AsuHd (Dose *et al*, 2016), to pull-down active RPD3/HDA1-class KDACs from leaf extracts and isolated chloroplasts, respectively, in comparison with the mini-Lys probe as background control (Table 2). The mini-AsuHd probe contains a hydroxamate moiety spaced with five carbon atoms to the peptide backbone, which chelates the catalytic Zn^{2+} ion of RPD3/HDA1 family-HDACs with nanomolar affinities (Dose *et al*, 2016). The mini-Lys probe contains a lysine residue instead. Three different *Arabidopsis* KDACs were identified in total leaf extracts (HDA5, 14, 15), while in isolated chloroplasts, only HDA14 was identified (Table 2). Although HDA14 contains a predicted target sequence for plastids, it was reported to be localized in the cytosol in a previous study (Tran *et al*, 2012). To confirm the plastid localization of HDA14, we fused GFP to the C-terminus of the HDA14 protein, instead of the N-terminus as in the previous study. Protoplasts of the stable transformed 35S:HDA14:GFP plants showed that the signal of the HDA14:GFP fusion protein was overlapping with the autofluorescence of the chlorophyll, as well as with a TMRM signal which visualizes

Table 2. KDAC pull-down with mini-AsuHd probe.

Majority protein IDs	Name	Peptides	MS/MS count	Log2-LFQ CP AsuHd	Log2-LFQ CP Lys	Log2 enrichment CP	Log2-LFQ LF AsuHd	Log2-LFQ LF Lys	Log2 enrichment LF
AT5G61060.1/.2	HDA5	6	11	n.d.	n.d.	n.d.	24.50 ± 0.12	n.d.	> 7[a]
AT4G33470.1	HDA14	7	22	24.74 ± 1.83	19.54 ± 0.05	5.2	26.20 ± 0.2	21.12 ± 0.49	5.1
AT3G18520.1/2	HDA15	2	2	n.d.	n.d.	n.d.	20.84 ± 0.08	n.d.	> 3[a]
ATCG00490.1	RBCL	28	545	33.65 ± 0.36	33.73 ± 0.13	−0.1	33.91 ± 0.09	33.75 ± 0.03	0.2

Selected proteins identified and quantified in pull-downs by LC-MS/MS analysis. Protein abundances are expressed as label free quantification (LFQ) values. Numbers indicate mean log2-transformed LFQ values from two biological replicates of *Arabidopsis* leaves (LF) and isolated chloroplasts (CP). Mini-Lys probes were used as pull-down controls to calculate relative enrichments of proteins. LFQ values for RuBisCO are indicated in all samples as background control.
[a]Estimated enrichment factor assuming a minimum Log2-LFQ threshold of 17.

mitochondria (Fig 4A). Hence, these results indicate a dual localization of HDA14 in mitochondria and chloroplasts. To further confirm the results, we performed a Western blot analysis with HDA14 antiserum on proteins from isolated chloroplasts and mitochondria from WT and stably expressing 35S:HDA14:GFP seedlings and detected the endogenous HDA14 as well as the HDA14-GFP fusion protein in the chloroplast stroma as well as in mitochondria (Appendix Fig S4).

HDA14 is a functional lysine deacetylase and is mainly inhibited by TSA *in vitro*

We produced a recombinant N-terminally His-tagged HDA14 protein, which lacks the first 45 amino acids of the predicted N-terminal signal peptide (Appendix Fig S5), to investigate the predicted KDAC activity of HDA14. The activity of the purified protein was tested in a colorimetric assay based on the deacetylation of a synthetic acetylated p53 peptide coupled to a chromophore (Dose *et al*, 2012). Using this assay, a deacetylase rate of 0.05/s (± 0.0032) was calculated for His-HDA14 at 100 µM substrate, which is active with both Zn^{2+} or Co^{2+} as cofactors (Fig 4B). Recent publications have shown that recombinant human HDAC8 is more active when the catalytic Zn^{2+} is replaced by Co^{2+} (Gantt *et al*, 2006). However, this is not the case for HDA14, but the enzyme is also active with Co^{2+}. Interestingly, apicidin acted only as a weak inhibitor for HDA14 even at concentrations of 100 µM. In contrast, TSA inhibited its activity by 80% at a concentration of 5 µM, the same concentration used in the leaf strip inhibitor experiments.

HDA14 regulates lysine acetylation levels of plastid proteins related to photosynthesis

To analyze the *in vivo* function of HDA14, a knock-out line (*hda14*) was obtained (Appendix Fig S6) and changes in lysine acetylation site and protein abundances between *hda14* and WT leaves were compared (Fig 5A–F). In total, 832 lysine acetylation sites were identified and quantified from leaves under normal light conditions (Fig 5A; Dataset EV3), and a further 425 lysine acetylation sites were identified from isolated thylakoids (Fig 5B, Dataset EV4), presumably associated with photosynthetic membrane proteins. While no major changes in protein abundances and plant growth were detected for *hda14* in comparison with WT (Fig 5D–F, Appendix Fig S6), 26 lysine acetylation sites on 26 protein groups

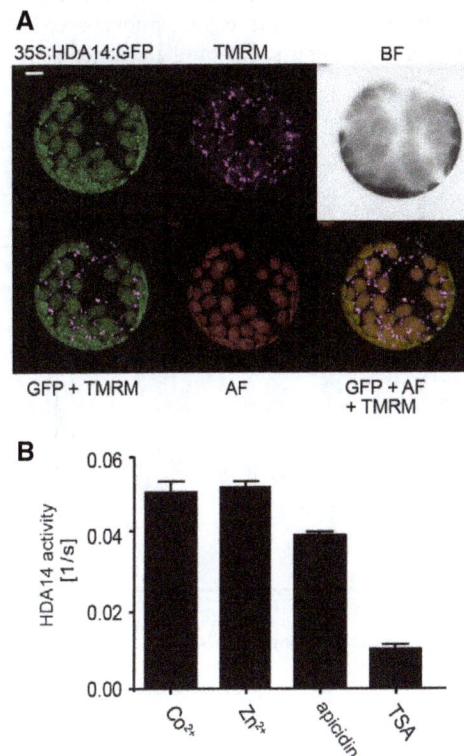

Figure 4. HDA14 protein localizes to the chloroplasts and mitochondria in *Arabidopsis*, and its activity is dependent on cofactors and can be inhibited by deacetylase inhibitors.

A GFP localization (green) of the HDA14-GFP fusion constructs in *Arabidopsis* protoplasts (35S:HDA14:GFP) from stable transformants. The mitochondrial marker TMRM is depicted in purple. GFP+TMRM shows the overlay image of 35S:HDA14:GFP and TMRM, AF indicates the chlorophyll autofluorescence and BF the bright-field image of the protoplast. GFP+AF+TMRM represents the overlay image of the three fluorescence channels. Scale bar: 10 µm.

B Deacetylase activity of the recombinant 6xHis-HDA14 protein using a colorimetric assay. Co^{2+} and Zn^{2+} were used as cofactors, apicidin (100 µM) and trichostatin A (5 µM) as deacetylase inhibitors (*n* = 5, ± SD).

were increased in between twofold and 80-fold in abundance in the mutant with a FDR < 5%. All of these lysine acetylation sites were detected on proteins localized in the plastid. Another 137 lysine acetylation sites from 122 protein groups were found significantly up-regulated in the mutant but with a lower confidence level

(P < 0.05). Of these 137 up-regulated lysine acetylation sites, 35 were uniquely identified in the thylakoid fraction and 13 sites were detected in both pull-downs. More than 90% of these proteins are annotated as plastid-localized and are involved in several biochemical processes according to a MapMan analysis (Thimm *et al*, 2004). While around 30% of the proteins have unknown functions, 24% are involved in photosynthesis, 12% in protein synthesis, degradation, and assembly, and around 5% each in lipid metabolism, redox regulation, regulation of transcription, and tetrapyrrole synthesis, as well as 1–3% each are involved in nucleotide metabolism, cell division, ABC transport, secondary metabolism, signaling, organic acid transformation, and amino acid metabolism. Eight of the HDA14 potential target proteins are encoded in the plastome, which further indicates that the deacetylation reaction is occurring within the chloroplast stroma. Among the eight plastome-encoded proteins affected in their acetylation status by the absence of HDA14, the alpha and beta-subunit of the ATP-synthase as well as several photosystem proteins, including the PSII reaction center protein D

and the PSI PsaA/PsaB protein, were identified. These results provide a further indication that HDA14 has a regulatory role in photosynthesis.

The regulation of photosynthesis by post-translational modifications such as phosphorylation and redox regulation is known to be of major importance at low light intensities, for example, during dawn and sunset, when the Calvin–Benson cycle becomes gradually activated or inactivated, respectively, due to changes in stromal pH, ATP, and NADPH levels (Carmo-Silva & Salvucci, 2013; Buchanan, 2016). Hence, we analyzed the acetylation status of the *hda14* plants in comparison with WT after the plants were transferred from normal light (100 µmol quanta m^2/s) to low light (20 µmol quanta m^2/s) intensities for 2 h. Under these conditions, 36 lysine acetylation sites on 32 protein groups showed a significant increase (P < 0.05, 2 to 100-fold), while the total protein abundances of these proteins were unchanged (Fig 5C and F, Dataset EV5). Twenty-six of these proteins are predicted to be localized in plastids. The MapMan analysis revealed that the biological process

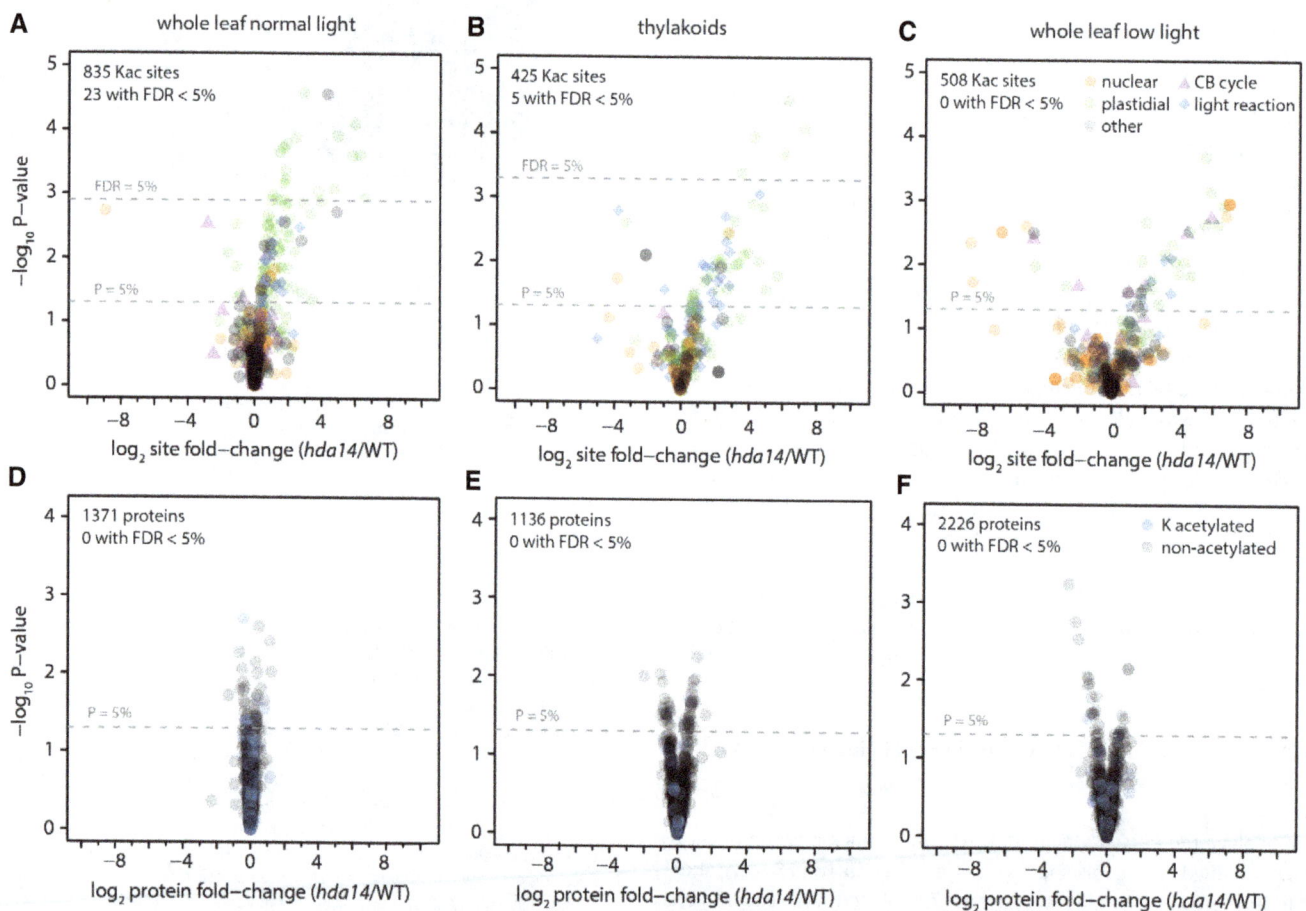

Figure 5. Differential lysine acetylation and protein expression in *hda14* versus wild-type leaves under normal light (A, D), in isolated thylakoids (B, E), and under low-light conditions (C, F).

A–F   Volcano plots depict lysine acetylation site ratios (A–C, top row) or protein ratios (D–F, bottom row) for mutant versus control, with P-values determined using the LIMMA package. Orange, protein with nuclear localization; green, protein with plastidial localization; purple triangles, proteins of the Calvin–Benson (CB) cycle; blue diamonds (top row), proteins of the light reaction; localization information according to SUBA4 database. Blue circles (bottom row), proteins with lysine acetylation sites identified. Dashed lines indicate significance thresholds of either uncorrected P-values < 5% or Benjamini–Hochberg corrected FDR < 5%. A missing FDR line indicates that the 5% threshold was not reached by any of the data points.

photosynthesis is significantly enriched among those regulated proteins, including RCA (K368/K438, log2-FC 5.89/4.85) as a master regulator of Calvin–Benson cycle activity and RuBisCO large subunit (K474, log2-FC 1.96) itself. RuBisCO catalyzes the carboxylation or alternatively oxygenation of ribulose-1,5-bisphosphate as the first step in either the Calvin–Benson cycle or photorespiration, and thereby enables the photoautotrophic lifestyle of plants. The RuBisCO enzyme is activated by carbamylation of the active site, a process that is dependent on pH, Mg^{2+}-ions and requires the removal of sugar-phosphate inhibitors that otherwise block the active site (Portis *et al*, 2008). The removal of these inhibitors requires specific conformational changes to RuBisCO that are induced by RCA, a AAA^+-ATPase enzyme. RCA is composed of redox-active alpha isoforms as well as redox-inactive beta-isoforms in *Arabidopsis* (Carmo-Silva & Salvucci, 2013). The RCA activity itself is inhibited under low light by rising ADP concentrations and remains inactive until the photosynthetic electron transport chain again raises the ATP/ADP ratio in response to higher irradiance. The RuBisCO activation state or initial activity, that is, the percentage of active sites free to perform catalysis, as well as total potential activity can be measured by rapid leaf protein extractions (Carmo-Silva *et al*, 2012). Hence, we determined the RuBisCO activity as well as the RuBisCO activation state of the *hda14* plants compared to WT. The results clearly demonstrate that the RuBisCO initial as well as total activity is significantly increased in the *hda14* mutant compared to WT (Fig 6A). While the total activity was increased on average by around 30%, the initial activity was more than doubled in *hda14* compared to WT (Fig 6A), leading to a significantly 90% increased RuBisCO activation state in the mutant under low light (Fig 6B).

Since the lysine acetylation site K438 of the RCA β1-isoform was also found increased after TSA treatment (but not K368), we performed a site-directed mutagenesis on this site in a N-terminally His-tagged RCA-β1 protein. Lysine 438 was exchanged to glutamine (K438Q) and arginine (K438R) to mimic and abolish the lysine acetylation status, respectively. The ATPase activities of the purified mutant RCA proteins were compared to the unmodified WT-like RCA-β1 protein (Fig 6C). Strikingly, the total activity of the K438Q mutant was not affected by this mutation, while the replacement of lysine to arginine led to a strongly diminished enzyme activity. Under low-light conditions, the increase in plastid ADP level plays an important role in the regulation of the RCA activity. Hence, we tested the level of ADP inhibition on the three RCA variants. While the WT-like isoform was inhibited by nearly 19% at an ATP:ADP ratio of 0.11, the activity of the K438Q mutant was only inhibited by about 8%. The K438R mutant, which mimics the non-acetylated state, showed an even stronger ADP inhibition of about 30% under these conditions (Fig 6C). Taken together, the results from this experiment further support that lysine acetylation at K438 leads to a higher RCA and thus RuBisCO activity under low light as observed in the *hda14* mutant.

Discussion

KDACs have important functions in plant development and acclimation of plants to environmental stresses (Shen *et al*, 2015). So far, these enzymes have mainly been studied with respect to their deacetylase function on histones in plants, despite the large number of different types of lysine-acetylated proteins detected in

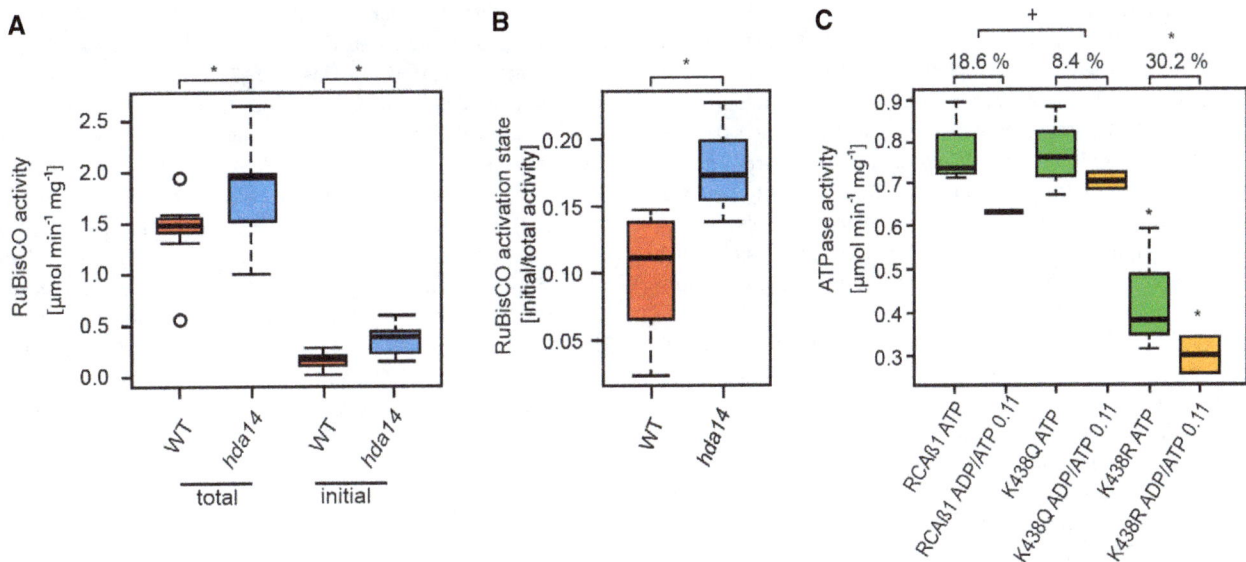

Figure 6. RuBisCO activity and RuBisCO activation state are increased in the *hda14* mutant under low-light conditions.

A RuBisCO initial and total activity in WT and hda14 in low-light-treated plants. Initial activity was measured directly upon extraction. For the total activity, samples were incubated with H_2CO_3 for 3 min to fully carbamylate the active site of RuBisCO ($n = 10$, *$P < 0.05$, t-test).

B RuBisCO activation state ($P < 0.05$, t-test).

C ATPase activity of recombinant 6x-HisRCAβ1 WT, K438Q and K438R with ATP and ADP/ATP = 0.11, respectively ($n = 3$, *$P < 0.05$, $^+P < 0.1$, t-test). Percentage values on top indicate percent ADP inhibition.

Data information: Boxes indicate lower and upper quartiles of data and whiskers indicate highest and lowest values. Small circles represent outliers. The bars across boxes indicate median values.

recent years (Hosp *et al*, 2016). In this work, we studied the proteome-wide putative targets of the RPD3/HDA1 class of lysine deacetylases in *Arabidopsis* by relative quantification of the changes in the lysine acetylome after inhibitor treatment of *Arabidopsis* leaves with apicidin and TSA. In total, we detected 2,152 lysine acetylation sites in 4-week-old *Arabidopsis* leaves when combining all experiments included in this study. The lysine acetylation sites were found on 1,022 protein groups from all different subcellular compartments and compartment-specific amino acid motifs surrounding the lysine acetylation sites were detected. Similar to human and *Drosophila* sequence motifs, glutamic acid and glycine can be frequently found at position −1 next to the lysine acetylation site also in *Arabidopsis*, while tyrosine, phenylalanine, but not proline, are also enriched in the *Arabidopsis* motifs at position +1, but with a lower frequency (Choudhary *et al*, 2009; Weinert *et al*, 2011). Generally, the acetylated lysines also occur in lysine-rich regions in *Arabidopsis* similar to those described for human and fly (Weinert *et al*, 2011).

In HeLa cells, apicidin was identified as an inhibitor of mainly the RPD3-like KDACs, while TSA inhibits enzymes from both RPD3/HDA1 classes (Scholz et al, 2015). Although we cannot exclude that apicidin and TSA have different specificities for *Arabidopsis* KDACs compared to humans, we observed that recombinant *Arabidopsis* HDA14, which is a HDA1-like KDAC, is not efficiently inhibited by apicidin but by TSA. This supports the notion that similar specificities of both inhibitors exist for the *Arabidopsis* KDACs as well. In *Arabidopsis*, four KDACs belong to the RPD3-like group, including HDA1/19, HDA6, HDA7, and HDA9 (Hollender & Liu, 2008). Two additional KDAC genes, *HDA10* and *HDA17*, are closely related to *HDA9*, but the predicted proteins lack a catalytic domain and therefore are probably inactive. Lysine acetylation sites on 91 protein groups were significantly ($P < 0.05$) up-regulated after apicidin treatment and are therefore most likely substrates of at least one of these four KDACs. The dominant nuclear localization among these proteins fits to the observed localizations of the RPD3-like KDACs in the nucleus. While HDA1/19, HDA6, and HDA9 were detected mainly in the nucleus, the localization of HDA7 has yet to be determined. Although the predicted HDA7 protein contains both a nuclear localization sequence as well as a nuclear export signal, it is unclear to what extent this protein is active due to its low expression level in most tissues.

From the 91 target protein groups identified upon inhibition with apicidin, only 14 are histone-like proteins. Hence, we identified 77 new candidate protein groups, which are potential substrates of the RPD3-like KDACs in *Arabidopsis*. This list of potential target proteins with the exact information on their acetylation sites can be regarded as valuable resource for future studies on the KDAC functions in plant stress response and development. Interestingly, a high mobility group box protein with ARID/BRIGHT DNA-binding domain (At1g76110) was identified as one of the substrate proteins, which was also regulated upon TSA treatment. These types of proteins have been identified as interaction partners of human HDAC1/2 (Joshi *et al*, 2013). Furthermore, a physical interaction was previously detected between *Arabidopsis* HDA6 and the histone H3.3 (At4g40030) (Earley *et al*, 2006). We identified several peptides of histone H3-like proteins that were up-regulated by more than 2-fold at the positions K9, 14, 18, 23, 27, 36, and 37 after apicidin treatment, but less so after treatment with TSA. Several of

these lysine acetylation sites on histone H3 are of great importance for chromatin regulation and remodeling [e.g., (Mahrez *et al*, 2016)]. For example, H3K9 acetylation was found to be associated with actively transcribed genes and has a strong impact on various developmental processes in plants (e.g., Ausín *et al*, 2004; Benhamed *et al*, 2006). Differences in the strength of TSA and apicidin inhibition could be explained by differences in the uptake of the inhibitors into the *Arabidopsis* cells, as well as by differences in the Ki values of the different *Arabidopsis* KDACs for these chemicals. Furthermore, our data indicate that TSA might not be effectively taken up into plastids, since the recombinant HDA14 protein was strongly inhibited by TSA, but only few plastid proteins were affected by TSA treatment of *Arabidopsis* leaves.

In addition to the many new KDAC inhibitor target proteins in the nucleus, there were also several interesting candidate proteins identified in the cytosol, such as the FRIENDLY protein (At3g52140), which is required for correct distribution of mitochondria within the cell. In a previous study, we already demonstrated that two lysine sites, which can be acetylated, regulate FRIENDLY function (El Zawily *et al*, 2014).

After TSA treatment, lysine acetylation sites from unique protein groups were regulated, which were not affected by apicidin treatment, indicating that those sites are specifically regulated by HDA1-type HDACs. These proteins included RCA-β1, photosystem I subunit D-2, ribosomal L6 family protein, S-adenosyl-L-methionine-dependent methyltransferases superfamily protein, the telomere repeat binding factor 1, and a histone H2B protein (Dataset EV2). Candidates of KDAC proteins from the HDA1-type group that might be responsible for the regulation of the lysine acetylation sites of these proteins include HDA5, 8, 14, 15, and 18, which cluster together with the human class 2 KDACs (Alinsug *et al*, 2009). By using a hydroxamate-based KDAC-probe, which allows the enrichment of active RPD3/HDA1-class KDACs from protein extracts, we were able to detect HDA5, 14, and 15 in total leaf extracts of 4-week-old *Arabidopsis* leaves. By using the same probe, we previously enriched all class 1 and class 2b KDACs from HeLa cells, indicating that the probe is able to bind all types of RPD3/HDA1-class KDACs (Dose *et al*, 2016). Hence, we conclude that HDA5, 14, and 15 are the most abundant KDACs in *Arabidopsis* leaves. Strikingly, both TSA and apicidin treatment resulted in an increased acetylation of plastid proteins involved in photosynthesis. Here, we identified HDA14 as the first organellar-localized RPD3/HDA1 class protein which is active as a KDAC and which has the majority of its candidate target proteins in the plastid stroma. At the concentration of apicidin used in our study, HDA14 was not significantly inhibited in its activity, which further supports the observation that apicidin mainly inhibits RPD3-like KDACs. Hence, the plastid target proteins, which showed mildly increased lysine acetylation after apicidin treatment, might be regulated by an unknown RPD3-like deacetylase. For example, we identified six lysine acetylation sites on the TROL protein, which is required for anchoring the ferredoxin-NADP reductase (FNR) to the thylakoid membranes and to sustain efficient linear electron flow in the light reactions of photosynthesis. While K328 and K348 of TROL were more than twofold increased in their acetylation level in the *hda14* loss-of-function mutant, only K337 showed a 1.2-fold increased acetylation after apicidin treatment. Lysine acetylation sites on FNR itself were not significantly regulated upon KDAC inhibition. Here, we identified four lysine acetylation sites on both FNR1 (K287, 290, 321, 325) and FNR2 (K90, 243, 299, 330) isoforms, of which only two have been previously reported (Lehtimaki et al, 2014).

By analyzing the acetylome of hda14 mutants, we were able to identify the unique substrate proteins of HDA14. HDA14 was previously identified as a nuclear/cytosolic protein based on enrichment in the microtubule fraction of a red fluorescent-tagged version of the protein (Tran et al, 2012). However, the N-terminal location of this tag would hinder the protein from entering the chloroplast. Moreover, the authentic N-terminus of HDA14 contains a clear signal sequence for the plastids as predicted by bioinformatical analysis (Alinsug et al, 2009). By using the Asu-Hd probe on isolated plastid fractions as well as by using C-terminal GFP-tagged fusion proteins, we were able to confirm the predicted plastid localization of HDA14. Furthermore, most of the candidate HDA14 substrate proteins identified reside in the plastids and are involved in metabolism and photosynthesis.

With the analysis of the HDA14-dependent acetylome, we found that the RCA-β1 site K438 is a substrate site of HDA14. Increased acetylation of this site reduced the ADP sensitivity of the RCA protein, which plays an important role for the Calvin–Benson cycle activation under low light intensities when the ADP/ATP ratio in plastids is still high (Carmo-Silva & Salvucci, 2013). Since the alpha-isoform of RCA in Arabidopsis is considerably more sensitive to inhibition by ADP than the beta-isoform (Carmo-Silva & Salvucci, 2013), the effect of acetylation of K438 of RCA-β1 in relieving ADP inhibition might be mediated by RCA-α through an effect of acetylation on subunit (i.e., alpha–beta) interaction.

Lysine acetylation on Arabidopsis RCA was detected in a previous study (Finkemeier et al, 2011), but on a different lysine residue and the functional consequences were not studied so far. In addition to increased acetylation, we also observed increased RuBisCO activity under low-light conditions in the hda14 mutant, which co-occurred with increased acetylation at K474 on the RuBisCO large subunit next to the strongly increased acetylation of K368 and K438 on RCA and on carbonic anhydrase (K269, At3g01500). Acetylation on all of these proteins might play an important role in fine-tuning of RuBisCO activity. In contrast to our results obtained in the hda14 mutant, two independent previous studies revealed that a decrease in RuBisCO acetylation resulted in a higher activity of the enzyme (Finkemeier et al, 2011; Gao et al, 2016). However, K474 on the RuBisCO large subunit (RBCL) was not detected in either of these studies, and hence, this site could have a different role than acetylation on any of the other 18 acetylation sites detected here and elsewhere.

In conclusion, in this study, we were able to define the heretofore-unknown acetylation candidate target proteins of RPD3/HDA1 class HDACs in Arabidopsis and specifically those of HDA14, as the first identified RPD3/HDA1 KDACs in organelles. Furthermore, our study revealed that about 10% of the detected lysine acetylation sites can be regulated by these types of KDACs in Arabidopsis leaves. Many sites might be specifically regulated under certain environmental or developmental conditions due to changes in KDAC activities, as we observed for low-light conditions for example. The activity of KDAC themselves might be regulated via post-translational modifications (Mengel et al, 2017) or by change in interaction partners that lead to the formation of different KDAC complexes (Dose et al, 2016). Future studies, with more detailed analyses of individual lysine sites in proteins and the analysis of further KDAC mutants and environmental conditions, will allow unraveling this complex network of fine-tuning of protein functions and interactions by lysine acetylation. Since lysine acetylation sites can act as molecular switches, they could be engineered in plant proteins to regulate cell-signaling cascades, the expression of certain genes, or to modulate the activities of metabolic enzymes. Furthermore, due to recent advances in advances in CRISPR/CAS technologies, lysine acetylation sites can be used for site-directed mutagenesis also in crop plants. Modifying these lysine residues to constitute acetylated or non-acetylated mimics ideally will allow a switching of metabolic activities and outputs that have the potential to enhance plant yields or direct metabolism in a way to enhance the accumulation of metabolic intermediates to increase the nutritional values of crops and thereby indirectly promote human health.

Materials and Methods

Plant material and growth conditions

Arabidopsis thaliana (Col-0) plants were grown for 4 weeks in a climate chamber using a 12-h light/12-h dark (21°C) photoperiod with a light intensity of 100 µmol quanta m^2/s and 50% relative humidity. For low-light treatments, plants were transferred to 20 µmol quanta m^2/s for 2 h before harvest. For growth on plates, Arabidopsis seeds were sterilized and transferred to half-strength Murashige–Skoog medium supplemented with 0.8% phytoagar. The hda14 line (SALK_144995C) was obtained from the Nottingham Arabidopsis stock center (NASC) and PCR screened according to Salk Institute Genomic Analysis Laboratory instructions (O'Malley et al, 2007) using the following primers: HDA14_LP 5'-GAAAC ATGTCACGCAAAAATG-3', HDA14_RP 5'-TTTTGTTGGTTTGCTTC TTCG-3', and the TDNA primer SALK-Lb1.3 5'-ATTTTGCCGA TTTCGGAAC-3'. PCR products were run on 1% agarose Tris–acetate (TAE: 40 mM Tris, 20 mM acetate, 1 mM EDTA, pH 8.0) and visualized by UV illumination upon ethidium bromide staining.

Trichostatin A and apicidin treatment

About 20 fully expanded leaves from 4-week-old Arabidopsis plants were pooled and cut into 2-mm-diameter leaf slices (for each biological replicate). After vacuum infiltration in effector solutions (three times for 5 min), the leaf slices were incubated at 100 µmol quanta m^2/s for 4 h. All solutions used for infiltrations were made in 1 mM MES pH 5.5 (KOH). All chemicals were purchased from Sigma-Aldrich (Gillingham, Dorset, UK). All stock solutions were dissolved in DMSO. Control experiments were then performed with DMSO added in same concentrations without effectors. Leaf material was briefly dried on tissues for harvest and flash-frozen in liquid nitrogen.

GFP fusion and plant transformation

Entry clones for Gateway cloning were generated with the pENTR/ SD/TOPO vector (Invitrogen™). The open reading frame of HDA14 (At4g33470) without stop codon from Arabidopsis (Col-0) was amplified from cDNA using the following primers: 5'-CACCATGTC CATGGCGCTAATTGT-3' and 5'-TAAGCAATGAATGCTTTTGGCTC

TC-3'. LR reactions were performed for recombination into the pK7GW2 vector (Karimi et al, 2007). The vector construct was verified by sequencing and transformed into *Agrobacterium tumefaciens* strain C58 followed by floral dip transformation of *Arabidopsis* (Col-0) plants (Clough & Bent, 1998). Transformants were selected by germination of seeds on MS-agar plates containing kanamycin (50 µg/ml). Resistant plants were transferred to soil and propagated.

RNA isolation and RT–PCR

Total RNA of *Arabidopsis* leaves was extracted using Trizol® (Invitrogen™) followed by chloroform extraction, and precipitation with isopropanol and subsequently $LiCl_2$. The quality and quantity of the RNA were confirmed on agarose gels and a UV-spectrometer. Complementary DNA (cDNA) was synthesized from DNase-treated RNA with SuperScriptIII reverse transcriptase (Invitrogen™) following the manufacturer's instruction and using dT_{20}. Real-time qPCR was carried out in triplicate in an iQ™5 Multicolor Real-Time PCR Detection System (Bio-Rad) using iQ™SYBR Green Super Mix (Bio-Rad) and gene-specific primers: HDA14-F 5'-ATCTGTGGCAGACT CGTTTCG-3', HDA14-R 5'-TCGCACCTTTCTCATTGGTTC-3'. Levels of selected transcripts in each sample were calculated using a standard curve method (Finkemeier et al, 2013). Expression levels of the HDA14 transcript were normalized to *ACTIN2* (At3g18780) transcript as housekeeping gene using the following primers: actin2-F 5'-CTGTACGGTAACATTGTGCTCAG-3' and actin2-R 5'-CCGATCCA GACACTGTACTTCC-3'.

Protoplast isolation and confocal laser scanning microscopy

Protoplast isolation was performed from 4-week-old *Arabidopsis* leaves after the tape-sandwich method (Wu et al, 2009). Staining with 20 nM TMRM (Sigma) was performed according to the manufacturer's protocol. Imaging was performed with a spectral TCS SP5 MP confocal laser scanning microscope (Leica Microsystems, Mannheim, Germany) using an argon and DPSS-Laser laser, respectively, at an excitation wavelength of 488 nm (eGFP) and 543 nm (TMRM). The water immersion objective lens HCX PL APO 20.0× 0.70 IMM UV was used for imaging in multitrack mode with line switching. eGFP fluorescence and TMRM fluorescence were measured at 500–530 and 565–615 nm, respectively.

Heterologous expression and purification of recombinant HDA14 protein

cDNAs were amplified by PCR excluding the coding region for the 45aa signal peptide using the following primers: HDA14p-F: 5'-TTTAGTACAGAGAAGAATCCTCTATTACCATCT-3' and 5'-TCAAA CAAATTCACCTTATAAGCAATG-3'. The PCR product was cloned into pEXP-5-NT/TOPO® TA (Invitrogen™), which allows expression and purification of the recombinant, N-terminally 6× His-tagged protein. Vector constructs were verified by sequencing. After transforming *E. coli* BL21(DE3) (Invitrogen™) with the expression vector, the recombinant protein was expressed using the EnPresso system (BioSilta, Germany) as described before (Jost et al, 2015): 500 ml of EnPresso medium was mixed with 12.5 µl $ZnCl_2$ (1 mM) solution and 25 µl of the "EnZ I'm" mix. Freshly prepared medium was

inoculated 1:100 with a 6-h-cultivated pre-culture at 37°C under gentle shaking (160 rpm) and incubated overnight. Subsequently, the temperature was reduced to 25°C and a "booster tablet" and 50 µl "EnZ I'm" mix were added to the culture medium followed by 500 µM IPTG. The culture was incubated for 24 h at 250 rpm. The cells were harvested by centrifugation (10 min at $3,000 \times g$), and the pellet was resuspended in PBS buffer (pH 8.0) and lysed with a homogenizer (EmulsiFlex-C5, Avestin) at 4°C. The cleared lysate (centrifugation for 20 min at $30,000 \times g$) was incubated with 1 ml of Ni-NTA Agarose slurry (Qiagen) for 2 h at 4°C. The resin was washed with 50 ml PBS, pH 8.0, 4°C, and the protein was eluted with 300 mM imidazole in HDAC buffer (8 mM KCl, 100 mM NaCl, 10 mM HEPES, pH 8.0). Pure fractions were combined and dialyzed against HDAC buffer containing 10 mM EDTA and subsequently against HDAC buffer supplied with 0.5 mM EDTA. The sample was concentrated with a centrifugation filter device with 10 kD MWCO (Amicon Ultra, Merck Millipore), supplied with 20% (v/v) glycerol and stored at −80°C until usage.

HDA14 activity assay

The deacetylation assays were performed with a previously described p53-derived peptide substrate containing the chromophore 5-amino-2-nitrobenzoic acid (p53-5,2-ANB) (Dose et al, 2012). To produce the apoenzyme, the purified HDA14 protein was first dialyzed against 10 mM EDTA and 1 mM DTT and in a second step against 0.5 mM EDTA to remove bound metal ions. For the enzyme asssay, HDA14 was supplied with either Zn^{2+} or Co^{2+} ions by incubating the enzyme solution with 1 mM of $ZnCl_2$ or $CoCl_2$ on ice for 30 min. Deacetylation assays were performed by incubating 1 µM of either Zn^{2+} or Co^{2+} supplied HDA14 with 100 µM p53-5,2-ANB substrate in HDAC reaction buffer (10 mM HEPES, 100 mM NaCl, 8 mM KCl, 10 µM BSA, pH 8.0) in a total volume of 50 µl at RT. The reaction was stopped after 10 min by adding 10 µl quenching solution (6.25 µM TSA in 0.1% (v/v) TFA) and developed by adding 10 µl of trypsin solution (6 mg/ml). After 30 min of trypsinization, the reaction mixture was supplied with 70 µl of HDAC reaction buffer, transferred into a 100-µl quartz cuvette, and the absorbance was monitored at 405 nm in a photometer (Helma, Germany). Studies with inhibitors apicidin (5 and 100 µM) and TSA (5 µM) were performed by adding these compounds to the assay before the reactions were started. All rates were normalized to the concentration of HDA14.

Western blot analyses

Proteins were separated on 12% SDS–polyacrylamide gels, blotted onto nitrocellulose membrane, and incubated overnight with the primary antibodies. The secondary IRDye 800CW antibody (LI-COR) was used in a 1:10,000 dilution and detected with the Odyssey reader (LI-COR).

RuBisCO activity measurements for activation state determination

For determination of the RubisCO activation state under low-light conditions, plants were transferred to low irradiation (20 µmol quanta m^2/s) for 5 h at 21°C, then harvested, and frozen in liquid nitrogen. RuBisCO initial and total activity were assayed by incorporation of $^{14}CO_2$ into acid-stable products (Salvucci, 1992). The leaves were homogenized in extraction medium [100 mM Tricine–NaOH pH 8.0, 1 mM EDTA, 5% polyvinylpyrrolidone (PVP-40), 5% polyethylene glycol 3350 (PEG3350), 5 mM (DTT), and protease inhibitor cocktail (Roche)]. Initial activities were measured immediately upon extraction, whereas total activities were measured after 3-min incubation in assays without RuBP to fully carbamylate the enzyme (Carmo-Silva et al, 2012). For each sample, assays were conducted in duplicate. Initial and total activities were used to calculate RuBisCO activation state, that is, (initial/total activity × 100) = % activation.

Purification and assay of RCA

The coding sequence of the RCAβ1 spliceform (At2g39730.2) was amplified from Arabidopsis cDNA using the following primers: 5′-CTCCGATATCTTACTTGCTGGGCTCCTTT-3′ and 5′-TTTTTGATA TCTCAAACCTCTGTTTTACC-3′ introducing SacI and EcoRV restriction sites for cloning into pCDFDuet-1 (Novagen®). Site-directed mutants of RCA K438R and K438Q were introduced with the Quik-Change Site-Directed Mutagenesis Kit (Agilent Technologies) using the following primers: RCAβ2-K438R 5′-GAACTTTCTACGGTAGAA CAGAGGAAAAGG-3′ and RCAβ2-K438Q 5′-GAACTTTCTACGGTCAA ACAGAGGAAAAGG-3′. The N-terminally 6-His-tagged protein was expressed and purified from Rosetta-gami cells (Novagen®) as described detail in (Barta et al, 2011). ATPase activity of 5 µg recombinant RCA was measured for 1 min at 23°C in 50 µl reaction buffer (100 mM HEPES-KOH (pH 8.0), 20 mM MgCl₂) containing 500 µM ATP and 500 µM ATP and 55 µM ADP, respectively. The reaction was heat inactivated at 95°C. The ATP consumption was determined using the KinaseGlo Max Luminescent Assay Kit (Promega) according to the manufacturer's protocol.

Isolation of intact chloroplasts and mitochondria

Chloroplasts were isolated from dark incubated (12 h) 5-week-old rosette leaves of Arabidopsis. Leaves were homogenized in ice-cold HB-buffer (0.45 M sorbitol, 20 mM Tricine-KOH pH 8.45, 10 mM EDTA, 10 mM NaHCO₃, 0.1% BSA, and 2 mM sodium ascorbate). Chloroplasts were purified on a Percoll gradient (40–80%) and resuspended in sorbitol buffer (0.3 M sorbitol, 20 mM Tricine-KOH pH 8.45, 2.5 mM EDTA, and 5 mM MgCl₂, 2 mM sodium ascorbate). Mitochondria were isolated as described previously (König et al, 2014a).

Isolation of thylakoids

Chloroplasts were lysed in 2 ml TMK buffer (50 mM HEPES/KOH pH 7.5, 0.1 M sorbitol, 5 mM MgCl₂, 10 mM NaF), and thylakoid membranes were sedimented at 14,000 × g.

Preparation of cell extracts and enrichment of active histone deacetylases

Leaves from 5-week-old Arabidopsis plants were homogenized in extraction buffer (50 mM Tris–KOH (pH 7.5), 150 mM NaCl, 10% [v/v] glycerol, 5 mM dithiothreitol (DTT), 1% [v/v] Triton X-100,

and protease inhibitor cocktail (Sigma-Aldrich). Homogenates were centrifuged at 14,000 × g, and protein concentration of the supernatant was determined with the Pierce 660 nm Protein Assay (Thermo Fisher Scientific). All protein extracts were desalted on PD-10 Desalting Columns (GE Life Sciences), and the samples were eluted with immunoprecipitation buffer (50 mM Tris, pH 7.5, 150 mM NaCl, 10% [v/v] glycerol).

The immobilized peptide probes, mini-AsuHd, and mini-Lys (Dose et al, 2016) were equilibrated with immunoprecipitation buffer two times and incubated with the protein extracts overnight at 4°C, under constant rotation. The next day, the beads were gently pelleted by centrifugation. The beads were transferred onto microcentrifugal filter system (Amchro GmbH) and washed five times with 1 ml immunoprecipitation buffer. Proteins bound on beads were subjected to on-bead digestion. Proteins were denatured in 6 M urea prepared in 0.1 M Tris–HCl (pH 8.0), 1 mM CaCl₂ and reduced with 5 mM DTT. Reduced cysteines were alkylated with 14 mM chloroacetamide for 30 min. Excess chloroacetamide was quenched with DTT. Proteins were trypsinated at a urea concentration of 1 M and a trypsin (Sigma-Aldrich) to protein ratio of 1:100 at 37°C. Resulting peptides were desalted on SDB-RPS and C18 Stage-Tips, respectively (Rappsilber et al, 2007; Kulak et al, 2014).

Protein extraction, peptide dimethyl labeling, and lysine-acetylated peptide enrichment

Frozen leaf material was ground to fine powder in liquid nitrogen and extracted using a modified filter-assisted sample preparation (FASP) protocol with 30k MWCO Amicon filters (Merck Millipore) as described in detail in Lassowskat et al (2017). Digested peptides were dimethyl-labeled on C18 Sep-Pak plus short columns (Waters) as described previously (Boersema et al, 2009; Lassowskat et al, 2017). Equal amounts of light and medium-labeled peptides (3–5 mg) were pooled for each replicate and the solvent evaporated in a vacuum centrifuge. The dried peptides were dissolved in 1 ml TBS buffer (50 mM Tris–HCl, 150 mM NaCl, pH 7.6), and pH was checked and adjusted where required. 15 µg peptide mixture was stored for whole proteome analysis. About 10 mg of the pooled labeled peptides was resuspended in 2 ml 95% solvent A (95% acetonitrile, 5 mM ammonium acetate) and 5% buffer B (5 mM ammonium acetate) and fractionated with a flow rate of 500 µl/min on a Sequant ZIC®-HILIC column (3.5 µm, Merck) using a segmented linear gradient of 0–60%. The fractions were combined to seven final fractions and dried in a vacuum centrifuge. Peptides were resuspended in IP buffer (50 mM Tris–HCl pH 7.6, 150 mM NaCl), and the concentration was determined on the spectrophotometer at 280 nm. Lysine-acetylated peptide enrichment was performed as previously described with 1 mg peptide per fraction (Hartl et al, 2015; Lassowskat et al, 2017). After enrichment, the eluted peptides were desalted using C18 StageTips and dried in a vacuum centrifuge.

LC-MS/MS

Dried peptides were redissolved in 2% ACN, 0.1% TFA for analysis. Total proteome samples were adjusted to a final concentration of 0.2 µg/µl. Samples were analyzed using an EASY-nLC 1000 (Thermo Fisher) coupled to a Q Exactive, Q Exactive Plus, and an Orbitrap Elite mass spectrometer (Thermo Fisher), respectively. Peptides were separated on 16 cm frit-less silica emitters (New Objective, 0.75 µm inner diameter), packed in-house with reversed-phase ReproSil-Pur C18 AQ 3 µm resin (Dr. Maisch). Peptides (5 µl) were loaded on the column and eluted for 120 min using a segmented linear gradient of 0% to 95% solvent B (solvent A 5% ACN, 0.5% FA; solvent B 100% ACN, 0.5% FA) at a flow rate of 250 nl/min. Parameters for the different machines are listed in Dataset EV1.

MS data analysis

Raw data were processed using MaxQuant software version 1.5.2.8 (http://www.maxquant.org/) (Cox & Mann, 2008). MS/MS spectra were searched with the Andromeda search engine against the TAIR10 database (TAIR10_pep_20101214; ftp://ftp.arabidopsis.org/home/tair/Proteins/TAIR10_protein_lists/). Sequences of 248 common contaminant proteins and decoy sequences were automatically added during the search. Trypsin specificity was required, and a maximum of two (proteome) or four missed cleavages (acetylome) were allowed. Minimal peptide length was set to seven amino acids. Carbamidomethylation of cysteine residues was set as fixed, oxidation of methionine, and protein N-terminal acetylation as variable modifications. Acetylation of lysines was set as variable modification only for the antibody-enriched samples. Light and medium dimethylation of lysines and peptide N-termini were set as labels. Peptide–spectrum matches and proteins were retained if they were below a false discovery rate of 1%, modified peptides were filtered for a score ≥ 35 and a delta score of ≥ 6. Match between runs and requantify options were enabled. Downstream data analysis was performed using Perseus version 1.5.5.3 (Tyanova *et al*, 2016). For proteome and acetylome, reverse hits and contaminants were removed, the site ratios log2-transformed, and flip-label ratios inverted. For quantitative lysine acetylome analyses, sites were filtered for a localization probability of ≥ 0.75. The "expand site table" feature of Perseus was used to allow separate analysis of site ratios for multiply acetylated peptides occurring in different acetylation states. Technical replicates were averaged, and proteins or sites displaying less than two out of three ratios were removed. The resulting matrices for proteome and acetylome, respectively, were exported and significantly differentially abundant protein groups and lysine acetylation sites were determined using the LIMMA package (Ritchie *et al*, 2015) in R 3.3.1 (R Core Team, 2016). Volcano plots were generated with R base graphics, plotting the non-adjusted *P*-values versus the log2 fold-change and marking data points below 5% FDR (i.e., adjusted *P*-values, Benjamini–Hochberg) when present.

Acknowledgements

We would like to thank Jon Nield (Queen Mary, University of London) for the light-reaction figure template, Julia Kimmel for cloning of 35S:HDA14-GFP constructs, and Anne Harzen and Gintaute Dailydaite (MPIPZ) as well as Anne Orwat (LMU Munich) for technical assistance. This work was supported by the Deutsche Forschungsgemeinschaft, Germany (Emmy Noether Programme FI-1655/1-1), and the Max Planck Gesellschaft.

Author contributions

MF, MH, IF, AB, MP, KK, MS, PJB, J-OJ, and JS performed research; JC performed computational analysis; MH, MS, and IF designed research; DS, DL, GU, GBGM, and MM provided reagents and analytical equipment; MF, MH, J-OJ, KK, and IF analyzed data; IF drafted the manuscript with input from MH, KK, DL, DS, and MES.

References

Alinsug MV, Yu CW, Wu K (2009) Phylogenetic analysis, subcellular localization, and expression patterns of RPD3/HDA1 family histone deacetylases in plants. *BMC Plant Biol* 9: 37

Allfrey VG, Faulkner R, Mirsky AE (1964) Acetylation and methylation of histones and their possible role in regulation of Rna synthesis. *Proc Natl Acad Sci USA* 51: 786–794

Ausín I, Alonso-Blanco C, Jarillo JA, Ruiz-García L, Martínez-Zapater JM (2004) Regulation of flowering time by FVE, a retinoblastoma-associated protein. *Nat Genet* 3: 162–166

Barta C, Carmo-Silva AE, Salvucci ME (2011) Purification of rubisco activase from leaves or after expression in *Escherichia coli*. In *Photosynthesis research protocols*, Carpentier R (ed.), pp 363–374. Totowa, NJ: Humana Press

Benhamed M, Bertrand C, Servet C, Zhou DX (2006) *Arabidopsis* GCN5, HD1, and TAF1/HAF2 interact to regulate histone acetylation required for light-responsive gene expression. *Plant Cell* 18: 2893–2903

Boersema PJ, Raijmakers R, Lemeer S, Mohammed S, Heck AJR (2009) Multiplex peptide stable isotope dimethyl labeling for quantitative proteomics. *Nat Protoc* 4: 484–494

Buchanan BB (2016) The carbon (formerly dark) reactions of photosynthesis. *Photosynth Res* 128: 215–217

Calfapietra C, Penuelas J, Niinemets U (2015) Urban plant physiology: adaptation-mitigation strategies under permanent stress. *Trends Plant Sci* 20: 72–75

Carmo-Silva AE, Gore MA, Andrade-Sanchez P, French AN, Hunsaker DJ, Salvucci ME (2012) Decreased CO2 availability and inactivation of Rubisco limit photosynthesis in cotton plants under heat and drought stress in the field. *Environ Exp Bot* 83: 1–11

Carmo-Silva AE, Salvucci ME (2013) The regulatory properties of rubisco activase differ among species and affect photosynthetic induction during light transitions. *Plant Physiol* 161: 1645–1655

Chen LT, Luo M, Wang YY, Wu K (2010) Involvement of *Arabidopsis* histone deacetylase HDA6 in ABA and salt stress response. *J Exp Bot* 61: 3345–3353

Choi SM, Song HR, Han SK, Han M, Kim CY, Park J, Lee YH, Jeon JS, Noh YS, Noh B (2012) HDA19 is required for the repression of salicylic acid biosynthesis and salicylic acid-mediated defense responses in *Arabidopsis*. *Plant J* 71: 135–146

Choudhary C, Kumar C, Gnad F, Nielsen ML, Rehman M, Walther TC, Olsen JV, Mann M (2009) Lysine acetylation targets protein complexes and co-regulates major cellular functions. *Science* 325: 834–840

Cigliano RA, Cremona G, Paparo R, Termolino P, Perrella G, Gutzat R, Consiglio MF, Conicella C (2013) Histone deacetylase AtHDA7 is required for female gametophyte and embryo development in *Arabidopsis*. *Plant Physiol* 163: 431–440

Clough SJ, Bent AF (1998) Floral dip: a simplified method for Agrobacterium-mediated transformation of *Arabidopsis thaliana*. *Plant J* 16: 735–743

Cox J, Mann M (2008) MaxQuant enables high peptide identification rates, individualized p.p.b.-range mass accuracies and proteome-wide protein quantification. *Nat Biotechnol* 26: 1367–1372

Dietz KJ (2015) Efficient high light acclimation involves rapid processes at multiple mechanistic levels. *J Exp Bot* 66: 2401–2414

Dose A, Jost JO, Spiess AC, Henklein P, Beyermann M, Schwarzer D (2012) Facile synthesis of colorimetric histone deacetylase substrates. *Chem Commun* 48: 9525–9527

Dose A, Sindlinger J, Bierlmeier J, Bakirbas A, Schulze-Osthoff K, Einsele-Scholz S, Hartl M, Essmann F, Finkemeier I, Schwarzer D (2016) Interrogating substrate selectivity and composition of endogenous histone deacetylase complexes with chemical probes. *Angew Chem Int Ed Engl* 55: 1192–1195

Drazic A, Myklebust LM, Ree R, Arnesen T (2016) The world of protein acetylation. *Bba-Proteins Proteom* 1864: 1372–1401

Earley K, Lawrence RJ, Pontes O, Reuther R, Enciso AJ, Silva M, Neves N, Gross M, Viegas W, Pikaard CS (2006) Erasure of histone acetylation by *Arabidopsis* HDA6 mediates large-scale gene silencing in nucleolar dominance. *Gene Dev* 20: 1283–1293

El Zawily AM, Schwarzlander M, Finkemeier I, Johnston IG, Benamar A, Cao Y, Gissot C, Meyer AJ, Wilson K, Datla R, Macherel D, Jones NS, Logan DC (2014) FRIENDLY regulates mitochondrial distribution, fusion, and quality control in *Arabidopsis*. *Plant Physiol* 166: 808–828

Fang XP, Chen WY, Zhao Y, Ruan SL, Zhang HM, Yan CQ, Jin L, Cao LL, Zhu J, Ma HS, Cheng ZY (2015) Global analysis of lysine acetylation in strawberry leaves. *Front Plant Sci* 6: 739

Finkemeier I, Laxa M, Miguet L, Howden AJ, Sweetlove LJ (2011) Proteins of diverse function and subcellular location are lysine acetylated in *Arabidopsis*. *Plant Physiol* 155: 1779–1790

Finkemeier I, König AC, Heard W, Nunes-Nesi A, Pham PA, Leister D, Fernie AR, Sweetlove LJ (2013) Transcriptomic analysis of the role of carboxylic acids in metabolite signaling in *Arabidopsis* leaves. *Plant Physiol* 162: 239–253

Gantt SL, Gattis SG, Fierke CA (2006) Catalytic activity and inhibition of human histone deacetylase 8 is dependent on the identity of the active site metal ion. *Biochemistry* 45: 6170–6178

Gao X, Hong H, Li WC, Yang L, Huang J, Xiao YL, Chen XY, Chen GY (2016) Downregulation of rubisco activity by non-enzymatic acetylation of RbcL. *Mol Plant* 9: 1018–1027

Hartl M, Finkemeier I (2012) Plant mitochondrial retrograde signaling: post-translational modifications enter the stage. *Front Plant Sci* 3: 253

Hartl M, König A-C, Finkemeier I (2015) Identification of lysine-acetylated mitochondrial proteins and their acetylation sites. In *Plant mitochondria: methods and protocols*, Whelan J, Murcha MW (eds), pp 107–121. New York, NY: Springer New York

He DL, Wang Q, Li M, Damaris RN, Yi XL, Cheng ZY, Yang PF (2016) Global proteome analyses of lysine acetylation and succinylation reveal the widespread involvement of both modification in metabolism in the embryo of germinating rice seed. *J Proteome Res* 15: 879–890

Heazlewood JL, Verboom RE, Tonti-Filippini J, Small I, Millar AH (2007) SUBA: the *Arabidopsis* subcellular database. *Nucleic Acids Res* 35: D213–D218

Hollender C, Liu ZC (2008) Histone deacetylase genes in *Arabidopsis* development. *J Integr Plant Biol* 50: 875–885

Hosp F, Lassowskat I, Santoro V, De Vleesschauwer D, Fliegner D, Redestig H, Mann M, Christian S, Hannah MA, Finkemeier I (2016) Lysine acetylation in mitochondria: from inventory to function. *Mitochondrion* 33 58–71

Huber SC (2007) Exploring the role of protein phosphorylation in plants: from signalling to metabolism. *Biochem Soc Trans* 35: 28–32

Johnova P, Skalak J, Saiz-Fernandez I, Brzobohaty B (2016) Plant responses to ambient temperature fluctuations and water-limiting conditions: a proteome-wide perspective. *Bba-Proteins Proteom* 1864: 916–931

Joshi P, Greco TM, Guise AJ, Luo Y, Yu F, Nesvizhskii AI, Cristea IM (2013) The functional interactome landscape of the human histone deacetylase family. *Mol Syst Biol* 9: 672

Jost JO, Hanswillemenke A, Schwarzer D (2015) A miniaturized readout strategy for endogenous histone deacetylase activity. *Mol BioSyst* 11: 1820–1823

Karimi M, Depicker A, Hilson P (2007) Recombinational cloning with plant gateway vectors. *Plant Physiol* 145: 1144–1154

Kleff S, Andrulis ED, Anderson CW, Sternglanz R (1995) Identification of a gene encoding a yeast histone H4 acetyltransferase. *J Biol Chem* 270: 24674–24677

König AC, Hartl M, Boersema PJ, Mann M, Finkemeier I (2014a) The mitochondrial lysine acetylome of *Arabidopsis*. *Mitochondrion* 19(Pt B): 252–260.

König AC, Hartl M, Pham PA, Laxa M, Boersema PJ, Orwat A, Kalitventseva I, Plöchinger M, Braun HP, Leister D, Mann M, Wachter A, Fernie AR, Finkemeier I (2014b) The *Arabidopsis* class II sirtuin is a lysine deacetylase and interacts with mitochondrial energy metabolism. *Plant Physiol* 164: 1401–1414

Kulak NA, Pichler G, Paron I, Nagaraj N, Mann M (2014) Minimal, encapsulated proteomic-sample processing applied to copy-number estimation in eukaryotic cells. *Nat Methods* 11: 319–U300

Lassowskat I, Hartl M, Hosp F, Boersema PJ, Mann M, Finkemeier I (2017) Dimethyl-labeling based quantification of the lysine acetylome and proteome of plants. *Methods Mol Biol*, photorespiration, AFeaE (ed), 1653: 65–81

Lehtimaki N, Koskela MM, Dahlstrom KM, Pakula E, Lintala M, Scholz M, Hippler M, Hanke GT, Rokka A, Battchikova N, Salminen TA, Mulo P (2014) Posttranslational modifications of FERREDOXIN-NADP(+) OXIDOREDUCTASE in *Arabidopsis* chloroplasts. *Plant Physiol* 166: 1764–U1917

Maddelein D, Colaert N, Buchanan I, Hulstaert N, Gevaert K, Martens L (2015) The iceLogo web server and SOAP service for determining protein consensus sequences. *Nucleic Acids Res* 43: W543–W546

Mahrez W, Arellano MST, Moreno-Romero J, Nakamura M, Shu H, Nanni P, Kohler C, Gruissem W, Hennig L (2016) H3K36ac is an evolutionary conserved plant histone modification that marks active genes. *Plant Physiol* 170: 1566–1577

Mengel A, Ageeva A, Georgii E, Bernhardt J, Wu K, Durner J, Lindermayr C (2017) Nitric oxide modulates histone acetylation at stress genes by inhibition of histone deacetylases. *Plant Physiol* 173: 1434–1452

Mignolet-Spruyt L, Xu EJ, Idanheimo N, Hoeberichts FA, Muhlenbock P, Brosche M, Van Breusegem F, Kangasjarvi J (2016) Spreading the news: subcellular and organellar reactive oxygen species production and signalling. *J Exp Bot* 67: 3831–3844

Nallamilli BRR, Edelmann MJ, Zhong XX, Tan F, Mujahid H, Zhang J, Nanduri B, Peng ZH (2014) Global analysis of lysine acetylation suggests the involvement of protein acetylation in diverse biological processes in rice (*Oryza sativa*). *PLoS ONE* 9: e89283

Newkirk TL, Bowers AA, Williams RM (2009) Discovery, biological activity, synthesis and potential therapeutic utility of naturally occurring histone deacetylase inhibitors. *Nat Prod Rep* 26: 1293–1320

Nunes-Nesi A, Fernie AR, Stitt M (2010) Metabolic and signaling aspects underpinning the regulation of plant carbon nitrogen interactions. *Mol Plant* 3: 973–996

O'Malley RC, Alonso JM, Kim CJ, Leisse TJ, Ecker JR (2007) An adapter ligation-mediated PCR method for high-throughput mapping of T-DNA inserts in the *Arabidopsis* genome. *Nat Protoc* 2: 2910–2917

Pandey R, Muller A, Napoli CA, Selinger DA, Pikaard CS, Richards EJ, Bender J, Mount DW, Jorgensen RA (2002) Analysis of histone acetyltransferase and histone deacetylase families of *Arabidopsis thaliana* suggests functional diversification of chromatin modification among multicellular eukaryotes. *Nucleic Acids Res* 30: 5036–5055

Parthun MR, Widom J, Gottschling DE (1996) The major cytoplasmic histone acetyltransferase in yeast: Links to chromatin replication and histone

Portis AR Jr, Li C, Wang D, Salvucci ME (2008) Regulation of Rubisco activase and its interaction with Rubisco. *J Exp Bot* 59: 1597–1604

R Core Team (2016) *R: A language and environment for statistical computing.* Vienna, Austria: R Foundation for Statistical Computing. https://www.R-project.org/

Rappsilber J, Mann M, Ishihama Y (2007) Protocol for micro-purification, enrichment, pre-fractionation and storage of peptides for proteomics using StageTips. *Nat Protoc* 2: 1896–1906

Ritchie ME, Phipson B, Wu D, Hu YF, Law CW, Shi W, Smyth GK (2015) limma powers differential expression analyses for RNA-sequencing and microarray studies. *Nucleic Acids Res* 43: e47

Salvucci ME (1992) Subunit Interactions of Rubisco Activase - Polyethylene-Glycol Promotes Self-Association, Stimulates ATPase and Activation Activities, and Enhances Interactions with Rubisco. *Arch Biochem Biophys* 298: 688–696

Scholz C, Weinert BT, Wagner SA, Beli P, Miyake Y, Qi J, Jensen LJ, Streicher W, McCarthy AR, Westwood NJ, Lain S, Cox J, Matthias P, Mann M, Bradner JE, Choudhary C (2015) Acetylation site specificities of lysine deacetylase inhibitors in human cells. *Nat Biotechnol* 33: 415–423

Shen Y, Wei W, Zhou DX (2015) Histone acetylation enzymes coordinate metabolism and gene expression. *Trends Plant Sci* 20: 614–621

Smith-Hammond CL, Hoyos E, Miernyk JA (2014) The pea seedling mitochondrial N-epsilon-lysine acetylome. *Mitochondrion* 19: 154–165

Thimm O, Blasing O, Gibon Y, Nagel A, Meyer S, Kruger P, Selbig J, Muller LA, Rhee SY, Stitt M (2004) MAPMAN: a user-driven tool to display genomics data sets onto diagrams of metabolic pathways and other biological processes. *Plant J* 37: 914–939

Tran HT, Nimick M, Uhrig RG, Templeton G, Morrice N, Gourlay R, DeLong A, Moorhead GBG (2012) *Arabidopsis thaliana* histone deacetylase 14 (HDA14) is an alpha-tubulin deacetylase that associates with PP2A and enriches in the microtubule fraction with the putative histone acetyltransferase ELP3. *Plant J* 71: 263–272

Tyanova S, Temu T, Sinitcyn P, Carlson A, Hein MY, Geiger T, Mann M, Cox J (2016) The Perseus computational platform for comprehensive analysis of (prote)omics data. *Nat Methods* 13: 731–740

Wagner GR, Payne RM (2013) Widespread and enzyme-independent N-is an element of-acetylation and N-is an element of-succinylation of proteins in the chemical conditions of the mitochondrial matrix. *J Biol Chem* 288: 29036–29045

Weinert BT, Wagner SA, Horn H, Henriksen P, Liu WSR, Olsen JV, Jensen LJ, Choudhary C (2011) Proteome-wide mapping of the *Drosophila* acetylome demonstrates a high degree of conservation of lysine acetylation. *Sci Signal* 4: ra48

Weinert BT, Iesmantavicius V, Wagner SA, Scholz C, Gummesson B, Beli P, Nystrom T, Choudhary C (2013) Acetyl-phosphate is a critical determinant of lysine acetylation in *E. coli. Mol Cell* 51: 265–272

Wu FH, Shen SC, Lee LY, Lee SH, Chan MT, Lin CS (2009) Tape-*Arabidopsis* Sandwich - a simpler *Arabidopsis* protoplast isolation method. *Plant Methods* 5: 16

Wu X, Oh MH, Schwarz EM, Larue CT, Sivaguru M, Imai BS, Yau PM, Ort DR, Huber SC (2011) Lysine acetylation is a widespread protein modification for diverse proteins in *Arabidopsis. Plant Physiol* 155: 1769–1778

Xiong YH, Peng XJ, Cheng ZY, Liu WD, Wang GL (2016) A comprehensive catalog of the lysine-acetylation targets in rice (Oryza sativa) based on proteomic analyses. *J Proteomics* 138: 20–29

Yang XJ, Seto E (2008) Lysine acetylation: codified crosstalk with other posttranslational modifications. *Mol Cell* 31: 449–461

Yuan H, Marmorstein R (2013) Histone acetyltransferases: rising ancient counterparts to protein kinases. *Biopolymers* 99: 98–111

Zhang YM, Song LM, Liang WX, Mu P, Wang S, Lin Q (2016) Comprehensive profiling of lysine acetylproteome analysis reveals diverse functions of lysine acetylation in common wheat. *Sci Rep* 6: 21069

Zhen SM, Deng X, Wang J, Zhu GR, Cao H, Yuan LL, Yan YM (2016) First comprehensive proteome analyses of lysine acetylation and succinylation in seedling leaves of *Brachypodium distachyon* L. *Sci Rep* 6: 31576

Zheng Y, Ding Y, Sun X, Xie SS, Wang D, Liu XY, Su LF, Wei W, Pan L, Zhou DX (2016) Histone deacetylase HDA9 negatively regulates salt and drought stress responsiveness in *Arabidopsis. J Exp Bot* 67: 1703–1713

Zhou CH, Zhang L, Duan J, Miki B, Wu KQ (2005) HISTONE DEACETYLASE19 is involved in jasmonic acid and ethylene signaling of pathogen response in *Arabidopsis. Plant Cell* 17: 1196–1204

5

Genomewide landscape of gene–metabolome associations in *Escherichia coli*

Tobias Fuhrer[†] [iD], Mattia Zampieri[†], Daniel C Sévin[†,‡], Uwe Sauer[*] [iD] & Nicola Zamboni [iD]

Abstract

Metabolism is one of the best-understood cellular processes whose network topology of enzymatic reactions is determined by an organism's genome. The influence of genes on metabolite levels, however, remains largely unknown, particularly for the many genes encoding non-enzymatic proteins. Serendipitously, genomewide association studies explore the relationship between genetic variants and metabolite levels, but a comprehensive interaction network has remained elusive even for the simplest single-celled organisms. Here, we systematically mapped the association between > 3,800 single-gene deletions in the bacterium *Escherichia coli* and relative concentrations of > 7,000 intracellular metabolite ions. Beyond expected metabolic changes in the proximity to abolished enzyme activities, the association map reveals a largely unknown landscape of gene–metabolite interactions that are not represented in metabolic models. Therefore, the map provides a unique resource for assessing the genetic basis of metabolic changes and conversely hypothesizing metabolic consequences of genetic alterations. We illustrate this by predicting metabolism-related functions of 72 so far not annotated genes and by identifying key genes mediating the cellular response to environmental perturbations.

Keywords functional genomics; GWAS; interaction network; metabolism; metabolomics
Subject Categories Genome-Scale & Integrative Biology; Metabolism; Methods & Resources

Introduction

Decades of *in vitro* biochemistry have established extensive enzyme-catalyzed networks of metabolite conversions, culminating in genome-scale reconstructions of bacterial metabolism with about 1,000 reactions (Feist *et al*, 2009; Orth *et al*, 2011). Largely unexplored is the even larger network of metabolites affecting general protein activities (Heinemann & Sauer, 2010; Link *et al*, 2013) and proteins influencing metabolism, mechanistically connected through direct or indirect relationships such as regulation processes or functional consequences, respectively. Scarce molecular knowledge of these interactions limits our ability to predict how genetic perturbations propagate throughout the multiple interlinked networks and affect the global metabolic state of a cell (Wang *et al*, 2010; Ghazalpour *et al*, 2014; Shin *et al*, 2014). Although gene–gene interactions exclusively based on growth phenotype measurements have provided valuable guidance in the form of genome-scale functional interaction maps (Costanzo *et al*, 2010; Nichols *et al*, 2011), the underlying mechanisms of such genetic interactions often remain obscure.

To resolve the complex traits by which abolished gene products can influence metabolism, we here exploit the multi-feature readout of non-targeted metabolomics to systematically map gene–metabolite associations at genome-scale in the bacterium *Escherichia coli*. To this end, we developed an experimental-computational approach to quantify the strength of gene–metabolite ion interactions from high-throughput, non-targeted metabolomics (Fuhrer *et al*, 2011) data obtained from two independent clones of each of the 3,807 mutants (Table EV1) in the *E. coli* single-gene deletion collection (Baba *et al*, 2006).

Analysis of our metabolome signatures revealed the presence of local effects caused by simple enzyme-reactant relationships but also numerous strong gene–metabolite associations that cannot be explained by metabolic proximity in classical stoichiometric models. The application potential of this comprehensive resource is demonstrated by predicting functionality of genes with unknown metabolic function and identifying genes involved in the cellular response to environmental perturbations solely on the basis of the metabolome. Hence, we believe the reported empirical associations between genes and metabolites to be a unique and powerful resource to support and inspire functional genomics studies.

Results

Metabolome profiling of *Escherichia coli* single knockout collection

Metabolome extracts were prepared from cultures growing exponentially in mineral salts medium containing glucose and amino acids

Institute of Molecular Systems Biology, ETH Zürich, Zürich, Switzerland
*Corresponding author. E-mail: sauer@imsb.biol.ethz.ch
†These authors contributed equally to this work
‡Present address: Cellzome, GlaxoSmithKline R&D, Heidelberg, Germany

and analyzed in technical duplicates by non-targeted mass spectrometry. To enable analysis of more than 34,000 injections, we used high-throughput, flow-injection analysis on an high-resolution, time-of-flight (TOF) instrument (Fuhrer *et al*, 2011). This is a chromatography-free system that is well suited for large-scale profiling of the polar metabolome but cannot resolve metabolites with similar molecular weight and is subject to misquantification or misannotation in regions of the measured spectrum that are densely crowded with peaks or in the presence of unknown metabolites. Spectral data processing identified 3,169 and 4,365 distinct mass-to-charge (*m/z*) features in negative and positive ionization mode, respectively.

Based on the measured accurate mass, a total of 3,130 ions with distinct *m/z* could be putatively matched to expectable ions of 1,432 of the 2,028 chemical formulas listed in the KEGG *E. coli* database (Kanehisa *et al*, 2012) (Table EV2). Since metabolites with equal molecular weight are not distinguishable, these 1,432 formulas theoretically match to 2,472 metabolites. As expected, most of the potentially detected compounds relate to abundant and polar metabolites such as intermediates of primary metabolism (Fig EV1). All data were condensed in a two-dimensional gene–metabolite ion association matrix that reports relative abundances of all detectable metabolite ions in all 3,807 analyzed deletion strains (Fig 1). Modified *z*-score normalization was applied to compare ion changes across all mutants independent of ionization mode and signal intensity. Ninety-nine percent of variability between biological replicates was estimated to be smaller than a *z*-score of 2.765 (Fig EV2A). Thus, the gene–metabolite association matrix can be directly queried for reproducibly changing ions in each mutant.

The overall rearrangements of steady-state metabolite concentrations in each mutant varied greatly across genotypes, regardless of mutant growth rate (Fig EV2B). Metabolites responded to deletion of genes from essentially all functional classes, including those with unknown function (Fig 2). The largest normalized metabolic changes in individual mutants were often located 1–2 metabolites upstream of deleted enzymes (e.g., $\Delta purK$ in Fig 1 and chorismate biosynthesis mutants in Fig EV3A), consistent with earlier observations (Ishii *et al*, 2007; Fendt *et al*, 2010). On the other hand, some gene deletions led to a widespread alteration of several metabolites (e.g., $\Delta pstS$ in Fig 1). Analogously, some metabolites responded only to a limited number of genetic ablations (e.g., enterobactin) while others varied in many mutants (e.g., xanthine) (Fig 1).

Proximity and similarity of metabolic changes across gene deletions

Generally, metabolites are expected to change in proximity of reactions that were directly affected by gene deletions because of perturbed metabolic fluxes (Fendt *et al*, 2010). We tested this

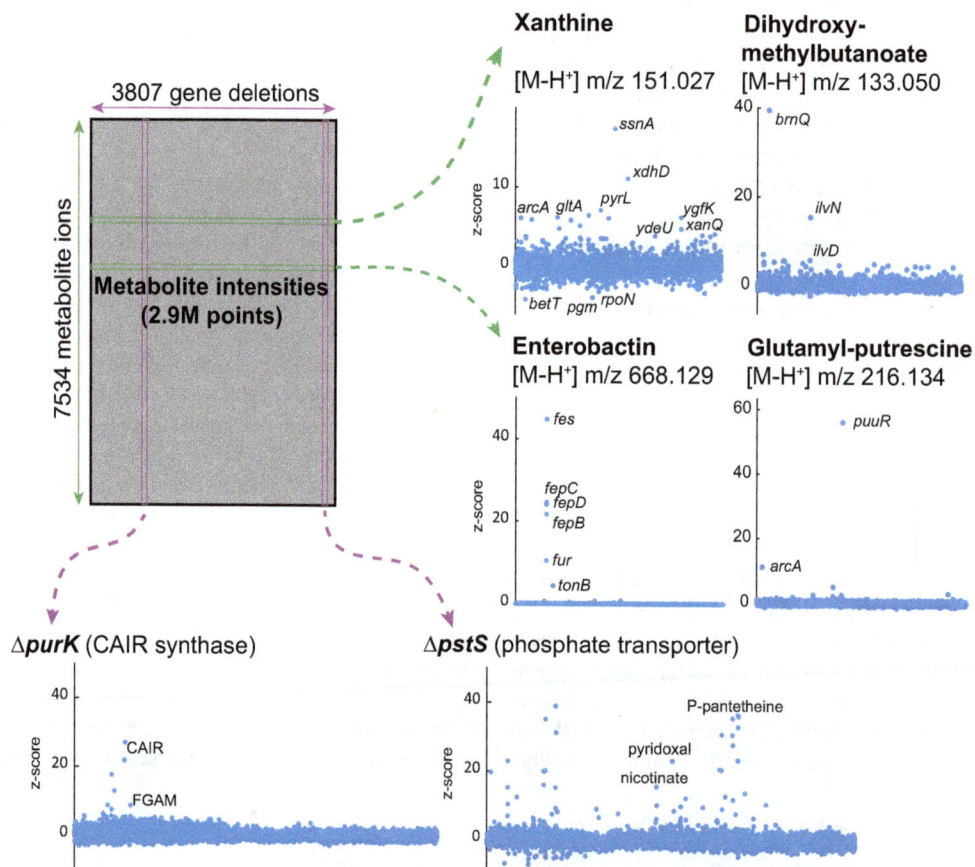

Figure 1. Gene–metabolite association matrix derived from metabolome analysis of single-gene deletion mutants.

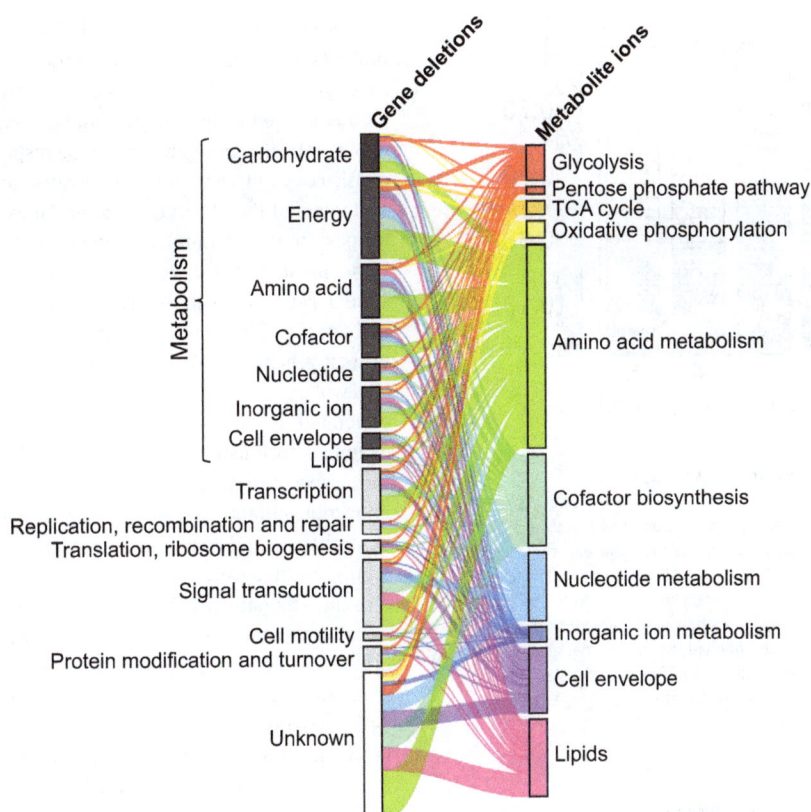

Figure 2. Gene–metabolite associations in the 0.1% most significant associations ranked by z-score, corresponding to an absolute z-score > ~5.
Genes were classified according to the Cluster of Orthologous Groups. Annotated ions were grouped according to the genome-scale metabolic model of *Escherichia coli* (Orth *et al*, 2011). Unknown ions were omitted. The ribbon width scales with the number of interactions.

functional association between metabolite changes and site of perturbation in mutants for which the metabolic effect can be predicted from known cellular networks. For enzyme deletions, we indeed observed a significant enrichment of metabolic changes in the immediate metabolic vicinity of the affected reactions (Figs 3, EV4 and EV5). We found a significant (P-value < 0.05) overrepresentation of enzyme deletions yielding the largest metabolic changes up to two enzymatic steps distance. Local metabolic changes are reabsorbed already at a distance of three, after which the reduced probability of finding the large metabolic changes remains constant.

Extending this locality analysis to larger models including the genome-scale metabolic network, the transcriptional regulatory network, and protein–protein interactions allowed us to probe the locality of measured metabolic responses of mutants lacking one of 166 transcription factors or 1,426 non-metabolic, non-enzymatic proteins, for example, those involved in regulation (e.g., PuuR) and membrane transport (e.g., BrnQ, Fig 1) (Andres Leon *et al*, 2009). In all cases, metabolite changes were enriched in the metabolic proximity of reactions that are known to be affected by the mutation (Fig EV6).

While this proximity analysis confirms the occurrence of local metabolic effects, a surprising result of this empirical genome–metabolome map is the many reproducible associations between genes and seemingly distant metabolites. Overall, genes involved in

coenzyme transport and metabolism, nucleotide transport and metabolism, signal transduction mechanisms, and transcription tend to induce widespread metabolic changes (Fig EV7). For example, we detected a strong interaction between malate and the *aro* and *pur* genes in chorismate and purine biosynthesis (Fig EV3B). Such distal changes could reflect functional interactions beyond the metabolic network topology. In most cases, the molecular links underlying such distal gene–metabolite associations remain elusive at this point, yet these interactions appear to be gene-specific rather than an unspecific consequences of mutant growth rate. One of the better understood molecular links is the strong association between the enterobactin-dependent iron uptake system (encoded by the *ent*, *fep*, and *fes* operons) and citrate/aconitate in the TCA cycle, highlighting the dichotomous role of iron as an essential modulator of aconitase activity (Varghese *et al*, 2003) and citrate as an iron chelator (Fig EV3B). These newly identified distal interactions provide leads to mechanisms of coordination across cellular pathways and functional modules.

The observed occurrence of widespread metabolic responses to specific genetic perturbations complicates the interpretation of metabolomics data, in particular from mutants lacking a gene with unknown function. We wondered whether the metabolome profiles are sufficient to characterize the genetic lesion regardless of prior information on network structure or metabolic proximity. We therefore determined the metabolome similarities between mutants

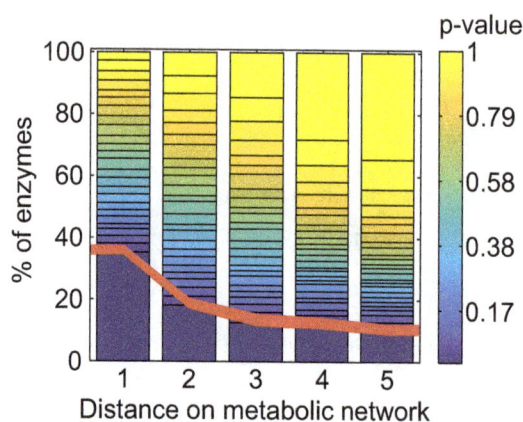

Figure 3. Locality analysis for enzyme deletions.
Distribution of empirical *P*-values (calculated from a permutation test) for enzyme deletions and the respective metabolites up to a distance of five enzymatic steps are plotted. In each enzyme-deletion mutant, the modified z-scores of metabolites at distance 1, 2, 3, 4 or 5 are compared to the average changes generated by selecting metabolites at random. For the five tested distances between enzyme and metabolites, the fraction of enzyme deletions yielding significant distance enriched metabolic changes are highlighted below the red line. For a substantial fraction of tested enzymes, the largest metabolic changes are observed within up to two enzymatic distance steps.

lacking genes encoding for (i) complementary partners of protein complexes or (ii) isoenzymes. Although not all genes are expected to be relevant under the tested condition, we were able to recover significant fractions of both types of functional relationships using the context likelihood of relatedness (CLR) algorithm (Faith *et al*, 2007) (Fig 4A). The best-reconstructed protein complexes belonged to multi-subunit enzymes (Fig 4B), but strong similarity was also found among functionally related processes, for example, between the *sdh* operon-encoded succinate dehydrogenase and the quinol oxidase (*cyo*) or between the NADH:ubiquinone oxidoreductase (*nuo*) and the fumarate reductase (*frd*) complexes. Thus, similar metabolic profiles indeed reflect functional dependencies between genes.

Our analyses demonstrate that metabolome profiles correctly capture known functional links between related genes and cellular processes, even if underlying gene–metabolite associations go beyond the canonical metabolic network. Consequently, our empirically constructed association map provides a unique resource for data-driven investigation of gene–metabolite interactions. In the following, we provide two representative applications to illustrate how the association map can predict metabolic function of orphan genes and identify potential genes mediating metabolic adaptations to external perturbations.

Prediction of orphan gene function

A particularly persisting problem in the post-genomic era is the 30–40% fraction of genes with unknown function (y-genes) even in the best-characterized species (Jaroszewski *et al*, 2009; Hanson *et al*, 2010). To demonstrate the use of the association map in detailing the roles of uncharacterized genes, we attempted to infer functions for the 1,274 y-genes (Hu *et al*, 2009) in our screen by searching for

similarities between the metabolome profiles of their deletion mutants and those of the 2,533 mutants lacking genes with known functions. Metabolic and other cellular functions were enriched among deleted genes eliciting similar responses for one quarter and about half of the y-gene mutants, respectively (Fig 5A). Based on consistency of enriched pathways among similar responding mutants and the observed differential ions in y-gene knockouts, we propose metabolism-related functions for 72 y-genes (Table EV3). Enrichment scores of enzymes or transcription factors among the similar genes were mutually exclusive, indicating that strong local and weaker global effects form two separate groups also among orphan genes (Fig 5B).

Two of our top y-gene predictions, *Yhhk* and *YgfY*, were annotated during the course of this study. These are representative examples to illustrate how functional hypotheses are derived from the association map and similarity analysis (Fig EV8A–C). *YhhK* knockout mutant exhibited strong similarity to several enzymes in pantothenate and coenzyme A biosynthesis (*panB*, *panC*, *panD*, and *panE*), with several differential metabolites in the coenzyme A biosynthesis pathway, including (R)-pantoate and (R)-pantothenate. Consistent with our data, YhhK was found to be an acetyltransferase activating PanD and recently annotated as PanZ (Nozaki *et al*, 2012; Stuecker *et al*, 2012). Similarly, metabolome profiling of *ygfY* deletion mutant featured common metabolic changes (Table EV3) to genes encoding subunits of succinate dehydrogenase (i.e., *sdhB* and *sdhC*). In addition, the top ranked differential ions observed in the *ygfY* mutant are dominated by succinate annotated ions (Table EV3). While enzymatic assays of YgfY using succinate as only substrate did not result in any detectable catalytic activity, when using crude cell lysate of the *ygfY* mutant instead of purified protein, we found succinate dehydrogenase activity to be drastically reduced in comparison with wild type (Fig EV8B). In addition, growth of the *ygfY* mutant was almost completely abolished when cells were grown on succinate as sole carbon source (Fig EV8C). Consistent with our findings, *ygfY* was recently reannotated to encoded sdhE, a FAD assembly factor for SdhA and FrdA (McNeil *et al*, 2012, 2013, 2014).

Different from the two previous examples, the functional role of YidK and YidR is still unknown. Metabolome-based predictions suggested a common role of these two genes in galactose and gluconate/galacturonate metabolism, respectively (Fig EV8D and E). Deletion mutants of functionally characterized genes with similar metabolic responses are mainly related to sugar catabolism (e.g., *treF*, *malS*, and *tktA*). Furthermore, most strongly affected metabolites include various sugar derivates including UDP-galactose, 3-deoxy-D-manno-2-octulosonate, or tagatose 6-phosphate (Table EV3). These observations are consistent with other large-scale datasets such as StringDB (Szklarczyk *et al*, 2011) suggesting a genomic context-based relation to galactose metabolism, or M3D showing good expression level correlation with genes involved in carbohydrate metabolism such as *fucR*, *fucI*, *fucK*, *xylAB*, and *rfaB* (Fig EV8E). Because of this metabolic similarity and because of its membrane-localized domain (Reizer *et al*, 1994), we hypothesized that YidK might be involved in the transport of some sugar-related compounds. Consistent with our prediction, Δ*yidK* mutant exhibited a growth phenotype when grown in galactose (Fig EV8D).

For *yidR* gene deletion, the strongest metabotype similarity was observed for *dgoT* (galactonate transporter), and we consistently

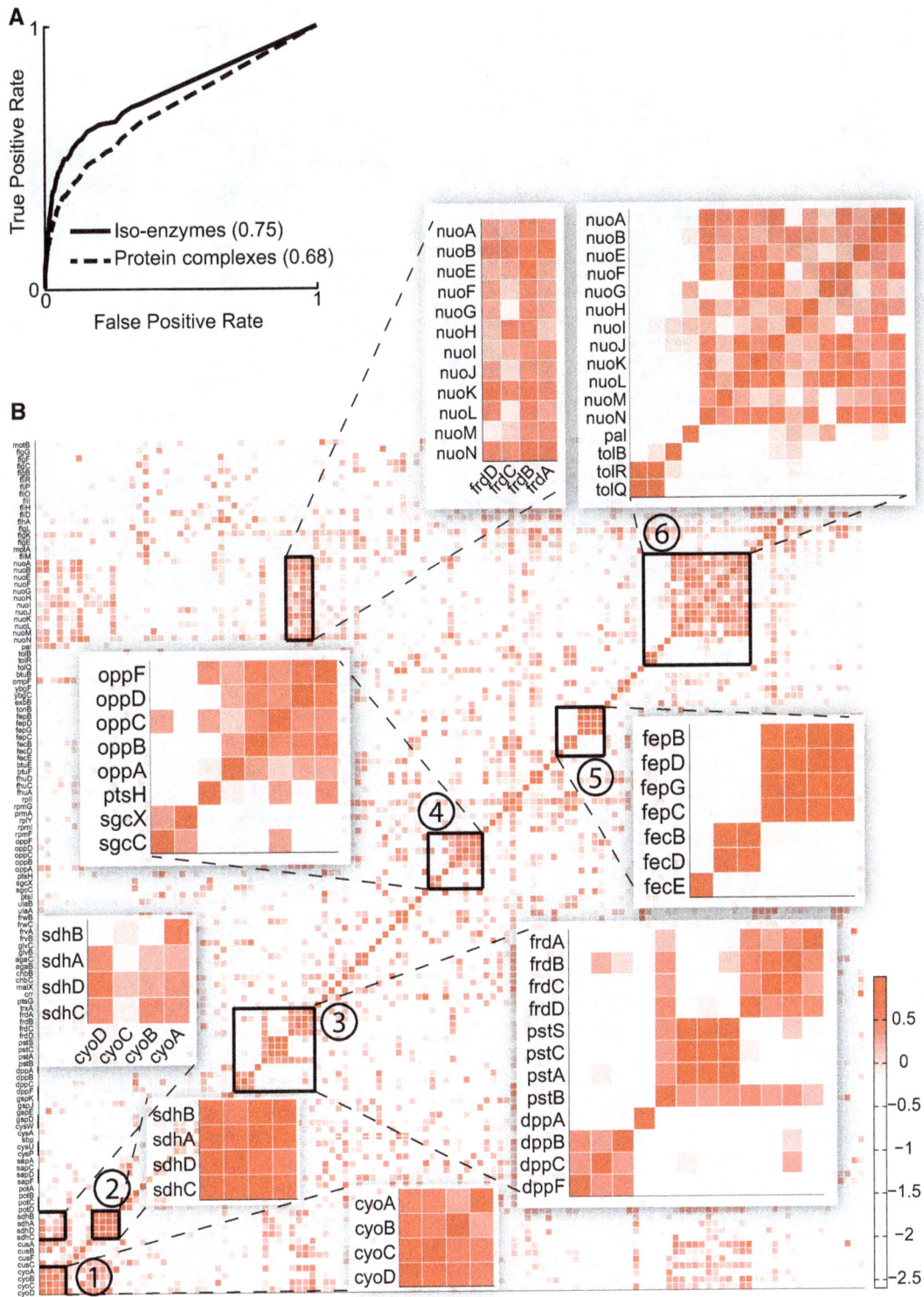

Figure 4. Network recovery for isoenzymes and protein complexes.

A   Recovery of enzyme function. Receiver operating characteristic curves obtained for the recovery of *Escherichia coli* isoenzymes and protein complexes based on the metabolome profiles recorded in single deletion mutants. The area under the curve (AUC) is reported in parentheses.

B   Consistent metabolic patterns in mutants of protein complex subunits. The heatmap shows the pair-wise similarity (e.g., CLR index) between metabolome response to gene deletions. Genes related to densely connected protein complexes consisting of at least three subunits are selected. We visualized the protein complex adjacency matrix, opportunely reordered. Magnified protein complexes are 1, succinate dehydrogenase; 2, cytochrome bo terminal oxidase; 3, fumarate reductase/phosphate ABC transporter/dipeptide ABC transporter; 4, murein tripeptide ABC transporter; 5, ferric enterobactin transport complex/ferric dicitrate transport system; 6, NADH: ubiquinone oxidoreductase/Tol–Pal cell envelope complex and high-scoring combinations thereof.

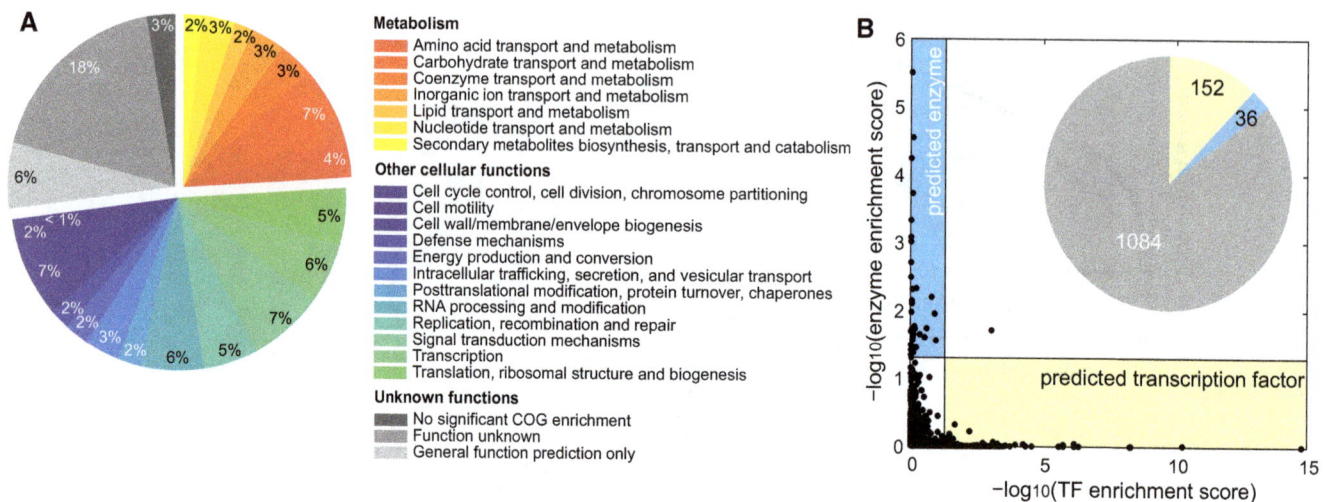

Figure 5. Enrichment of metabolic functions for orphan genes.

A Enrichment of metabolic functions (defined by Clusters of Orthologous Groups, COG) for each y-gene based on genes of known function with similar metabolome profiles, as determined by CLR.

B Mutually exclusive function prediction of orphan genes as either enzymes or transcription factors (TF). The inset represents the number of genes predicted to be TFs (yellow), enzymes (blue), or neither (gray). One gene was predicted to be both a TF and an enzyme.

observed changes in different metabolites related to carbohydrate metabolism, such as sedoheptulose-7-phosphate or D-sorbitol 6-phosphate. Additionally, four genes of the D-galactonate degradation pathway (*dgoT, dgoD, dgoR,* and *dgoK*) were among the top ten correlating genes in the M3D database (Fig EV8E). Growth phenotyping revealed that, indeed, the *yidR* deletion mutant displays a growth defect specifically on gluconate and galacturonate (Fig EV8D), confirming that *yidR* is functionally relevant for the assimilation of these compounds. The association map thus informs on metabolism-related gene functions and can provide leads for further functional genomics investigations. In most cases, however, the prediction relates to pathways because of the complex and widespread metabolic changes we observed. More specific prediction on specific reactions or enzyme class is only possible if the immediate substrates are detectable and characterized by a particularly strong response.

Predicting genes mediating the metabolic response to environmental perturbations

The rapidly growing number of large-scale metabolomics studies poses a serious challenge to data interpretation, in particular when complicated patterns emerge (Sevin *et al*, 2015). Here, we investigated whether complex metabolic patterns in response to an external perturbations can be functionally interpreted with our gene–metabolite association map. To this end, we recorded the metabolome response of wild-type *E. coli* to a variety of naturally relevant nutrient limitations (e.g., phosphate, sulfur, oxygen, or iron limitation) and stresses (osmotic and deoxycholate). By comparing the metabolic response to an external perturbing agent to the metabolome profiles of individual gene deletions, we tested the ability to recover genes mediating the adaptive response of *E. coli* to sudden environmental changes (Fig EV9). In agreement with our expectations, for nutritional limitations, we consistently identified

genes directly related to the utilization of the corresponding limiting nutrient, such as iron uptake and cysteine biosynthesis in the case of iron and sulfur limitations, respectively. Deoxycholate is found in bile acids and represents a common stress factor for bacteria in the gut, such as *E. coli* (Merritt & Donaldson, 2009). However, little is known about the underlying metabolic response. Based on our genomewide metabolome map, we identified seven single-gene deletion mutants in the compendium eliciting similar metabolic changes to those observed upon exposure to deoxycholate (Fig 6A). Notably, we found that six out of these seven mutants were substantially more resistant to deoxycholate inhibition compared to the wild-type *E. coli* strain, confirming the predicted functional role of these genes in mediating the stress response to deoxycholate (Fig 6B). Five of these beneficial mutants were disrupted in the uptake of ferric enterobactin (*fepB, fepC, fepD,* and *fepG*) or the release of iron into the cytosol (*fes*). While the function of these genes might suggest a direct role of iron, growth experiments under iron limitation and deoxycholate stress demonstrated that iron depletion *per se* is not sufficient to confer deoxycholate resistance (Fig EV10A). A common metabolic feature between these mutants and deoxycholate-stressed cells was the intracellular accumulation of enterobactin (Fig 1A). While enterobactin supplementation (Fig EV10B) did not affect deoxycholate sensitivity in the wild type (Fig 6C), mutants with disrupted enterobactin biosynthesis (Δ*entF*, Δ*entB*, Δ*entC*) were surprisingly insensitive to deoxycholate stress in minimal medium containing exogenous enterobactin (Fig 6C). Hence, while the underlying mechanism remains to be elucidated, these results support the predicted functional link between enterobactin biosynthesis and deoxycholate tolerance. Overall, interrogating the metabolic response to complex perturbations using our compendium of gene deletion profiles can reveal new and non-obvious hypothesis on the molecular players that mediate the adaptive response of *E. coli* to an external stimulus.

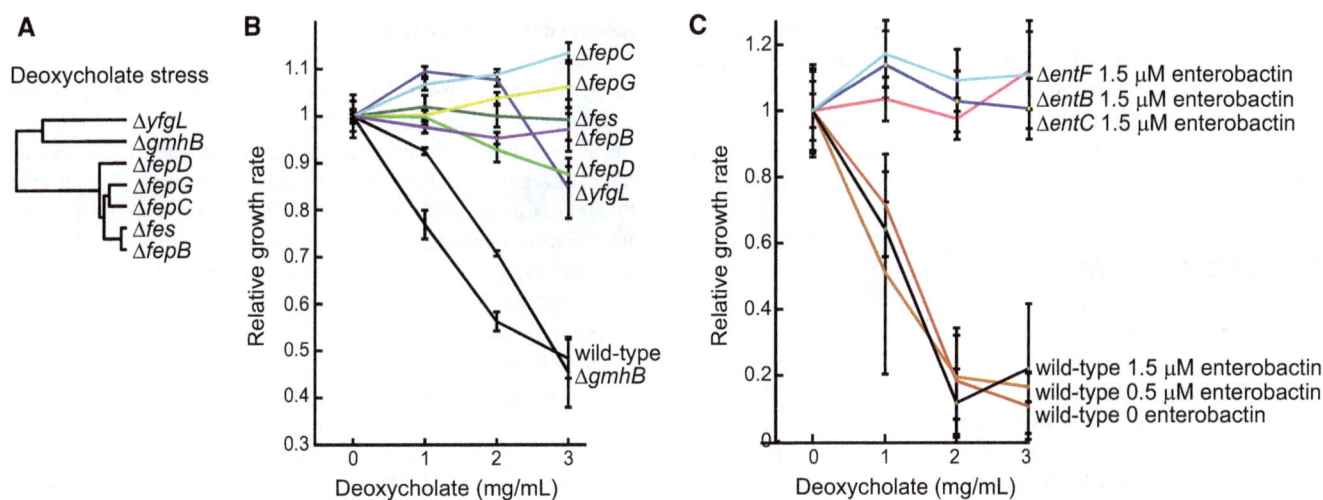

Figure 6. Predicting genes mediating the metabolic response to environmental perturbations.

A   Dendrogram representing genes with significant overlap of differential metabolites in the respective knockout and during growth in the presence of 10 mg/ml deoxycholate in wild-type *Escherichia coli*. Genes are hierarchically clustered based on their topological distance assessed by the minimum number of connecting reactions in the metabolic network.

B   Relative growth rates of wild-type *E. coli* and deletion mutants in glucose minimal medium supplemented with casein hydrolysate and deoxycholate. Error bars represent standard deviations from three biological replicates.

C   Relative growth rates of wild-type *E. coli* and enterobactin biosynthesis mutants in glucose minimal medium supplemented with enterobactin and deoxycholate. Error bars represent standard deviations from three biological replicates.

Discussion

Large-scale phenotypic screening of single- and double-gene deletion mutants has proven powerful in understanding gene functions, gene–gene interactions, and condition-dependent gene essentiality (Costanzo *et al*, 2010; Nichols *et al*, 2011). However, screening of large mutant libraries typically comes at the cost of measuring only one or few phenotypic traits (e.g., growth rate or viability). The implicit lack of fine-grained molecular or intracellular information, in turn, hinders attaining a detailed understanding on how interactions between genes, or between genes and the environment are established. Increasing the space of features that characterize functional consequences of genetic deletions on a molecular level improves the ability to disentangle the interplay between genes and to discern the regulatory architecture within cells (van Wageningen *et al*, 2010).

For the investigation of metabolism and its regulation, metabolomics provides a direct functional readout that convolutes the cell-wide interplay of enzyme activity and metabolites. Regulatory events that modify enzyme properties such as their abundance, localization, or kinetic properties eventually affect metabolite levels. Metabolomics offers the possibility of quickly profiling hundreds of metabolites involved in primary metabolism and thus characterizes the response of all key pathways that sustain growth and energy production. To this end, we generated the first comprehensive empirical map of gene–metabolite interactions by systematically measuring the relative changes in the abundance of hundreds of metabolite (ions) in 3,807 *E. coli* single-gene deletion mutants (Baba *et al*, 2006). This comprehensive compendium can be searched to find gene deletions that have the largest impact on a metabolite of interest, or vice versa, to find which metabolites are affected upon a specific gene deletion. This gene–metabolite interaction map complements classical genomewide phenotypic screens and is a valuable resource to mechanistically interpret macroscopic phenotypes.

An important result of our analysis of the gene–metabolite map is that marked metabolite changes can occur distant from the genetic lesion, even when enzymes with known catalytic functions were deleted and no global growth defect was detected. Hence, the topology and connectivity defined by the metabolic network are not sufficient to explain and predict the impact of gene deletions on the overall cell/metabolic phenotype. While these distal gene–metabolite interactions remain largely elusive to explain at this point, they can be interpreted as metabolic fingerprints of gene function and in our study were used to (i) predict the enzymatic functions of y-genes even in the absence of growth phenotypes and (ii) establish a metabolic link between gene and their functional role in mediating the cell response to environmental perturbations.

Altogether, we hypothesized metabolic function for 72 y-genes, of which YidK and YidR were experimentally validated. Our metabolome compendium also provides a key to interpret metabolic changes induced upon perturbations other than gene deletions. By establishing an indirect link between common metabolic changes induced by gene deletions and external perturbations, we predicted non-obvious mediators of the cellular response to deoxycholate treatment, which involved genes in iron metabolism.

The two above examples of functional gene annotation and predicting genes involved in cellular responses demonstrate the utility of the association map for generating novel lead hypotheses on the functional roles of genes in metabolism. Because of the large number of tested genetic lesions and covered metabolites, the map constitutes a unique resource to inspire new and less conventional approaches to predict the mode of action of genetic and environmental perturbations. Moreover, the large number of newly revealed gene–metabolite associations paves the road to explore so far unknown functional and regulatory interactions beyond those represented in current genome-scale metabolic models (Feist *et al*, 2009; Orth *et al*, 2011).

Materials and Methods

Chemicals

Water, methanol and 2-propanol, all CHROMASOLV LC-MS grade, buffer additives for online mass referencing, media, and sample preparation chemicals at the highest available purity were purchased from Sigma-Aldrich and Agilent Technologies. Pure water for extraction and resuspension with an electric resistance greater than 16 MΩ was obtained from a NANOpure purification unit (Barnstead, Dubuque, IA, USA).

Biological samples

Escherichia coli wild-type and 4,320 deletion mutants (Table EV1) from the KEIO knockout collection (Baba *et al*, 2006) were grown on glucose minimal medium supplemented with casein hydrolysate containing (per liter): 4 g glucose, 2 g N-Z Case Plus, 7.52 g $Na_2HPO_4\cdot2H_2O$, 3 g KH_2PO_4, 0.5 g NaCl, 2.5 g $(NH_4)_2SO_4$, 14.7 mg $CaCl_2\cdot2H_2O$, 246.5 mg $MgSO_4\cdot7H_2O$, 16.2 mg $FeCl_3\cdot6H_2O$, 180 µg $ZnSO_4\cdot7H_2O$, 120 µg $CuCl_2\cdot2H_2O$, 120 µg $MnSO_4\cdot H_2O$, 180 µg $CoCl_2\cdot6H_2O$, 1 mg thiamine·HCl. Culture volumes of 1 ml were incubated in 96-deep well plates at 37°C with shaking at 300 rpm. Growth was followed via absorbance at 600 nm measured at four time-points, and all samples were harvested during mid-exponential growth phase by centrifugation for 10 min at 0°C and 2,200 *g*. Cell pellets were immediately extracted with 150 µl preheated water containing 2 µM reserpine and 2 µM taurocholic acid for 10 min at 80°C and occasional vortexing. This extraction broth was centrifuged for 10 min at 0°C and 2,200 *g*, and supernatants were stored at −80°C until further analysis.

Flow-injection analysis—TOF MS

The analysis was performed on a platform consisting of an Agilent Series 1100 LC pump coupled to a Gerstel MPS2 autosampler and an Agilent 6520 Series Quadrupole TOF mass spectrometer (Agilent, Santa Clara, CA, USA) as described previously (Fuhrer *et al*, 2011). The flow rate was 150 µl/min of mobile phase consisting of isopropanol:water (60:40, v/v) buffered with 5 mM ammonium carbonate at pH 9 for negative mode and methanol:water (60:40, v/v) with 0.1% formic acid at pH 3 for positive mode. For online mass axis correction, 2-propanol (in the mobile phase) and taurocholic acid or reserpine were used for negative mode or for positive mode, respectively. Mass spectra were recorded in profile mode from m/z 50 to 1,000 with a frequency of 1.4 s for 2 × 0.48 min (double injection) using the highest resolving power (4 GHz HiRes). Source temperature was set to 325°C, with 5 l/min drying gas and a nebulizer pressure of 30 psig. Fragmentor, skimmer, and octopole voltages were set to 175 V, 65 V, and 750 V, respectively.

Spectral data processing and annotation

All steps of mass spectrometry data processing and analysis were performed with MATLAB (The Mathworks, Natick, MA, USA) using functions embedded in the Bioinformatics, Statistics, Database, and Parallel Computing toolboxes as described previously (Fuhrer *et al*, 2011). Peak picking was done for each sample once on the total profile spectrum obtained by summing all single scans recorded over time, and using wavelet decomposition as provided by the Bioinformatics toolbox. In this procedure, we applied a cutoff to filter peaks of less than 500 ion counts (in the summed spectrum) to avoid detection of features that are in any case too low to deliver statistically meaningful insights. Centroid lists from samples were then merged to a single matrix by binning the accurate centroid masses within the tolerance given by the instrument resolution (about 0.002 amu at m/z 300). The resulting matrix lists the intensity of each mass peak in each analyzed sample. An accurate common m/z was recalculated with a weighted average of the values obtained from independent centroiding. Because mass axis calibration is applied online during acquisition, no m/z correction was applied during processing to correct for potential drifts. After merging, 3,169 and 4,365 common ions were obtained for negative and positive mode, respectively, which were annotated based on accurate mass using 3 mDa tolerance (Tables EV1 and EV2). Annotation was based on assumption that $-H^+$, $+OH^-$, and $+Cl^-$ are the possible ionization options for negative mode, and $+H^+$, $+K^+$, and $+Na^+$ for positive mode. Additional commonly observed adducts were considered as described previously in detail (Fuhrer *et al*, 2011): $^{12}C_1\text{-}^{13}C_1$, $^{12}C_2\text{-}^{13}C_2$, $-H^++Na^+$, $-H^++K^+$, $-H_2O$, $-CO_2$, $-NH_3$, $-HPO_3$, $-H_3PO_4$, $+H_2PO_4Na$, $+H_2PO_4K$, $+HPO_4Na_2$, $+HPO_4K_2$, $+(H_2PO_4Na)_2$, $+(H_2PO_4K)_2$, $+(H_2PO_4)_2NaH$, and $+(H_2PO_4)_2KH$. Putative coverage of *E. coli* metabolism is shown in Fig EV1.

Physiology

Growth rates were calculated as the slopes of linear fits to log-transformed experimentally determined OD values. Approximate harvest OD values were estimated by the calculated growth rates and harvest time. Identification of extremely sick mutants was based on a 0.1 OD cutoff on the last two OD values as the respective calculated growth rates were not reliable. Of the 3,847 unique gene knockouts, 21 extremely sick mutants were omitted from data analysis: *aceF, carA, eptB, frmA, guaA, guaB, hflD, lpd, parC, purA, rfaI, rfaP, rfaS, rplA, ybaB, ybeY, yiQ, yjjG, yqaB, yrdA,* and *ytfE*. Furthermore, 16 mutants were removed due to injection errors during mass spectrometry analysis: *ybdG, stfQ, ybiB, mntR, ybgJ, ybaK, ybjJ, yahD, ylaB, yqcA, wcaJ, ybhR, yliJ, yaiB, ybhN,* and *ykgF*. The remaining 3,810 unique gene knockouts were further analyzed.

Data normalization and calculation of differential ions

Preprocessing of raw mass spectrometry data consists of four steps: (i) correction of intensity drift throughout sequential injections within one plate by a low pass filter over a moving window of five injections, (ii) correction of intensity drifts across plates due to regular instrument cleaning procedures, extraction effects or long-term ionization drifts by normalizing the mean of every ion to be equal over all plates, (iii) correction of intensities for harvest OD dependencies using a locally weighted scatterplot smoothing (LOWESS). This procedure was used to perform a model-free estimation of the ion-specific dependency with harvest ODs. The local least square regression was employed to normalize for OD effects as follows:

$$\text{intensity}_{\text{ion i}} = \text{raw intensity}_{\text{ion i}} - \text{trend}_{\text{ion i}} + \text{median intensity}_{\text{ion i}}$$

(iv) a modified z-score is then used to select for significantly and specifically affected ions according to:

$$z\text{-score}_{(i,j)} = \frac{\text{intensity}_{(i,j)} - \text{median}_{(i,\text{all})}}{\text{std}_{(i,\text{all})}}$$

where i and j denote ion and samples, respectively, and median as well as standard deviation (std) refers to all intensities of ion i in the entire dataset. Modified z-scores referring to technical and biological replicates are summarized into a unique median modified z-score, and three additional mutants were removed from the dataset having inconsistent z-score among the replicates (flhC, hrpB, and yhfW), resulting in a final data set with 3,807 mutants. Notably, no internal standards were used because they are not suited to normalize thousands of mostly unknown chemical entities.

Reproducibility between biological replicates

For each knockout mutant, two different clones contained on separate plates in the library were separately processed on different days. We assessed reproducibility of the metabolome profiles between the two biological replicates by calculating the absolute difference between z-scores (Fig EV2A). The resulting distribution of z-score differences shows that above a z-score of 2.765, we have 1% probability of having false positives.

Network recovery from pair-wise similarity

Pair-wise similarity among normalized metabolome profiles for each tested mutant was calculated by the CLR approach (Faith et al, 2007), which estimates a similarity score for each pair of gene profiles by comparing a joint likelihood measure based on mutual information. Isoenzymes were identified by the Orth genome-scale model (Orth et al, 2011). Protein complexes were obtained from EciD (Andres Leon et al, 2009). We then evaluated the overlap between the above known interaction graphs and the inferred network of similarities derived from metabolome profiles using receiver operating characteristic (ROC) curves. Briefly, this framework allows us to span the entire range of cutoff values for the estimated pair-wise similarities to describe the trade-off between sensitivity and the false-positive rate (FPR) in the network reconstruction. Sensitivity is defined as the fraction of the known interactions, which were also inferred by our similarity score (true positive, TP) and the total number of known interactions, given by the sum of TP and true negatives (TN) [sensitivity = TP/(TP + FN)], while FPR is the fraction of incorrect inferred relationships (false positive, FP) over the sum of all known negative interactions (given by the sum of FP and true negative, TN) [FPR = FP/(FP + TN)]. Finally, the area under the curve (AUC) was used to give an estimate of the quality of the reconstruction. For Fig 4A, the

analysis was performed by restricting the calculation of TP and FP interactions only among 46% of genes with at least one significant metabolic change (i.e., silent genes were excluded, Fig EV7C).

Locality analysis

A genome-scale network model of E. coli metabolism was used to determine the distance between each enzyme–metabolite pair. The resulting pair-wise distance matrix between metabolic enzymes and metabolites was estimated by means of the minimum number of reactions separating the two in a non-directional network. All highly connected metabolites were removed prior to calculation. To assess whether largest metabolic changes are statistically more probable in the proximity of the deleted enzymes (Fig 3), we used a permutation test. For each enzyme deletion, we calculated the sum of absolute changes of metabolites directly linked to the enzyme (i.e., substrates/products) corresponding to distance 1, up to metabolites at five enzymatic steps away from deleted gene. For each tested enzyme, the observed statistic (S_{obs}) was compared with a permuted one (S_{perm}) obtained by randomizing 1,000 times the original distance matrix. P-value was empirically estimated as follows:

$$P\text{-value}_{g,d} = \frac{(S_{\text{perm}} \geq S_{\text{obs}})}{1,000}$$

where g is the selected gene and d is the distance (from 1 to 5).

This first statistical analysis revealed a tendency of enzyme deletions, often within amino acids biosynthetic pathways, to exhibit larger metabolic changes within one to two enzymatic steps (Figs 3 and EV4). This result enabled us to generalize and extend our analysis to non-metabolic genes using a locality scoring function as follows.

For transcription factors (TFs), we augmented the metabolic network with the connections between TFs and their known metabolic enzyme targets extracted from RegulonDB (Salgado et al, 2013). Similarly, to calculate the distance between non-metabolic proteins and detectable compounds, we included known protein–enzyme interactions from the EciD database (Andres Leon et al, 2009) (Fig EV6). The distance for all pairs of genes and detectable compounds was calculated as the minimum number of edges (regulatory links) connecting the non-metabolic proteins to enzymes, and enzymes to metabolites (reactions). We then implemented a locality scoring function weighted for distance as follows:

$$S(g_i) = \frac{\sum_{m=1}^{M} D_{i,m}^{-2} \cdot (1 - z.pval_{i,m}) \cdot |Z_{i,m}|}{\sum_{m=1}^{M} D_{i,m}^{-2} \cdot (1 - z.pval_{i,m})}$$

$$S_{\text{rand}}^{k}(g_i) = \frac{\sum_{m=1}^{M} D_{\text{rand}_{i,m}}^{k-2} \cdot (1 - z.pval_{i,m}) \cdot |Z_{i,m}|}{\sum_{m=1}^{M} D_{\text{rand}_{i,m}}^{k-2} \cdot (1 - z.pval_{i,m})}$$

$$P.value(g_i) = \frac{\sum_{k} (S_{\text{rand}}^{k} \geq S)}{K}$$

where for each gene (encoding for enzymes, TFs, or proteins) g_i, a weighted mean over corresponding $Z_{i,m}$ (Z-scores of metabolome profiles corresponding to gene i and metabolite m) is computed. Weights are a function of the inverse of the squared distance between the gene i and the metabolite m and the z-test P-value associated with the ion measurements (measure of confidence of metabolite measurements). We performed a permutation test with randomly shuffling of the distance matrix D (D_{rand}) K times (i.e., $K = 100$) to assign a significance value to each gene.

Pathway enrichment analysis

Enzyme deletions and annotated metabolites were grouped according to the corresponding metabolic pathways as defined in the genome-scale model of Orth *et al* (2011). Only the largest 0.1% of metabolic changes were considered. The statistical significance of enzymes belonging to a metabolic pathway to elicit metabolic changes in each different metabolic pathway (Fig EV5) was assessed by a permutation test, where the metabolite and gene orders are randomly shuffled and the number of randomly assigned metabolite with a relative change in absolute abundance (modified z-score) > 5 is counted. Significance of the observed statistics was assessed by counting how many times randomly assigned metabolites exceed the originally observed metabolic changes in each metabolic pathway.

Categorization of mutants based on number of differential ions and enrichment of cellular functions

We applied stringent absolute 0.1 percentile (0.1%) cutoff to the distribution of the z-scores to select significantly differential ions. Choosing a typical P-value cutoff of 0.05 for 0.1 percentile cutoff leads to four distinct scenarios: 0 hits for silent mutations, a region where we get less hits than expected (1–4 hits), a region with significant probability to get hits (5–10 hits), and a region where we get more hits than expected (> 10 hits) (Fig EV7B). Based on these distributions, we classified the strains as mutants being silent, or having rare, moderate, and global effects, respectively (Fig EV7C). Enrichment significance (P-value) for of Cluster of Orthologous Groups (Tatusov *et al*, 2003; Hu *et al*, 2009) within the different categories silent, rare, moderate, and global was derived by hypergeometric probability density function for each cellular function category for the 0.1 percentile cutoff (Fig EV7D).

Potential function predictions for orphan genes—calculation of prediction score

Metabolome profile similarity based on CLR enabled us to relate genes with similar catalytic activity (Fig 4A). Hence, we envisaged the possibility to employ such similarity patterns to infer potential catalytic properties of functionally uncharacterized proteins. Moreover, the observed locality of metabolic responses can suggest potential reactants of eventually catalyzed reactions. In order to predict the enzymatic activity of a functionally uncharacterized protein, we thus combined the information of both CLR and differential annotated ions. To allow for contribution of more subtly affected differential ions, the percentile cutoff on the z-score was relaxed to 0.5% (corresponding to 0.01 P-value based on the probability distribution estimated in Fig EV2A). First, for a particular candidate gene, we selected among the top similar gene based on CLR (Table EV3—CLR hits, also see legend for prediction table

supplied as Word file) the annotated enzymes and extrapolated from overrepresented linked metabolites (i.e., their substrates and products enriched with P-value < 0.01, Table EV3—MET prediction hits, top hit and top score) when at least two enzymes were significantly similar. The genes with the highest similarity scores (i.e., highest CLR index) were selected using a threshold corresponding to a 5% FPR in the recovery of iso-enzyme network (Fig 4A). Columns CLR—top hit and top score in Table EV3 report the most similar gene by CLR and the corresponding CLR index value, respectively. Second, we tested for each annotated metabolite how likely it is that perturbations yielding most significant changes of the annotated ions were associated with deletion of those enzymes directly capable of converting the selected metabolite. For this test, we used a ROC curve analysis and ranked deletion mutants based on the z-score matrix. From this analysis, we derived for each ion-annotation an estimate of the AUC index representing annotation confidence (Table EV4). The changes in ion abundance (i.e., z-score) were then weighted by corresponding AUC indices (Table EV3, DIFF IONS hits, top hit, and top score). Third, we identified overrepresented metabolites linked to annotated enzymes either as substrates or products of catalyzed reactions and determined their intersection with metabolites identified as annotated differential ions (based on 0.5% percentile cutoff, Table EV3, MET Prediction overlap). Finally, the sum of AUC weighted z-scores of overlapping metabolites (Table EV3, CLR—prediction score) was used to rank all y-genes according to consistency between CLR-based predictions and observed differential annotated ions to prioritize further validations. Table EV3 also includes enrichment of KEGG pathways using hypergeometric probability tests of annotated similar genes based on differential ions, similar genes, or both (KEGG PATHWAYS by CLR, hits, top hits and top score). We also provide hyperlinks to three databases (i.e., Ecocyc, KEGG, and PortEco, Table EV3—links), and of enriched COG terms among similar genes (Table EV3—COG hits, top hit, and top score).

Overexpression and purification of proteins encoded by y-genes

A genomewide *E. coli* gene overexpression library (Kitagawa *et al*, 2005) served as resource for expression and purification of proteins for functional validation. This library consists of 4,267 clones of *E. coli* K-12 AG1 each containing a plasmid encoding one ORF as a N-terminal His$_6$-fusion under control of the isopropyl-β-D-thiogalactopyranoside (IPTG)-inducible lac-promoter as well as chloramphenicol-acetyl-transferase providing chloramphenicol resistance as dominant selection marker. The selected expression strains were grown in 500 ml shake flasks (50 ml culture volume) on an orbital shaker for 16 h at 37°C and 300 rpm. Overexpression was performed in LB medium (10 g/l yeast extract, 10 g/l tryptone, 5 g/l NaCl at pH 7.5) supplemented with 5 g/l glucose, 20 µg/ml chloramphenicol, and 100 µg/ml IPTG. Cells were harvested by centrifugation at 2,200 g and 4°C for 10 min. The supernatant was discarded, and the pellet was washed once with 500 µl (2 ml cultures) or 10 ml (50 ml cultures) 0.9% NaCl + 1 mM MgCl$_2$. For the 50 ml cultures, cells were resuspended in 4 ml of 100 mM Tris-buffer at pH 7.5 supplemented with 5 mM MgCl$_2$, 20 mM imidazole, 2 mM DTT, and 4 mM PMSF and lysed by three passages through a French pressure cell (Thermo Fisher) at 1,000 psig. Proteins were purified by immobilized metal-ion affinity chromatography using a sepharose resin with nitriotriacetic acid groups chelating Ni^{2+} ions. Imidazole and salts from the elution buffer were removed by ultrafiltration using 10 kDa MWCO centrifugal filters in 4 ml tube (50 ml cultures) or 96-well plate format (2 ml cultures) (Millipore). Proteins were resuspended in 2 mM Tris pH 7.4 and 1 mM $MgCl_2$ and stored at 4°C until further usage. Expression and purity was checked by standard Bradford assays and SDS–PAGE analysis.

Predicting gene mediators of environmental perturbations

For deoxycholate, the IC_{50} (concentration inhibiting growth by 50%) of 10 mg/l was determined by automated microtiter-scale cultivations with a gradient of 0–20 mg/l deoxycholate. Limiting concentrations of 35 μM sulfate ($MgSO_4$), 100 nM ferric iron ($FeCl_3$), and 120 μM phosphate (KH_2PO_4) was determined analogously. Perturbation experiments were performed in 5 ml cultures in glucose-M9 minimal medium without casein hydrolysate at 37°C and 300 rpm. Anaerobic cultures were performed at 37°C under N_2 atmosphere in N_2 sparged septum flasks. 1 ml of cells at OD 0.5 was harvested in triplicates by fast filtration and extracted in 4 ml 40:40:20 acetonitrile:MeOH: ddH$_2$O (vol %) for 2 h at −20°C. Extracts were dried under vacuum and resuspended in 150 μl ddH$_2$O. Metabolomics measurements were performed by flow-injection TOF-MS as described previously (Fuhrer *et al*, 2011), and for osmotic stress (500 mM NaCl), previously published data were used (Sevin & Sauer, 2014).

Metabolites undergoing significant changes upon an environmental perturbations compared to mock treated cells were defined as passing an absolute fold-change cutoff of 2 at a false-discovery rate of 0.01 relative to unperturbed cells. Subsequently, the overlap between metabolites significantly affected by an environmental perturbation and metabolites influenced by single-gene deletions (considering only the largest 0.1% of observed metabolic changes) was systematically determined. For each gene deletion, the sum of relative changes among the set of metabolites (Ω) commonly affected by each environmental perturbation (E) was calculated according to:

$$S_g = \sum_{\Omega \cap E} Z\text{-score}_i$$

Significance of the overlap was estimated by means of P-values obtained using a permutation test:

$$S_g^{\text{perm}} = \sum_{\Omega_{\text{perm}} \cap E} Z\text{-score}_i$$

$$P\text{-value}_g = \frac{(S_g^{\text{perm}} \geq S_g)}{1,000}$$

where S_{perm} is the permuted score obtained by selecting at random the set of metabolites affected upon gene deletion (Ω_{perm}). Randomization was performed 1,000 times for P-value estimation.

Growth assay of predicted gene knockouts under deoxycholate stress

Escherichia coli wild-type strain and single-gene deletion mutants with a metabolic phenotype consistent with that of cells under deoxycholate stress (Fig 6A) were cultivated at microtiter scale at

different deoxycholate concentrations (0, 1, 2, 3 mg/ml) in glucose-M9 minimal medium with casein hydrolysate (Fig 6B). Maximum growth rates during exponential growth phase were calculated from triplicate cultivations.

Growth assay of *Escherichia coli* wild-type strain and enterobactin biosynthesis mutant strains with enterobactin supplementation and deoxycholate stress

Escherichia coli wild-type and mutant strains were cultivated at microtiter scale at different deoxycholate concentrations (0, 1, 2, 3 mg/ml) with varying enterobactin concentration (0, 0.5, 1.5 μM) in glucose-M9 minimal medium (Figs 6C and EV10B). Maximum growth rates during exponential growth phase were calculated from triplicate cultivations.

Growth assay of *Escherichia coli* wild-type strain under iron limitation and deoxycholate stress

Escherichia coli wild-type strain was cultivated at microtiter scale at different deoxycholate concentrations (0, 1, 2, 3 mg/ml) with varying iron concentration (0.05, 0.15, 0.5, 1, 5, 10, 25, 50 μM) in glucose minimal medium (Fig EV10A). Maximum growth rates during exponential growth phase were calculated from triplicate cultivations.

Acknowledgements

This work was in part supported by an ETH postdoctoral fellowship to MZ.

Author contributions

TF, DCS, and MZ conceived and designed the study, performed the experimental work and computational analysis, and wrote the manuscript. NZ performed computational analysis, supervised the study, and wrote the manuscript. US supervised the study and wrote the manuscript. All authors read and approved the final paper.

References

Andres Leon E, Ezkurdia I, Garcia B, Valencia A, Juan D (2009) EcID. A database for the inference of functional interactions in *E. coli*. *Nucleic Acids Res* 37: D629–D635

Baba T, Ara T, Hasegawa M, Takai Y, Okumura Y, Baba M, Datsenko KA, Tomita M, Wanner BL, Mori H (2006) Construction of *Escherichia coli* K-12 in-frame, single-gene knockout mutants: the Keio collection. *Mol Syst Biol* 2: 2006.0008

Costanzo M, Baryshnikova A, Bellay J, Kim Y, Spear ED, Sevier CS, Ding H, Koh JL, Toufighi K, Mostafavi S, Prinz J, St Onge RP, VanderSluis B, Makhnevych T, Vizeacoumar FJ, Alizadeh S, Bahr S, Brost RL, Chen Y, Cokol M *et al* (2010) The genetic landscape of a cell. *Science* 327: 425–431

Faith JJ, Hayete B, Thaden JT, Mogno I, Wierzbowski J, Cottarel G, Kasif S, Collins JJ, Gardner TS (2007) Large-scale mapping and validation of *Escherichia coli* transcriptional regulation from a compendium of expression profiles. *PLoS Biol* 5: e8

Feist AM, Herrgard MJ, Thiele I, Reed JL, Palsson BO (2009) Reconstruction of biochemical networks in microorganisms. *Nat Rev Microbiol* 7: 129–143

Fendt SM, Buescher JM, Rudroff F, Picotti P, Zamboni N, Sauer U (2010)

Tradeoff between enzyme and metabolite efficiency maintains metabolic homeostasis upon perturbations in enzyme capacity. *Mol Syst Biol* 6: 356

Fuhrer T, Heer D, Begemann B, Zamboni N (2011) High-throughput, accurate mass metabolome profiling of cellular extracts by flow injection-time-of-flight mass spectrometry. *Anal Chem* 83: 7074–7080

Ghazalpour A, Bennett BJ, Shih D, Che N, Orozco L, Pan C, Hagopian R, He A, Kayne P, Yang WP, Kirchgessner T, Lusis AJ (2014) Genetic regulation of mouse liver metabolite levels. *Mol Syst Biol* 10: 730

Hanson AD, Pribat A, Waller JC, de Crecy-Lagard V (2010) 'Unknown' proteins and 'orphan' enzymes: the missing half of the engineering parts list - and how to find it. *Biochem J* 425: 1–11

Heinemann M, Sauer U (2010) Systems biology of microbial metabolism. *Curr Opin Microbiol* 13: 337–343

Hu P, Janga SC, Babu M, Diaz-Mejia JJ, Butland G, Yang W, Pogoutse O, Guo X, Phanse S, Wong P, Chandran S, Christopoulos C, Nazarians-Armavil A, Nasseri NK, Musso G, Ali M, Nazemof N, Eroukova V, Golshani A, Paccanaro A *et al* (2009) Global functional atlas of *Escherichia coli* encompassing previously uncharacterized proteins. *PLoS Biol* 7: e96

Ishii N, Nakahigashi K, Baba T, Robert M, Soga T, Kanai A, Hirasawa T, Naba M, Hirai K, Hoque A, Ho PY, Kakazu Y, Sugawara K, Igarashi S, Harada S, Masuda T, Sugiyama N, Togashi T, Hasegawa M, Takai Y *et al* (2007) Multiple high-throughput analyses monitor the response of E. coli to perturbations. *Science* 316: 593–597

Jaroszewski L, Li Z, Krishna SS, Bakolitsa C, Wooley J, Deacon AM, Wilson IA, Godzik A (2009) Exploration of uncharted regions of the protein universe. *PLoS Biol* 7: e1000205

Kanehisa M, Goto S, Sato Y, Furumichi M, Tanabe M (2012) KEGG for integration and interpretation of large-scale molecular data sets. *Nucleic Acids Res* 40: D109–D114

Kitagawa M, Ara T, Arifuzzaman M, Ioka-Nakamichi T, Inamoto E, Toyonaga H, Mori H (2005) Complete set of ORF clones of *Escherichia coli* ASKA library (a complete set of E. coli K-12 ORF archive): unique resources for biological research. *DNA Res* 12: 291–299

Link H, Kochanowski K, Sauer U (2013) Systematic identification of allosteric protein-metabolite interactions that control enzyme activity in vivo. *Nat Biotechnol* 31: 357–361

McNeil MB, Clulow JS, Wilf NM, Salmond GP, Fineran PC (2012) SdhE is a conserved protein required for flavinylation of succinate dehydrogenase in bacteria. *J Biol Chem* 287: 18418–18428

McNeil MB, Iglesias-Cans MC, Clulow JS, Fineran PC (2013) YgfX (CptA) is a multimeric membrane protein that interacts with the succinate dehydrogenase assembly factor SdhE (YgfY). *Microbiology* 159: 1352–1365

McNeil MB, Hampton HG, Hards KJ, Watson BN, Cook GM, Fineran PC (2014) The succinate dehydrogenase assembly factor, SdhE, is required for the flavinylation and activation of fumarate reductase in bacteria. *FEBS Lett* 588: 414–421

Merritt ME, Donaldson JR (2009) Effect of bile salts on the DNA and membrane integrity of enteric bacteria. *J Med Microbiol* 58: 1533–1541

Nichols RJ, Sen S, Choo YJ, Beltrao P, Zietek M, Chaba R, Lee S, Kazmierczak KM, Lee KJ, Wong A, Shales M, Lovett S, Winkler ME, Krogan NJ, Typas A, Gross CA (2011) Phenotypic landscape of a bacterial cell. *Cell* 144: 143–156

Nozaki S, Webb ME, Niki H (2012) An activator for pyruvoyl-dependent l-aspartate alpha-decarboxylase is conserved in a small group of the gamma-proteobacteria including *Escherichia coli*. *Microbiologyopen* 1: 298–310

Orth JD, Conrad TM, Na J, Lerman JA, Nam H, Feist AM, Palsson BO (2011) A comprehensive genome-scale reconstruction of *Escherichia coli* metabolism–2011. *Mol Syst Biol* 7: 535

Reizer J, Reizer A, Saier MH Jr (1994) A functional superfamily of sodium/solute symporters. *Biochim Biophys Acta* 1197: 133–166

Salgado H, Peralta-Gil M, Gama-Castro S, Santos-Zavaleta A, Muniz-Rascado L, Garcia-Sotelo JS, Weiss V, Solano-Lira H, Martinez-Flores I, Medina-Rivera A, Salgado-Osorio G, Alquicira-Hernandez S, Alquicira-Hernandez K, Lopez-Fuentes A, Porron-Sotelo L, Huerta AM, Bonavides-Martinez C, Balderas-Martinez YI, Pannier L, Olvera M *et al* (2013) RegulonDB v8.0: omics data sets, evolutionary conservation, regulatory phrases, cross-validated gold standards and more. *Nucleic Acids Res* 41: D203–D213

Sevin DC, Sauer U (2014) Ubiquinone accumulation improves osmotic-stress tolerance in *Escherichia coli*. *Nat Chem Biol* 10: 266–272

Sevin DC, Kuehne A, Zamboni N, Sauer U (2015) Biological insights through nontargeted metabolomics. *Curr Opin Biotechnol* 34: 1–8

Shin SY, Fauman EB, Petersen AK, Krumsiek J, Santos R, Huang J, Arnold M, Erte I, Forgetta V, Yang TP, Walter K, Menni C, Chen L, Vasquez L, Valdes AM, Hyde CL, Wang V, Ziemek D, Roberts P, Xi L *et al* (2014) An atlas of genetic influences on human blood metabolites. *Nat Genet* 46: 543–550

Stuecker TN, Hodge KM, Escalante-Semerena JC (2012) The missing link in coenzyme A biosynthesis: PanM (formerly YhhK), a yeast GCN5 acetyltransferase homologue triggers aspartate decarboxylase (PanD) maturation in *Salmonella enterica*. *Mol Microbiol* 84: 608–619

Szklarczyk D, Franceschini A, Kuhn M, Simonovic M, Roth A, Minguez P, Doerks T, Stark M, Muller J, Bork P, Jensen LJ, von Mering C (2011) The STRING database in 2011: functional interaction networks of proteins, globally integrated and scored. *Nucleic Acids Res* 39: D561–D568

Tatusov RL, Fedorova ND, Jackson JD, Jacobs AR, Kiryutin B, Koonin EV, Krylov DM, Mazumder R, Mekhedov SL, Nikolskaya AN, Rao BS, Smirnov S, Sverdlov AV, Vasudevan S, Wolf YI, Yin JJ, Natale DA (2003) The COG database: an updated version includes eukaryotes. *BMC Bioinformatics* 4: 41

Varghese S, Tang Y, Imlay JA (2003) Contrasting sensitivities of *Escherichia coli* aconitases A and B to oxidation and iron depletion. *J Bacteriol* 185: 221–230

van Wageningen S, Kemmeren P, Lijnzaad P, Margaritis T, Benschop JJ, de Castro IJ, van Leenen D, Groot Koerkamp MJ, Ko CW, Miles AJ, Brabers N, Brok MO, Lenstra TL, Fiedler D, Fokkens L, Aldecoa R, Apweiler E, Taliadouros V, Sameith K, van de Pasch LA *et al* (2010) Functional overlap and regulatory links shape genetic interactions between signaling pathways. *Cell* 143: 991–1004

Wang K, Li M, Hakonarson H (2010) Analysing biological pathways in genome-wide association studies. *Nat Rev Genet* 11: 843–854

6

Pharmacoproteomic characterisation of human colon and rectal cancer

Martin Frejno[1,2] (iD), Riccardo Zenezini Chiozzi[2,3], Mathias Wilhelm[2], Heiner Koch[2,4,5], Runsheng Zheng[2] (iD), Susan Klaeger[2,4,5], Benjamin Ruprecht[2,6], Chen Meng[2], Karl Kramer[2], Anna Jarzab[2], Stephanie Heinzlmeir[2,4,5], Elaine Johnstone[1], Enric Domingo[1,7], David Kerr[8], Moritz Jesinghaus[9], Julia Slotta-Huspenina[9], Wilko Weichert[9], Stefan Knapp[10], Stephan M Feller[11,12,*] (iD) & Bernhard Kuster[2,4,6,13,**] (iD)

Abstract

Most molecular cancer therapies act on protein targets but data on the proteome status of patients and cellular models for proteome-guided pre-clinical drug sensitivity studies are only beginning to emerge. Here, we profiled the proteomes of 65 colorectal cancer (CRC) cell lines to a depth of > 10,000 proteins using mass spectrometry. Integration with proteomes of 90 CRC patients and matched transcriptomics data defined integrated CRC subtypes, highlighting cell lines representative of each tumour subtype. Modelling the responses of 52 CRC cell lines to 577 drugs as a function of proteome profiles enabled predicting drug sensitivity for cell lines and patients. Among many novel associations, MERTK was identified as a predictive marker for resistance towards MEK1/2 inhibitors and immunohistochemistry of 1,074 CRC tumours confirmed MERTK as a prognostic survival marker. We provide the proteomic and pharmacological data as a resource to the community to, for example, facilitate the design of innovative prospective clinical trials.

Keywords CPTAC; CRC65; drug response; patient stratification; proteomics
Subject Categories Cancer; Pharmacology & Drug Discovery; Post-translational Modifications, Proteolysis & Proteomics

Introduction

Owing to the high molecular heterogeneity of human cancers, profiling technologies such as genomics and transcriptomics have been employed for some time to identify entity-specific molecular subtypes of tumours that can be used for diagnostic refinement, prediction of disease prognosis or to stratify patients for therapy (McDermott et al, 2011). More recently, this concept has been extended to the measurement of cancer proteomes and their post-translational modification status as exemplified by the NCI's Clinical Proteomic Tumor Analysis Consortium (CPTAC; Mertins et al, 2016; Zhang et al, 2016). The first published CPTAC study was on the proteome of colon and rectal cancer (CRC; Zhang et al, 2014), for which extensive transcriptional profiling data had previously been used to define consensus molecular subtypes (CMS) of CRC (Guinney et al, 2015). The comparison of subtype information at both transcript and proteome level showed general concordance but also significant discrepancies, with many features only detectable at the protein level. In parallel with molecular profiling of cancers, phenotypic drug screening campaigns in large panels of (mostly) genomically well-characterised cancer cell lines representing many tumour entities have been performed (Barretina et al, 2012; Medico et al, 2015; Iorio et al, 2016; Rees et al, 2016). These studies aimed at identifying effective drugs or combinations thereof in cellular model systems that recapitulate the genomic alterations found in human tumours and may thus also show efficacy in humans. However, although genomics might play a role in determining drug

1 Department of Oncology, University of Oxford, Oxford, UK
2 Chair of Proteomics and Bioanalytics, Technical University of Munich, Freising, Germany
3 Department of Chemistry, Sapienza – Università di Roma, Rome, Italy
4 German Cancer Consortium (DKTK), Munich, Germany
5 German Cancer Research Center (DKFZ), Heidelberg, Germany
6 Center for Integrated Protein Science (CIPSM), Munich, Germany
7 Wellcome Trust Centre for Human Genetics (WTCHG), University of Oxford, Oxford, UK
8 Nuffield Division of Clinical Laboratory Sciences (NDCLS), University of Oxford, Oxford, UK
9 Institute of Pathology, Technical University of Munich, Munich, Germany
10 Institute of Pharmaceutical Chemistry, Goethe University, Frankfurt am Main, Germany
11 Weatherall Institute of Molecular Medicine, University of Oxford, Oxford, UK
12 Institute of Molecular Medicine, Martin-Luther-University, Halle, Germany
13 Bavarian Biomolecular Mass Spectrometry Center (BayBioMS), Freising, Germany
    *Corresponding author. E-mail: stephan.feller@uk-halle.de
    **Corresponding author. E-mail: kuster@tum.de

sensitivity, given that the majority of drugs act on protein targets, it appears logical to correlate protein expression with drug sensitivity. Recent proteome profiling of the NCI60 cancer cell line panel and a panel of 20 breast cancer cell lines showed that protein signatures predicting drug sensitivity or resistance can be found (Gholami et al, 2013; Lawrence et al, 2015). Despite these previous efforts, the number of cell lines for any given cancer entity in these panels was limited, impairing the analysis of drug sensitivity for tumour subtypes and possible translation to human patients. Using published transcriptomics data, Medico et al (2015) assigned 151 CRC cell lines to different molecular subtypes of CRC patients, in order to identify model systems amenable for drug sensitivity screens in CRC but to date, no comprehensive proteomic dataset on CRC cell lines has been published that would allow for the direct discovery of proteomic signatures of drug sensitivity and resistance in different CRC subtypes.

In this study, we measured the proteomes of a panel of 65 well-characterised human colorectal cancer cell lines (Emaduddin et al, 2008) to a depth of > 10,000 proteins and integrated this data with the proteome profiles of 90 CRC patients (Zhang et al, 2014) and matched transcriptome profiles (Appendix Supplementary Methods) to define integrated proteomic subtypes of CRC. Integration with drug sensitivity data available for 52 CRC cell lines allowed us to predict the drug sensitivity of cell lines and patients towards 577 drugs or combinations thereof. The analysis revealed that, for example, high MAP2K1 (MEK1) expression renders small-molecule EGFR inhibitors less effective and also identified the kinase MERTK to confer partial resistance to MEK1/2 inhibitors. Analysis of 1,074 CRC patients from the QUASAR 2 trial (Kerr et al, 2016) proved that high MERTK expression is a prognostic marker for poor survival, defining this receptor tyrosine kinase as an attractive potential target for therapeutic intervention. We are making the proteomic data and our full analysis available to the scientific community to provide a rich resource for aiding in the design of future prospective clinical studies in CRC based on multi-omics molecular tumour data and phenotypic drug sensitivity data.

Results

Proteome profiles of CRC cell lines and patients

We devised a multi-omics data integration strategy to determine integrated proteomic subtypes of human colorectal cancer cell lines and patient samples in order to predict their sensitivity towards a variety of clinical and pre-clinical drugs and combinations thereof (Figs 1A and EV1, Appendix Supplementary Methods). To accomplish this, we first used LC-MS/MS-based shotgun proteomics to measure the proteomes of 65 CRC lines and to quantify their expressed kinomes using Kinobeads (Bantscheff et al, 2007; Medard et al, 2015). This led to the identification of a total of 11,796 protein groups (median/cell line = 9,447) representing 10,951 genes (median/cell line = 9,068) and 235 human kinases (median/cell line = 155, from Kinobeads experiments; Fig 2A; Table EV1A and B). We next re-analysed the proteomic profiles of 90 CRC patients published by the CPTAC (Zhang et al, 2014) using the same analysis pipeline and identified 7,005 protein groups (median/patient = 4,980) representing 6,727 gene groups (median/

patient = 4,901; Fig 2C; Table EV1C, Appendix Supplementary Methods). The CRC65 data contained most proteins of the CPTAC CRC data (Fig 2B), but we noted that gene groups unique to the CPTAC dataset are enriched in extracellular matrix proteins and IgGs as one might expect for tissue samples containing blood vessels and connective tissue. For quantification of full proteomes, we used a modified version of the intensity-based absolute quantitation (iBAQ) approach (Schwanhausser et al, 2011) termed gene-centric iBAQ (giBAQ, Appendix Supplementary Methods), while kinomes were quantified using the label-free quantification (LFQ) intensities provided by MaxQuant (Cox et al, 2014). For all subsequent data analyses, we used the proteomic data aggregated at the gene group level annotated with gene symbols as identifiers in order to be able to compare proteomics and transcriptomics data (referred to as protein expression/abundance throughout the manuscript). An overview of the data integration pipeline is depicted in Fig EV1.

Integration of multiple mRNA datasets reveals consensus molecular subtypes of the CRC65 panel

A number of studies have reported CRC patient subtypes based on mRNA measurements, and these were recently consolidated into consensus molecular subtypes (CMS) and applied to patients from the CPTAC study (Guinney et al, 2015). On this basis, we sought to determine the CMS membership of CRC cell lines in order to identify cell lines representative of CRC tumour subtypes. We downloaded 10 public mRNA datasets (Appendix Supplementary Methods), eight for the cell lines and two for the patients (Fig 1B), analysed them from the data level closest to raw data available to us and aggregated the datasets using a scheme similar to the one used by Guinney et al. This involved the selection of a reference dataset, followed by the selection of a probe set for each gene based on a consistency criterion and subsequent cross-dataset normalisation (Appendix Supplementary Methods). This resulted in a combined expression matrix of 9,737 transcripts (7,113 after filtering for low abundance transcripts) across 145 non-redundant cell lines (including the CRC65 panel) and 89 tumours (Table EV1D). We adapted the single-sample CMS classifier reported by Guinney et al to accept gene symbols as identifiers (rather than Entrez IDs; Appendix Supplementary Methods) and predicted the CMS for cell lines and patients based on 382 of the 692 classifier genes contained in the combined expression matrix. The correct classification of 65 out of 81 patients (80%, using the original CMS assignment as the ground truth) provided confidence that cell lines can be placed into CMSs with good accuracy and the resulting subtype labels for the CRC65 cell lines and the CPTAC patients are shown in Fig 1B. A subtype-resolved evaluation of the prediction accuracy using a confusion matrix and a table containing a variety of commonly used metrics for evaluating classification performance can be found in Table EV2E.

Integrated proteomic subtypes of CRC cell lines and tumours

Despite the fairly deep proteomic measurements, the quantification of proteins across many cell lines (and patients) suffered from an increasing number of missing values for proteins of decreasing abundance (Fig EV2A). We addressed this frequently encountered issue by mRNA-guided and minimum-guided missing value

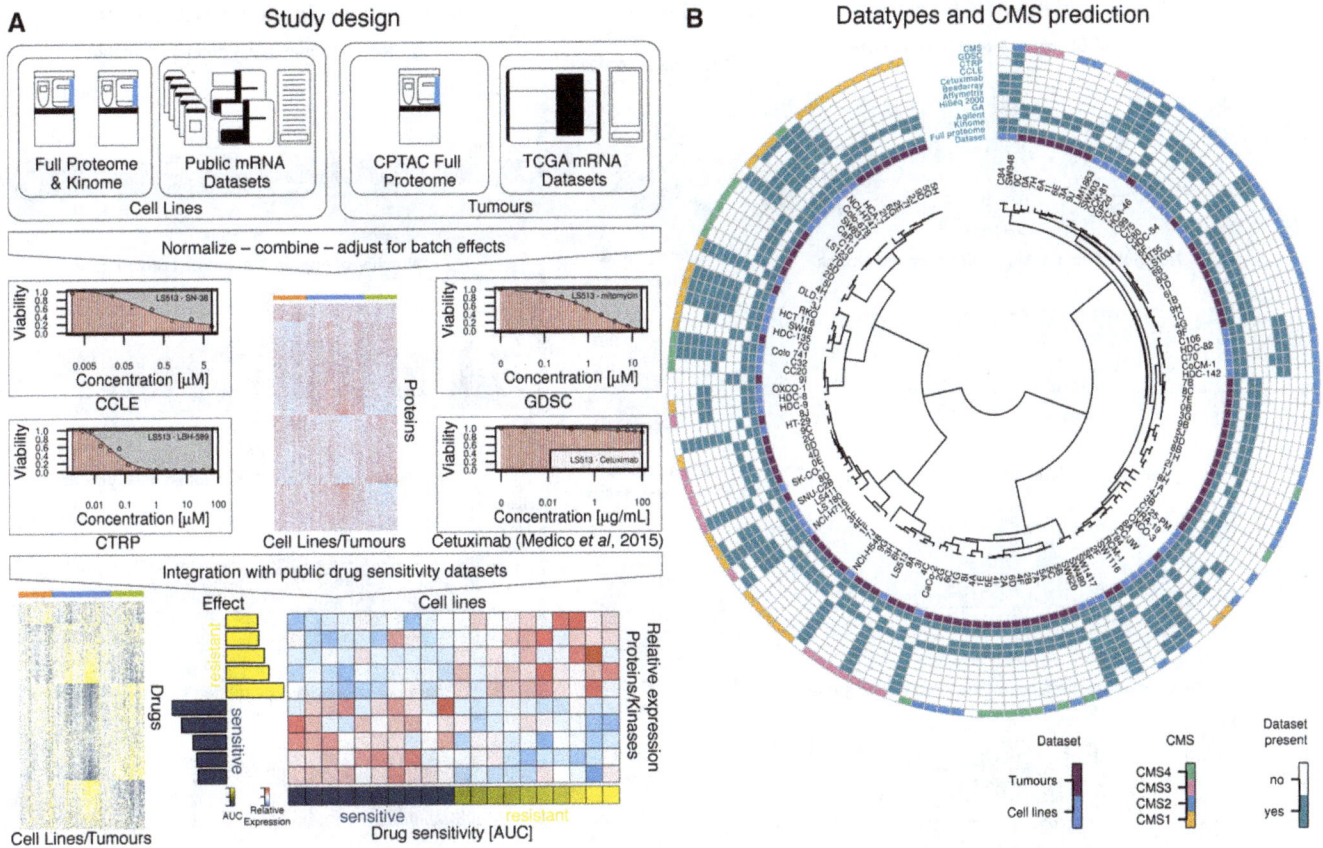

Figure 1. Study design, datatypes & CMS prediction for cell lines.

A Overview of the multi-omics data integration workflow followed in this study. Proteomic and transcriptomic data generated as part of this study or available from the literature for colorectal cancer (CRC) cell lines and patients were integrated in order to identify cell lines and tumours forming proteomic subtypes. Four published drug sensitivity datasets (abbreviated CCLE, CTRP, GDSC and cetuximab (Medico *et al*, 2015); one dose–response plot for each data source in shown) were overlaid onto the proteomic data to identify protein signatures associated with sensitivity or resistance. An example of an effect-size heat map for one drug, ten proteins and 20 cell lines is shown at the bottom-right (see main text and Appendix Supplementary Methods for details).

B Circos plot visualising the different datasets that were integrated in this study. The dendrogram in the centre hierarchically ordered tumours (violet "Dataset" label) and cell lines (blue "Dataset" label) based on the mRNA expression of classifier genes from Guinney *et al* (2015). A set of rings around the dendrogram indicates which proteomics data (full proteome, kinome), mRNA technologies (Agilent microarray, Genome-Analyser-based mRNA-Seq, HiSeq-2000-based mRNA-Seq, Affymetrix microarrays or Illumina Beadarrays) and drug sensitivity datasets (cetuximab, CCLE, CTRP or GDSC) were included in this study. The outermost ring indicates the membership of cell lines/tumours in a consensus molecular subtype ("CMS"). Undetermined CMS class labels or unavailable data were left white (see main text and Appendix Supplementary Methods for details). See also Fig EV1.

imputation on the peptide level to generate one complete protein expression matrix consisting of 59 cell lines, 81 tumours and 6,254 proteins (Fig EV2, Table EV1E), of which 323 were contained in the CMS classifier by Guinney *et al* (CMSgene in Fig 3A; see Appendix Supplementary Methods for details). In order to estimate protein levels from mRNA levels, we removed systematic differences (Fig EV3A and B) between proteomics and transcriptomics data using MComBat (Stein *et al*, 2015; see Appendix Supplementary Methods). This increased the protein/mRNA correlation for both the CRC65 and CPTAC datasets (Figs EV2B and EV3C), enabling mRNA-guided missing value imputation. After imputing missing values separately for both datasets (Fig EV2D and E, Appendix Supplementary Methods), we accounted for differences in their proteomic depth (Fig EV2C, Appendix Supplementary Methods) using ComBat (Johnson *et al*, 2007) before we merged the two protein expression matrices. Using the combined protein expression

matrix and consensus clustering (Appendix Supplementary Methods), we identified three integrated Full Proteome Subtypes (short: FPSs, namely FPA, FPB and FPC; Fig 3A). Each subtype consisted of cell lines as well as patients in a ratio of 28/34 for FPA, 22/12 for FPB and 9/26 for FPC, indicating that indeed there are cell lines, which are molecularly more similar to tumours than they are to other cell lines. We measured the association of these FPSs with previously published subtypes, as well as genomic and epigenomic features using Fisher's exact test (Table EV2A), and found good overall concordance but with some differences in detail (see discussion). Interestingly, FPA was associated with TP53 mutations, while FPB was associated with ATM mutations, suggesting that p53 signalling in response to DNA damage is perturbed through distinct mechanisms in these two subtypes. FPC showed association with mutations in BCL9L, a transcriptional activator of b-catenin activity, RNF43, an E3 ubiquitin-protein ligase, which acts as a negative

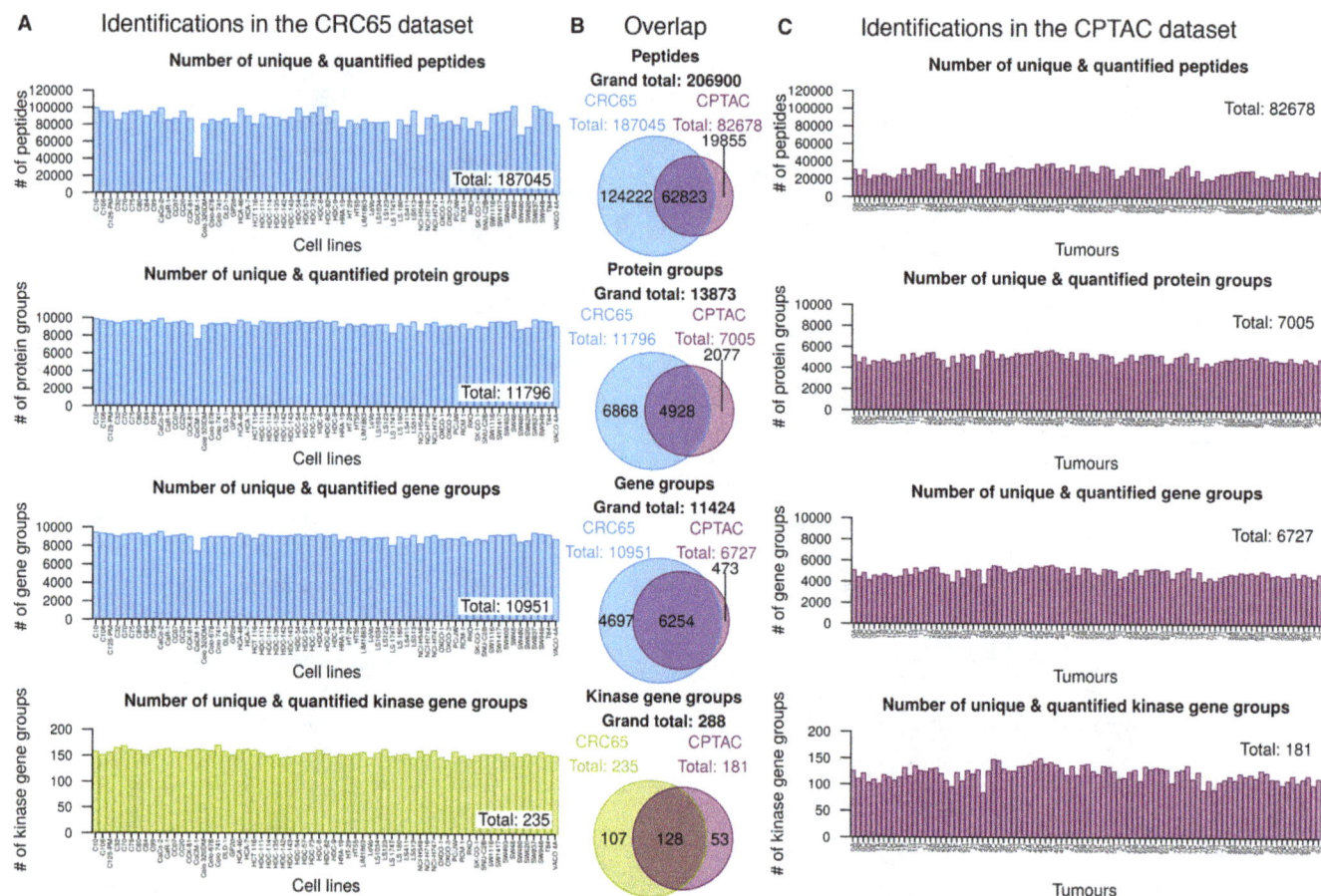

Figure 2. LC-MS/MS-based identifications.

A Bar charts visualising the number of unique identified and quantified peptides, protein groups and gene groups (full proteomes), as well as kinase gene groups (Kinobeads), across the CRC65 cell line panel (n = 65 cell lines). Insets indicate the corresponding number of identifications across the entire dataset.

B Venn diagrams showing the intersection and complements with respect to identifications in the aforementioned categories across both the CRC65 cell line and CPTAC patient dataset (n = 89 tumours). Kinase identifications in the CPTAC dataset were extracted from the full proteome data.

C Same as (A) for the CPTAC dataset. The different proteomic datasets were colour-coded (green = Kinobeads, blue = CRC65 full proteomes and purple = CPTAC full proteomes).

regulator of Wnt signalling, as well as with mutations in the histone acetyltransferase EP300, all pointing towards deregulated WNT signalling in conjunction with aberrant histone acetylation. Enrichment analysis of functional classifications using MetaCore (Fig EV4, Appendix Supplementary Methods) revealed that proteins significantly under- or overrepresented in the different FPSs fall into specific categories. Briefly, the analysis suggests that FPA is characterised by a "high metabolism, low cell cycle, microsatellite stable (MSI−)" signature, FPB harbours a "high immune response, low metabolism, microsatellite instable (MSI+)" signature and FPC shows a "low immune response, low inflammation, low adhesion (invasion)" signature.

Integration of drug sensitivity data reveals protein signatures of drug response

Since the response to drug treatment cannot be determined as quickly and as broadly as would be desirable in clinical studies, we took advantage of the fact that CRC cell lines recapitulate the main

molecular subtypes in CRC and were recently extensively characterised for drug sensitivity as part of major screening efforts. On the basis of the proteomic data, we used elastic net regression (Zou & Hastie, 2005) to predict the response (sensitivity/resistance) of CPTAC patients and CRC65 cell lines to 577 drugs or combinations thereof (Fig 3B, Appendix Supplementary Methods). Briefly, common cell lines (52 cell lines overlap in total; median overlap per drug is 27) between the CRC65 panel and three small-molecule drug sensitivity screens (Barretina *et al*, 2012; Garnett *et al*, 2012; Rees *et al*, 2016) and one study investigating sensitivity of CRC cell lines towards cetuximab (Medico *et al*, 2015) served as the training set (Fig 1, Appendix Supplementary Methods). We used significance analysis of microarrays (SAM; median FDR of 0.001; Appendix Supplementary Methods) to identify drugs which showed specificity for certain FPSs and performed target space enrichment by measuring the association between these subtypes and recurrent annotated targets of these drugs using Fisher's exact test (Fig 3B, Table EV2B). Cetuximab, for example, was more effective in FPA than in the other FPSs, with EGFR target enrichment reaching statistical significance

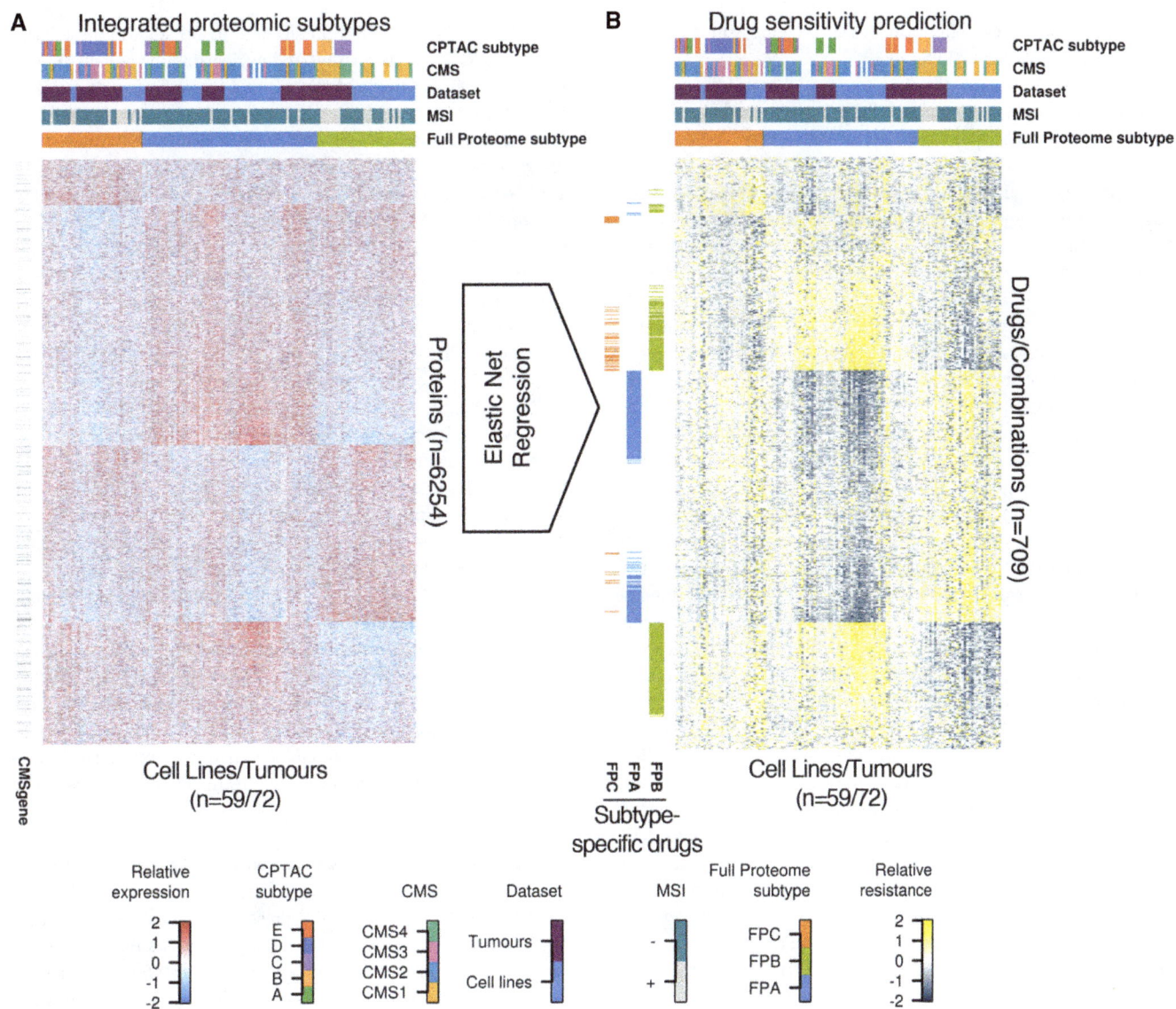

Figure 3. From integrated proteomic subtypes to drug sensitivity prediction.

A Heat map of standardised, log2-transformed and median-centred giBAQ protein quantification (z-scores) across the combined CRC65/CPTAC dataset. Cell lines/patients are displayed as columns and proteins are shown as rows. Black bars to the left of the heat map indicate the presence of the respective protein in the CMS classifier. Annotation bars on top of the heat map visualise the membership of the different cell lines/patients in five annotation categories.

B Elastic net regression was used to model drug sensitivity as a function of protein profile with the CCLE, CTRP, GDSC and cetuximab datasets as input. The resulting models were used to predict the drug sensitivity of cell lines/patients (see also Appendix Supplementary Methods). In the heat map of standardised predicted AUC values (z-scores), cell lines/patients are displayed as columns with the same annotations and ordering as in (A), while drugs or combinations thereof are shown as rows. Coloured bars to the left of the heat map indicate drugs, which are more effective in a respective subtype (see main text and Appendix Supplementary Methods for details). See also Figs EV2–EV4.

($P = 0.05$). FPA was also significantly (multiple-test adjusted $P < 0.05$) associated with drugs targeting the kinases MAP2K1 and MAP2K2 (MEK1 and 2; mainly selumetinib/AZD6244 and its combinations with other compounds), as well as with drugs and combinations targeting the bromodomain containing protein BRDT (mainly by the compound JQ-1), histone deacetylases or HDACs (vorinostat among others) and the nicotinamide phosphoribosyltransferase NAMPT (daporinad among these). Even though more drugs were differentially effective in FPB than in FPA (205 versus 170), most of the former did not show significant enrichment in their target space, making it difficult to link the drug phenotype to the presumed mechanism of action. However, FPB did show positive association with drugs targeting inhibitors of apoptosis (IAPs; all these compounds are mimetics of second mitochondrial activator of caspases, SMAC) as well as negative association with drugs targeting EGFR, while FPC showed association with drugs targeting dihydroxyfolate reductase (DHFR, including methotrexate). The complete list of predicted drug sensitivities for the patients and cell lines can be found in

Table EV3A, while mean effect sizes of all proteins from elastic net regression can be found in Table EV3B.

High MERTK expression in CRC cell lines is predictive of resistance to MEK1/2 inhibitors

Since many targeted cancer drugs act on kinases, we performed independent experiments focussing on the expressed kinome of the CRC65 cell line panel to identify kinases associated with drug sensitivity or resistance. Pulling down protein kinases using immobilised kinase inhibitors (Kinobeads; Bantscheff *et al*, 2007; Medard *et al*, 2015) led to the reproducible (average $R = 0.91$ between replicates; Fig EV5A) quantification of 138 kinases by mass spectrometry (Appendix Supplementary Methods), which correlated well ($R = 0.94$; Fig EV5B) with data obtained by Western blotting and densitometry for the kinases EPHA4 (Fig EV5C; variable expression across the panel) and ABL1 (Fig EV5D; relatively stable expression across the CRC65 panel). While kinase expression as measured using Kinobeads correlated reasonably well with measurements in full proteomes, an enrichment of kinases was clearly visible in the Kinobeads data (Fig EV5E) and led to the identification and quantification of more than 50 kinases not detected in the deep proteome analysis (Fig EV5F). Consensus clustering identified three Kinobeads Subtypes (KSs) KA, KB and KC (Fig EV5G) that fully recapitulated the Full Proteome Subtypes and showed significant association with other subtype classifications (Table EV2C).

By searching for differentially expressed kinases between the different Kinobeads Subtypes using SAM (median FDR of 0.01), we identified proteins frequently mutated in microsatellite instable (MSI+) tumours with concomitant decrease in expression of proteins such as ACVR2A and TGFBR2 (Kim *et al*, 2013), as well as the receptor tyrosine kinase EPHA2, which was overexpressed in KC (Table EV4). Overexpression of EPHA2 is known to predict resistance towards cetuximab in CRC (Strimpakos *et al*, 2013). Elastic net regression was used to identify kinases associated with drug sensitivity or resistance and confirmed the strong association of EPHA2 with resistance towards cetuximab (Fig EV5H, Table EV3C and D). We also noted that overexpression of MAP2K1 (MEK1) was associated with resistance to two-thirds (12/18) of all inhibitors targeting EGFR (Fig 4 for example). Activating mutations in K57 of MAP2K1 were previously shown to be a potential mechanism of primary resistance towards EGFR-targeted therapy (Bertotti *et al*, 2015), and our data suggest that high expression can have a similar effect. SAM also identified MERTK —a receptor tyrosine kinase correlating with disease progression in melanoma (Schlegel *et al*, 2013) and not detected in the full proteome measurements—to be differentially expressed between the different Kinobeads Subtypes (*q*-value < 0.01). MERTK is down-regulated in KRAS-mutant CRC (Watanabe *et al*, 2011) and up-regulated in pancreatic cancer cell lines resistant towards selumetinib (Beech & Kelly, 2014). Interestingly, we also recurrently found high MERTK expression associated with resistance towards MEK1/2 inhibitors as well as other drugs from multiple drug sensitivity datasets (Figs 5A and B, and EV6A, Appendix Supplementary Methods). To validate hypotheses arising from the above predictions, we performed a series of *in vitro* drug treatment experiments.

On the basis of MERTK expression, we selected three cell lines (CC07, HDC-143 and SK-CO-1) predicted to be sensitive to two MEK1/2 inhibitors (RDEA119 and PD-0325901) as well as three cell lines (C10, CaCo-2 and T84) predicted to be resistant to these drugs (Fig EV6B). Cell viability assays and nuclei counting of cells (Fig 6C and D) showed that the predictions could be confirmed, as cell lines expected to be resistant did not respond as well as cell lines expected to be sensitive ($P \leq 0.05$, one-sided Mann–Whitney test). Similar results were obtained for the predicted association of ACVR2A expression with drug sensitivity to AUY922 (an HSP90 inhibitor) and BAY 61-3606 (a designated SYK inhibitor) in LS-180, OXCO-1 and RKO cells (sensitive) or C125-PM, HDC-111 and HT55 cells (resistant; Fig 5C and D). Since ACVR2A expression is reduced in microsatellite instable CRC, these tumours might benefit from treatment with these drugs.

We next asked whether knockout of MERTK in MEK1/2-inhibitor-resistant C10 cells could re-sensitise them to treatment with RDEA119 (MEK1/2 inhibitor). A Western blot-based validation of the CRISPR/Cas9-mediated knockout is depicted in Fig 5E. We also confirmed the knockout by sequencing of the targeted genomic region, which showed two insertions: one in exon 7 (FN3 domain) and one in exon 14 (kinase domain), inducing frameshifts. As depicted in Fig 5F, C10$^{MERTKN^{KO}}$ was more sensitive to RDEA119 than the parental C10$^{MERTKN^{WT}}$ cell line. We next tested whether the combination of MEK (by RDEA119 or PD-0325901) and MERTK inhibition (UNC569) could re-sensitise MEK1/2-inhibitor-resistant cell lines. As shown in Fig EV6C–F, co-treatment indeed significantly ($P \leq 0.05$, one-sided Mann–Whitney test) reduced the viability of MEK1/2 inhibitor-resistant cell lines to levels comparable to sensitive cell lines. However, this reduction appeared to be mainly due to UNC569 on its own, since both MEK1/2 inhibitor-sensitive and resistant cell lines show comparable response to the co-treatment when treated with UNC569 alone.

MERTK is a prognostic survival marker in CRC patients

In order to evaluate the general expression status as well as the clinical impact of MERTK expression in CRC, we quantified its abundance in 1,074 patients enrolled in the QUASAR 2 trial by immunohistochemistry (IHC). After establishing cell-based positive (CaCo-2) and negative (RKO) controls with high and low expression of MERTK, respectively (Fig 6A and B), tissue microarrays (TMAs) were stained for MERTK and the fraction of positive cells, as well as the staining intensity, were quantified by pathologists. While MERTK expression was found primarily at the membrane in CaCo-2 cells, the TMA data showed combined cytoplasmic and membrane staining. Therefore, TMAs were categorised to have either high or low MERTK expression (more or < 5% cytoplasm/membrane-positive tumour cells, Appendix Supplementary Methods) without distinguishing between these compartments. Figure 6C shows three representative TMA samples alongside their quantification of MERTK, also highlighting the strong variability of MERTK expression in primary tumours. Modelling the outcome variables of the QUASAR 2 trial as a function of MERTK expression showed that high MERTK expression was prognostic of worse 5-year overall survival (OS), disease-free survival (DFS) and recurrence-free survival (RFS) in both univariate and multivariate Cox proportional hazards regression (Table EV5, Fig 6D). Notably, MERTK expression is more informative in predicting these outcome variables than the patients' treatment status, since its coefficient was significant in the final multivariate Cox proportional hazards model, while the

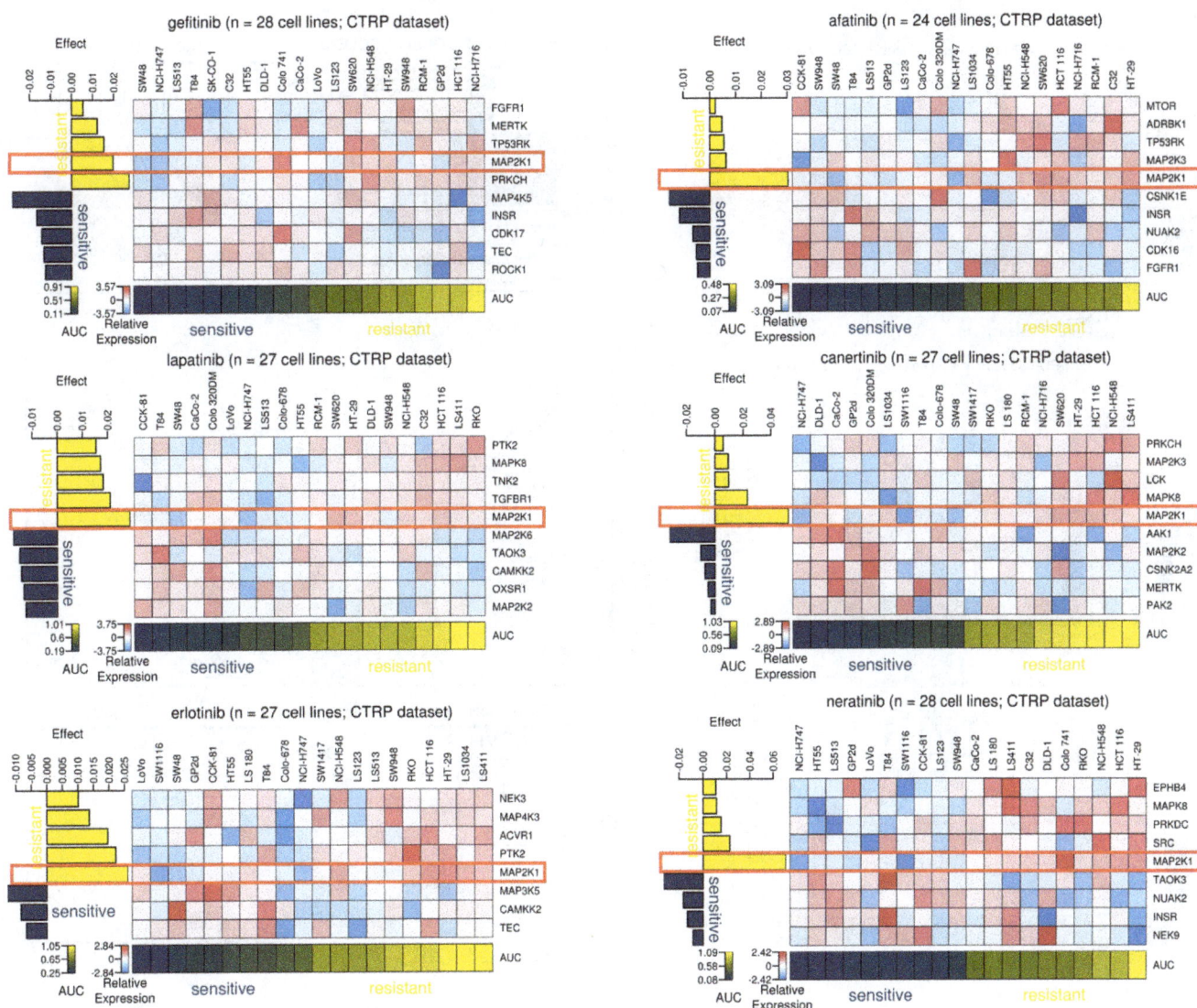

Figure 4. MAP2K1 is a predictive marker for inhibitors targeting EGFR.

Effect-size heat maps of six drugs (see titles of panels) targeting EGFR. It is evident that the different drugs showed different profiles but also that high MAP2K1 expression (blue/red gradient across cell lines) was consistently associated with drug resistance (dark blue/yellow gradient across cell lines; AUC: area under the curve; see main text and Appendix Supplementary Methods for details). See also Fig EV5.

coefficient of the treatment variable was not (see also Appendix Supplementary Methods). Comparing tumours with high versus low MERTK expression resulted in a multivariate hazard ratio (HR) of 1.61 for OS ($CI_{0.95}$ = 1.1–2.36; significant coefficient for MERTK with P = 0.015; Wald test), 1.70 for DFS ($CI_{0.95}$ = 1.24–2.33; significant coefficient for MERTK with P = 0.00094; Wald test) and 1.77 for RFS ($CI_{0.95}$ = 1.26–2.48; significant coefficient for MERTK with P = 0.00087; Wald test), respectively. High expression of MERTK was also significantly associated with BRAF[V600E] mutations (12% of the CRC65 cell line panel and 13% of the QUASAR 2 cohort carry the mutation, Fig EV5D) and stage T4 tumours, while low cytoplasmic/membranous expression was associated with wild-type BRAF and stage T2/T3 tumours (two-sided Fisher's exact test P < 0.05, Table EV2D). We note that more work is needed in order to determine the direction of causality for these observations and whether

MERTK might qualify as a drug target in CRC in addition to being a prognostic survival marker in CRC patients.

Discussion

Recent landmark studies established resources like the "Genomics of Drug Sensitivity in Cancer" (GDSC; Iorio *et al*, 2016), the "Cancer Cell Line Encyclopedia" (CCLE; Barretina *et al*, 2012) and the "Cancer Therapeutics Response Portal" (CTRP; Rees *et al*, 2016), all of which focus on the identification of genomic and transcriptomic associations with drug sensitivity. However, since drugs almost always target proteins, it appears obvious to incorporate proteome-wide measurements into drug sensitivity association studies. Here, we overlaid these large-scale drug sensitivity datasets and data on

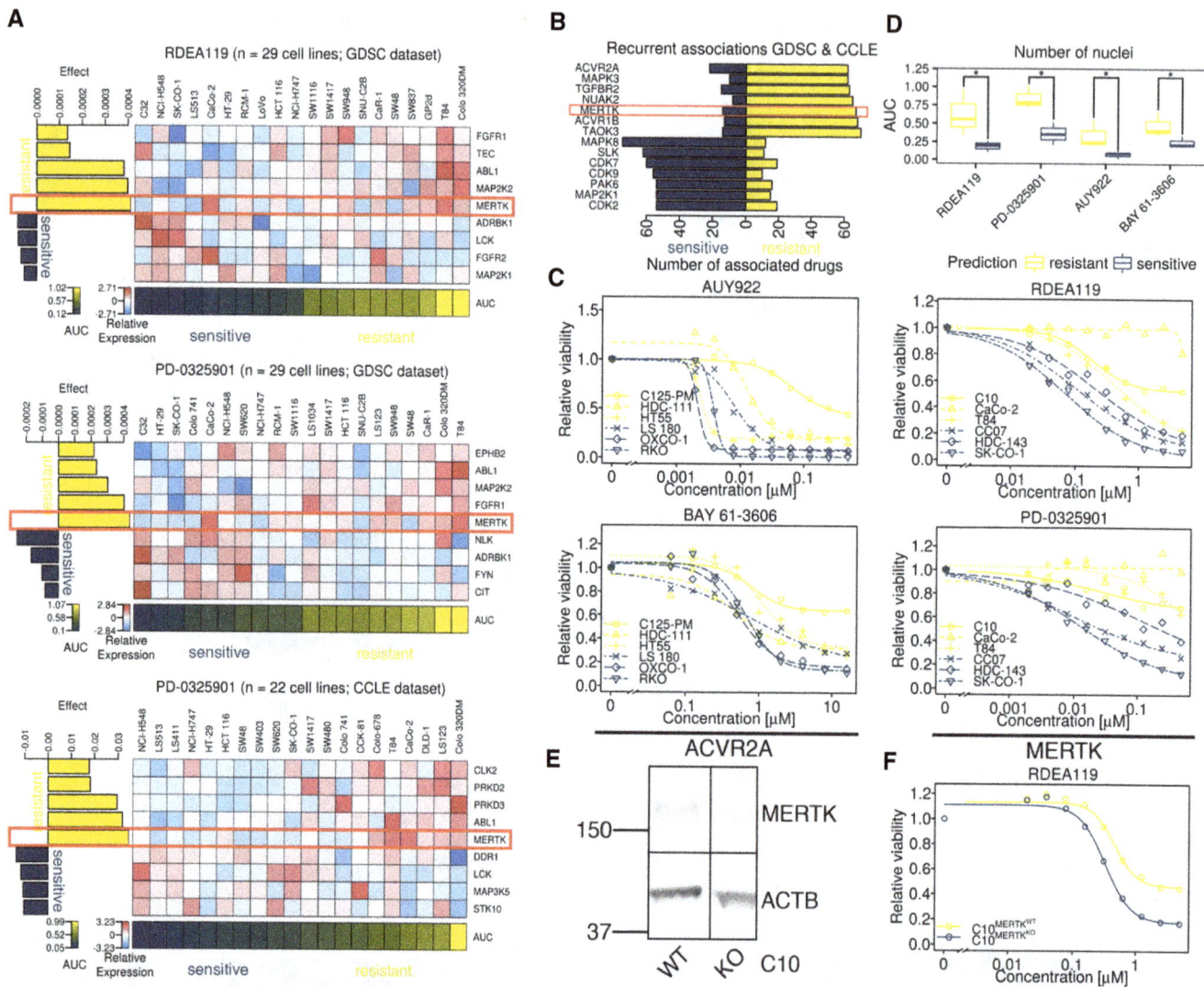

Figure 5. MERTK is a predictive marker for inhibitors targeting MEK1/2 in CRC cell lines.

A Effect-size heat maps of two drugs (one from two different drug sensitivity screens) targeting MEK1/2 show consistent association of high MERTK expression with drug resistance. The colour scheme is the same as in Fig 4.

B Bar chart visualising the top kinases recurrently associated (absolute effect size > 0) with drug resistance (top seven bars) and sensitivity (bottom seven bars) in the GDSC and CCLE drug sensitivity datasets.

C Dose–response curves of two drugs for which high MERTK (left panels) or ACVR2A (right panels) expression was predicted to confer drug resistance. For each drug, three cell lines predicted to be sensitive (dark blue) and three cell lines predicted to be resistant (yellow) were assayed for viability. The experimental data validated that cell lines predicted to be more sensitive to a drug indeed showed this phenotype (data represent the average of three technical replicates; see Appendix Supplementary Methods for details).

D Boxplots summarising the data shown in panel (C) using the area under the curve (AUC) as a measure for drug sensitivity. The whiskers extend to the minimum and maximum AUC for a given drug and sensitivity prediction, while the median AUC is marked with a bold horizontal line inside a box spanning the interquartile range (IQR) from the 25% quantile (lower horizontal line) to the 75% quantile (upper horizontal line). High AUC values indicate drug resistance. Again, it is evident that cell lines predicted to be more resistant to a certain drug were in fact significantly (*$P \leq 0.05$, one-sided Mann-Whitney test) more resistant than cell lines predicted to be more sensitive.

E Western blot of $C10^{MERTKN^{WT}}$ and $C10^{MERTKN^{KO}}$ cells, visualising successful knockout by CRISPR/Cas9.

F Dose–response curve of RDEA119 (MEK1/2 inhibitor) in $C10^{MERTKN^{WT}}$ and $C10^{MERTKN^{KO}}$ cells. The knockout is more sensitive to MEK1/2 inhibition than the wild type. See also Fig EV6.

cetuximab (Medico *et al*, 2015) onto extensive proteomic profiles of 65 colon cancer cell lines, created as part of this study and 90 published human colon cancer proteomes (Zhang *et al*, 2014; reprocessed for this study) in order to identify molecular subtypes of colorectal cancer and molecular markers of drug response with translational potential to the human disease.

It was shown before that cell lines included in the CRC65 panel might capture core molecular subtypes of CRC (Mouradov *et al*, 2014; Medico *et al*, 2015); however, the recently established consensus molecular subtype CMS of these cell lines was not known as of yet. By combining and re-analysing multiple public mRNA datasets acquired using a variety of technologies across a number of

A Western blot of MERTK in control cell lines

B Immunohistochemistry of MERTK in control cell lines

RKO (negative control)   CaCo-2 (positive control)

D Survival analysis of QUASAR 2, stratified by MERTK expression

C Immunohistochemistry of three representative tissue microarrays from QUASAR 2

Intensity: 2
Fraction of MERTK positive tumour cells: 10%

Intensity: 2
Fraction of MERTK positive tumour cells: 30%

Intensity: 3
Fraction of MERTK positive tumour cells: 100%

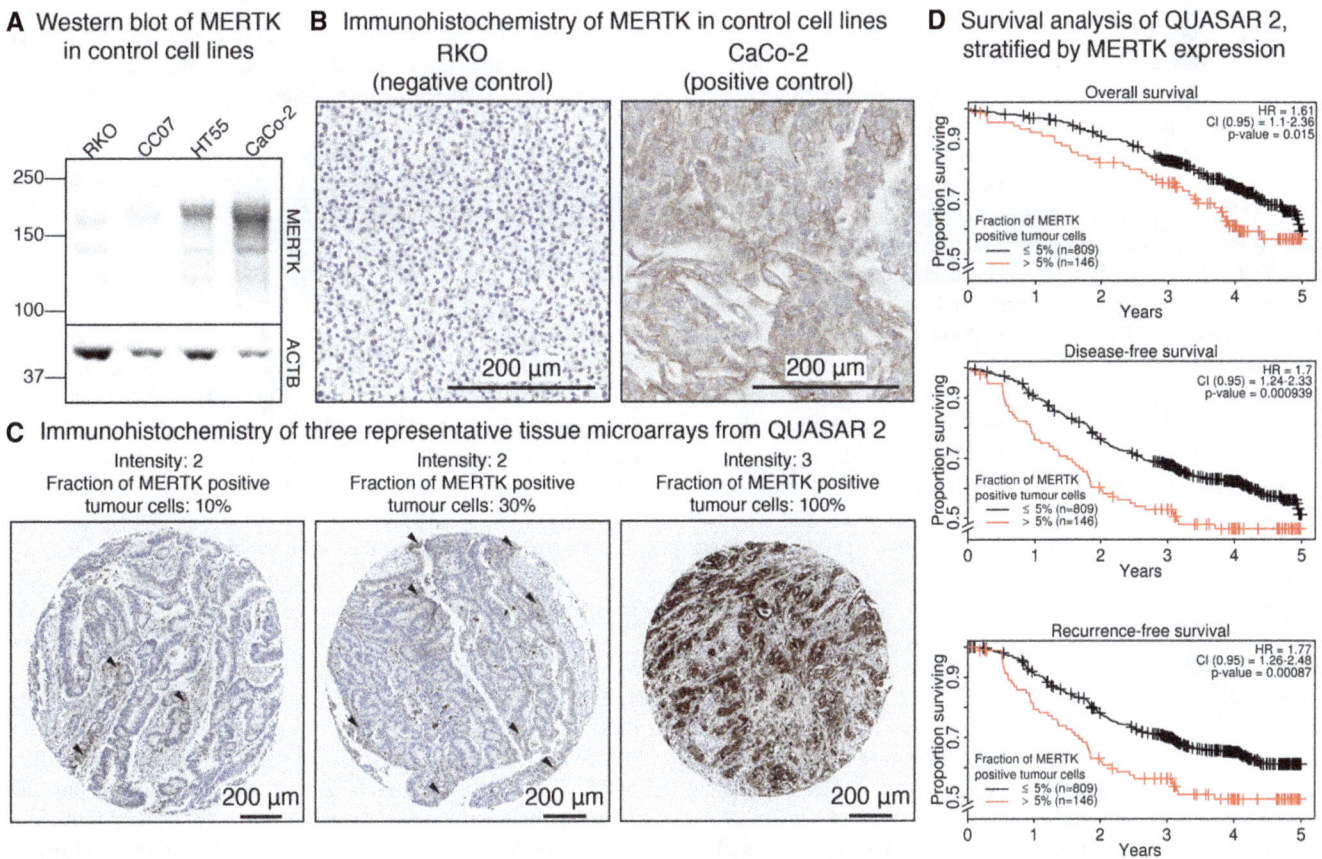

Figure 6. MERTK is a prognostic marker in CRC patients.

A   Western blot visualising MERTK expression in two negative (RKO and CC07) and two positive control cell lines (CaCo-2 and HT55).

B   IHC staining of MERTK expression in RKO (negative control) and CaCo-2 (positive control) cells.

C   IHC staining of three representative TMAs from patients enrolled in the QUASAR 2 clinical trial and the corresponding quantification of the signal by pathologists. Arrowheads indicate tumour cells positive for MERTK.

D   Kaplan–Meier plots showing worse overall, disease-free and recurrence-free survival of patients with high cytoplasmic/membranous expression of MERTK. P-values indicate the significance of MERTK as a predictor based on the Wald statistic.

laboratories involving several different batches of cell lines cultivated under a diverse set of conditions, we were able to assign 42 of them to a defined CMS subtype using the single-sample classifier developed by Guinney et al (2015), confirming that these cell lines indeed represent the main molecular subtypes of CRC. Dunne et al recently noted that patients might be misclassified by the CMS classifier due to regional heterogeneity in tumours arising from stromal contributions to measured gene expression (Dunne et al, 2016), highlighting the need for independent unsupervised class discovery while searching for representative cell lines. This might also explain why the cluster in Fig 1B containing most of the CMS4 tumours did not contain cell lines annotated as CMS4, since cell lines only capture the epithelial component of a tumour. The transcriptional profiles of the cell lines were consistent across the different datasets if systematic differences between them were accounted for.

Combining the CRC65 and CPTAC datasets then enabled the identification of integrated Full Proteome Subtypes of CRC (consisting of cell lines and patients), which were associated with previously published mRNA-based and proteomics-based subtypes. Since the CMS represent the current consensus in the field, we note that they could only

be partially recapitulated from proteomics data. The data analysis grouped CMS1 and CMS4 together into Full Proteome Subtype FPB, which might be explained by the fact that mesenchymal genes overrepresented in CMS4 are mainly of stromal rather than epithelial origin and will therefore largely evade detection in cell culture experiments (Calon et al, 2015; Isella et al, 2015). Despite this loss in granularity, the data identified clear proteomic signatures characterised by "high metabolism, low cell cycle, MSI−" for FPA, "high immune response, low metabolism, MSI+" for FPB and "low immune response, low inflammation, low adhesion (invasive)" for FPC.

Due to the substantial overlap of our CRC65 full proteome data on cell lines with the CPTAC patient data, we were able to predict the drug sensitivity of both cell lines and patients towards 577 drugs or combinations thereof as a function of protein profile. This allowed us to confirm a number of known and suggest a number of novel promising drugs and targets specific to the integrated Full Proteome Subtypes identified in this project. One example of a known subtype-specific drug is cetuximab, which was more effective in FPA than in FPB or FPC. Since FPA was associated with (i.e. is similar to) the transit-amplifying (TA) subtype from the subtype model by Sadanandam et al

(Table S2A), our analysis confirmed previous findings on cell lines and patients where cetuximab sensitivity was recurrently observed in the TA subtype (Sadanandam *et al*, 2013). Interestingly, modelling drug sensitivity as a function of kinase protein expression also confirmed previous findings by Strimpakos *et al* (2013), as high expression of EPHA2 was associated with resistance towards cetuximab. Since the Kinobeads-based expression level of EPHA2 was significantly lower in KB/FPA than in the other subtypes, one might speculate that treatment with EPHA2-targeted drugs might re-sensitise cetuximab-resistant patients and cell lines to the antibody. Koch *et al* previously made similar observations for non-small-cell lung cancer and proposed a model in which EGFR and EPHA2 functionally interact to mediate resistance to EGFR inhibition (Koch *et al*, 2015). Target space enrichment analysis then facilitated the identification of proteins, which are recurrently targeted by drugs specific to certain FPSs. This analysis enabled the suggestion of target-based treatment options for each FPS, while the drug sensitivity predictions themselves provide more detailed information for the selection of rational treatment regimens as part of, for example, prospective clinical trials. For example, selumetinib (MEK inhibitor), lapatinib (EGFR inhibitor) and vorinostat (HDAC inhibitor) were specific to FPA, SN-38 (topoisomerase inhibitor) and sorafenib (a RAF inhibitor) were specific to FPB, while AZD7762 (checkpoint kinase inhibitor) and methotrexate (DHFR inhibitor) were specific to FPC. Together with our target space enrichment analysis, this suggested that patients of the FPA subtype might profit most from drugs targeting the classical MAPK signalling cascade or acetylation-based epigenetic modifications, patients of the FPB subtype might benefit most from drugs promoting apoptosis and patients of the FPC subtype might be best served by DHFR-targeted drugs.

The kinase-centric drug sensitivity analysis found high expression of MAP2K1 (MEK1) frequently associated with resistance towards EGFR inhibitors, and high expression of MERTK recurrently associated with resistance towards MEK1/2 inhibitors. While activating mutations of K57 in MAP2K1 are known to play a role in the development of resistance towards EGFR-targeted treatments (Bertotti *et al*, 2015), to our knowledge, MERTK was so far only shown to be up-regulated on the mRNA level in pancreatic cancer cell lines resistant towards the MEK1/2 inhibitor selumetinib (Beech & Kelly, 2014). We were able to confirm the predictive potential of MERTK protein expression for CRC cell lines *in vitro*. In addition, exploration of MERTK expression levels in tumours using immunohistochemistry on tissue microarrays (TMAs) showed that high MERTK expression is a biomarker for worse overall, disease-free and recurrence-free survival in CRC patients. Aberrant expression of MERTK was observed previously in a variety of cancers (Graham *et al*, 2014). We showed that high MERTK expression was associated with stage T4 tumours. Tavazoie *et al* (2008) suggested that MERTK is a target of miR-335-dependent post-transcriptional regulation. Since miR-335 was shown to be down-regulated in aggressive CRC and breast cancer (Tavazoie *et al*, 2008; Sun *et al*, 2014), one could speculate that aberrant expression of MERTK in CRC may be due to down-regulation of miR-335. Given that MERTK is a receptor tyrosine kinase, the protein might also be an actionable target for the treatment of advanced CRC, possibly in combination with MEK1/2 inhibitors. However, further work is needed to evaluate whether MERTK inhibition alone or combination treatments have clinical potential in CRC. In the light of the fact that, for example,

the approved drugs crizotinib and sunitinib are also potent MERTK inhibitors and that the designated compound MRX-2843 will soon enter clinical trials, such investigations are becoming feasible in future.

Materials and Methods

Cell lines

All cell lines in the CRC65 panel apart from SW480 (kind gift from Ulrike Stein from the MDC, Berlin) were collected by the laboratory of Prof Sir Walter Bodmer FRS at the University of Oxford and were previously HLA-typed and characterised for other genetic changes to determine whether they are derived from cancers of different donors (Emaduddin *et al*, 2008). Information on the CRC65 cell lines and the CPTAC tumour samples like MSI status, original sources (cell lines) or subtype membership is compiled in Table EV6A.

Patient samples

We analysed the expression of MERTK in 1,074 patients from the QUASAR 2 clinical trial cohort, with approval from the West Midlands Research Ethics Committee (Edgbaston, Birmingham, UK; REC reference: 04/MRE/11/18). All participants provided written informed consent for treatment, and separate consent was obtained regarding the use of tumour tissue. QUASAR 2 is a phase III international randomised controlled trial, which collected data on toxicity, overall survival (OS), disease-free survival (DFS) and recurrence-free survival (RFS) for 1,941 stage II/III CRC patients, with the aim to determine the efficacy of adjuvant capecitabine±bevacizumab after resection of the primary tumour. A biobank comprising 1,350 FFPE blocks was established, and tissue microarrays (TMAs) were generated from 1.2-mm cores.

Cell culture and lysis

Slightly altering the culture conditions described by Emaduddin *et al* (2008), cells were grown in high glucose Dulbecco's modified Eagle medium (DMEM, including GlutaMAX and pyruvate; PAA) containing 1% Pen-Strep (penicillin at 100 units/ml and streptomycin at 100 μg/ml final concentration, PAA) and 10% foetal bovine serum (FBS, SLI EU-000F Batch 503005) in a humidified incubator at 37°C and 10% CO_2 using T-175 flasks (Corning). Hereafter, this medium composition is referred to as "culture medium" and the culture conditions apart from the culture vessel are referred to as "standard culture conditions". Adherent cells were harvested at ~80–90% confluency. Suspension cells (e.g. HDC-135) growing in 250 ml of culture medium were harvested by centrifugation in 250-ml centrifuge tubes (Corning) at 300 ×*g* and 4°C for 5 min. We used RIPA100 buffer (20 mM Tris–HCl pH 7.5, 1 mM EDTA, 100 mM NaCl, 1% Triton X-100, 0.5% sodium deoxycholate and 0.1% SDS) containing protease (Complete™ mini with EDTA; Roche) and phosphatase inhibitors (Phosphatase Inhibitor cocktail 1 and 2; Sigma Aldrich) at 2× and 5× the final concentration recommended by the manufacturer, respectively, and lysed the cells for 30 min at 4°C. After clearing the total cell lysates (TCLs) at 22,000 ×*g* for 30 min at 4°C, the protein concentration was determined using a Coomassie-based protein assay kit (Thermo Fisher) and the TCLs were stored at −80°C until further use.

Cell viability assays

We performed *in vitro* cell viability assays in order to test our *in silico* predictions. Assays were performed as described previously (Garnett et al, 2012), with minor modifications. Cell viability assays were carried out in technical triplicates in order to generate 10-point dose–response curves for drugs with the resistance of which either ACVR2A or MERTK was associated. For each drug, we selected three cell lines predicted to be resistant and three cell lines predicted to be sensitive towards the respective drug and seeded them in 100 μl culture medium at their optimal seeding density on day zero (C10 = 4,000, CaCo-2 = 2,000, CC07 = 4,000, HDC-143 = 12,000, SK-CO-1 = 8,000, T84 = 11,000, RKO = 4,000, LS 180 = 2,000, HDC-111 = 4,000, HT55 = 12,000, OXCO-1 = 8,000, C125-PM = 11,000). Following an overnight incubation under standard culture conditions, 50 μl of fresh medium containing either 1% DMSO or the respective drug in a 9-point twofold dilution series in 1% DMSO was added to the corresponding wells. This resulted in a final DMSO concentration of 0.33%, while the highest final drug concentration was 10 μM for UNC569 (Merck), as well as 0.5 μM for PD-0325901, 5 μM for RDEA119, 0.5 μM for AUY922 (all from Cambridge Bioscience) and 16 μM for BAY 61-3606 (Insight Biotechnology), respectively. For drug co-treatments, the respective compounds were combined at constant ratios over the entire concentration range used for the single-agent treatments, keeping the final DMSO concentration in the assay at 0.33%. After 72 h of incubation time under standard culture conditions, the medium was either replaced with 150 μl of fresh culture medium containing 1 μM of Hoechst 33342 (Thermo Fisher #H3570) or 10 μl Alamar-Blue (Thermo Fisher #88952) added to each well (only C10^{MERTKNKO} and C10^{MERTKNWT}). Cells were incubated under standard culture conditions for 1 h or 4 h, respectively, before they were either imaged using an In Cell Analyzer 6000 automated confocal microscope (GE Healthcare) with four fields of view (FOVs) per well or before AlamarBlue fluorescence was quantified using a FLUOstar Omega plate reader (BMG Labtech). The Hoechst 33342 channel of all images was subsequently analysed with Columbus v2.6.0 (PerkinElmer) using two of the built-in algorithms "B" and "C" to automatically segment nuclei. For each well, we then counted the number of nuclei satisfying standard quality control criteria in all four FOVs. Subsequently, the nuclei count or AlamarBlue fluorescence was normalised to the mean of the corresponding DMSO controls, followed by dose–response modelling and parameter extraction (Appendix Supplementary Methods).

CRISPR/Cas9 targeting of MERTK in C10 cells

Knockout of MERTK in C10 cells was installed by CRISPR/Cas9 gene targeting as described previously (Ran et al, 2013). Guide RNA sequences were selected by using the CRISPR Design Tool (http://www.genome-engineering.org/crispr/?page_id=41). Three guide RNAs targeting exons 7, 8 (encoding the FNIII domain) and 14 (encoding the kinase domain) of MERTK were obtained, which were designed to induce double-strand breaks at position 982, 1,273 and 1,882 bp. The guide RNA sequences were cloned into pSpCas9

(BB)-2A-GFP (the vector was a gift from Feng Zhang; Addgene plasmid #48138) by standard Golden Gate Assembly using the BbsI site, followed by transformation of chemically competent DH5α cells. Three colonies of each sgRNA transformation were picked and the correct insertion confirmed by sequencing. C10 cells were transfected with a mixture of all three sgRNAs using Lipofectamine 3000 (Invitrogen) according to the manufacturer's instructions. Transfected cells were microscopically identified by expression of GFP and sub-cloned 3 days post-transfection using cloning rings. After 2–4 weeks, individual cell clones were tested for successful knockout of MERTK by Western blot and sequencing.

Western blots

Total cell lysates were diluted to 1 mg/ml protein concentration and 1× final sample buffer concentration with 4× sample buffer (70 mM Tris pH 6.8, 5% v/v 2-mercaptoethanol, 40% v/v glycerol, 3% w/v SDS, 0.05% w/v bromophenol blue) and stored at −80°C until further use. Samples were separated using 4-12% NuPAGE Bis-Tris mini/midi gels (70 μg/14 μg per sample) and subsequently blotted to Hybond-P PVDF (Amersham) or nitrocellulose (iBlot Transfer Stack, Thermo Fisher) membranes according to the manufacturer's instructions. We used primary antibodies against EPHA4 (ab157588, 1:500, Abcam), ABL1 (OP20, 1:1,000, Merck), BRAF^{V600E} (E19290, 1:500, Spring Bioscience) and ERK1/2 (#4695, 1:1,000, Cell Signaling Technology) for PVDF and primary antibodies against MERTK (ab52968, 1:2,000, Abcam) and ACTB (sc-47778, 1:1,000, Santa Cruz) for nitrocellulose membranes. Membranes were incubated in the dark with appropriate fluorophore-coupled secondary antibodies (IRDye 800CW goat anti-mouse IgG #926-32210, IRDye 680RD donkey anti-rabbit IgG #926-68023, IRDye 800CW goat anti-rabbit IgG #926-32211 and IRDye 680RD goat anti-mouse IgG #925-68070). Membranes were measured using an Odyssey near-infrared scanner (LI-COR). Densitometry of Western blots was carried out using ImageStudioLite v5.2.5 (LI-COR), expressing EPHA4 and ABL1 expression relative to the respective ERK1/2 signal, followed by dividing all expression values by the expression value of OXCO-1 (present on each gel) and log2-transformation.

Immunohistochemistry

In order to evaluate MERTK expression in tissue microarrays, a staining protocol was developed using positive and negative control cell lines for MERTK expression, selected based on the expression level as measured by LC-MS/MS. For each of the four control cell lines (negative: RKO, CC07; positive: HT55, CaCo-2), we grew one T-175 flask under standard culture conditions in culture medium to ~70% confluency, washed the cells once with 10 ml ice-cold PBS and subsequently scraped them into 2 ml 4% PFA in PBS, followed by centrifugation at 400 ×g for 10 min and fixation overnight. Afterwards, the cell pellets were embedded in paraffin using standard procedures. Immunohistochemistry (IHC) of QUASAR 2 TMAs and FFPE control cell pellets was carried out as already described (Schlegel et al, 2013), with minor modifications. Stainings quantifying MERTK expression were performed using the Bond-MAX automated IHC system (Leica Biosystems) at room temperature if not specified otherwise. After deparaffinising and re-hydrating the slides, heat-induced epitope retrieval (HIER) was performed for 10 min at

100°C with epitope retrieval solution 1 (AR9961, pH 6). MERTK was detected using the Bond Polymer Refine Detection Kit (DS9800) together with a primary antibody against MERTK (ab52968, 1:1,000, Abcam) according to the manufacturer's instructions, followed by counterstaining nuclei with haematoxylin solution (< 0.1% haematoxylin). Slides were washed, dehydrated with an increasing alcohol series (30 s in 50, 70, 100 and 100% EtOH), cleared with two 30-s washes in 100% xylene and subsequently mounted using DPX mountant (Sigma Aldrich). Finally, slides were scanned at 40× magnification using a ScanScope (Aperio).

Sample processing for mass spectrometry

Full proteomes

Acetone precipitation and re-solubilisation TCLs were first acetone-precipitated overnight using four volumes of pre-cooled (−40°C) acetone to remove detergents, followed by two additional washing steps with 1 ml fresh, cold acetone. In between precipitation and washing steps, samples were centrifuged for 10 min at 13,000 ×g and 4°C. After the final washing step, the supernatant was taken off and the samples were left to dry in a fume hood at room temperature. Following acetone precipitation, the samples were re-suspended in urea buffer (40 mM Tris–HCl pH = 7.6, 8 M urea) containing protease (Complete mini without EDTA; Roche) and phosphatase inhibitors (Phosphatase Inhibitor cocktail 1 and 2; Sigma Aldrich) at 1× and 5× the final concentration recommended by the manufacturer, respectively, as well as 20 nM calyculin A. In order to ensure proper re-solubilisation of proteins, the precipitates were first mixed thoroughly by pipetting up and down, followed by sonication of each sample for 5 min on ice using an HF generator GM mini20, equipped with an ultrasonic converter UW mini20 and a microtip MS 2.5 sonotrode (Bandelin), which was set to 3-s pulses at 30% intensity with 3-s pause in between. After re-solubilisation, the protein concentration of the lysate was determined again using a Coomassie-based protein assay kit (Thermo Fisher).

In-solution digestion For in-solution digestion of proteins, 3.5 mg of TCL per sample was reduced with 10 mM DTT and subsequently alkylated using 55 mM chloroacetamide. Afterwards, samples were diluted with 40 mM Tris–HCl pH 7.6 to reduce the urea concentration to 1.5 M, 1.5 µl of $CaCl_2$ was added to each sample and proteins were digested overnight at 37°C and 700 rpm in a thermomixer using trypsin (Roche) at a protease-to-protein ratio of 1:50.

Desalting Desalting of peptide mixtures was carried out at room temperature using Sep-Pak cartridges (50 mg sorbent per cartridge, Waters) and a vacuum manifold according to the manufacturer's instructions. Desalted samples were stored at −80°C until further use.

hSAX chromatography For hydrophilic strong anion exchange (hSAX) chromatography, peptide solutions were first dried down in a Speed-Vac, re-solubilised in hSAX solvent A (5 mM Tris–HCl, pH 8.5) to a concentration of 2.73 µg/µl peptide and then centrifuged for 30 s at 5,000 ×g to spin down insoluble debris. Chromatography was carried out using a Dionex Ultimate 3000 HPLC system (Thermo Fisher), which was equipped with an IonPac AG24 guard column (Thermo Fisher), as well as an IonPac AS24 strong anion exchange column (Thermo Fisher). Chromatography was performed

at 30°C and a flow rate of 250 µl/min. Following the injection of 100 µl sample and 3 min of equilibration with 100% hSAX solvent A, peptides were eluted using a two-step linear gradient from 0 to 25% hSAX solvent B (5 mM Tris–HCl, pH 8.5, 1 M NaCl) in 24 min and from 25 to 100% solvent B in 13 min. Solvent B was kept at 100% for another 4 min to flush the column before returning to 0% in 1 min and additional 5 min of equilibration with 100% hSAX solvent A. We started collecting 48 fractions of 250 µl after 2 min of gradient time, which were subsequently combined to form 24 fractions after consulting the 214-nm chromatography trace. For that, fractions 5–7 were pooled to form fraction 5, fractions 24–25 formed fraction 22, fractions 26–28 formed fraction 23 and fractions 29–48 formed fraction 24, respectively.

Post-hSAX desalting All 24 fractions were desalted using StageTips as described earlier (Rappsilber *et al*, 2007), with minor modifications. Briefly, we used five C18 discs (3M Empore) of about 1.5 mm diameter in 200-µl pipette tips. After wetting (MeOH followed by 5% TFA in 80% ACN) and equilibration of stage tips (0.1% TFA), acidified samples (pH 2) were loaded twice onto the StageTips to ensure proper binding. Following two washes with 0.1% TFA, samples were eluted using 200 µl 0.1% TFA in 60% ACN. Eluates were transferred to 96-well plates, evaporated to dryness using a Speed-Vac and subsequently stored at −20°C until further use.

Kinobeads

Kinobeads gamma (KBγ) pulldowns (biological triplicates) were performed in 96-well filter plates (Porvair Sciences) as described elsewhere (Medard *et al*, 2015), with minor modifications. For each pulldown, 3 ml of 2.2 mg/ml TCL was cleared by ultracentrifugation at 167,000 ×g and 4°C for 20 min. Following washing (CP buffer: 50 mM Tris–HCl pH 7.5, 5% glycerol, 1.5 mM $MgCl_2$, 150 mM NaCl, 1 mM Na_3VO_4) and equilibration (CP buffer supplemented with 0.4% Igepal CA-630) of 35 µl settled KBγ per TCL, ~1.8 ml (equivalent of 4 mg of protein) of each TCL was transferred to its corresponding well. After 60-min incubation at 4°C on a head-over-end shaker, the beads were washed thrice with CP buffer containing 0.4% Igepal CA-630 and twice with CP buffer containing 0.2% Igepal CA-630. Subsequently, proteins were eluted by incubating the beads for 30 min at 50°C and 700 rpm in a thermomixer with 60 µl of 2× NuPAGE LDS sample buffer (Thermo Fisher) per well. After collecting eluates by centrifugation, samples were alkylated using 55 mM chloroacetamide. Finally, detergents and salts were removed from samples by running a short electrophoresis (~0.5 cm) using 4−12% NuPAGE gels (Thermo Fisher), followed by tryptic in-gel digestion according to the standard procedures.

LC-MS/MS data acquisition

Full proteomes

Reverse-phase gradient Full proteome fractions were measured using nanoflow LC-MS/MS by directly coupling an Eksigent nanoLC-Ultra 1D+ (Eksigent) to an Orbitrap Velos mass spectrometer (Thermo Fisher). Peptides were dissolved in 20 µl buffer A (0.1% FA), injecting 5 µl per measurement. Using a flow rate of 5 µl/min, peptides were then loaded onto a 2-cm trap column

(100 μm i.d., ReproSil-Pur 120 ODS-3 5 μm, Dr. Maisch) and washed for 10 min with 100% buffer A. Subsequently, peptides were separated on a 40-cm analytical column (75 μm i.d., ReproSil-Gold 120 C18 3 μm, Dr. Maisch) using a flow rate of 300 nl/min and a gradient from 2% buffer B (0.1% FA and 5% DMSO in ACN) to 4% in 2 min (buffer A now also contained 5% DMSO) and from 4 to 32% in 96 min. Buffer B was then ramped from 32 to 80% in 1 min and the column was flushed with 80% buffer B for 4 min, before returning to 2% buffer B in 2 min and a final equilibration step with 2% buffer B for 5 min. This resulted in a turnaround time of 120 min per full proteome fraction, totalling 130 days of measurement time for the entire CRC65 cell line panel.

Acquisition parameters The eluate from the analytical column was sprayed via stainless steel emitters (Thermo) at a source voltage of 2.6 kV towards the orifice of the mass spectrometer; a transfer capillary heated to 275°C. The Orbitrap Velos was set to data-dependent acquisition in positive ion mode, automatically selecting the top 10 most intense precursor ions from the preceding full MS (MS1) spectrum with an isolation width of 2.0 Th for fragmentation using higher-energy collisional dissociation (HCD) at 30% normalised collision energy and subsequent identification by MS/MS (MS2). MS1 (360–1,300 m/z) and MS2 (precursor-dependent m/z range, starting at m/z 100) spectra were acquired in the Orbitrap using a resolution of 30,000 and 7,500 at m/z 400, with an automatic gain control (AGC) target value of 1×10^6 and 3×10^4 charges and a maximum injection time of 100 and 200 ms, respectively. Dynamic exclusion was set to 60 s.

Kinobeads

Reverse-phase gradient Kinobeads eluates were measured using nanoflow LC-MS/MS by directly coupling an Eksigent nanoLC-Ultra 1D+ (Eksigent) to an Orbitrap Elite mass spectrometer (Thermo Fisher). Chromatography as well as data acquisition was similar to the settings used for full proteome fractions; hence, we only describe differences between the two set-ups. We injected 10 μl per measurement instead of 5 μl and used a slightly steeper gradient from 2% buffer B to 4% in 2 min and from 4 to 32% in 88 min instead of 96 min. The rest of the gradient was kept the same, resulting in a turnaround time of 110 min per Kinobeads pulldown, totalling ~15 days of measurement time for the entire CRC65 cell line panel in biological triplicates.

Acquisition parameters The source voltage was at 2.2 kV instead of 2.6 kV and the Orbitrap Elite selected the top 15 most intense precursor ions for MS2 instead of the top 10 most intense ones because of its higher scanning speed. While the MS1 m/z range, resolution, AGC target value and maximum injection time were the same as for the Orbitrap Velos, the Orbitrap Elite acquired MS2 spectra at a resolution of 15,000 at m/z 400 with an AGC target value of 2×10^4 charges and a maximum injection time of 100 ms. Dynamic exclusion was set to 20 s instead of 60 s and the Orbitrap Elite also made use of a global kinase peptide inclusion list, which contained precursor ions and retention times from frequently observed kinase peptides.

Processing of LC-MS/MS raw data

Full proteomes and CPTAC patient data

MaxQuant v.1.5.3.30 (Cox & Mann, 2008) was used to search our LC-MS/MS raw data, as well as the raw data from the original CPTAC publication on human colon and rectal cancer (Zhang et al, 2014) against the UniProtKB human reference proteome (v25.11.2015; 92,011 sequences), concatenated with a list of common contaminants supplied by MaxQuant (245 sequences) in two separate runs with identical settings. Therefore, some data used in this publication were generated by the Clinical Proteomic Tumor Analysis Consortium (NCI/NIH). We set the digestion mode to fully tryptic, allowing for cleavage before proline (Trypsin/P) and a maximum of two missed cleavages. Carbamidomethylation of cysteines was set as a fixed modification and oxidation of methionines and acetylation of protein N-termini were set as variable modifications, allowing for a maximum number of five modifications per peptide. Candidate peptides were required to have a length of at least seven amino acids, with a maximum peptide mass of 4,600 Da. The fragment ion tolerance was set to 20 ppm for FTMS (CRC65) and 0.4 Da for ITMS spectra (CPTAC), respectively. A first search with a precursor ion tolerance of 20 ppm was used to recalibrate raw data based on all peptide spectrum matches (PSMs) without filtering using hard score cut-offs. After recalibration, the data were searched with a precursor ion tolerance of 4.5 ppm, while chimeric spectra were searched a second time using MaxQuant's "Second peptides" option to identify co-fragmented peptide precursors. We used "Match between runs" with an alignment time window of 30 min and a match time window of 1.1 min to transfer identifications between raw files of the same and neighbouring fractions (± 1 fraction). Using the classical target-decoy approach with a concatenated database of reversed peptide sequences, data were filtered using a PSM and protein false discovery rate (FDR) of 1%. Protein groups were required to have at least one unique or razor peptide, with each razor peptide being used only once during the calculation of the protein FDR. No score cut-offs were applied in addition to the target-decoy FDR.

Kinobeads

Raw files from triplicate Kinobeads pulldowns were processed using a pipeline similar to the one employed for the analysis of full proteomes, adapting some of the parameters described hereafter. The fragment ion tolerance was set to 120 ppm for FTMS spectra instead of 20 ppm, since we observed systematic fragment mass deviations, which were linearly dependent on the m/z of the fragment ions in ppm space. Because this was likely due to problems during data acquisition, we had to compensate for it during the processing of raw data. Since Kinobeads pulldowns were not fractionated, "Match between runs" was used with the same parameters as described above to transfer identifications between all raw files.

Quantification, statistical analysis and multi-omics data integration

Proteins detected in proteomics experiments were quantified based on MaxQuant output data, which was subsequently integrated with transcriptomics data from various CRC samples (see Appendix Supplementary Methods for details). All statistical analyses were carried out using R v3.2.4 (R Core Team, 2016).

Code availability

Modified MComBat and computeAUC functions (Appendix Supplementary Methods) can be downloaded from https://github.com/mfrejno/pharmacoproteomics_crc.

Acknowledgements

We thank Marcus Green and Helen Scott for help with the immunohistochemistry of TMAs, Guillaume Médard for fruitful discussions and Michaela Krötz-Fahning, Andrea Hubauer and Andreas Klaus for laboratory assistance. This work was supported in part by the Medical Research Council, the Department of Oncology of the University of Oxford and a Scatcherd European Scholarship (to M.F.).

Author contributions

Conceptualisation: MF, SKn, SMF and BK; Methodology: MF, MW, BR, CM and SMF; Software: MF, MW and CM; Formal Analysis: MF; Investigation: MF, RZC, HK, RZ, SKl, KK, AJ, SH, MJ, JS-H and WW; Resources: EJ, ED, DK, WW, SKn, SMF and BK; Data curation: MF: Writing—original draft: MF, BK; writing—review and editing, MF, SMF and BK; Visualisation: MF; Supervision: WW, SKn, SMF and BK; Project administration: BK; Funding acquisition, MF, SKn, SMF and BK.

References

Bantscheff M, Eberhard D, Abraham Y, Bastuck S, Boesche M, Hobson S, Mathieson T, Perrin J, Raida M, Rau C, Reader V, Sweetman G, Bauer A, Bouwmeester T, Hopf C, Kruse U, Neubauer G, Ramsden N, Rick J, Kuster B et al (2007) Quantitative chemical proteomics reveals mechanisms of action of clinical ABL kinase inhibitors. Nat Biotechnol 25: 1035–1044

Barretina J, Caponigro G, Stransky N, Venkatesan K, Margolin AA, Kim S, Wilson CJ, Lehar J, Kryukov GV, Sonkin D, Reddy A, Liu M, Murray L, Berger MF, Monahan JE, Morais P, Meltzer J, Korejwa A, Jane-Valbuena J, Mapa FA et al (2012) The cancer cell line encyclopedia enables predictive modelling of anticancer drug sensitivity. Nature 483: 603–607

Beech J, Kelly K (2014) Abstract 5594: JNK1 and MERTK are markers of MEK inhibitor resistance and new targets for therapy. Can Res 73: 5594

Bertotti A, Papp E, Jones S, Adleff V, Anagnostou V, Lupo B, Sausen M, Phallen J, Hruban CA, Tokheim C, Niknafs N, Nesselbush M, Lytle K, Sassi F, Cottino F, Migliardi G, Zanella ER, Ribero D, Russolillo N, Mellano A et al (2015) The genomic landscape of response to EGFR blockade in colorectal cancer. Nature 526: 263–267

Calon A, Lonardo E, Berenguer-Llergo A, Espinet E, Hernando-Momblona X, Iglesias M, Sevillano M, Palomo-Ponce S, Tauriello DV, Byrom D, Cortina C, Morral C, Barcelo C, Tosi S, Riera A, Attolini CS, Rossell D, Sancho E, Batlle E (2015) Stromal gene expression defines poor-prognosis subtypes in colorectal cancer. Nat Genet 47: 320–329

Cox J, Mann M (2008) MaxQuant enables high peptide identification rates, individualized p.p.b.-range mass accuracies and proteome-wide protein quantification. Nat Biotechnol 26: 1367–1372

Cox J, Hein MY, Luber CA, Paron I, Nagaraj N, Mann M (2014) Accurate proteome-wide label-free quantification by delayed normalization and maximal peptide ratio extraction, termed MaxLFQ. Mol Cell Proteomics 13: 2513–2526

Dunne PD, McArt DG, Bradley CA, O'Reilly PG, Barrett HL, Cummins R, O'Grady T, Arthur K, Loughrey MB, Allen WL, McDade SS, Waugh DJ, Hamilton PW, Longley DB, Kay EW, Johnston PG, Lawler M, Salto-Tellez M, Van Schaeybroeck S (2016) Challenging the cancer molecular stratification dogma: intratumoral heterogeneity undermines consensus molecular subtypes and potential diagnostic value in colorectal cancer. Clin Cancer Res 22: 4095–4104

Emaduddin M, Bicknell DC, Bodmer WF, Feller SM (2008) Cell growth, global phosphotyrosine elevation, and c-Met phosphorylation through Src family kinases in colorectal cancer cells. Proc Natl Acad Sci USA 105: 2358–2362

Garnett MJ, Edelman EJ, Heidorn SJ, Greenman CD, Dastur A, Lau KW, Greninger P, Thompson IR, Luo X, Soares J, Liu Q, Iorio F, Surdez D, Chen L, Milano RJ, Bignell GR, Tam AT, Davies H, Stevenson JA, Barthorpe S et al (2012) Systematic identification of genomic markers of drug sensitivity in cancer cells. Nature 483: 570–575

Gholami AM, Hahne H, Wu Z, Auer FJ, Meng C, Wilhelm M, Kuster B (2013) Global proteome analysis of the NCI-60 cell line panel. Cell Rep 4: 609–620

Graham DK, DeRyckere D, Davies KD, Earp HS (2014) The TAM family: phosphatidylserine sensing receptor tyrosine kinases gone awry in cancer. Nat Rev Cancer 14: 769–785

Guinney J, Dienstmann R, Wang X, de Reynies A, Schlicker A, Soneson C, Marisa L, Roepman P, Nyamundanda G, Angelino P, Bot BM, Morris JS, Simon IM, Gerster S, Fessler E, De Sousa EMF, Missiaglia E, Ramay H, Barras D, Homicsko K et al (2015) The consensus molecular subtypes of colorectal cancer. Nat Med 21: 1350–1356

Iorio F, Knijnenburg TA, Vis DJ, Bignell GR, Menden MP, Schubert M, Aben N, Goncalves E, Barthorpe S, Lightfoot H, Cokelaer T, Greninger P, van Dyk E, Chang H, de Silva H, Heyn H, Deng X, Egan RK, Liu Q, Mironenko T et al (2016) A landscape of pharmacogenomic interactions in cancer. Cell 166: 740–754

Isella C, Terrasi A, Bellomo SE, Petti C, Galatola G, Muratore A, Mellano A, Senetta R, Cassenti A, Sonetto C, Inghirami G, Trusolino L, Fekete Z, De Ridder M, Cassoni P, Storme G, Bertotti A, Medico E (2015) Stromal contribution to the colorectal cancer transcriptome. Nat Genet 47: 312–319

Johnson WE, Li C, Rabinovic A (2007) Adjusting batch effects in microarray expression data using empirical Bayes methods. Biostatistics 8: 118–127

Kerr RS, Love S, Segelov E, Johnstone E, Falcon B, Hewett P, Weaver A, Church D, Scudder C, Pearson S, Julier P, Pezzella F, Tomlinson I, Domingo E, Kerr DJ (2016) Adjuvant capecitabine plus bevacizumab versus capecitabine alone in patients with colorectal cancer (QUASAR 2): an open-label, randomised phase 3 trial. Lancet Oncol 17: 1543–1557

Kim TM, Laird PW, Park PJ (2013) The landscape of microsatellite instability in colorectal and endometrial cancer genomes. Cell 155: 858–868

Koch H, Busto ME, Kramer K, Medard G, Kuster B (2015) Chemical proteomics uncovers EPHA2 as a mechanism of acquired resistance to small molecule EGFR kinase inhibition. J Proteome Res 14: 2617–2625

Lawrence RT, Perez EM, Hernandez D, Miller CP, Haas KM, Irie HY, Lee SI, Blau CA, Villen J (2015) The proteomic landscape of triple-negative breast cancer. Cell Rep 11: 630–644

McDermott U, Downing JR, Stratton MR (2011) Genomics and the continuum of cancer care. N Engl J Med 364: 340–350

Medard G, Pachl F, Ruprecht B, Klaeger S, Heinzlmeir S, Helm D, Qiao H, Ku X, Wilhelm M, Kuehne T, Wu Z, Dittmann A, Hopf C, Kramer K, Kuster B (2015) Optimized chemical proteomics assay for kinase inhibitor profiling. J Proteome Res 14: 1574–1586

Medico E, Russo M, Picco G, Cancelliere C, Valtorta E, Corti G, Buscarino M, Isella C, Lamba S, Martinoglio B, Veronese S, Siena S, Sartore-Bianchi A, Beccuti M, Mottolese M, Linnebacher M, Cordero F, Di Nicolantonio F, Bardelli A (2015) The molecular landscape of colorectal cancer cell lines unveils clinically actionable kinase targets. *Nat Commun* 6: 7002

Mertins P, Mani DR, Ruggles KV, Gillette MA, Clauser KR, Wang P, Wang X, Qiao JW, Cao S, Petralia F, Kawaler E, Mundt F, Krug K, Tu Z, Lei JT, Gatza ML, Wilkerson M, Perou CM, Yellapantula V, Huang KL et al (2016) Proteogenomics connects somatic mutations to signalling in breast cancer. *Nature* 534: 55–62

Mouradov D, Sloggett C, Jorissen RN, Love CG, Li S, Burgess AW, Arango D, Strausberg RL, Buchanan D, Wormald S, O'Connor L, Wilding JL, Bicknell D, Tomlinson IP, Bodmer WF, Mariadason JM, Sieber OM (2014) Colorectal cancer cell lines are representative models of the main molecular subtypes of primary cancer. *Cancer Res* 74: 3238–3247

R Core Team (2016) *R: A language and environment for statistical computing.* Vienna, Austria: R Foundation for Statistical Computing. URL http://www.R-project.org/

Ran FA, Hsu PD, Wright J, Agarwala V, Scott DA, Zhang F (2013) Genome engineering using the CRISPR-Cas9 system. *Nat Protoc* 8: 2281–2308

Rappsilber J, Mann M, Ishihama Y (2007) Protocol for micro-purification, enrichment, pre-fractionation and storage of peptides for proteomics using StageTips. *Nat Protoc* 2: 1896–1906

Rees MG, Seashore-Ludlow B, Cheah JH, Adams DJ, Price EV, Gill S, Javaid S, Coletti ME, Jones VL, Bodycombe NE, Soule CK, Alexander B, Li A, Montgomery P, Kotz JD, Hon CS, Munoz B, Liefeld T, Dancik V, Haber DA et al (2016) Correlating chemical sensitivity and basal gene expression reveals mechanism of action. *Nat Chem Biol* 12: 109–116

Sadanandam A, Lyssiotis CA, Homicsko K, Collisson EA, Gibb WJ, Wullschleger S, Ostos LC, Lannon WA, Grotzinger C, Del Rio M, Lhermitte B, Olshen AB, Wiedenmann B, Cantley LC, Gray JW, Hanahan D (2013) A colorectal cancer classification system that associates cellular phenotype and responses to therapy. *Nat Med* 19: 619–625

Schlegel J, Sambade MJ, Sather S, Moschos SJ, Tan AC, Winges A, DeRyckere D, Carson CC, Trembath DG, Tentler JJ, Eckhardt SG, Kuan PF, Hamilton RL, Duncan LM, Miller CR, Nikolaishvili-Feinberg N, Midkiff BR, Liu J, Zhang W, Yang C et al (2013) MERTK receptor tyrosine kinase is a therapeutic target in melanoma. *J Clin Invest* 123: 2257–2267

Schwanhausser B, Busse D, Li N, Dittmar G, Schuchhardt J, Wolf J, Chen W, Selbach M (2011) Global quantification of mammalian gene expression control. *Nature* 473: 337–342

Stein CK, Qu P, Epstein J, Buros A, Rosenthal A, Crowley J, Morgan G, Barlogie B (2015) Removing batch effects from purified plasma cell gene expression microarrays with modified ComBat. *BMC Bioinformatics* 16: 63

Strimpakos A, Pentheroudakis G, Kotoula V, De RW, Kouvatseas G, Papakostas P, Makatsoris T, Papamichael D, Andreadou A, Sgouros J, Zizi-Sermpetzoglou A, Kominea A, Televantou D, Razis E, Galani E, Pectasides D, Tejpar S, Syrigos K, Fountzilas G (2013) The prognostic role of ephrin A2 and endothelial growth factor receptor pathway mediators in patients with advanced colorectal cancer treated with cetuximab. *Clin Colorectal Cancer* 12: 267–274.e262

Sun Z, Zhang Z, Liu Z, Qiu B, Liu K, Dong G (2014) MicroRNA-335 inhibits invasion and metastasis of colorectal cancer by targeting ZEB2. *Med Oncol* 31: 982

Tavazoie SF, Alarcon C, Oskarsson T, Padua D, Wang Q, Bos PD, Gerald WL, Massague J (2008) Endogenous human microRNAs that suppress breast cancer metastasis. *Nature* 451: 147–152

Vizcaino JA, Csordas A, del-Toro N, Dianes JA, Griss J, Lavidas I, Mayer G, Perez-Riverol Y, Reisinger F, Ternent T, Xu QW, Wang R, Hermjakob H (2016) 2016 update of the PRIDE database and its related tools. *Nucleic Acids Res* 44: D447–D456

Watanabe T, Kobunai T, Yamamoto Y, Matsuda K, Ishihara S, Nozawa K, Iinuma H, Ikeuchi H, Eshima K (2011) Differential gene expression signatures between colorectal cancers with and without KRAS mutations: crosstalk between the KRAS pathway and other signalling pathways. *Eur J Cancer* 47: 1946–1954

Zhang B, Wang J, Wang X, Zhu J, Liu Q, Shi Z, Chambers MC, Zimmerman LJ, Shaddox KF, Kim S, Davies SR, Wang S, Wang P, Kinsinger CR, Rivers RC, Rodriguez H, Townsend RR, Ellis MJ, Carr SA, Tabb DL et al (2014) Proteogenomic characterization of human colon and rectal cancer. *Nature* 513: 382–387

Zhang H, Liu T, Zhang Z, Payne SH, Zhang B, McDermott JE, Zhou JY, Petyuk VA, Chen L, Ray D, Sun S, Yang F, Chen L, Wang J, Shah P, Cha SW, Aiyetan P, Woo S, Tian Y, Gritsenko MA et al (2016) Integrated proteogenomic characterization of human high-grade serous ovarian cancer. *Cell* 166: 755–765

Zou H, Hastie T (2005) Regularization and variable selection via the elastic net. *J Roy Stat Soc B* 67: 301–320

7

Pervasive coexpression of spatially proximal genes is buffered at the protein level

Georg Kustatscher[1],* (iD), Piotr Grabowski[2] (iD) & Juri Rappsilber[1,2],** (iD)

Abstract

Genes are not randomly distributed in the genome. In humans, 10% of protein-coding genes are transcribed from bidirectional promoters and many more are organised in larger clusters. Intriguingly, neighbouring genes are frequently coexpressed but rarely functionally related. Here we show that coexpression of bidirectional gene pairs, and closeby genes in general, is buffered at the protein level. Taking into account the 3D architecture of the genome, we find that co-regulation of spatially close, functionally unrelated genes is pervasive at the transcriptome level, but does not extend to the proteome. We present evidence that non-functional mRNA coexpression in human cells arises from stochastic chromatin fluctuations and direct regulatory interference between spatially close genes. Protein-level buffering likely reflects a lack of coordination of post-transcriptional regulation of functionally unrelated genes. Grouping human genes together along the genome sequence, or through long-range chromosome folding, is associated with reduced expression noise. Our results support the hypothesis that the selection for noise reduction is a major driver of the evolution of genome organisation.

Keywords gene expression noise; genome organisation; proteomics; regulatory interference; transcriptomics

Subject Categories Chromatin, Epigenetics, Genomics & Functional Genomics; Genome-Scale & Integrative Biology

Introduction

The position of genes in the human genome is not random (Hurst et al, 2004). Genes are often found in pairs or larger clusters that tend to be coexpressed (Caron et al, 2001; Lercher et al, 2002; Trinklein et al, 2004). Some of these coordinate transcription of genes with related functions, for example histone genes and other clusters resulting from gene duplication. However, the majority of closeby, coexpressed human genes appear not to have a higher functional similarity than random gene pairs (Hurst et al, 2004; Williams & Bowles, 2004; Li et al, 2006; Purmann et al, 2007; Michalak, 2008; Xu et al, 2012). For example, 35 DNA repair genes are transcribed from bidirectional promoters, but none of their paired genes is involved in DNA repair (Xu et al, 2012). This raises intriguing questions: Why are functionally unrelated genes clustered in the genome and how can the cell tolerate their coexpression?

Pioneering work in yeast identified the selection for reduced gene expression noise as a key driver for the evolution of chromosome organisation (Batada & Hurst, 2007; Wang et al, 2011). A major cause of gene expression noise is thought to be the random fluctuation of chromatin domains between an active and inactive state, causing mRNAs to be synthesised in short, stochastic bursts (Raj et al, 2006). Clusters of active genes may mutually reinforce their open chromatin state, minimising stochastic chromatin remodelling, and thereby reduce expression noise (Batada & Hurst, 2007; Wang et al, 2011). Similarly, genes flanking bidirectional promoters have lower expression noise than other genes, even if one of the divergent partners is a noncoding RNA (Wang et al, 2011). Noise-sensitive genes, such as those encoding protein complex subunits, are enriched among bidirectional pairs, but neither in yeast nor in human do any of these pairs encode two subunits of the same protein complex (Li et al, 2006; Wang et al, 2011). Consequently, it has been suggested that bidirectional promoters may drive noise reduction rather than the coexpression of functionally related genes (Wang et al, 2011).

The noise reduction model not only provides a potential explanation for the occurrence of clusters of functionally unrelated genes, but also predicts that such genes may be coexpressed (Wang et al, 2011). In yeast, chromatin-modifying enzymes are major contributors to gene expression noise (Newman et al, 2006) and chromatin remodelling drives the incidental coexpression of neighbouring, functionally unrelated genes (Batada et al, 2007). This coexpression may be due to a passive mechanism, whereby random transitions between open and closed chromatin simultaneously expose all genes within a chromatin domain to the transcriptional machinery. Alternatively, for very close genes such as those with bidirectional promoters, the up- or downregulation of one gene may directly affect the transcriptional status of its neighbour (Wang et al, 2011). Indeed, such a "ripple effect" of transcriptional activation has been observed in yeast and humans (Ebisuya et al, 2008). The noise and

1 Wellcome Trust Centre for Cell Biology, University of Edinburgh, Edinburgh, UK
2 Chair of Bioanalytics, Institute of Biotechnology, Technische Universität Berlin, Berlin, Germany
   *Corresponding author. E-mail: georg.kustatscher@ed.ac.uk
   **Corresponding author. E-mail: juri.rappsilber@ed.ac.uk

expression levels of transgenes also vary with their insertion site, as a result of both domain-wide effects and interference with individual neighbouring genes (Gierman *et al*, 2007; Chen & Zhang, 2016). Transgenes can also affect the mRNA expression levels of endogenous genes located close to the insertion site (Akhtar *et al*, 2013).

If the transcription of noise-reduced, clustered genes is unduly influenced by their neighbours, how can individual genes reach their optimal expression levels? Notably, gene expression is usually measured at the mRNA level. However, protein levels are buffered against certain transcript fluctuations (Liu *et al*, 2016), such as those caused by stochastic transcription initiation (Raj *et al*, 2006; Gandhi *et al*, 2011) and genetic variation between individuals (Battle *et al*, 2015) and species (Khan *et al*, 2013). The abundance of some proteins can also be buffered against gene copy number variations (Geiger *et al*, 2010; Stingele *et al*, 2012; Dephoure *et al*, 2014). We therefore speculated that protein abundances may also be buffered against regulatory interference between genes in close spatial proximity.

Results

Coexpression of bidirectional gene pairs is buffered at the protein level

We investigated the expression of 4,188 genes across 60 different human lymphoblastoid cell lines (LCLs), for which mRNA (Pickrell *et al*, 2010) and protein abundances (Battle *et al*, 2015) have been reported (Fig 1A, Dataset EV1). These genes are highly expressed in all human tissues and their promoters are in active chromatin states (Appendix Fig S1). Although constitutively active, expression levels of these "housekeeping" genes vary between LCLs, as a result of genetic and other differences, including age and growth conditions (Akey *et al*, 2007; Stark *et al*, 2014; Yuan *et al*, 2015). The LCL cell line panel has been instrumental in identifying expression quantitative trait loci, that is DNA sequence variants that specifically influence the expression level of one or more genes (Albert & Kruglyak, 2015). Here, instead of assessing how a gene's expression level depends on the genotype, we analyse how it is influenced by the expression of other, closeby genes. LCLs are a valuable test system as their genome structure and regulatory elements have been mapped at unparalleled resolution (Lieberman-Aiden *et al*, 2009; Ernst *et al*, 2011; ENCODE Project Consortium, 2012; Rao *et al*, 2014).

First, we analysed gene pairs that are transcribed from bidirectional promoters. These are commonly defined as genes that are found in head-to-head orientation with < 1 kb between their transcription start sites (TSSs) (Trinklein *et al*, 2004). Out of 167 such gene pairs in this dataset, the mRNA abundances of 31 (19%) are strongly and significantly co-regulated across LCLs (Pearson's correlation coefficient, PCC > 0.5, BH-adjusted *P*-value < 0.001). However, protein co-regulation is attenuated or buffered for 28 of these (Fig 1B, Appendix Table S1). Literature analysis revealed that the buffered gene pairs generally have unrelated biological functions, in contrast to the three gene pairs whose co-regulation is sustained at the protein level (Appendix Table S1).

We next considered the 929 non-bidirectional gene pairs with up to 50 kb between their TSSs, regardless of their orientation (Dataset EV2). Although these pairs do not share a promoter region, we find that 22% have co-regulated mRNA abundances (PCC > 0.5, BH-adjusted *P* < 0.001). However, only 3% are also co-regulated at the protein level (Fig 1B).

Genes with similar functions have co-regulated mRNA and protein abundances

To confirm that the different impact of gene proximity on mRNA and protein abundances reflects a biological phenomenon, rather than simply a difference in data quality, we assessed the co-regulation of genes with known functional links, irrespective of their genomic position. We analysed subunits of the same protein complex, enzymes catalysing consecutive reactions in metabolic pathways and proteins with identical subcellular localisations. In all cases, we observe strong co-regulation on mRNA and protein levels, but co-regulation of proteins is significantly stronger than that of mRNAs (Fig EV1, $P < 3 \times 10^{-16}$). Therefore, data quality appears not to be limiting. Instead, the observed differences between mRNA and protein co-regulation indicate that post-transcriptional processes eliminate co-regulation of genes which are related spatially, but not functionally.

A fraction of closeby genes is enriched for similar functions

Our observation that only 3% of closeby genes have co-regulated protein abundances appears to contrast with the fact that genes in close genomic proximity are enriched for similar functions (Thévenin *et al*, 2014). However, functional enrichment does not exclude the possibility that the bulk of closeby gene pairs does not share similar functions. For example, we find that co-regulation of transcripts and proteins from closeby genes is more common than for random protein pairs (Fig 1B), and this enrichment is highly significant (3% versus 0.4%, $P < 4 \times 10^{-14}$).

To analyse the relationship between gene distance and function more systematically, we assessed functional associations between our gene pairs using the STRING database (Szklarczyk *et al*, 2017). We considered gene pairs to be functionally associated if their STRING score, that is the likelihood of the association to be biologically meaningful, specific and reproducible, was > 0.7. Using this comprehensive definition, we find that 4.5% of closeby gene pairs, that is those with < 50 kb between their TSSs, are related functionally (Fig EV2A). As observed by Thévenin *et al*, we find this to be a significant enrichment over gene pairs that are farther apart. Likewise, gene pairs from the same chromosome are enriched for similar functions relative to those from different chromosomes. Nevertheless, the extent of mRNA co-regulation (22%) strongly exceeds co-function, and mRNA co-regulation of most closeby gene pairs is not sustained at the protein level (Fig EV2A).

Notably, a similar analysis in yeast has shown that adjacent genes tend to have correlated mRNA expression and are statistically enriched for similar functions (Cohen *et al*, 2000). However, in striking agreement with our observations, only about 2% of these coexpressed neighbouring gene pairs have related functions (Batada *et al*, 2007) and only for these is gene order evolutionarily conserved (Hurst *et al*, 2002). Coexpression of neighbouring genes has also been observed in *Arabidopsis thaliana*, but only a fraction

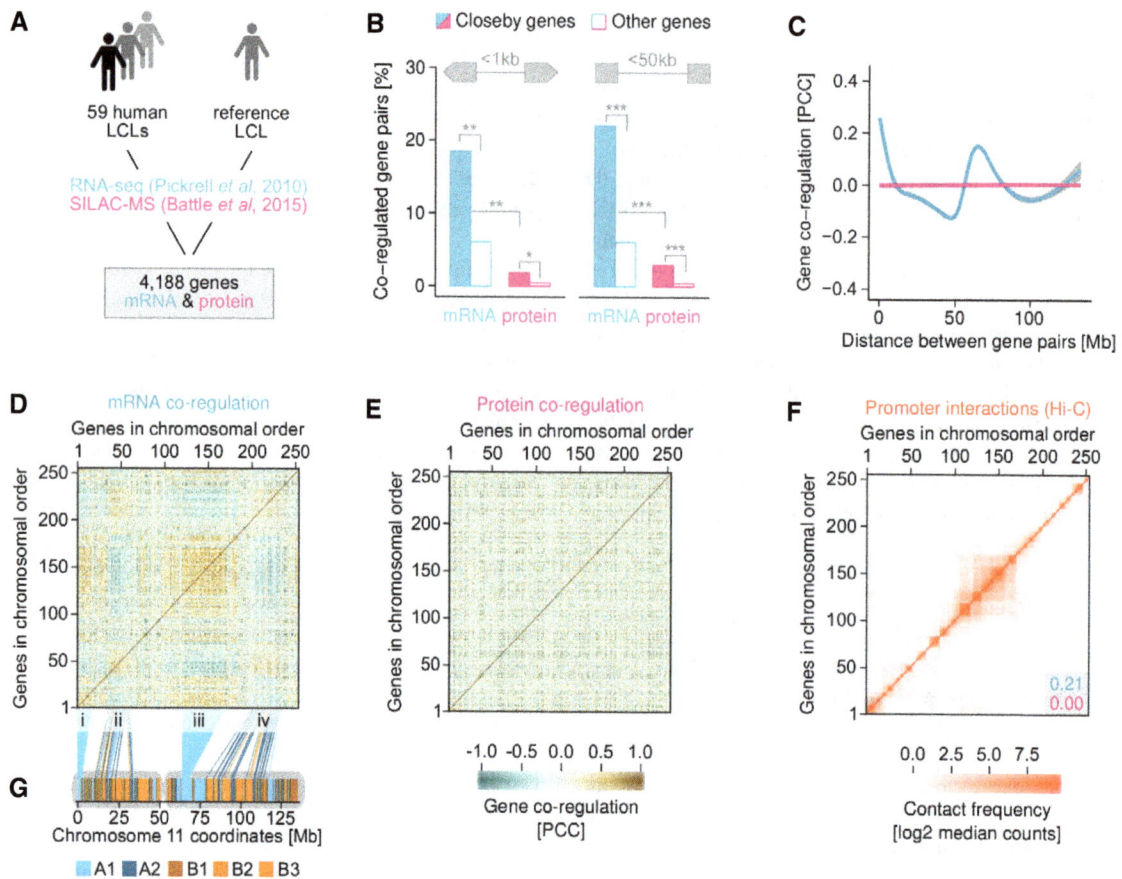

Figure 1. Spatial proximity of genes affects mRNA but not protein regulation.

A We analysed previously reported mRNA and protein abundances in 59 lymphoblastoid cell lines (LCLs), relative to a reference sample.

B Genes transcribed from bidirectional promoters frequently have co-regulated mRNA abundances, but only a fraction of these also have co-regulated protein abundances (left). The same is true for non-bidirectional gene pairs whose transcription start sites (TSS) are < 50 kb apart, irrespective of their orientation (right) (*$P < 0.05$, **$P < 2 \times 10^{-7}$, ***$P < 4 \times 10^{-14}$ based on Fisher's exact test).

C mRNA co-regulation of gene pairs on chromosome 11 decreases with chromosomal distance over many megabases, but not monotonously. Protein co-regulation is unaffected by genomic distance.

D mRNA co-regulation map for chromosome 11 showing large patches of co-regulated (brown) and anti-regulated (blue) gene pairs. Four large, co-regulated patches are highlighted (i–iv).

E No regulation patches exist on the protein level.

F mRNA co-regulation patches partially coincide with physical associations between genes derived from Hi-C data (Rao et al, 2014). Numbers in grey box show the Pearson correlation between the Hi-C map and mRNA (blue) or protein (red) co-regulation maps.

G Patches i, iii and ii, iv broadly coincide with genome subcompartments A1 and A2, respectively.

of the observed cases could be explained through a shared function (Williams & Bowles, 2004).

Long-range gene co-regulation leads to coordinated mRNA but not protein expression

The influence of gene distance on co-regulation of transcripts is not limited to genes in close proximity. As seen in the example of chromosome 11, mRNA co-regulation extends over many megabases but does not affect protein abundances (Fig 1C). Although co-regulation generally declines with increasing gene distance, such long-range effects are unlikely to result from transcriptional interference *in cis*. A major co-regulation peak of genes that are more than 50 Mb apart on chromosome 11 suggests that long-range chromosome folding

may be involved. In agreement with this, all chromosomes have distinct co-regulation curves (Appendix Fig S2).

The co-regulation map of chromosome 11 shows large patches of genes whose transcripts are coordinately up- and downregulated (Fig 1D). Importantly, no corresponding co-regulation is observed on the protein level (Fig 1E). However, the mRNA co-regulation map shows a striking similarity to physical associations observed for our gene set, as extracted from existing Hi-C data (Rao et al, 2014; Fig 1F). The Hi-C contact matrix of chromosome 11 is correlated with the mRNA co-regulation map (PCC 0.21, $P < 2 \times 10^{-318}$), but not the protein map (PCC 0.00, $P = 0.4$). Similar mRNA co-regulation patches can be observed on other chromosomes (Fig EV3) as well as between different chromosomes (Fig EV4). Generally, both intra- and interchromosomal co-regulation patches correspond to areas with increased Hi-C contacts (Appendix Table S2). Some chromosomes have more prominent patches than others (Fig EV3). Chromosome 19, which is short but exceptionally gene-dense, is unique in forming a single large co-regulation patch (Fig EV3C). Importantly, none of these mRNA co-regulation patches are reflected at the protein level (Figs EV3 and EV4, Appendix Fig S2). This suggests that regulatory interference between genes that are close in 3D could be associated with similar non-functional mRNA co-regulation as observed for neighbouring genes in the genome sequence.

We next sought to determine which structural features of the genome give rise to mRNA co-regulation patches. Four large mRNA co-regulation patches can be observed on chromosome 11 (labelled i–iv in Fig 1D). Co-regulation patches differ widely in size but often span many megabases, likely reflecting broad architectural features. Notably, promoters and enhancers typically interact on a smaller scale, within topologically associated domains (Gibcus & Dekker, 2013). However, co-regulated groups of genes are more reminiscent of genome compartments. Genome compartments were first identified on the basis of long-range interactions mapped by Hi-C, which showed that open and closed chromatin spatially segregate into two genome-wide compartments (Lieberman-Aiden *et al*, 2009). The compartments containing active and repressive chromatin were designated A and B, respectively. A high-resolution Hi-C map of the genome in LCLs subsequently identified that these compartments segregate further into six subcompartments: A1-2 and B1-4 (Rao *et al*, 2014). Genomic loci within each subcompartment tend to be associated with each other more often than with loci from other subcompartments, that is they are in closer spatial proximity. We find that co-regulation patches i and iii of chromosome 11 align with subcompartment A1 and patches ii and iv align with subcompartment A2 (Fig 1G). These are the two subcompartments of the genome formed by transcriptionally active chromatin, which is expected given that we analyse housekeeping genes. Interestingly, genes across patches i and iii are co-regulated, as are genes across patches ii and iv, suggesting that co-localisation in subcompartments may contribute to the existence of these patches.

Genes with co-regulated mRNAs co-localise in genome subcompartments

To assess systematically the overlap of co-regulated gene groups with genome compartments, we clustered genes by co-regulation. We found four transcriptome regulation groups T1-4 (Fig 2A and Dataset EV3), explaining more than 50% of the total variance (Appendix Fig S3). Transcripts within each group are co-regulated (Fig 2A and B). Genes from T1 and T2 are strongly enriched for subcompartments A2 and A1, respectively (Fig 2C). Curiously, they are anti-correlated, that is when T1 genes are upregulated, T2 genes tend to be downregulated, and vice versa (Fig 2B). Co-regulated genes of the T3 and T4 groups are also enriched for A1 and A2 subcompartments, respectively. However, they are independent of T1 and T2, that is there is neither a positive nor a negative correlation between T1/T2 and T3/T4 (Fig 2B). Therefore, while subcompartments A1 and A2 are strongly related to transcriptome regulation groups, they are not sufficient to explain them.

Genome compartments and subcompartments were defined solely based on their physical interaction patterns, but also have

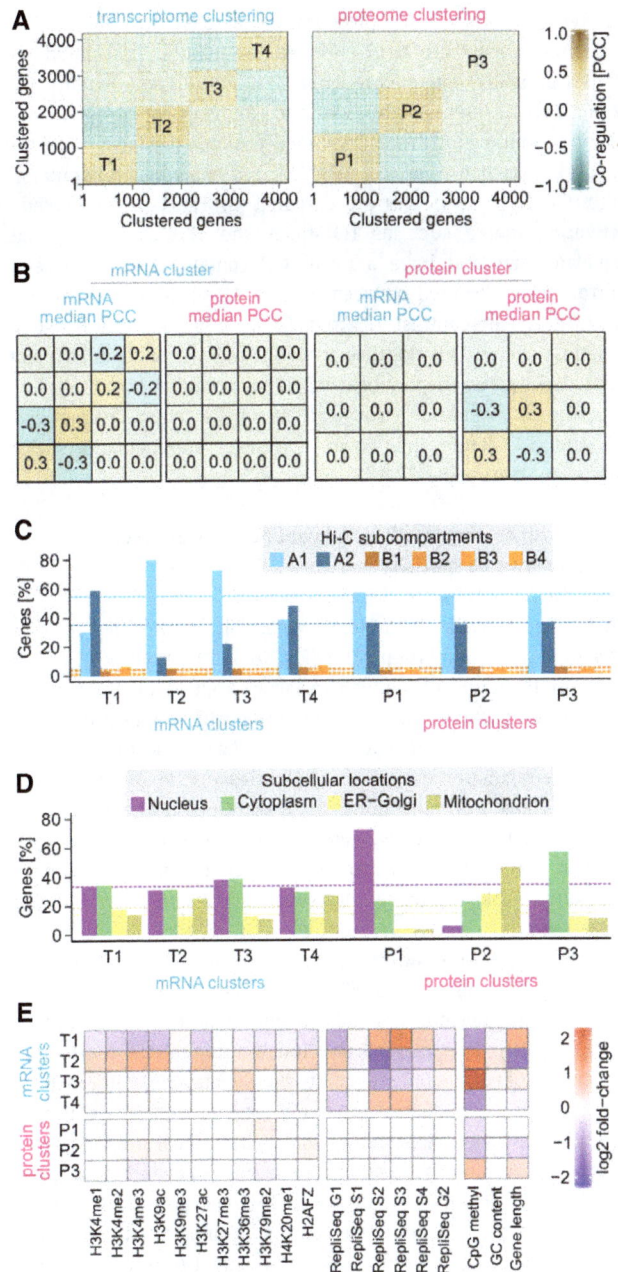

Figure 2. Transcriptome and proteome regulation are driven by different factors.

A *k*-means clustering of genes based on their mRNA or protein abundance changes across LCLs.

B Median Pearson's correlation coefficients (PCCs) for each transcriptome and proteome *k*-means cluster. Genes assigned to different *k*-means clusters can either be anti-regulated (e.g. T1 and T2) or not correlated (e.g. T1 and T3). *k*-means clusters formed by genes that are co-regulated at the mRNA level are not generally co-regulated at the protein level, and vice versa.

C Transcriptome clusters are strongly enriched for subcompartment A1 or A2. Dashed lines indicate the percentage of genes expected if subcompartments were evenly distributed across clusters.

D Proteome clusters are mainly composed of proteins from distinct subcellular locations. Dashed lines indicate the percentage of genes expected if subcellular locations were evenly distributed across clusters.

E Genomic and epigenomic features enriched in each cluster relative to the whole dataset.

different genomic and epigenomic characteristics. A1 and A2 subcompartments are both enriched for features associated with transcriptionally active chromatin, but to different extents (Rao *et al*, 2014). Interestingly, we also found clear differences in histone modifications and DNA methylation associated with transcriptome regulation groups (Fig 2E). For example, in comparison with T2, T1 gene bodies are enriched for H3K9me3, depleted in activating marks such as H3K4me3 and H3K27ac, are longer, replicate later and have a lower GC content. These differences mirror those observed between A2 and A1 subcompartments (Rao *et al*, 2014). In contrast, T3 and T4 do not show these features despite preferentially localising to A1 and A2 subcompartments. Instead, T3 genes display heavy CpG methylation, which is almost an order of magnitude stronger than for T4 genes. Consequently, T3 and T4 define their own epigenetic subpopulation within A-type compartments.

Genes with co-regulated protein abundances are related functionally, not spatially

Clustering analysis of protein expression profiles led to three proteome regulation groups P1-3 (Fig 2A and Dataset EV3), explaining more than 50% of the total variance (Appendix Fig S3). Neither genome compartments nor epigenomic signatures appear to be associated with proteome regulation groups (Fig 2C and E). In contrast, proteome regulation groups broadly correspond to subcellular locations: nucleus (P1), mitochondria, ER and Golgi (P2) and cytoplasm (P3) (Fig 2D). They are also enriched for biological processes taking place in these subcellular locations (Appendix Fig S4). In contrast, T1-4 only weakly coincide with subcellular locations or biological processes.

Intriguingly, T1-4 and P1-3 are independent of each other, that is genes that are clustered based on their transcript expression signature are generally not co-regulated on the protein level, and vice versa (Fig 2B). This suggests that much of the mRNA coexpression of genes from the same subcompartment may be non-functional. Note that as for sequence proximity (see above), this appears to contrast with a previous report that genes which are close in 3D nuclear space often have similar functions (Thévenin *et al*, 2014). However, we also find significant enrichment of functional associations between genes from the same subcompartment (Fig EV2B). Nevertheless, in quantitative terms, the extent of mRNA co-regulation strongly exceeds co-function as well as protein co-regulation. For example, while 11% of gene pairs in the same (intrachromosomal) subcompartment have co-regulated mRNAs, < 1% have similar functions according to STRING and are co-regulated at the protein level (Fig EV2B).

Gene clustering within but not between chromosomes associates with reduced expression noise

In yeast, clustering of genes in the genome sequence is associated with reduced expression noise (Batada & Hurst, 2007; Wang *et al*, 2011). However, the situation is more complex when considering the 3D structure of the genome. Highly transcribed gene clusters tend to form fewer contacts with other chromosomes, and genomic loci with more interchromosomal contacts tend to have higher expression noise (McCullagh *et al*, 2010; Sandhu, 2012).

We tested whether gene clustering has a similar effect in human cells. For each gene in our dataset, we calculated a clustering degree, defined as the average distance to its three nearest neighbouring genes along the DNA sequence. We then compared the expression noise of the 5% most and least clustered genes, respectively. As observed in yeast, we find that gene expression noise in LCLs is significantly reduced for genes in gene-dense areas (Fig 3A). The noise-reducing effect is much more significant on the mRNA than the protein level.

In a second step, we investigated whether gene clustering in nuclear space has a similar noise-reducing effect. In principle, gene-dense regions may interact with each other in 3D to benefit from further noise reduction by forming "super-clusters". The three human histone gene clusters on chromosome 6, for example, converge in 3D to form such a super-cluster (Sandhu *et al*, 2012). Therefore, we calculated a second clustering degree for each gene, defined as the average distance to its three nearest neighbours in 3D, using Hi-C contacts. To capture long-range interactions resulting from chromosome folding, we only considered neighbouring genes that were on the same chromosome, but at least 500 kb up- or downstream in terms of DNA sequence. There is a positive correlation between the clustering degree in 1D and 3D (PCC 0.32, $P < 6 \times 10^{-97}$), suggesting that genes clustered along the sequence are also more densely packed in the 3D structure of a chromosome. Moreover, this gene clustering due to chromosome folding is also associated with a significant reduction of gene expression noise, albeit not as strongly as sequence-based clusters (Fig 3A).

Next, we investigated clusters that genes from different chromosomes may form in nuclear space, calculating a third clustering degree based on interchromosomal Hi-C contacts. As shown in yeast (McCullagh *et al*, 2010; Sandhu, 2012), we find a negative correlation between sequence-based and interchromosomal clustering (PCC -0.1, $P < 5 \times 10^{-11}$). This suggests that gene-dense regions, while forming long-range, noise-reducing interactions within the same chromosome, are less likely to interact with gene clusters on a different chromosome. Moreover, genes forming interchromosomal clusters are associated with higher expression noise than those with fewer interactions (Fig 3A). This difference is not statistically significant but is in agreement with earlier findings in yeast (McCullagh *et al*, 2010; Sandhu, 2012).

Coexpression of closeby genes is driven by stochastic epigenetic fluctuations and regulatory interference

How can gene proximity lead to mRNA coexpression? Many incidents of coexpressed genes that are close in sequence have been linked to stochastic alternation between an active and inactive chromatin state (Batada *et al*, 2007). Such chromatin fluctuations can lead to coordinated transcriptional bursts of all genes within a chromatin domain (Raj *et al*, 2006). We first compared the chromatin environment of genes that are co-regulated with their sequence neighbours with genes that show no such co-regulation ("neighbours" being defined as genes whose TSSs are < 50 kb away). We find that genes which are coexpressed with their neighbours are more often flanked by heterochromatin, upstream of their transcription start site (Fig 3B). This is consistent with mRNA coexpression driven by stochastic spreading of the adjacent heterochromatin domain into the active locus, silencing all genes therein. This is

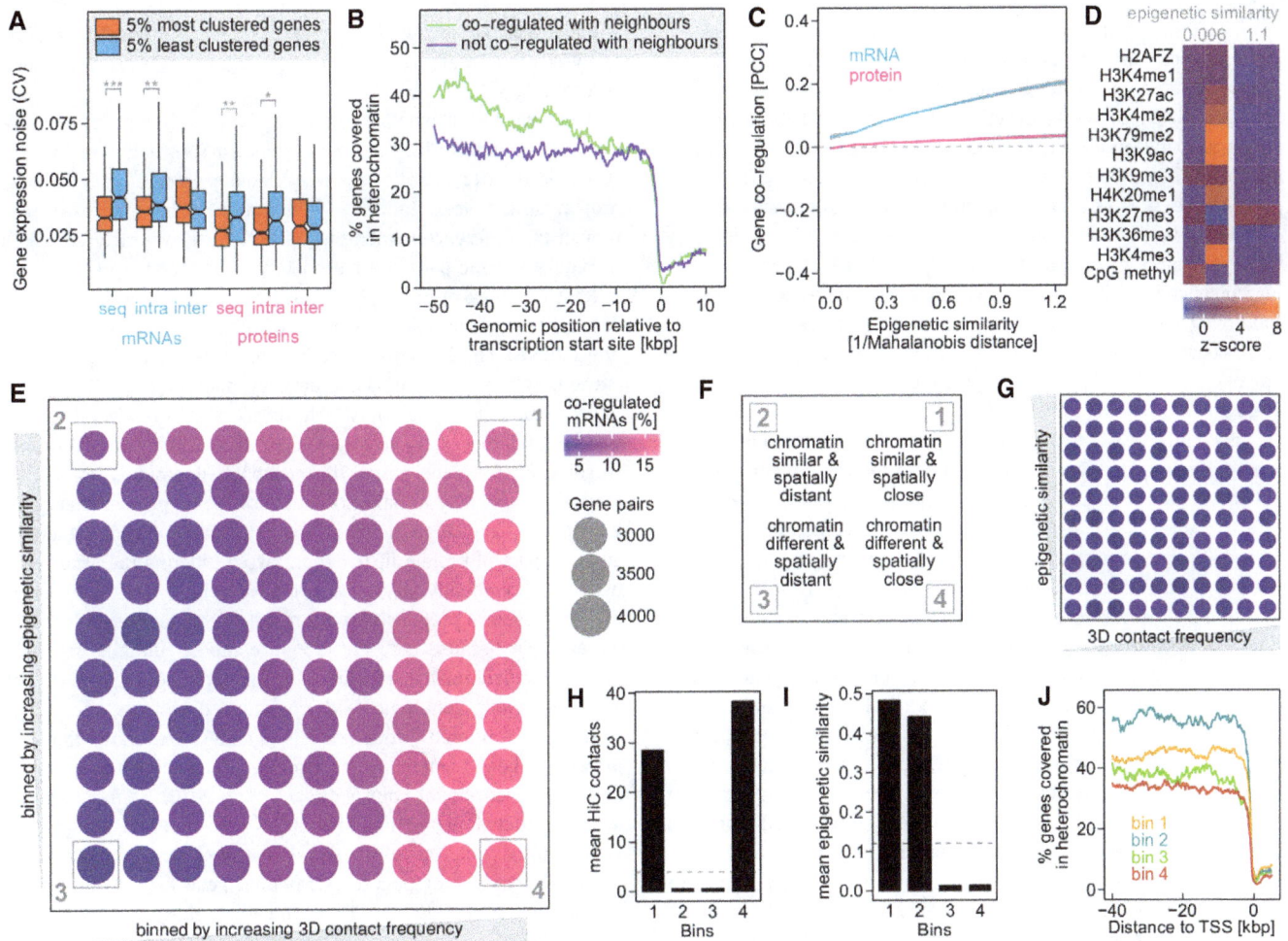

Figure 3. mRNA coexpression of neighbouring genes is driven by chromatin fluctuations and regulatory interference.

A   Intrachromosomal gene clustering reduces gene expression noise. We determined the expression noise (coefficient of variation, CV) of the most and least densely clustered genes, considering three different types of clustering: in terms of sequence proximity (seq), using long-range Hi-C contacts (> 500 kb) within the same chromosome (intra) and using interchromosomal Hi-C contacts (inter). Expression noise is reduced for clustered genes, except for genes forming more interchromosomal contacts (*P < 0.01, **P < 0.002, ***P < 5 × 10⁻⁶ based on Kolmogorov–Smirnov test). Boxplot drawn in the style of Tukey, that is box limits indicate the first and third quartiles, central lines the median, whiskers extend 1.5 times the interquartile range from the box limits. Notches indicate the 95% confidence interval for comparing medians.

B   The upstream region of genes that are co-regulated with their neighbours, that is other genes within 50 kb, is more likely to be occupied by heterochromatin than that of genes showing no such co-regulation. Heterochromatin regions in LCLs have been reported previously (Ernst *et al*, 2011).

C   Epigenetic similarity calculated on the basis of histone marks and CpG methylation is a strong general predictor of mRNA co-regulation. Curves are fitted to all intrachromosomal gene pairs irrespective of their genomic distance.

D   Two randomly picked gene pairs exemplifying low and high epigenetic similarity, respectively. Each column represents a gene and each row an epigenetic feature. Colours show the standardised, average abundance of each mark across the gene body.

E   mRNA co-regulation requires epigenetic similarity or spatial proximity, but not both. Intrachromosomal gene pairs were binned by epigenetic similarity and spatial proximity (Hi-C contacts), and the percentage of co-regulated mRNAs is shown in colour. Note bins 2 and 4 are both enriched for co-regulated mRNAs despite containing gene pairs that are spatially distant and epigenetically different, respectively.

F   Description of bins highlighted in panel (E).

G   Gene pairs binned as in (E) but colour showing percentage of co-regulated proteins. Protein co-regulation does not depend on epigenetic similarity or spatial proximity.

H   On average, gene pairs in bins 1 and 4 have many more Hi-C contacts than those in bins 2 and 3, that is they are spatially closer. Dashed line shows average Hi-C contacts between genes in the dataset.

I   On average, gene pairs in bins 1 and 2 are epigenetically much more similar than those in bins 3 and 4. Dashed line shows average epigenetic similarity between genes in the dataset.

J   Heterochromatin profile for genes in bins 1–4.

reminiscent of subtelomeric regions in yeast, which are hot spots for expression noise (Batada & Hurst, 2007) due to transient spreading of telomeric heterochromatin (Anderson *et al*, 2014).

Notably, chromatin fluctuations may lead to mRNA coexpression that is not restricted to genes in close spatial proximity. Chromatin factors play a key role in creating gene expression noise (Newman *et al*, 2006). Fluctuating expression levels of, for example, a histone-modifying enzyme may simultaneously affect all its target chromatin domains in the genome. To test for such a global chromatin-mediated co-regulation effect, we determined the epigenetic similarity between all genes in our dataset. We defined "epigenetic similarity" based on the abundance of various histone marks within gene bodies. We used the Mahalanobis distance to measure similarity, as this takes into account that some histone marks are strongly co-dependent, for example H3K9ac and H3K4me3. Genes with similar epigenetic profiles are targeted by a similar set of chromatin-modifying factors, and are therefore expected to respond similarly to stochastic fluctuations of these factors. Indeed, we find that the epigenetic similarity is a strong predictor of non-functional mRNA co-regulation (Fig 3C and D).

This chromatin fluctuation scenario is a passive mechanism where genes simply respond to changes in their chromatin domain. However, on a local scale, transcriptional changes of one gene may directly affect the transcription of its neighbours, if chromatin remodelling or transcription factors spill over to adjacent genomic regions (Ebisuya *et al*, 2008; Wang *et al*, 2011). This "regulatory interference" model crucially depends on spatial proximity, but does not require co-regulated genes to be part of the same chromatin domain. To compare the impact of chromatin and gene distance on non-functional mRNA coexpression, we grouped gene pairs based on epigenetic similarity as well as based on Hi-C contact frequency. We then observed which groups contain co-regulated mRNAs (Fig 3E). This shows that gene pairs which are far apart both spatially and epigenetically are rarely co-regulated (bin 3 in Fig 3E and F). Gene pairs with similar histone marks tend to be co-regulated, even if they are spatially distant (Fig 3E and H). Co-regulation of such genes is consistent with the passive chromatin fluctuation model, but not the transcriptional interference model. Importantly, spatially close gene pairs can be co-regulated even if their histone marks show no similarity (bin 4 in Fig 3E and I). This type of coexpression is not consistent with the passive chromatin fluctuation model, since the epigenetic differences between the gene pairs suggest that, in steady state, they occupy distinct chromatin domains. These genes are also the least likely to be flanked by heterochromatin (Fig 3J). However, the behaviour of gene pairs in bin 4 is consistent with the regulatory interference model, where fluctuations in one gene affect the chromatin and transcriptional state of its neighbours, in sequence and 3D. Note that this effect is buffered at the protein level (Fig 3G), which is in agreement with this type of coexpression being not functional.

Buffering of non-functional mRNA coexpression tends to be a non-selective process

Finally, we asked which post-transcriptional mechanisms might buffer the coexpression of genes that are spatially close, but functionally unrelated. In principle, this could be a selective process that specifically targets closeby genes and disentangles their expression patterns. Alternatively, buffering could be a neutral process, where the lack of coordination between post-transcriptional mechanisms prevents the mRNA coexpression to be propagated to the protein level. In this case, a selective process would need to exist to ensure that functionally related genes do in fact have co-regulated protein

abundances. To distinguish between these two possibilities, we analysed five measures of post-transcriptional gene expression control (Fig 4).

First, we tested whether gene pairs with sustained protein co-regulation are more likely to have similar mRNA half-lives in LCLs (Duan *et al*, 2013), relative to co-regulated gene pairs with buffered protein abundances. Indeed, we find this to be the case, even though the difference is modest (Fig 4A). Next, we analysed which co-regulated gene pairs are more likely to be targeted by the same miRNA (Helwak *et al*, 2013). Again, gene pairs that are also co-regulated on the protein level are enriched for pairs sharing at least one miRNA. Third, as an indication for translation-related effects, we took into account ribosome profiling data for the LCL cell line panel (Battle *et al*, 2015), which reflect both the abundance of mRNAs and the extent to which they are occupied by ribosomes (Ingolia, 2014). Gene pairs with coexpressed proteins are almost three times as likely to have correlated ribosome profiles than pairs which only have co-regulated mRNA abundances. Then, we looked at the impact of protein degradation, by considering the occurrence of non-exponentially degraded proteins (NEDs) (McShane *et al*, 2016). These are proteins that are rapidly degraded after synthesis, for example because they are protein complex subunits produced in super-stoichiometric amounts. Again, we find that NEDs are enriched among gene pairs with co-regulated proteins rather than those with buffered protein levels. Finally, we show that the protein sequence length, which strongly correlates with the extent of post-transcriptional control (Vogel *et al*, 2010), is more similar for co-regulated than buffered proteins. Proximity in the genome seemed to have no impact on the similarity of gene pairs in any of the five measures of post-transcriptional gene expression control investigated here (Fig 4B). Taken together, these results suggest that buffering of co-regulated closeby genes may occur via a neutral mechanism, with buffered gene pairs consistently lacking the extent of shared post-transcriptional processing observed for functionally related gene pairs. If mRNA coexpression is functionally relevant, multiple layers of post-transcriptional control appear to work together to ensure that this is propagated to the protein level.

Discussion

Genes are not randomly distributed across the sequence and structure of the genome, forming clusters that tend to be coexpressed but do not generally have a shared function. Gene expression noise is detrimental to cell fitness, especially for housekeeping genes (Fraser *et al*, 2004). Clusters of actively transcribed genes have low expression noise, which may drive the evolution of non-random gene order (Batada & Hurst, 2007). The coexpression of functionally unrelated neighbouring genes may then be a side effect of the selection for noise reduction. However, such coexpression is not necessarily deleterious. As we show here, non-functional co-regulation is frequently observed at the mRNA level, but is largely buffered at the protein level. Consequently, non-functional coexpression is unlikely to offset the benefit of noise reduction.

The expression profiles of genes in a cluster co-evolve, such that the evolutionary change in expression of one gene on average predicts changes in its neighbours (Ghanbarian & Hurst, 2015).

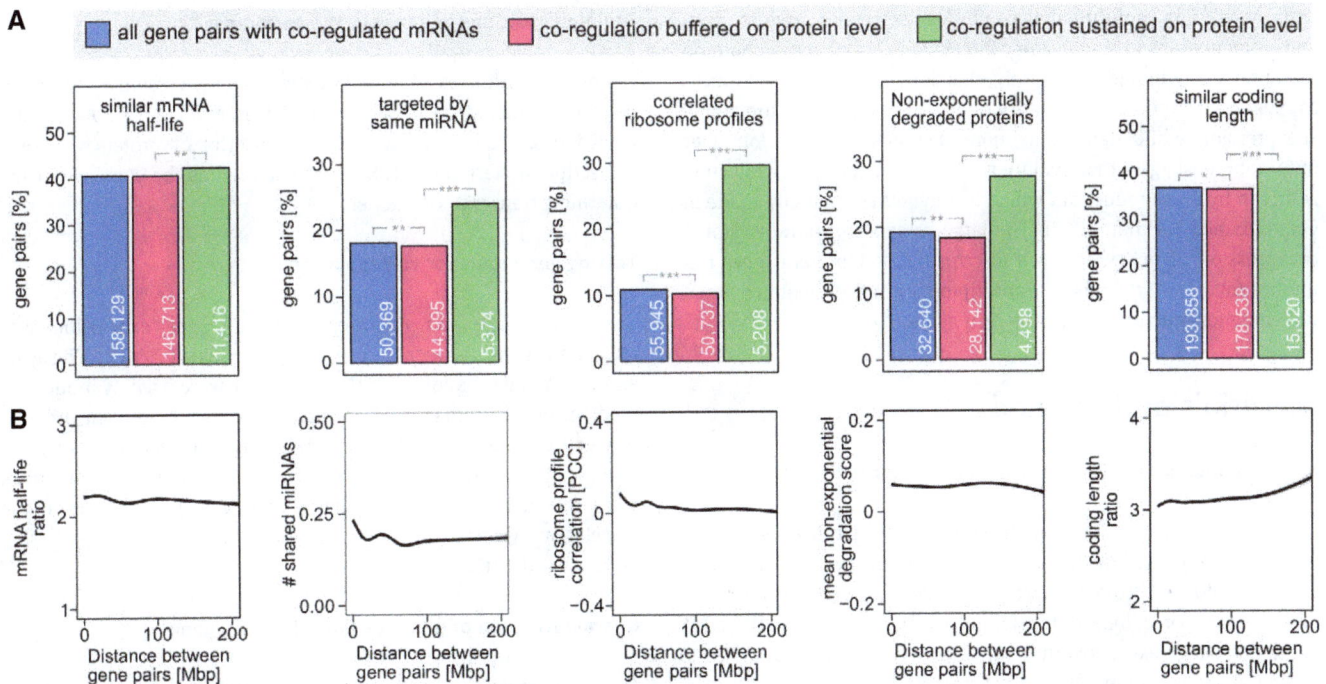

Figure 4. Buffering of non-functional mRNA co-regulation likely is a passive process.

A Percentage of gene pairs with coordinated post-transcriptional regulation, irrespective of genomic distance. Gene pairs with sustained protein co-regulation consistently stand out as more likely to share similar aspects of post-transcriptional control. Genes were considered to have a similar mRNA half-life if the half-life ratio between the more and less stable gene was < 1.5. For miRNAs, all gene pairs targeted by at least one shared miRNA were considered. Gene pairs were said to have correlated ribosome profiles if their ribosome occupancy correlated with PCC > 0.5 (BH adj. P < 0.001) across LCLs. For the non-exponentially degraded proteins (NEDs) barchart, gene pairs containing at least one NED were counted. Coding length was considered similar if the longer protein was < 1.5-fold longer than the shorter protein. Numbers of gene pairs are shown inside the bars. Statistical significance was calculated using Fisher's exact test (*P < 0.01, **P < 1 × 10^{-6}, ***P < 3 × 10^{-27}).

B No striking relationship between gene distance and the extent to which gene pairs show similar post-transcriptional regulation. Note that the small increase of similar ribosome occupancy towards closeby genes may be explained by the fact that ribosome profiles partially reflect mRNA abundance.

Nevertheless, it is still unclear whether expression clusters are the result of natural selection. In yeast, only the most highly coexpressed neighbours are conserved as a pair, but these also tend to be functionally related (Hurst *et al*, 2002). Neighbouring gene pairs that separate tend to show interchromosomal co-localisation (Dai *et al*, 2014). In *Drosophila*, highly coexpressed neighbouring gene pairs are less likely to be conserved than expected (Weber & Hurst, 2011). In mammals, although some coexpression clusters are evolutionarily maintained (Sémon & Duret, 2006), natural selection generally tends to separate gene pairs that show a strong position-related coexpression effect (Liao & Zhang, 2008) or that involve tissue-specific expression (Lercher *et al*, 2002). This indicates that non-functional coexpression can affect cell fitness under some circumstances, possibly if it becomes so strong that it persists through the uncoordinated post-transcriptional processes.

The existence of coexpression clusters may also reflect the way new genes originate. For example, highly transcribed chromatin regions are more susceptible to retroposition (Hurst *et al*, 2004). Recently, it has been proposed that the large number of human gene pairs in head-to-head orientation may arise from divergent transcription of single genes, when initially noncoding, antisense transcripts evolve into new protein-coding genes (Wu & Sharp, 2013). In both of these cases, new genes would have no sequence homology with their neighbours, and would therefore be unlikely to share

their function. However, some of the most well-known coexpression clusters, such as histone gene clusters, arose by gene duplication. Gene duplicates could potentially explain why some gene clusters are functionally related. There are 30 gene pairs in our dataset that are located within 50 kb from each other and are coexpressed on both the mRNA and the protein level. Of these, 10 (33%) are classified as paralogues by Ensembl, a strong enrichment considering that paralogues account for only 1.5% of these closeby gene pairs overall. However, 20 (66%) of the clustered gene pairs with co-regulated protein abundances show no evidence for paralogy, suggesting that functionally relevant clusters need not necessarily arise by gene duplication.

Our analysis focussed on housekeeping genes, because comparable data for tissue- or condition-specific genes were not available. Housekeeping genes constitute about half of all human genes (Uhlén *et al*, 2015). They have a higher tendency to cluster than other genes (Lercher *et al*, 2002), presumably because they are more sensitive to gene expression noise (Fraser *et al*, 2004). Interestingly, post-transcriptional expression control is particularly important for housekeeping genes (Gandhi *et al*, 2011; Jovanovic *et al*, 2015). Notably, transcriptional activation of induced genes can also lead to co-activation of functionally unrelated neighbouring genes (Spitz *et al*, 2003; Ebisuya *et al*, 2008). However, it remains to be seen if such co-activation is also buffered at the protein level.

In conclusion, non-functional mRNA coexpression, due to chromatin fluctuations and regulatory interference, is far more common than previously thought. Generally, this does not hamper cell fitness as post-transcriptional regulatory mechanisms enforce functional coexpression while dampening non-functional coexpression. Our observations suggest that evolution of human genome organisation is driven by noise reduction, which is a hypothesis initially made in yeast (Batada & Hurst, 2007). The large presence of non-functional coexpression of genes at the transcript but not protein level has implications for the fields of transcriptomics and proteomics when screening for functional links between genes.

Materials and Methods

mRNA abundances in human lymphoblastoid cell lines

RNA-sequencing data for human lymphoblastoid cell lines (LCLs) have been reported (Pickrell et al, 2010). Counts per mapped reads were downloaded from http://eqtl.uchicago.edu and converted to log2 "reads per kilobase transcript per million mapped reads" (RPKMs). Genes expressed in < 30 LCLs were removed. In order to make mRNA measurements comparable to proteomics data, expression levels needed to be analysed relative to the same reference LCL. To do so, log2 RPKMs values from the reference cell line GM19238 were subtracted from all other LCLs.

Protein abundances in human lymphoblastoid cell lines

Protein abundances in LCLs have also been reported (Battle et al, 2015). They have been measured by mass spectrometry and quantified relative to the reference cell line GM19238, using stable isotope labelling by amino acids in cell culture (SILAC) (Ong et al, 2002). Mass spectrometry raw files were downloaded from the PRIDE repository (Vizcaíno et al, 2016) (project identifier PXD001406) and re-processed using MaxQuant 1.5.2.8 (Cox & Mann, 2008). Raw files tagged as "run2" were omitted. Mass spectra were searched against human Swiss-Prot sequences downloaded from Uniprot (UniProt Consortium, 2015). To facilitate combining mRNA and protein datasets, no protein isoforms were considered. We used non-normalised SILAC ratios obtained by MaxQuant with at least two ratio counts. Because the internal standard had been used as heavy SILAC sample, heavy/light (H/L) SILAC ratios were inverted to obtain L/H ratios (i.e. test LCLs / reference LCL). Proteins that could not be unambiguously mapped to a single gene were removed, as were proteins detected in 30 LCLs or less. SILAC ratios were also log2-transformed.

Combining mRNA and protein expression data

To combine mRNA and protein data, ENSEMBL gene IDs from RNA sequencing were mapped to Uniprot IDs using Uniprot's webtool (UniProt Consortium, 2015). Genes with ambiguous mappings were removed. We also only considered LCLs for which both mRNA and protein data were available. The resulting file contains mRNA and protein abundances for 4,188 human genes in 59 LCLs, relative to the GM19238 reference sample (Dataset EV1). It contains 0.1 and 6.7% missing values for mRNA and protein measurements, respectively.

Defining positions of genes in the genome

Genomic coordinates of human genes (dataset version GRCh38.p5) were downloaded from ENSEMBL (Yates et al, 2016). As we are considering genes but not specific transcript or protein isoforms, transcription start sites (TSSs) were defined as the start site of the outermost transcript of a gene.

Testing gene pairs for co-regulation

Coordinated up- and downregulation of gene expression was measured using Pearson's correlation coefficient (PCC). The gene expression datasets for LCLs (Dataset EV1) were used as input. The median log2 fold change of each LCL was set to zero, in order to prevent correlations reflecting irrelevant data features such as uneven mixing of light and heavy SILAC samples. Gene pairs were considered to be co-regulated at PCC > 0.5, but only if the correlation was significant (Benjamini and Hochberg-adjusted P-values < 0.001).

Characterisation of genes as housekeeping genes

To demonstrate that the 4,188 genes in the LCL dataset belong to the constitutively expressed core proteome, we performed a number of tests:

Chromatin states of gene promoters
Chromatin states of the genome of the GM12878 lymphoblastoid cell line were determined previously (Ernst et al, 2011). They were downloaded as hg19 genome coordinates from the USCS genome browser (Rosenbloom et al, 2015) and converted to GRCh38 coordinates using the liftOver command line tool (available at https://genome-store.ucsc.edu/). Genomic regions with conflicting chromatin state annotations, resulting from the genome coordinates update, were removed. For each gene in our dataset, the chromatin state mapping to its transcription start site was determined.

GO term enrichment
A statistical overrepresentation test was performed using the PANTHER classification system (Mi et al, 2016) according to the reported protocol (Mi et al, 2013). Overrepresentation of Gene Ontology Biological Process (slim) terms was assessed for our 4,188 genes compared to the entire human genome. Only significantly enriched terms (more than twofold; $P < 0.05$ after Bonferroni correction) were considered.

mRNA tissue expression data
mRNA expression levels in different human tissues have been assessed using RNA sequencing (Uhlén et al, 2015). Transcripts detected with FPKM ≥ 1 were considered to be expressed.

Protein tissue expression data
Protein expression levels in different human tissues have been assessed using mass spectrometry (Wilhelm et al, 2014) (available at www.proteomicsdb.org). To avoid bias due to the incomplete nature of current proteome maps, only tissues with expression values for more than 6,000 proteins were considered.

Defining pairs of genes with related functions (focussed on accuracy)

To test whether genes with related functions are co-regulated across LCLs, we defined three sets of functionally linked gene pairs. Functional associations in these test sets are as accurate—not as comprehensive—as possible.

Gene pairs from same protein complexes

Human protein–protein interaction pairs based on Reactome pathways (Fabregat et al, 2016) were downloaded from www.reactome. org (homo_sapiens.interactions.txt file; March 2016). They were filtered for physical interactions of the "direct_complex" category. Gene pairs belonging to more than one complex and homodimeric interactions were removed.

Gene pairs encoding enzymes from consecutive metabolic reactions

As for protein complexes, human protein–protein interaction pairs based on Reactome pathways (Fabregat et al, 2016) were downloaded from www.reactome.org (homo_sapiens.interactions.txt file; March 2016). They were filtered for interactions of the "neighbouring_reactions" category. These are interactions where one gene/protein produces the input or catalyst for the second reaction. Any gene pairs known to interact also physically, that is belonging to the "direct_complex" or "indirect_complex" categories, were removed. In addition, gene pairs were filtered for those involved in *metabolic* pathways, as opposed to, for example, the cell cycle pathway which would contain irrelevant reactions such as "Mis18 complex binds the centromere". To do so, we first inferred all pathways mapping to the metabolism root pathway, using the pathway hierarchy relationship file (ReactomePathwaysRelation.txt, available on www.reactome.org). Enzymatic reactions belonging to each metabolic pathway were then identified using another interaction file available from Reactome (homo_sapiens.mitab.interactions.txt). Finally, to avoid "trivial" consecutive reactions such as those involving ubiquitous metabolites like NAD^+, we removed metabolic reactions with more than ten neighbouring reactions.

Gene pairs from identical subcellular locations

Subcellular localisations of human proteins were downloaded from Uniprot (UniProt Consortium, 2015). Proteins localising to more than one subcellular location were removed. To avoid trivial localisations such as "cytoplasm", only subcellular compartments with 200 or less known protein components were considered.

Defining pairs of genes with related functions (focussed on completeness)

To estimate an upper limit for how many coexpressed neighbouring genes may be functionally related, we defined a separate test set based on the STRING database (Szklarczyk et al, 2017). Functional associations in this test set are as comprehensive as possible. Protein network data for *Homo sapiens* were downloaded from http://string-db.org. We considered all functional associations with a combined STRING score > 0.7. This score integrates various types of evidence and indicates the likelihood of the association to be biologically meaningful, specific and reproducible.

Testing functionally related gene pairs for co-regulation

Correlation coefficients were obtained for every gene pair in our three test sets (protein complexes, consecutive metabolic reactions, subcellular locations) and their distribution was displayed in histograms. As a control, gene pairs were randomly shuffled to break the link between the pairs. For example, gene pairs encoding subunits of the same protein complexes were shuffled such that the same genes were paired randomly, in which case most gene pairs encode subunits of different protein complexes. The Kolmogorov–Smirnov test was used to assess whether PCC distributions of relevant gene pairs were significantly different from those obtained with randomised pairs.

Chromosome co-regulation mapping

PCCs were calculated for all relevant gene combinations, as described for histograms above. For chromosome co-regulation curves, PCCs were plotted against the genomic distance between transcription start sites, with curves fitted by a generalised additive model. For chromosome co-regulation maps, genes were plotted in their chromosomal order and PCCs between all gene combinations were represented by a colour scale.

Hi-C interactions for our gene set

Hi-C contact matrices for a lymphoblastoid cell line (Rao et al, 2014) were downloaded from NCBI GEO database (accession GSE63525). An unpublished script from Liz Ing-Simmons (available at https://github.com/liz-is/readhic) was adapted (available at https://github.com/Rappsilber-Laboratory/readhic) and then used to import the Hi-C contact matrices into R, using 10-kb resolution and "KRnorm" normalisation for intrachromosomal pairs and 50-kb resolution and "INTERKRnorm" normalisation for interchromosomal pairs. All reads used passed the MAPQ>0 filter. Hi-C data are based on GRCh37 genome coordinates. GRCh37 transcription start sites for all genes were obtained using the biomaRt R package (Durinck et al, 2009), considering only the TSS of the outermost transcript of each gene. The GenomicInteractions R package (Harmston et al, 2015) was used to determine the contact frequency between the genes in our dataset, considering the median read count of all Hi-C pixels in a range $\pm$ 40 kb around the TSS of each gene.

Analysis of genome subcompartments

Nuclear subcompartments A1, A2, B1, B2, B3 and B4 have been defined previously (Rao et al, 2014). A genome-wide mapping of subcompartments in a lymphoblastoid cell line is available via the NCBI GEO database (accession GSE63525). Subcompartment annotations were lifted from hg19/b37 to GRCh38 genome coordinates using the UCSC genome browser service (Rosenbloom et al, 2015).

k-means clustering of transcript and protein expression changes

k-means clustering was performed using the default algorithm and settings in R (R Core Team, 2016), with $k = 4$ (mRNAs) or $k = 3$ (proteins) and five random start sets. Values of k were chosen such that the clusters explain at least 50% of the total variance.

Analysis of cluster features

Subcellular locations

To get a broad understanding of subcellular locations enriched in k-means clusters, we downloaded all Uniprot entries mapping to the locations Nucleus (Uniprot subcellular location ID: SL-0191), Endoplasmic reticulum (SL-0095), Golgi apparatus (SL-0132), Mitochondrion (SL-0173) and Cytoplasm (SL-0086) (UniProt Consortium, 2015). Proteins localising to the Endoplasmic reticulum and/or the Golgi apparatus were combined as "ER-Golgi". Proteins mapping to more than one organelle were removed.

GO term enrichment

A statistical overrepresentation test was performed using the PANTHER classification system (Mi *et al*, 2016) according to the reported protocol (Mi *et al*, 2013). Overrepresentation of Gene Ontology Biological Process (complete) terms in each cluster, relative to other clusters, was assessed. Using PANTHER's GO hierarchy annotation, we reported only the most specific GO terms and omitted any co-enriched parent terms for clarity. All reported GO terms were significantly enriched ($P < 0.05$ after Bonferroni correction).

Genomic and epigenomic features

Raw signals of ChIP-seq experiments for lymphoblastoid cells were downloaded from ENCODE (ENCODE Project Consortium, 2012) in hg19 genomic coordinates. ENCODE accessions were ENCFF000ARW (H2AZ), ENCFF000ARZ (H3K4me1), ENCFF000ATL (H3K4me2), ENCFF001EXX (H3K4me3), ENCFF000ASJ (H3K27ac), ENCFF000ATX (H3K79me2), ENCFF000AUF (H3K9ac), ENCFF000AUL (H3K9me3), ENCFF000AUS (H4K20me1), ENCFF001EXC (H3K27me3), ENCFF001EXP (H3K36me3), ENCFF001GNK (RepliSeq G1b), ENCFF001GNN (RepliSeq G2), ENCFF001GNR (RepliSeq S1), ENCFF001GNT (RepliSeq S2), ENCFF001GNX (RepliSeq S3) and ENCFF001GOA (RepliSeq S4). These bigWig files were converted to bedGraph files, lifted over to GRCh38 coordinates, cleared of any resulting overlaps and converted back to bigWig files using command line tools from the UCSC genome browser (Rosenbloom *et al*, 2015) (tools available at https://genome-store.ucsc.edu/). GC percentage over 5-bp windows was downloaded from the UCSC genome browser (Rosenbloom *et al*, 2015). Average signals over gene bodies were calculated with the UCSC bigWigAverageOverBed command line utility, using the coordinates of our genes as bed files. CpG methylation from reduced representation bisulphite sequencing of a lymphoblastoid cell line was also available from ENCODE (ENCODE Project Consortium, 2012) (experiment ENCSR000DFT; file accession ENCFF001TLQ). After lifting the hg19 bedMethyl file over to GRCh38 genomic coordinates, the mean percentage of CpG methylation in gene bodies was calculated using an R script. For each epigenomic or genomic feature, the median enrichment for genes in each k-means cluster, compared to all genes in our dataset, was calculated and plotted as log2 ratio in a heatmap.

Calculation of gene expression noise

Gene expression noise at the mRNA and protein levels was calculated as the coefficient of variation (CV; standard deviation divided by the mean) of log2-transformed RPKM and SILAC ratios, respectively. To avoid dividing by zero (for unchanged genes with a log2 ratio of zero), a constant value of 10 was added to all mRNA and protein log2 ratios before calculating the noise.

Calculating the clustering degree

To define local gene density in a manner that can be applied to both the sequence and the 3D structure of the genome, we determined the average distance of a gene to its three nearest neighbouring genes. We calculated three such "clustering degrees" for each gene in our dataset. For the sequence-based clustering degree, the distance to neighbouring genes was calculated in base pairs. For intrachromosomal clustering in 3D, gene distance was calculated based on Hi-C counts. However, we only considered "nearest" neighbours which were at least 500 kb away in terms of DNA sequence, to catch long-range interactions and avoid replicating the sequence-based clustering degree. For interchromosomal clustering, we considered the three nearest neighbours on other chromosomes, based on interchromosomal Hi-C contacts.

Heterochromatin profiles of upstream regions

Chromatin states throughout the LCL genome were previously described (Ernst *et al*, 2011). To simplify the analysis, we combined the five inactive chromatin states defined by Ernst *et al* ("Heterochromatin", "Repressed", "Repetitive", "Poised Promoter" and "Insulator") into one "heterochromatin" state. We then scanned the promoter region of test genes for the presence of heterochromatin, moving in 100-bp intervals from $-50,000$ bp to $+10,000$ bp relative to their transcription start site.

Calculating epigenetic similarity

Epigenetic similarity was calculated on the basis of the histone mark abundance within gene bodies (see section "Analysis of cluster features" for processing of ChIP-seq data). For this analysis, we considered H2AFZ, H3K4me1, H3K4me2, H3K4me3, H3K27ac, H3K79me2, H3K9ac, H3K9me3, H4K20me1, H3K27me3, H3K36me3 and CpG methylation, but not GC content, gene length and replication timing. For every pair of genes, we then determined how similar or dissimilar they are regarding the abundance of these epigenetic features. This was calculated using the Mahalanobis distance measure, which takes into account that some histone marks strongly covary.

Analysis of post-transcriptional mechanisms

mRNA half-lives in seven different LCLs were previously reported (Duan *et al*, 2013). We first calculated the average half-life of each mRNA in these LCLs. We considered two mRNAs to have a similar stability if the half-life of the more stable one was < 1.5-fold longer than the less stable one. mRNA targets of human miRNAs were also described previously (Helwak *et al*, 2013). Ribosome occupancy profiles for the LCL cell line panel were recently published (Battle *et al*, 2015). We considered ribosome profiles for 57 LCLs and 4,033 genes for which we had matching mRNA and protein measurements. We calculated Pearson correlation coefficients (PCCs) for ribosome profiles between all gene pairs. Two genes were said to have correlated ribosome profiles at PCC > 0.5 (BH-adjusted P-value < 0.001). Proteins subjected to non-exponential degradation in human RPE-1 cells were also described recently (McShane *et al*, 2016). Finally, protein sequence lengths were downloaded from Uniprot (UniProt Consortium, 2015).

Human paralogous genes

Human gene duplicates were downloaded from ENSEMBL (Yates *et al*, 2016). We only considered paralogues with at least 25% sequence identity.

General data processing and plotting

Data processing was performed in R (R Core Team, 2016), unless indicated otherwise. Plots were created using the ggplot2 package (Wickham, 2009).

Acknowledgements

This work was supported by the Wellcome Trust through a Senior Research Fellowship to JR (grant number 103139). The Wellcome Trust Centre for Cell Biology is supported by core funding from the Wellcome Trust (grant number 203149).

Author contributions

PG analysed Hi-C contact frequencies between the genes in our dataset. GK and JR designed the study, analysed the data and wrote the paper.

References

Akey JM, Biswas S, Leek JT, Storey JD (2007) On the design and analysis of gene expression studies in human populations. *Nat Genet* 39: 807–808; author reply 808–9

Akhtar W, de Jong J, Pindyurin AV, Pagie L, Meuleman W, de Ridder J, Berns A, Wessels LFA, van Lohuizen M, van Steensel B (2013) Chromatin position effects assayed by thousands of reporters integrated in parallel. *Cell* 154: 914–927

Albert FW, Kruglyak L (2015) The role of regulatory variation in complex traits and disease. *Nat Rev Genet* 16: 197–212

Anderson MZ, Gerstein AC, Wigen L, Baller JA, Berman J (2014) Silencing is noisy: population and cell level noise in telomere-adjacent genes is dependent on telomere position and sir2. *PLoS Genet* 10: e1004436

Batada NN, Hurst LD (2007) Evolution of chromosome organization driven by selection for reduced gene expression noise. *Nat Genet* 39: 945–949

Batada NN, Urrutia AO, Hurst LD (2007) Chromatin remodelling is a major source of coexpression of linked genes in yeast. *Trends Genet* 23: 480–484

Battle A, Khan Z, Wang SH, Mitrano A, Ford MJ, Pritchard JK, Gilad Y (2015) Genomic variation. Impact of regulatory variation from RNA to protein. *Science* 347: 664–667

Caron H, van Schaik B, van der Mee M, Baas F, Riggins G, van Sluis P, Hermus MC, van Asperen R, Boon K, Voûte PA, Heisterkamp S, van

Kampen A, Versteeg R (2001) The human transcriptome map: clustering of highly expressed genes in chromosomal domains. *Science* 291: 1289–1292

Chen X, Zhang J (2016) The genomic landscape of position effects on protein expression level and noise in yeast. *Cell Syst* 2: 347–354

Cohen BA, Mitra RD, Hughes JD, Church GM (2000) A computational analysis of whole-genome expression data reveals chromosomal domains of gene expression. *Nat Genet* 26: 183–186

Cox J, Mann M (2008) MaxQuant enables high peptide identification rates, individualized p.p.b.-range mass accuracies and proteome-wide protein quantification. *Nat Biotechnol* 26: 1367–1372

Dai Z, Xiong Y, Dai X (2014) Neighboring genes show interchromosomal colocalization after their separation. *Mol Biol Evol* 31: 1166–1172

Dephoure N, Hwang S, O'Sullivan C, Dodgson SE, Gygi SP, Amon A, Torres EM (2014) Quantitative proteomic analysis reveals posttranslational responses to aneuploidy in yeast. *Elife* 3: e03023

Duan J, Shi J, Ge X, Dölken L, Moy W, He D, Shi S, Sanders AR, Ross J, Gejman PV (2013) Genome-wide survey of interindividual differences of RNA stability in human lymphoblastoid cell lines. *Sci Rep* 3: 1318

Durinck S, Spellman PT, Birney E, Huber W (2009) Mapping identifiers for the integration of genomic datasets with the R/Bioconductor package biomaRt. *Nat Protoc* 4: 1184–1191

Ebisuya M, Yamamoto T, Nakajima M, Nishida E (2008) Ripples from neighbouring transcription. *Nat Cell Biol* 10: 1106–1113

ENCODE Project Consortium (2012) An integrated encyclopedia of DNA elements in the human genome. *Nature* 489: 57–74

Ernst J, Kheradpour P, Mikkelsen TS, Shoresh N, Ward LD, Epstein CB, Zhang X, Wang L, Issner R, Coyne M, Ku M, Durham T, Kellis M, Bernstein BE (2011) Mapping and analysis of chromatin state dynamics in nine human cell types. *Nature* 473: 43–49

Fabregat A, Sidiropoulos K, Garapati P, Gillespie M, Hausmann K, Haw R, Jassal B, Jupe S, Korninger F, McKay S, Matthews L, May B, Milacic M, Rothfels K, Shamovsky V, Webber M, Weiser J, Williams M, Wu G, Stein L *et al* (2016) The reactome pathway knowledgebase. *Nucleic Acids Res* 44: D481–D487

Fraser HB, Hirsh AE, Giaever G, Kumm J, Eisen MB (2004) Noise minimization in eukaryotic gene expression. *PLoS Biol* 2: e137

Gandhi SJ, Zenklusen D, Lionnet T, Singer RH (2011) Transcription of functionally related constitutive genes is not coordinated. *Nat Struct Mol Biol* 18: 27–34

Geiger T, Cox J, Mann M (2010) Proteomic changes resulting from gene copy number variations in cancer cells. *PLoS Genet* 6: e1001090

Ghanbarian AT, Hurst LD (2015) Neighboring genes show correlated evolution in gene expression. *Mol Biol Evol* 32: 1748–1766

Gibcus JH, Dekker J (2013) The hierarchy of the 3D genome. *Mol Cell* 49: 773–782

Gierman HJ, Indemans MHG, Koster J, Goetze S, Seppen J, Geerts D, van Driel R, Versteeg R (2007) Domain-wide regulation of gene expression in the human genome. *Genome Res* 17: 1286–1295

Harmston N, Ing-Simmons E, Perry M, Barešić A, Lenhard B (2015) GenomicInteractions: an R/Bioconductor package for manipulating and investigating chromatin interaction data. *BMC Genom* 16: 963

Helwak A, Kudla G, Dudnakova T, Tollervey D (2013) Mapping the human miRNA interactome by CLASH reveals frequent noncanonical binding. *Cell* 153: 654–665

Hurst LD, Williams EJB, Pál C (2002) Natural selection promotes the conservation of linkage of co-expressed genes. *Trends Genet* 18: 604–606

Hurst LD, Pál C, Lercher MJ (2004) The evolutionary dynamics of eukaryotic gene order. *Nat Rev Genet* 5: 299–310

Ingolia NT (2014) Ribosome profiling: new views of translation, from single codons to genome scale. *Nat Rev Genet* 15: 205–213

Jovanovic M, Rooney MS, Mertins P, Przybylski D, Chevrier N, Satija R, Rodriguez EH, Fields AP, Schwartz S, Raychowdhury R, Mumbach MR, Eisenhaure T, Rabani M, Gennert D, Lu D, Delorey T, Weissman JS, Carr SA, Hacohen N, Regev A (2015) Immunogenetics. Dynamic profiling of the protein life cycle in response to pathogens. *Science* 347: 1259038

Khan Z, Ford MJ, Cusanovich DA, Mitrano A, Pritchard JK, Gilad Y (2013) Primate transcript and protein expression levels evolve under compensatory selection pressures. *Science* 342: 1100–1104

Lercher MJ, Urrutia AO, Hurst LD (2002) Clustering of housekeeping genes provides a unified model of gene order in the human genome. *Nat Genet* 31: 180–183

Li Y-Y, Yu H, Guo Z-M, Guo T-Q, Tu K, Li Y-X (2006) Systematic analysis of head-to-head gene organization: evolutionary conservation and potential biological relevance. *PLoS Comput Biol* 2: e74

Liao B-Y, Zhang J (2008) Coexpression of linked genes in Mammalian genomes is generally disadvantageous. *Mol Biol Evol* 25: 1555–1565

Lieberman-Aiden E, van Berkum NL, Williams L, Imakaev M, Ragoczy T, Telling A, Amit I, Lajoie BR, Sabo PJ, Dorschner MO, Sandstrom R, Bernstein B, Bender MA, Groudine M, Gnirke A, Stamatoyannopoulos J, Mirny LA, Lander ES, Dekker J (2009) Comprehensive mapping of long-range interactions reveals folding principles of the human genome. *Science* 326: 289–293

Liu Y, Beyer A, Aebersold R (2016) On the dependency of cellular protein levels on mRNA abundance. *Cell* 165: 535–550

McCullagh E, Seshan A, El-Samad H, Madhani HD (2010) Coordinate control of gene expression noise and interchromosomal interactions in a MAP kinase pathway. *Nat Cell Biol* 12: 954–962

McShane E, Sin C, Zauber H, Wells JN, Donnelly N, Wang X, Hou J, Chen W, Storchova Z, Marsh JA, Valleriani A, Selbach M (2016) Kinetic analysis of protein stability reveals age-dependent degradation. *Cell* 167: 803–815.e21

Mi H, Muruganujan A, Casagrande JT, Thomas PD (2013) Large-scale gene function analysis with the PANTHER classification system. *Nat Protoc* 8: 1551–1566

Mi H, Poudel S, Muruganujan A, Casagrande JT, Thomas PD (2016) PANTHER version 10: expanded protein families and functions, and analysis tools. *Nucleic Acids Res* 44: D336–D342

Michalak P (2008) Coexpression, coregulation, and cofunctionality of neighboring genes in eukaryotic genomes. *Genomics* 91: 243–248

Newman JRS, Ghaemmaghami S, Ihmels J, Breslow DK, Noble M, DeRisi JL, Weissman JS (2006) Single-cell proteomic analysis of *S. cerevisiae* reveals the architecture of biological noise. *Nature* 441: 840–846

Ong S-E, Blagoev B, Kratchmarova I, Kristensen DB, Steen H, Pandey A, Mann M (2002) Stable isotope labeling by amino acids in cell culture, SILAC, as a simple and accurate approach to expression proteomics. *Mol Cell Proteomics* 1: 376–386

Pickrell JK, Marioni JC, Pai AA, Degner JF, Engelhardt BE, Nkadori E, Veyrieras J-B, Stephens M, Gilad Y, Pritchard JK (2010) Understanding mechanisms underlying human gene expression variation with RNA sequencing. *Nature* 464: 768–772

Purmann A, Toedling J, Schueler M, Carninci P, Lehrach H, Hayashizaki Y, Huber W, Sperling S (2007) Genomic organization of transcriptomes in mammals: coregulation and cofunctionality. *Genomics* 89: 580–587

R Core Team (2016) *R: a language and environment for statistical computing.* Vienna, Austria: R Core Team. Available at: https://www.R-project.org/

Raj A, Peskin CS, Tranchina D, Vargas DY, Tyagi S (2006) Stochastic mRNA synthesis in mammalian cells. *PLoS Biol* 4: e309

Rao SSP, Huntley MH, Durand NC, Stamenova EK, Bochkov ID, Robinson JT, Sanborn AL, Machol I, Omer AD, Lander ES, Aiden EL (2014) A 3D map of the human genome at kilobase resolution reveals principles of chromatin looping. *Cell* 159: 1665–1680

Rosenbloom KR, Armstrong J, Barber GP, Casper J, Clawson H, Diekhans M, Dreszer TR, Fujita PA, Guruvadoo L, Haeussler M, Harte RA, Heitner S, Hickey G, Hinrichs AS, Hubley R, Karolchik D, Learned K, Lee BT, Li CH, Miga KH *et al* (2015) The UCSC genome browser database: 2015 update. *Nucleic Acids Res* 43: D670–D681

Sandhu KS (2012) Did the modulation of expression noise shape the evolution of three dimensional genome organizations in eukaryotes? *Nucleus* 3: 286–289

Sandhu KS, Li G, Poh HM, Quek YLK, Sia YY, Peh SQ, Mulawadi FH, Lim J, Sikic M, Menghi F, Thalamuthu A, Sung WK, Ruan X, Fullwood MJ, Liu E, Csermely P, Ruan Y (2012) Large-scale functional organization of long-range chromatin interaction networks. *Cell Rep* 2: 1207–1219

Sémon M, Duret L (2006) Evolutionary origin and maintenance of coexpressed gene clusters in mammals. *Mol Biol Evol* 23: 1715–1723

Spitz F, Gonzalez F, Duboule D (2003) A global control region defines a chromosomal regulatory landscape containing the HoxD cluster. *Cell* 113: 405–417

Stark AL, Hause RJ Jr, Gorsic LK, Antao NN, Wong SS, Chung SH, Gill DF, Im HK, Myers JL, White KP, Jones RB, Dolan ME (2014) Protein quantitative trait loci identify novel candidates modulating cellular response to chemotherapy. *PLoS Genet* 10: e1004192

Stingele S, Stoehr G, Peplowska K, Cox J, Mann M, Storchova Z (2012) Global analysis of genome, transcriptome and proteome reveals the response to aneuploidy in human cells. *Mol Syst Biol* 8: 608

Szklarczyk D, Morris JH, Cook H, Kuhn M, Wyder S, Simonovic M, Santos A, Doncheva NT, Roth A, Bork P, Jensen LJ, von Mering C (2017) The STRING database in 2017: quality-controlled protein-protein association networks, made broadly accessible. *Nucleic Acids Res* 45: D362–D368

Thévenin A, Ein-Dor L, Ozery-Flato M, Shamir R (2014) Functional gene groups are concentrated within chromosomes, among chromosomes and in the nuclear space of the human genome. *Nucleic Acids Res* 42: 9854–9861

Trinklein ND, Aldred SF, Hartman SJ, Schroeder DI, Otillar RP, Myers RM (2004) An abundance of bidirectional promoters in the human genome. *Genome Res* 14: 62–66

Uhlén M, Fagerberg L, Hallström BM, Lindskog C, Oksvold P, Mardinoglu A, Sivertsson Å, Kampf C, Sjöstedt E, Asplund A, Olsson I, Edlund K, Lundberg E, Navani S, Szigyarto CA-K, Odeberg J, Djureinovic D, Takanen JO, Hober S, Alm T *et al* (2015) Proteomics. Tissue-based map of the human proteome. *Science* 347: 1260419

UniProt Consortium (2015) UniProt: a hub for protein information. *Nucleic Acids Res* 43: D204–D212

Vizcaíno JA, Csordas A, del-Toro N, Dianes JA, Griss J, Lavidas I, Mayer G, Perez-Riverol Y, Reisinger F, Ternent T, Xu Q-W, Wang R, Hermjakob H (2016) 2016 update of the PRIDE database and its related tools. *Nucleic Acids Res* 44: D447–D456

Vogel C, Abreu Rde S, Ko D, Le S-Y, Shapiro BA, Burns SC, Sandhu D, Boutz DR, Marcotte EM, Penalva LO (2010) Sequence signatures and mRNA concentration can explain two-thirds of protein abundance variation in a human cell line. *Mol Syst Biol* 6: 400

Wang G-Z, Lercher MJ, Hurst LD (2011) Transcriptional coupling of neighboring genes and gene expression noise: evidence that gene orientation and noncoding transcripts are modulators of noise. *Genome Biol Evol* 3: 320–331

Weber CC, Hurst LD (2011) Support for multiple classes of local expression clusters in *Drosophila melanogaster*, but no evidence for gene order conservation. *Genome Biol* 12: R23

Wickham H (2009) *Ggplot2 elegant graphics for data analysis*. New York: Springer

Wilhelm M, Schlegl J, Hahne H, Moghaddas Gholami A, Lieberenz M, Savitski MM, Ziegler E, Butzmann L, Gessulat S, Marx H, Mathieson T, Lemeer S, Schnatbaum K, Reimer U, Wenschuh H, Mollenhauer M, Slotta-Huspenina J, Boese J-H, Bantscheff M, Gerstmair A *et al* (2014) Mass-spectrometry-based draft of the human proteome. *Nature* 509: 582–587

Williams EJB, Bowles DJ (2004) Coexpression of neighboring genes in the genome of *Arabidopsis thaliana*. *Genome Res* 14: 1060–1067

Wu X, Sharp PA (2013) Divergent transcription: a driving force for new gene origination? *Cell* 155: 990–996

Xu C, Chen J, Shen B (2012) The preservation of bidirectional promoter architecture in eukaryotes: what is the driving force? *BMC Syst Biol* 6 (Suppl 1): S21

Yates A, Akanni W, Amode MR, Barrell D, Billis K, Carvalho-Silva D, Cummins C, Clapham P, Fitzgerald S, Gil L, Girón CG, Gordon L, Hourlier T, Hunt SE, Janacek SH, Johnson N, Juettemann T, Keenan S, Lavidas I, Martin FJ *et al* (2016) Ensembl 2016. *Nucleic Acids Res* 44: D710–D716

Yuan Y, Tian L, Lu D, Xu S (2015) Analysis of genome-wide RNA-sequencing data suggests age of the CEPH/Utah (CEU) lymphoblastoid cell lines systematically biases gene expression profiles. *Sci Rep* 5: 7960

8

High-throughput CRISPRi phenotyping identifies new essential genes in *Streptococcus pneumoniae*

Xue Liu[1,2] ⓘ, Clement Gallay[1] ⓘ, Morten Kjos[1,3] ⓘ, Arnau Domenech[1] ⓘ, Jelle Slager[1] ⓘ, Sebastiaan P van Kessel[1], Kèvin Knoops[4], Robin A Sorg[1], Jing-Ren Zhang[2] ⓘ & Jan-Willem Veening[1,5,*] ⓘ

Abstract

Genome-wide screens have discovered a large set of essential genes in the opportunistic human pathogen *Streptococcus pneumoniae*. However, the functions of many essential genes are still unknown, hampering vaccine development and drug discovery. Based on results from transposon sequencing (Tn-seq), we refined the list of essential genes in *S. pneumoniae* serotype 2 strain D39. Next, we created a knockdown library targeting 348 potentially essential genes by CRISPR interference (CRISPRi) and show a growth phenotype for 254 of them (73%). Using high-content microscopy screening, we searched for essential genes of unknown function with clear phenotypes in cell morphology upon CRISPRi-based depletion. We show that SPD_1416 and SPD_1417 (renamed to MurT and GatD, respectively) are essential for peptidoglycan synthesis, and that SPD_1198 and SPD_1197 (renamed to TarP and TarQ, respectively) are responsible for the polymerization of teichoic acid (TA) precursors. This knowledge enabled us to reconstruct the unique pneumococcal TA biosynthetic pathway. CRISPRi was also employed to unravel the role of the essential Clp-proteolytic system in regulation of competence development, and we show that ClpX is the essential ATPase responsible for ClpP-dependent repression of competence. The CRISPRi library provides a valuable tool for characterization of pneumococcal genes and pathways and revealed several promising antibiotic targets.

Keywords bacterial cell wall; competence; DNA replication; gene essentiality; teichoic acid biosynthesis

Subject Categories Chromatin, Epigenetics, Genomics & Functional Genomics; Genome-Scale & Integrative Biology; Microbiology, Virology & Host Pathogen Interaction

Introduction

Streptococcus pneumoniae (pneumococcus) is a major cause of community-acquired pneumonia, meningitis, and acute otitis media and, despite the introduction of several vaccines, remains one of the leading bacterial causes of mortality worldwide (Prina *et al*, 2015). The main antibiotics used to treat pneumococcal infections belong to the beta-lactam class, such as amino-penicillins (amoxicillin, ampicillin) and cephalosporines (cefotaxime). These antibiotics target the penicillin binding proteins (PBPs), which are responsible for the synthesis of peptidoglycan (PG) that plays a role in the maintenance of cell integrity, cell division, and anchoring of surface proteins (Sham *et al*, 2012; Kocaoglu *et al*, 2015). The pneumococcal cell wall furthermore consists of teichoic acids (TA), which are anionic glycopolymers that are either anchored to the membrane (lipo TA) or covalently attached to PG (wall TA) and are essential for maintaining cell shape (Brown *et al*, 2013; Massidda *et al*, 2013). Unfortunately, resistance to most beta-lactam antibiotics remains alarmingly high. For example, penicillin non-susceptible pneumococcal strains colonizing the nasopharynx of children remain above 40% in the United States (Kaur *et al*, 2016), despite the effect of the pneumococcal conjugate vaccines. Furthermore, multidrug resistance in *S. pneumoniae* is prevalent and antibiotic resistance determinants and virulence factors can readily transfer between strains via competence-dependent horizontal gene transfer (Chewapreecha *et al*, 2014; Johnston *et al*, 2014; Kim *et al*, 2016). For these reasons, it is crucial to understand how competence is regulated and to identify and characterize new essential genes and pathways. Interestingly, not all proteins within the pneumococcal PG and TA biosynthesis pathways are known (Massidda *et al*, 2013), leaving room for discovery of new potential antibiotic targets. For instance, not all enzymes in the biosynthetic route to lipid II, the precursor of PG, are known and annotated in *S. pneumoniae*. The pneumococcal TA biosynthetic pathway is even more enigmatic, and it is unknown which genes code for the enzymes responsible for polymerizing TA precursors (Denapaite *et al*, 2012).

Several studies using targeted gene knockout and depletion/overexpression techniques as well as transposon sequencing (Tn-seq)

1 Molecular Genetics Group, Groningen Biomolecular Sciences and Biotechnology Institute, Centre for Synthetic Biology, University of Groningen, Groningen, The Netherlands
2 Center for Infectious Disease Research, School of Medicine, Tsinghua University, Beijing, China
3 Department of Chemistry, Biotechnology and Food Science, Norwegian University of Life Sciences, Ås, Norway
4 Molecular Cell Biology, Groningen Biomolecular Sciences and Biotechnology Institute, University of Groningen, Groningen, The Netherlands
5 Department of Fundamental Microbiology, Faculty of Biology and Medicine, University of Lausanne, Lausanne, Switzerland
*Corresponding author. E-mail: Jan-Willem.Veening@unil.ch

have aimed to identify the core pneumococcal genome (Thanassi *et al*, 2002; Song *et al*, 2005; van Opijnen *et al*, 2009; van Opijnen & Camilli, 2012; Zomer *et al*, 2012; Mobegi *et al*, 2014; Verhagen *et al*, 2014). These genome-wide studies revealed a core genome of around 400 genes essential for growth either *in vitro* or *in vivo*. Most of the essential pneumococcal genes can be assigned to a functional category on the basis of sequence homology or experimental evidence. However, per the most recent gene annotation of the commonly used *S. pneumoniae* strain D39 (NCBI, CP000410.1, updated on 31-JAN-2015), approximately one-third of the essential genes belong to the category of "function unknown" or "hypothetical" and it is likely that several unknown cell wall synthesis genes, such as the TA polymerase, are present within this category.

To facilitate the high-throughput study of essential genes in *S. pneumoniae* on a genome-wide scale, we established CRISPRi (clustered regularly interspaced short palindromic repeats interference) for this organism. CRISPRi is based on expression of a nuclease-inactive *Streptococcus pyogenes* Cas9 (dCas9), which together with expression of a single-guide RNA (sgRNA) targets the gene of interest (Bikard *et al*, 2013; Qi *et al*, 2013; Peters *et al*, 2016). When targeting the non-template strand of a gene by complementary base-pairing of the sgRNA with the target DNA, the dCas9-sgRNA-DNA complex acts as a roadblock for RNA polymerase (RNAP) and thereby represses transcription of the target genes (Qi *et al*, 2013; Peters *et al*, 2016) (Fig 1A). Note that *S. pneumoniae* does not contain an endogenous CRISPR/Cas system, consistent with interference with natural transformation and thereby lateral gene transfer that is crucial for pneumococcal host adaptation (Bikard *et al*, 2012).

Using Tn-seq and CRISPRi, we refined the list of genes that are either essential for viability or for fitness in *S. pneumoniae* strain D39 (Avery *et al*, 1944). To identify new genes involved in pneumococcal cell envelope homeostasis, we screened for essential genes of unknown function (as annotated in NCBI), with a clear morphological defect upon CRISPRi-based depletion. This identified SPD_1416 and SPD_1417 as essential peptidoglycan synthesis proteins (renamed to MurT and GatD, respectively) and SPD_1198 and SPD_1197 as essential proteins responsible for precursor polymerization in TA biosynthesis (hereafter called TarP and TarQ, respectively). Finally, we demonstrate the use of CRISPRi to unravel gene regulatory networks and show that ClpX is the ATPase subunit that acts together with the ClpP protease as a repressor for competence development.

Results

Identification of potentially essential genes in *S. pneumoniae* strain D39

While several previous studies have identified many pneumococcal genes that are likely to be essential, the precise contribution to pneumococcal biology has remained to be defined for most of these genes. Here, we aim to characterize the functions of these proteins in the commonly used *S. pneumoniae* serotype 2 strain D39 by the CRISPRi approach. Therefore, we performed Tn-seq on *S. pneumoniae* D39 grown in C+Y medium at 37°C, our standard laboratory

condition (see Materials and Methods). We included all genes that we found to be essential in our Tn-seq study, and added extra genes that were found to be essential by previous Tn-seq studies with a serotype 4 strain TIGR4 (van Opijnen *et al*, 2009; van Opijnen & Camilli, 2012) in the CRISPRi library (see below). Finally, 391 potentially essential genes were selected, and the genes are listed in Dataset EV1.

CRISPRi enables tunable repression of gene transcription in *S. pneumoniae*

To develop the CRISPR interference system, we first engineered the commonly used LacI-based isopropyl β-D-1-thiogalactopyranoside (IPTG)-inducible system for *S. pneumoniae* (see Materials and Methods). The *dcas9* gene was placed under control of this new IPTG-inducible promoter, named P_{lac}, and was integrated into the chromosome via double crossover (Fig 1A and B). To confirm the reliability of the CRISPRi system, we tested it in a reporter strain expressing firefly luciferase (*luc*), in which an sgRNA targeting *luc* was placed under the constitutive P3 promoter (Sorg *et al*, 2015) and integrated at a non-essential locus (Fig 1B). To obtain high efficiency of transcriptional repression, we used the optimized sgRNA sequence as reported previously (Chen *et al*, 2013) (Fig EV1A).

Induction of dCas9 with 1 mM IPTG resulted in quick reduction in luciferase activity; ~30-fold repression of luciferase expression was obtained within 2 h without substantial impact on bacterial growth (Fig 1C). Furthermore, the level of repression was tunable by using different concentrations of IPTG (Fig 1C). To test the precision of CRISPRi in *S. pneumoniae*, we determined the transcriptome of the sgRNA*luc* strain (strain XL28) by RNA-Seq in the presence or absence of IPTG. The data were analyzed using Rockhopper (McClure *et al*, 2013) and T-REx (de Jong *et al*, 2015). The RNA-Seq data showed that the expression of dCas9 was stringently repressed by LacI without IPTG and was upregulated ~600-fold upon addition of 1 mM IPTG after 2.5 h. Upon dCas9 induction, the *luc* gene was significantly repressed (~84-fold) (Fig 1D). Our RNA-Seq data showed that the genes (*spd_0424*, *spd_0425*, *lacE-1*, *lacG-1*, *lacF-1*) that are downstream of *luc*, which was driven by a strong constitutive promoter without terminator, were significantly repressed as well (Appendix Fig S1A). This confirms the reported polar effect of CRISPRi (Qi *et al*, 2013). In addition, induction of dCas9 in the sgRNA-deficient strain XL29 (Fig EV1B) led to no repression of the target gene (Fig EV1C). By comparing strains with or without sgRNA*luc*, we found that repression in our CRISPRi system is stringently dependent on the expression of both dCas9 and the sgRNA, and detected no basal level repression (Fig EV1C). Furthermore, we compared the transcriptome of *luc* reporter strains with sgRNA*luc* (strain XL28) and without sgRNA*luc* (strain XL29) both grown in the presence of 1 mM IPTG. This showed that *galT-2*, *galK*, and *galR* were upregulated in both strains, indicating that these genes are activated in response to the inducer IPTG and not by the CRISPRi system itself (Dataset EV2). We also noted a slight repression of several competence genes in both XL28 and XL29 with 1 mM IPTG (Dataset EV2). Since this repression does not rely on the presence of a functional CRISPRi system, we anticipate that these changes are due to the noisy character of the competence system (Aprianto *et al*, 2016; Prudhomme *et al*, 2016). Taken together, the IPTG-inducible CRISPRi system is highly specific.

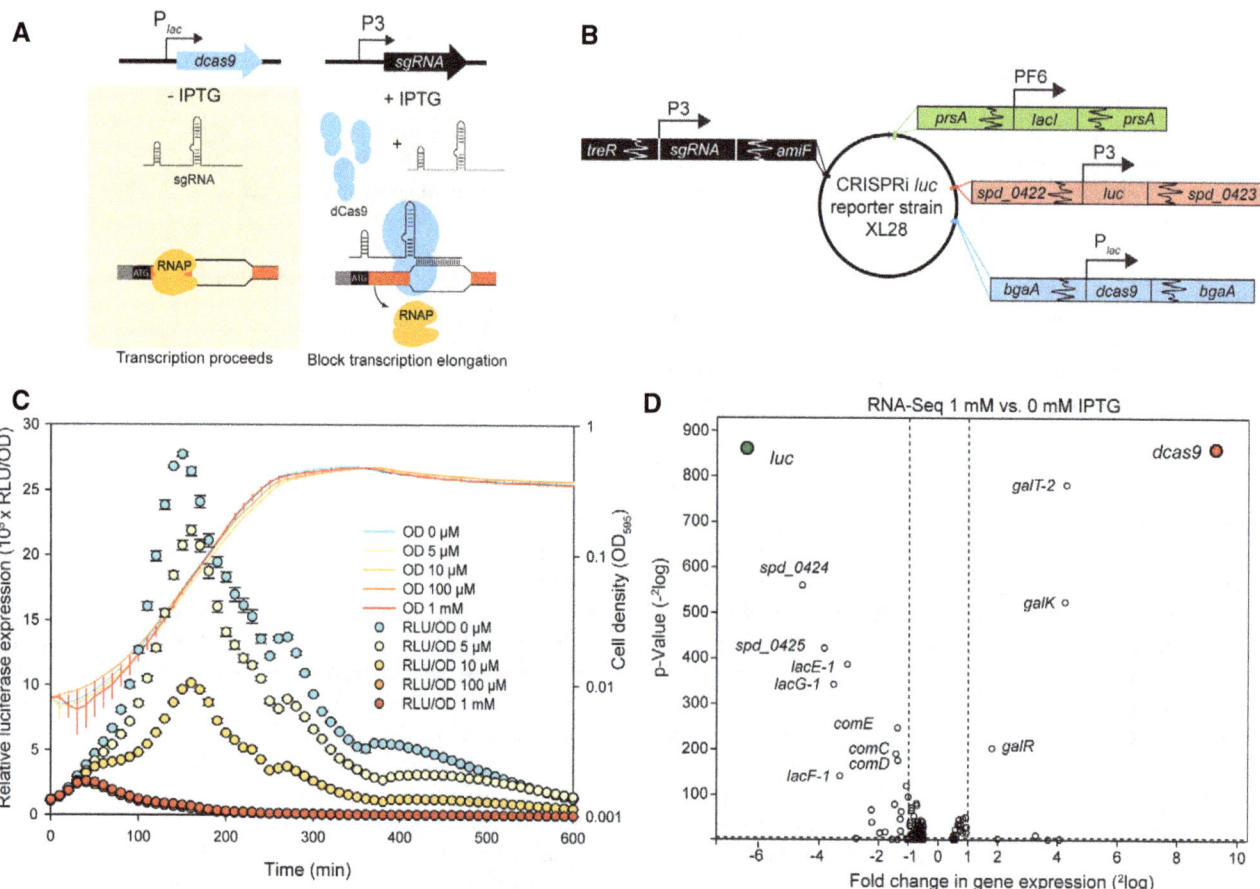

Figure 1. An IPTG-inducible CRISPRi system for tunable repression of gene expression in *S. pneumoniae*.

A   *dcas9* and sgRNA sequences were chromosomally integrated at two different loci, and expression was driven by an IPTG-inducible promoter (P_{lac}) and a constitutive promoter (P3), respectively. With addition of IPTG, dCas9 is expressed and guided to the target site by constitutively expressed sgRNA. Binding of dCas9 to the 5′ end of the coding sequence of its target gene blocks transcription elongation. In the absence of IPTG, expression of dCas9 is tightly repressed, and transcription of the target gene can proceed smoothly.

B   Genetic map of CRISPRi *luc* reporter strain XL28. To allow IPTG-inducible expression, the *lacI* gene, driven by the constitutive PF6 promoter, was inserted at the non-essential *prsA* locus; *luc*, encoding firefly luciferase, driven by the constitutive P3 promoter was inserted into the intergenic sequence between gene loci *spd_0422* and *spd_0423*; *dcas9* driven by the IPTG-inducible P_{lac} promoter was inserted into the *bgaA* locus; sgRNA-*luc* driven by the constitutive P3 promoter was inserted into the CEP locus (between *treR* and *amiF*).

C   The CRISPRi system was tested in the *luc* reporter strain XL28. Expression of *dcas9* was induced by addition of different concentrations of IPTG. Cell density (OD_{595}) and luciferase activity (shown as RLU/OD) of the bacterial cultures were measured every 10 min. The values represent averages of three replicates with SEM.

D   RNA-Seq confirms the specificity of the CRISPRi system in *S. pneumoniae*. RNA sequencing was performed on the *luc* reporter strain XL28 (panel B) with or without 1 mM IPTG. The *dcas9* and *luc* genes are highlighted. Data were analyzed with T-REx and plotted as a volcano plot. *P*-value equals 0.05 is represented by the horizontal dotted line. Two vertical dotted lines mark the twofold changes.

Construction and growth analysis of the CRISPRi library

We next used the CRISPRi system to construct an expression knock-down library of pneumococcal essential genes. An sgRNA to each of the 391 potentially essential genes was designed as described previously (Larson *et al*, 2013) (Dataset EV3). Based on the sgRNA*luc* plasmid (Fig 2A), we tested two different cloning strategies to introduce the unique 20-nt base-pairing region for each gene: infusion cloning and inverse PCR (Ochman *et al*, 1988; Irwin *et al*, 2012; Larson *et al*, 2013) (Fig EV2A). For infusion cloning, we synthesized two complementary primers consisting of the 20-nt base-pairing region flanked by 15-nt overlap sequences. The two complementary primers were then annealed to form a duplex DNA

fragment and cloned into the vector by the infusion reaction, followed by direct transformation into *S. pneumoniae* D39 strain DCI23. With inverse PCR, we used a phosphorylated universal primer, together with a gene-specific primer to fuse the 20-nt base-pairing region into the vector by PCR, followed by blunt-end ligation and direct transformation into *S. pneumoniae* D39 strain DCI23. We compared the efficiency of the two methods by creating sgRNA strains targeting the known essential gene *folA* (*spd_1401*). Depletion of *folA* causes a clear growth defect, which could thus be used to test the functionality of sgRNA*folA* in transformants. We found that 79% of the transformants produced by infusion cloning had a growth defect upon dCas9 induction with IPTG (38 out of 48 colonies), whereas 26% of the transformants generated by inverse PCR showed a phenotype (12/46). Sequencing validated that transformants with a growth defect contained the correct sgRNA sequence. Considering the convenience and efficiency, we adopted the infusion cloning strategy for sgRNA cloning in this study. All sgRNA constructs were sequence verified, and we considered them genetically functional when the sgRNA did not contain more than 1 mismatch to the designed sgRNA and no mismatches in the first 14-nt prior to the PAM. Using this approach, after a single round of cloning and sequencing, we successfully constructed 348 unique sgRNA strains (see Materials and Methods). Note that we are still in the process of constructing the remaining 43 sgRNA strains, the failure of which is likely caused by technical reasons (e.g., incorrect oligonucleotides, poor oligo annealing, low transformation).

To examine the effects of CRISPRi-based gene silencing, growth was assayed both in the presence and absence of 1 mM IPTG for 18 h in real time by microtiter plate assays. Two types of growth phenotypes were defined and identified: a growth defect and increased lysis (Fig EV2B–E). As shown in Fig 2B, CRISPRi-based repression of transcription led to a growth defect in 230 genes, 48 genes showed increased lysis, including 24 that demonstrated both a growth defect and increased lysis, and 94 genes showed no defect (see Dataset EV1). In total, 254 out of 348 target genes (about 73%) repressed by CRISPRi showed growth phenotypes. Comparing the optical densities between the uninduced and induced cells at the time point at which uninduced cells reached an OD_{595} of ~0.1, 174 genes repressed by CRISPRi displayed a more than fourfold growth defect, and 254 genes showed a more than twofold growth defect (Fig 2C). To further validate the specificity of the CRISPRi system, CRISPRi strains targeting eight genes identified as essential and eight genes as dispensable by Tn-seq were included in the growth analysis. The selected dispensable genes are present as a monocistron or are in an operon with other

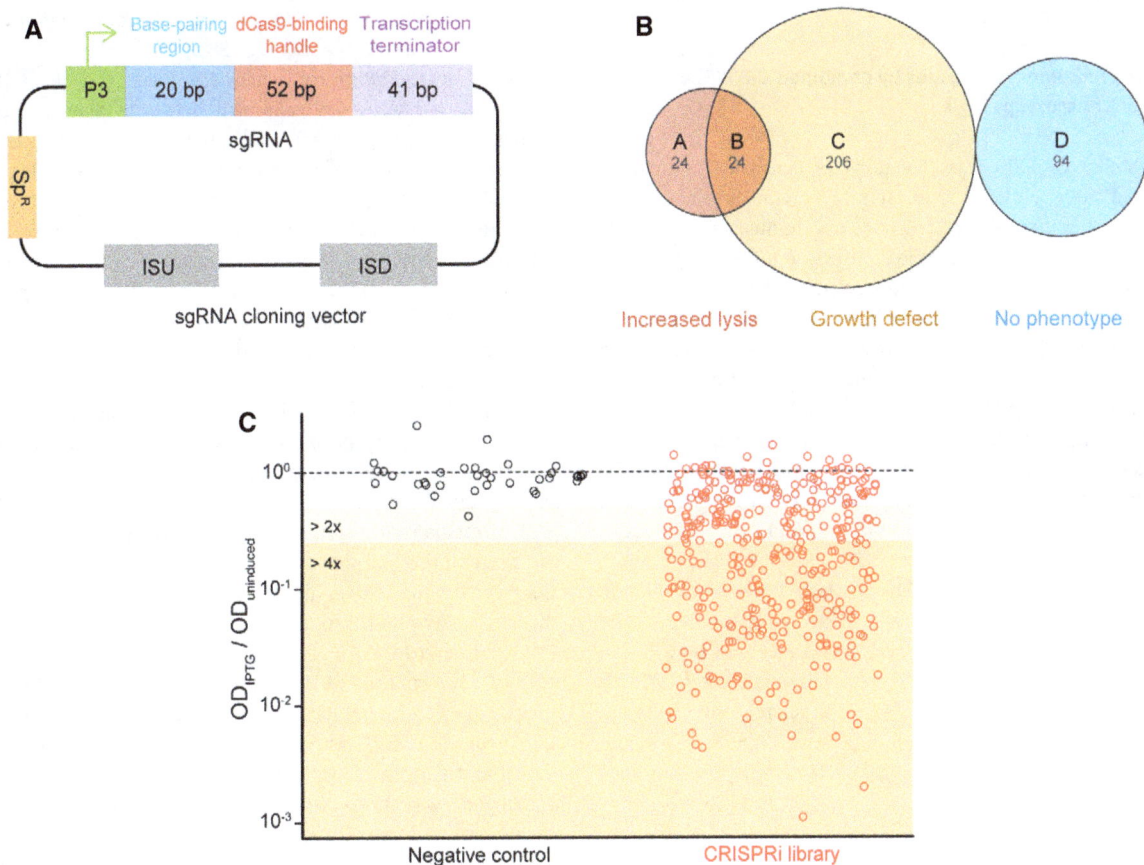

Figure 2. Construction and growth analysis of the CRISPRi library.

A   The plasmid map of the sgRNA cloning vector (pPEPX-P3-sgRNA*luc*). The sgRNA expression vector is a *S. pneumoniae* integration vector. It contains a constitutive P3 promoter, a spectinomycin-selectable marker (Sp^R), two homology sequences (ISU and ISD) for double crossover integration at the CEP locus (Sorg *et al*, 2015), and the sgRNA sequence. The sgRNA chimera contains a base-pairing region (blue), dCas9-binding handle (red), and the *S. pyogenes* transcription terminator (purple).

B, C   Growth analysis of the whole library. (B) Classification of the 348 genes targeted by the CRISPRi library according to growth analysis: A represents the 24 strains that only showed increased autolysis; B represents the 24 strains showing both increased autolysis and growth defects; C represents the 206 strains that showed only growth defects; D represents the 94 strains with no phenotype. Criteria for determination of a growth defect and increased lysis are demonstrated in Fig EV2B–E. (C) Comparison of the OD_{595} of IPTG-induced cells (OD_{IPTG}) to the OD_{595} of uninduced cells ($OD_{uninduced}$) at a time point. The time point at which uninduced cells have an optical density (595 nm) closest to 0.1 was selected for the plotting. *y*-axis represents the value of OD_{IPTG} divided by $OD_{uninduced}$. The red data points in the dark orange area (174/348 strains) correspond to strains displaying a strong growth defect (more than fourfold); points in the light orange area demonstrate a moderate growth defect of twofold to fourfold (71/348 strains). The same type of analysis was performed on 36 negative control strains, shown as the black data points.

non-essential genes. As shown in Fig EV3A, no apparent growth defects could be observed when these non-essential genes were targeted by CRISPRi, while repression of essential genes led to strong growth defects (Fig EV3B).

It should be noted that CRISPRi repression of dispensable genes that are cotranscribed with essential genes can lead to growth phenotypes (Appendix Fig S1), which is due to polar effect of CRISPRi system (Qi et al, 2013). Thus, some of the genes may be targeted multiple times in the CRISPRi library (in case of more than one essential gene within the operon). We also observed that after a lag phase, most CRISPRi knockdowns with growth phenotypes eventually grow out to the same final OD (Fig EV4A). Re-culturing these cells showed the absence of sensitivity to IPTG, indicative of the presence of suppressor mutations (Fig EV4A). Indeed, by sequencing the two key elements of the CRISPRi system, the sgRNA and dcas9, we found that most of the suppressor strains contain loss-of-function mutation in the dcas9 coding sequence (Fig EV4B). This is similar to observations made for the CRISPRi system in Bacillus subtilis (Zhao et al, 2016).

Phenotyping pneumococcal genes by combined CRISPRi and high-content microscopy

To test whether CRISPRi was able to place genes in a functional category and thereby allow us to identify previously uncharacterized genes with a function in cell envelope homeostasis, we first analyzed the effects of CRISPRi-based repression on cell morphology using 68 genes. These genes were selected as they represent different functional pathways and have been identified as essential or crucial for normal pneumococcal growth by Tn-seq studies (van Opijnen et al, 2009; van Opijnen & Camilli, 2012) and by displaying strong growth phenotypes in our CRISPRi assay (Fig 2B and C). The selected genes have been associated with capsule synthesis (three genes), transcription (four genes), cell division (six genes), translation (seven genes), teichoic acid biosynthesis (nine genes), cell membrane synthesis (11 genes), chromosome biology (14 genes), and peptidoglycan synthesis (14 genes) (Table 1). High-content microscopy of the CRISPRi knockdowns showed a good correlation between reported gene function and observed phenotype. The common features of the morphological changes caused by CRISPRi repression of genes belonging to the same functional categories are summarized in Table 1. Growth analysis and microscopy phenotyping of a representative gene of each pathway, CRISPRi repression of which showed typical morphological changes of its pathway, were included in Fig 3. Morphological changes of CRISPRi repression of the other genes of the pathways are shown in Appendix Figs S2–S9. For instance, compared with the control strain (Fig 3A, XL28), repression of transcription of genes involved in chromosome biology caused, as expected, appearance of anucleate cells or cells with aberrant chromosomes (Fig 3B, dnaA; Appendix Fig S2). Cells with repression of genes involved in transcription showed a significant growth defect, and no obvious morphological changes were observed (Fig 3C, rpoC; Appendix Fig S3). Repression of genes involved in translation showed heterogeneous cell shapes and condensed nucleoids (Fig 3D, infC; Appendix Fig S4), in line with our previous observations (Sorg & Veening, 2015) and observations made in Escherichia coli showing that inhibition of protein synthesis by antibiotics leads to nucleoid

condensation (Morgan et al, 1967; Zusman et al, 1973; Roggiani & Goulian, 2015).

In S. pneumoniae, the fatty acid biosynthesis genes are all located in a single cluster (Lu & Rock, 2006) (Appendix Fig S5A), and two promoters in front of fabT and fabK are regulated by the transcriptional repressor FabT (Jerga & Rock, 2009). It was shown that fabT and fabH are cotranscribed (Lu & Rock, 2006), but the transcription pattern of the other genes is still unknown, which makes functional study of these genes with CRISPRi very difficult due to polar effects of the block of transcription elongation (Qi et al, 2013). Nevertheless, repression of transcription of genes involved in cell membrane synthesis caused diverse patterns of morphological changes: repression of fabH, acpP, fabK, fabD, and fabG led to a spotty Nile red pattern and irregular cell shapes including more pointy cells (Fig 3E, fabK; Appendix Fig S5B), as was shown previously (Kuipers et al, 2016); repression of fabF, accB, fabZ, and accD led to chaining of cells, heterogeneous cell sizes and irregular cell shapes; repression of acpS resulted in elongated and enlarged cells, whereas repression of cdsA caused cell rounding with heterogeneous cell sizes (Appendix Fig S5B).

When transcription of genes involved in cell division was repressed, we observed cells with irregular shapes and heterogeneous sizes (Appendix Fig S6). Interestingly, repression of ftsZ and ftsL caused similar morphological changes (Fig 3F, ftsZ; Appendix Fig S6), consistent with the reported function of FtsL on regulating FtsZ ring (Z-ring) dynamics in B. subtilis (Kawai & Ogasawara, 2006). Cells with repression of ezrA formed twisting chains and contained multiple septa, some of which formed at cell poles instead of midcell. Indeed, it was reported that B. subtilis EzrA can modulate the frequency and position of the Z-ring formation (Chung et al, 2004).

Repression of genes involved in capsule synthesis caused aggregation of cells (Appendix Fig S7), which may be due to the reduction in the negatively charged capsule that can provide a repelling electrostatic force preventing cell aggregation (Li et al, 2013).

Repression of transcription of genes involved in cell wall synthesis caused different phenotypes, depending on which step in peptidoglycan synthesis was interrupted. S. pneumoniae is oval-shaped, and it displays both septal and peripheral growth (Massidda et al, 2013; Pinho et al, 2013). Peptidoglycan synthesis of S. pneumoniae starts from formation of UDP-MurNAc-pentapeptides. Repression of expression of genes playing roles in these very first steps, including glmU, alr, ddl, murI, murC, murD, murE, and murF, will block both septal and peripheral peptidoglycan synthesis. Consistent with this prediction, we observed severe changes in cell shape and size, including heterogeneous cell sizes, exploding cells, defective septa, round cells, and cells demonstrating a coccus-to-rod transition (Appendix Fig S8). MraY and MurG play roles in formation of lipid II, and they are thus also involved in both peripheral and septal peptidoglycan synthesis. CRISPRi strains repressing mraY or murG led to a mix of elongated cells and short cells (Appendix Fig S8). FtsW and RodA are members of SEDS (shape, elongation, division, and sporulation) proteins (Meeske et al, 2016) and were first identified in E. coli (Ikeda et al, 1989). Inactivation of FtsW in E. coli blocks cell division without an effect on cell elongation (Khattar et al, 1994), and FtsW is suggested to act as a lipid II flippase (Mohammadi et al, 2011). FtsW of S. pneumoniae was believed to have a conserved function with E. coli (Maggi et al, 2008), is

Table 1. Cellular pathways selected for CRISPRi phenotyping.

Pathway	Phenotype	Gene[a]	Pathway	Phenotype	Gene[a]
Chromosome replication	Anucleate cells; longer chains; uneven distribution of chromosomes; heterogeneous cell size	*dnaA* **(SPD_0001)**	Cell division	Exploding cells; heterogeneous cell size; defective septa; twisting chains	*ftsL (SPD_0305)*
		dnaN (SPD_0002)			*gpsB (SPD_0339)*
		gyrB (SPD_0709)			*ftsE (SPD_0659)*
		parE (SPD_0746)			*ftsX (SPD_0660)*
		parC (SPD_0748)			*ezrA (SPD_0710)*
		dnaX (SPD_0760)			***ftsZ (SPD_1479)***
		ftsK (SPD_0774)	Capsule synthesis	Cell aggregation; heterogeneous cell size	*cps2E (SPD_0319)*
		dnaG (SPD_0957)			*cps2I (SPD_0324)*
		xerS (SPD_1023)			*cps2L (SPD_0328)*
		gyrA (SPD_1077)	Peptidoglycan biosynthesis	Heterogeneous cell size; coccus-to-rod transition; round cells; elongated cells; enlarged cells; defective septa	*uppS (SPD_0243)*
		dnaI (SPD_1521)			***pbp2X (SPD_0306)***
		priA (SPD_1546)			*mraY (SPD_0307)*
		parB (SPD_2069)			*uppP (SPD_0417)*
		dnaC (SPD_2030)			*murD (SPD_0598)*
Transcription	No strong morphological phenotype	*rpoA (SPD_0218)*			*murG (SPD_0599)*
		rpoD (SPD_0958)			*rodA (SPD_0706)*
		rpoC (SPD_1758)			*glmU (SPD_0874)*
		rpoB (SPD_1759)			*ftsW (SPD_0952)*
Translation	Condensed nucleoids; short cells; heterogeneous cell size	*rpsJ (SPD_0192)*			*murE (SPD_1359)*
		rplD (SPD_0194)			*murF (SPD_1483)*
		rplV (SPD_0198)			*ddl (SPD_1484)*
		rpsC (SPD_0199)			*alr (SPD_1508)*
		efp (SPD_0395)			*murI (SPD_1661)*
		infC (SPD_0847)	Teichoic acid biosynthesis	Longer chains; elongated cells; enlarged cells; heterogeneous cell size; defective septa	*SPD_0099*
		tsf (SPD_2041)			*licC (SPD_1123)*
Cell membrane biosynthesis	Spotty membrane staining; irregular cell shape; heterogeneous cell size	*cdsA (SPD_0244)*			*licB (SPD_1124)*
		fabH (SPD_0380)			*licA (SPD_1125)*
		acpP (SPD_0381)			*tarJ (SPD_1126)*
		fabK (SPD_0382)			*tarI (SPD_1127)*
		fabD (SPD_0383)			*SPD_1200*
		fabG (SPD_0384)			***licD3 (SPD_1201)***
		fabF (SPD_0385)			*SPD_1620*
		accB (SPD_0386)			
		fabZ (SPD_0387)			
		accD (SPD_0389)			
		acpS (SPD_1509)			

[a]The genes highlighted in bold were included in Fig 3.

co-localized with septal HMW (high molecular weight) PBPs (Morlot *et al*, 2004), and is thus predicted to be involved in septal peptidoglycan synthesis. By morphological analysis, we provided experimental evidence to support this prediction: FtsW and Pbp2X are responsible for septal peptidoglycan synthesis, and elongated cells and coccus-to-rod transition were observed with CRISPRi repression of *ftsW* or *pbp2X* (Fig 3G, *pbp2X*; Appendix Fig S8, *ftsW*). RodA of *S. pneumoniae* shows 26% identity with RodA of

E. coli (Noirclerc-Savoye *et al*, 2003), which is required for cell elongation. Studies of RodA in *B. subtilis* also support its function on elongation of the lateral wall (Henriques *et al*, 1998; Meeske *et al*, 2016). RodA of *S. pneumoniae* was predicted to be a lipid II flippase responsible for peripheral peptidoglycan synthesis (Massidda *et al*, 2013). *Streptococcus pneumoniae* cells with repressed *rodA* expression by CRISPRi are consistently shorter (Appendix Fig S8), indicating a defect in cell elongation.

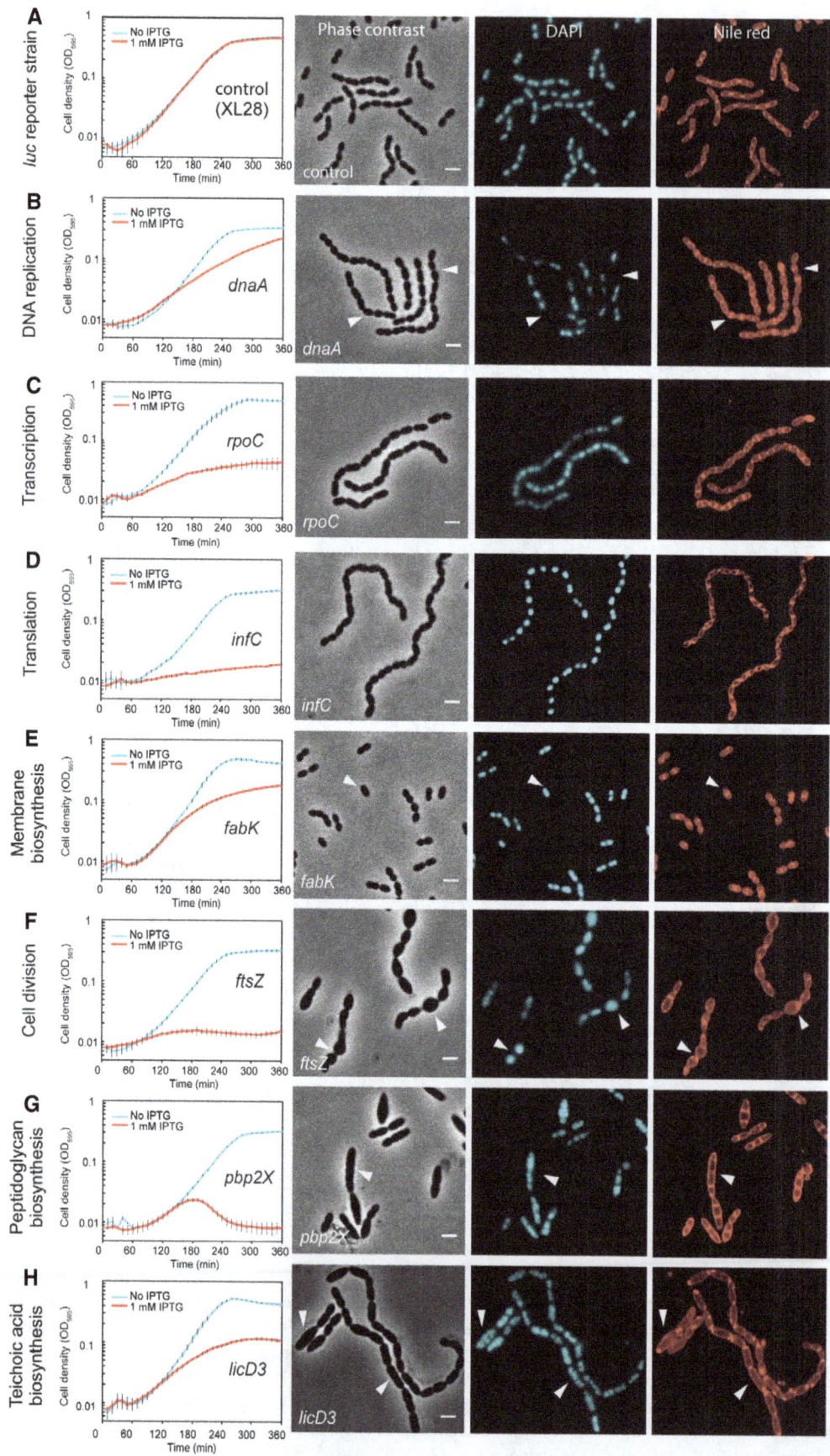

Figure 3.

Figure 3. Growth profiles and morphological changes of CRISPRi strains with sgRNA targeting genes of different functional pathways.

A–H   Growth of *S. pneumoniae* strains was performed in C+Y medium with (red) or without (cyan) 1 mM IPTG. Cell densities were measured every 10 min. The values represent averages of three replicates with SEM. Morphological changes were examined with fluorescence microscopy, and representative micrographs are shown. Phase contrast, DAPI staining, and Nile red staining are displayed. Scale bars = 2 μm. *Streptococcus pneumoniae* D39 reporter strain XL28 expresses firefly luciferase (*luc*) from a constitutive promoter and contains an sgRNA targeting the *luc* gene, and serves as a control strain without growth defects. White arrows point to specific (morphological) changes. For *dnaA*, arrows point to anucleate cells (B); *fabK*, non-uniform, spotty membrane staining (E); *ftsZ*, ballooning cells (F); *pbp2X*, elongated cells (G); *licD3*, elongated and enlarged cells (H). Repression of a transcription-related gene, *rpoC*, no strong morphological changes were observed (C); a translation-related gene, *infC*, led to generally condensed chromosomes as shown in the DAPI staining image (D). One gene of each pathway was presented in this figure. Additional information related to this figure can be found in Table 1 and Appendix Figs S2–S9, which show microscopy images of more genes of each pathway.

Repression of genes involved in teichoic acid (TA) biosynthesis led to morphological changes, including formation of longer chains and cells of heterogeneous sizes, mostly enlarged or elongated (Fig 3H, *licD3*; Appendix Fig S9). Growth of *S. pneumoniae* depends on exogenous choline, which is an essential molecule for the synthesis of pneumococcal TA, and the chaining phenotype caused by repression of genes involved in TA synthesis is in line with *S. pneumoniae* growing in medium without choline (Damjanovic *et al*, 2007).

In summary, by morphological analysis of CRISPRi strains for repression of transcription of genes with known function from different pathways, we established links between genotypes and phenotypes. Importantly, repression of transcription of genes known to be involved in cell envelope homeostasis, such as *ftsZ*, *ftsL*, *ftsW*, *rodA*, *pbp2X*, *glmU*, *murC*, *murF*, *tarI*, *tarJ*, *licA*, *licB*, *licC*, and *licD3*, caused severe changes in cell shape and size, including heterogeneous cell size, ballooning cells, defective septa, short cells, round cells, cells in chains, and cells demonstrating a coccus-to-rod transition. These observations provide a useful platform for the functional identification of hypothetical genes, especially genes involved in cell envelope homeostasis.

Functional verification and annotation of *pcsB* (*spd_2043*), *vicR* (*spd_1085*), *divIC* (*spd_0008*), and *rafX* (*spd_1672*)

We next analyzed 44 strains in the CRISPRi library that target genes that are annotated as hypothetical in the *S. pneumoniae* D39 genome in the NCBI database (CP000410.1, updated on 31-JAN-2015). From this approach, we were able to verify the function and annotate several genes, whose function had been studied in pneumococci before but have not been properly annotated in the D39 genome. For example, repression of genes (*spd_0008*, *spd_1085*, and *spd_2043*) led to significant growth defects and cell shape and cell size changes (Appendix Fig S11A and B). Knocking down *spd_2043* and *spd_1085* led to almost the same morphological changes, which included irregular cell shape, heterogeneous cell sizes, and appearance of ballooned cells, suggesting that these two genes might be functionally associated and play roles in peptidoglycan synthesis or cell division. By literature mining and BLAST searches, we recognized *spd_1085* as *vicR* and *spd_2043* as *pcsB* (Ng *et al*, 2003). Consistent with the observed phenotypes in the CRISPRi strains, *pcsB* was shown to be essential for cell wall separation and its expression relies on the response regulator encoded by *vicR* (Reinscheid *et al*, 2001; Ng *et al*, 2003; Sham *et al*, 2011; Bartual *et al*, 2014). Similarly, the morphological changes suggested a potential role of SPD_0008 in cell wall synthesis or cell division. In line with this, SPD_0008 was identified as DivIC, which was reported to form a trimeric complex with DivIB and FtsL and

colocalized at division sites of *S. pneumoniae* strain R6 (Noirclerc-Savoye *et al*, 2005).

CRISPRi knockdown strain targeting *spd_1672* showed no significant growth defect at exponential phase, but cells lysed quicker in the stationary phase (Appendix Fig S11C). Microscopy showed that bacterial cells with CRISPRi-repressed *spd_1672* formed significantly longer chains (Appendix Fig S11D). Chained cells displayed irregular shapes and heterogeneous cell sizes. These phenotypes are very similar to the morphological changes caused by repression of genes involved in the biosynthesis of teichoic acid (Appendix Fig S9). Actually, *spd_1672* has been studied in *S. pneumoniae* R6 and was shown to contribute to the biosynthesis of wall teichoic acid and was named *rafX* (Wu *et al*, 2014). The reported *spd_1672* knockout strain of *S. pneumoniae* R6 also displayed a reduced stationary phase with similar cell shape and cell size defects. Inconsistent with our study, longer chains were not observed by TEM (transmission electron microscopy) imaging in the Wu *et al* study. To exclude the possible polar effect of CRISPRi repression, we made a *spd_1672* knockout in *S. pneumoniae* D39, and the *spd_1672* knockout strain also showed longer chains. Thus, the mismatch in phenotypes between the studies may be due to the different genetic background of *S. pneumoniae* D39 and R6, or may be caused by the process of sample preparation for TEM examination.

Annotation and characterization of chromosome replication genes *dnaB* (*spd_1522*), *dnaD* (*spd_1405*), and *yabA* (*spd_0827*)

High-content microscopy screening of the CRISPRi library showed that repression of *spd_1405*, *spd_1522,* and *spd_0827* led to significant growth defects and generation of anucleate cells (Appendix Fig S10). Appearance of anucleate cells is an important sign of a defect in chromosome biology, thus suggesting that these three genes are involved in chromosome replication or segregation. SPD_0827 shows 33% identity with initiation control protein YabA of *Bacillus subtilis*, which interacts with DnaN and DnaA, and acts as a negative regulator of replication initiation (Noirot-Gros *et al*, 2002; Goranov *et al*, 2009). We thus named SPD_0827 to YabA. To test the function of *yabA* in *S. pneumoniae*, a deletion mutant was made by erythromycin marker replacement. The *yabA* deletion (Δ*yabA*) showed a significantly reduced growth rate compared to the wild type (Appendix Fig S12A) and displayed longer chains with frequent anucleate cells (Appendix Fig S12C). To test whether *S. pneumoniae* YabA is also a negative regulator of initiation of DNA replication, we determined the *oriC-ter* ratio using real-time quantitative PCR (qPCR). As shown in Appendix Fig S12D, the *oriC-ter* ratio was significantly higher in Δ*yabA* indicative of over-initiation, strongly suggesting a similar function as *B. subtilis* YabA.

When making a list of known genes involved in pneumococcal chromosome biology (Table 1), we noticed that *dnaB* and *dnaD*, two known bacterial DNA replication proteins (Smits *et al*, 2011; Briggs *et al*, 2012), are not annotated in *S. pneumoniae* D39. BlastP analyses showed that *spd_1405* and *spd_1522* might be coding for DnaD and DnaB, respectively. SPD_1405 showed 30% identity with DnaD of *B. subtilis*, and thus, we named *spd_1405* to *dnaD*. SPD_1522 has 389 amino acids (aa), and the N-terminal 1–149 aa-long domain showed 19.8% identity with domain I of DnaB of *B. subtilis*, whereas aa 206–379 showed 45.7% identity with domain II. DnaB of *B. subtilis* (472 aa) is longer than SPD_1522 of *S. pneumoniae* D39 (389 aa), because the former contains a degenerated middle DDBH2 domain (Briggs *et al*, 2012). Additionally, the arrangement of the neighboring genes of *S. pneumoniae dnaB* (*spd_1522*), *dnaI*, and *nrdR* is the same in *B. subtilis*. Based on these observations, we named *spd_1522* to *dnaB*.

It was reported that DnaD and DnaB are recruited to the chromosome by DnaA and play important roles in chromosome replication initiation in *B. subtilis* (Smits *et al*, 2011). To test the function of *S. pneumoniae* DnaD and DnaB, we constructed Zn^{2+}-inducible depletion strains (P_{Zn}-*dnaD*; P_{Zn}-*dnaB*), because efforts to make deletion mutants failed. In the absence of 0.1 mM Zn^{2+}, the depletion strains showed significant growth defects (Appendix Fig S12A), confirming their essentiality. If DnaB and DnaD indeed play a role in replication initiation, repression of them should lead to a decrease in the *oriC-ter* ratio. Indeed, the *oriC-ter* ratio of cells in absence of Zn^{2+} was significantly lower than in the presence of Zn^{2+} (Appendix Fig S12D). Together, we identified and annotated *yabA*, *dnaD*, and *dnaB* and confirmed their function in pneumococcal DNA replication.

SPD_1416 and SPD_1417 are involved in peptidoglycan precursor synthesis

We found that CRISPRi strains with sgRNA targeting hypothetical genes *spd_1416* or *spd_1417* showed significant growth retardation and morphological abnormality, such as heterogeneous cell size and elongated and enlarged cells with multiple incomplete septa (Appendix Fig S10). These manifestations mirrored what we observed upon inhibiting the expression of genes known to be involved in peptidoglycan (PG) synthesis (Appendix Fig S8). Consistent with the essentiality of these two genes as suggested by Tn-seq, we were unable to obtain deletion mutants of *spd_1416* or *spd_1417* after multiple attempts. To confirm that these genes are essential for pneumococcal growth, we constructed merodiploid strains of *spd_1416* and *spd_1417* by inserting a second copy of each gene fused to *gfp* (encoding a monomeric superfolder GFP) at their N-terminus (referred as *gfp-spd_1416* or *gfp-spd_1417*) or C-terminus (referred as *spd_1416-gfp* or *spd_1417-gfp*). These *gfp* fusions were integrated at the ectopic *bgaA* locus under the control of the zinc-inducible promoter, P_{Zn}. In the presence of Zn^{2+}, we could delete the native *spd_1416* or *spd_1417* gene by allelic replacement in the P_{Zn}-*gfp-spd_1417* or P_{Zn}-*gfp-spd_1416* genetic background. When transforming in the P_{Zn}-*spd_1417-gfp* genetic background, we did not obtain erythromycin resistant colonies, indicating that the C-terminal GFP fusion of SPD_1417 is not functional. Note that we could not replace *spd_1416* or *spd_1417* in the wild type in the presence of Zn^{2+}. While both the *spd_1416* and

spd_1417 mutants behaved normally in the presence of Zn^{2+}, severe growth retardation was observed in the absence of Zn^{2+} (Fig 4A). Together, these lines of evidence demonstrate that both *spd_1416* and *spd_1417* are essential genes.

Morphological analysis by light microscopy of bacterial cells upon depletion of *gfp-spd_1416* or *gfp-spd_1417* confirmed the morphological changes as observed in the CRISPRi knockdowns (Fig 4B). The *gfp-spd_1416* or *gfp-spd_1417* cells were further analyzed using freeze-substitution electron microscopy (Fig 4C). This showed the presence of elongated cells and the frequent formation of multiple septa per cell, in contrast to wild-type D39 cells which showed the typical diplococcal shape. Note that the mild sample preparation used in our freeze-substitution EM protocol also preserved the capsule, which can be readily lost during traditional EM sample preparation (Hammerschmidt *et al*, 2005). BlastP analysis shows that SPD_1416 contains a Mur-ligase domain with 36% sequence identity with MurT of *Staphylococcus aureus*, whereas SPD_1417 possesses a glutamine amidotransferase domain with 40% sequence identity with GatD of *S. aureus*. MurT and GatD, two proteins involved in staphylococcal cell wall synthesis (Figueiredo *et al*, 2012; Munch *et al*, 2012), form a complex to perform the amidation of the D-glutamic acid in the stem peptide of PG. It was previously reported that recombinant MurT/GatD of *S. pneumoniae* R6, purified from *E. coli*, indeed can amidate glutamate lipid II into iso-glutamine lipid II *in vitro* (Zapun *et al*, 2013). Therefore, we named *spd_1416* to *murT* and *spd_1417* to *gatD*. It is interesting to note that while MurT or GatD depletion strains in *S. aureus* showed reduced growth, cells exhibited normal cell morphologies (Figueiredo *et al*, 2012), in contrast to the strong morphological defects observed in *S. pneumoniae* D39.

MurT and GatD contain no membrane domain or signal peptide, and are thus predicted to be cytoplasmic proteins. However, fluorescence micyopy of the N-terminal GFP fused to MurT or GatD showed that they are partially membrane localized (Fig 4D). In-gel fluorescence imaging showed that GFP-MurT and GFP-GatD were correctly expressed without any detectable proteolytic cleavage (Appendix Fig S13). Since *in vitro* assays demonstrated that glutamate lipid II, which is anchored to the membrane by the bactoprenol hydrocarbon chain of lipid II, is a substrate of the MurT/GatD amidotransferase complex, it is reasonable to assume that membrane localization of MurT or GatD is caused by recruitment to the membrane-bound substrate. Indeed, amidation of the glutamic acid at position 2 of the peptide chain most likely occurs after formation of lipid-linked PG precursors (Rajagopal & Walker, 2016).

CRISPRi revealed novel pneumococcal genes involved in teichoic acid biosynthesis

CRISPRi-based repression of hypothetical essential genes *spd_1197* and *spd_1198* led to significant growth defects, and microscopy revealed chained cells with abnormal shape and size (Appendix Fig S10). Some of the cells were elongated and enlarged. These phenotypes are consistent with the typical morphological changes caused by repression of genes in teichoic acid (TA) biosynthesis (Appendix Fig S9). In accordance with this, analysis of the genetic context of *spd_1197* and *spd_1198* showed that they are in the *lic3* region, which was predicted to be a pneumococcal TA gene cluster

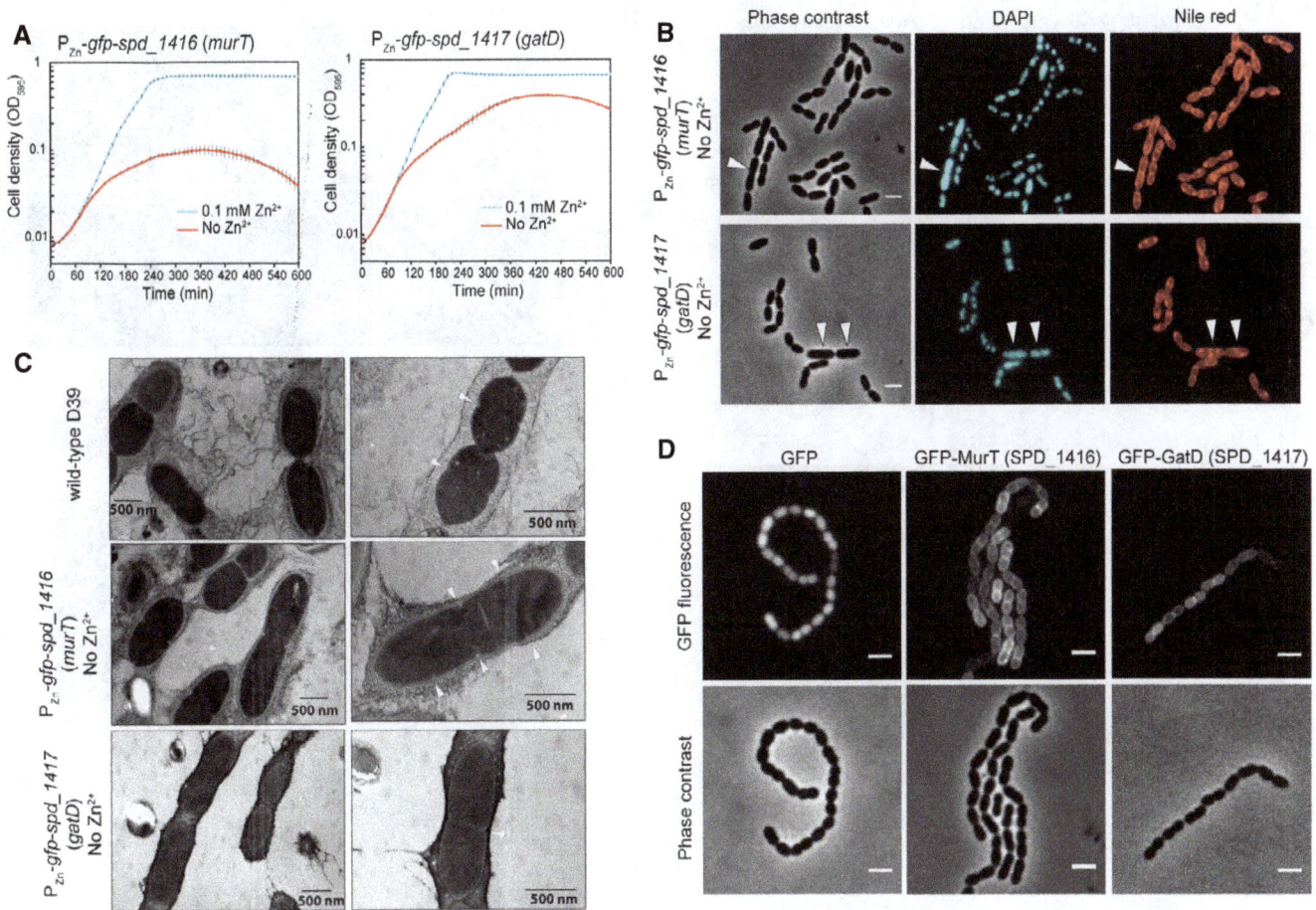

Figure 4. Identification of peptidoglycan synthesis genes *spd_1416* (*murT*) and *spd_1417* (*gatD*).

A   Growth curves of depletion strains P_{Zn}-*gfp-spd_1416* (*murT*) and P_{Zn}-*gfp-spd_1417* (*gatD*), in C+Y medium with (cyan) or without (red) 0.1 mM Zn^{2+}. The values represent averages of three replicates with SEM.

B   Microscopy of cells from panel (A) after incubating in C+Y medium without Zn^{2+} for 2.5 h. Representative micrographs of phase contrast, DAPI, and Nile red are shown. Scale bars = 2 μm. White arrows point to elongated and enlarged cells.

C   Electron micrographs of the same samples as in panel (B) and wild-type *S. pneumoniae* D39 as reference. Note that depletion of *spd_1416* or *spd_1417* resulted in elongated cells. Septa are pointed with white arrows.

D   Localization of GFP-MurT and GFP-GatD. Micrographs of GFP signal (upper panel) and phase contrast (lower panel) are shown. Scale bars = 2 μm. *Streptococcus pneumoniae* D39 with free GFP showing cytoplasmic localization was included as reference.

(Kharat *et al*, 2008; Denapaite *et al*, 2012). Similar to the approach described above, we generated Zn^{2+}-inducible C-terminal GFP fusions to SPD_1197 and SPD_1198, integrated these ectopically at the *bgaA* locus, and then deleted the native *spd_1197* or *spd_1198* genes in the presence of Zn^{2+}. Plate reader assays showed strong growth impairment in the absence of Zn^{2+} (Fig 5A), suggesting their essentiality. In line with this, we were unable to replace these genes with an erythromycin resistance marker in the wild-type background in either the absence or presence of Zn^{2+}. Consistent with the phenotypes of the CRISPRi screen, the zinc-depletion strains showed similar morphological defects with cells in chains and elongated or enlarged cell shape and size (Fig 5B). EM analysis of depleted cells also revealed uneven distribution of multiple septa within a single cell, increased extracellular material and a rough cell surface (Fig 5C).

SPD_1198 contains 11 predicted transmembrane (TM) helices, while SPD_1197 has 2 predicted TM segments with a C-terminal extracytoplasmic tail. In-gel fluorescence showed that SPD_1197-GFP was mainly produced as a full-length product. The SPD_1198-GFP fusion, however, showed clear signs of protein degradation (Appendix Fig S13). Nevertheless, we performed fluorescence microscopy to determine their localizations. In agreement with the prediction, SPD_1197-GFP and SPD_1198-GFP are clearly localized to the membrane (Fig 5D).

Phosphorylcholine is an essential component of pneumococcal TA, and for this reason, a phosphorylcholine antibody is frequently used to detect *S. pneumoniae* TA (Vollmer & Tomasz, 2001; Wu *et al*, 2014). To explore whether SPD_1197 and SPD_1198 indeed play a role in TA synthesis, we performed Western blotting to detect phosphorylcholine-decorated TA using whole-cell lysates (Fig 5E). Cells of strains P_{Zn}-*spd_1197-gfp* and P_{Zn}-*spd_1198-gfp* were grown in the presence or absence of 0.1 mM Zn^{2+}. As controls, we depleted expression of three genes involved in PG synthesis (*murT*, *gatD*, and *pbp2x*). As shown in Fig 5E, Zn^{2+} did

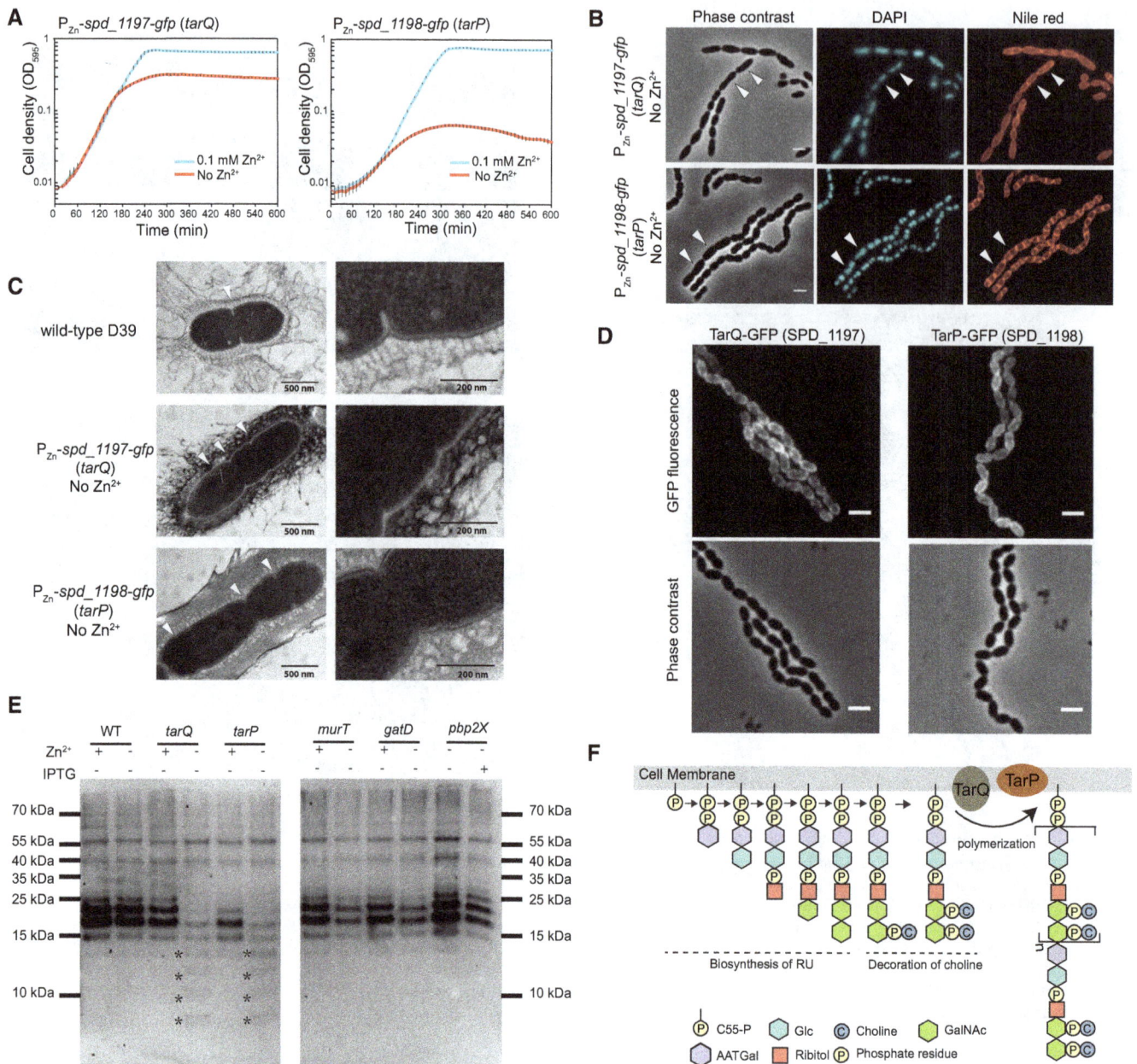

Figure 5. Newly identified genes of the teichoic acid biosynthesis pathway: *spd_1198* (*tarP*) and *spd_1197* (*tarQ*) are involved in precursor polymerization.

A Growth curves of depletion strains P_{Zn}-*spd_1197-gfp* (*tarQ*) and P_{Zn}-*spd_1198-gfp* (*tarP*) in C+Y medium with (cyan) or without (red) 0.1 mM Zn^{2+}. The values represent averages of three replicates with SEM.

B Microscopy of strains as in panel (A) after incubation in C+Y medium without Zn^{2+} for 2.5 h. Representative micrographs are shown. Scale bars = 2 μm. White arrows point to elongated and enlarged cells. Note that depletion of *spd_1197* or *spd_1198* led to long-chain formation of cells.

C Electron micrographs of the same samples as in panel (B) with wild-type *S. pneumoniae* D39 as reference. Arrowheads point to the septa of cells.

D Localization of TarQ-GFP, TarP-GFP, with C-terminal fused monomeric GFP. GFP signal (upper panel) and phase contrast (lower panel) are shown. Scale bars = 2 μm.

E Western blotting to detect phosphorylcholine-containing molecules of *S. pneumoniae*. Whole-cell lysates were separated with SDS–PAGE, and phosphorylcholine-containing molecules were detected by phosphorylcholine antibody TEPC-15. Smaller bands caused by depletion of *tarQ* (*spd_1197*) or *tarP* (*spd_1198*) are indicated by asterisks. Note that for *tarQ*, *tarP*, *murT*, and *gatD*, Zn^{2+}-inducible strains were used, and for *pbp2X*, a CRISPRi strain was used.

F Model for TarP/TarQ function in precursor polymerization of the teichoic acid biosynthesis pathway in *S. pneumoniae*. Steps of biosynthesis of repeat units (RU), decoration of RU with choline, and polymerization of the precursor are shown.

not influence TA synthesis of the *S. pneumoniae* D39 wild-type (WT) strain, and the four main TA bands are clearly visible, migrating in the range between 15 and 25 kDa consistent with

previous reports (Wu *et al*, 2014). In contrast, cells depleted for SPD_1197 or SPD_1198 displayed a different pattern and the 4 main bands around 15 and 25 kDa were missing or much weaker,

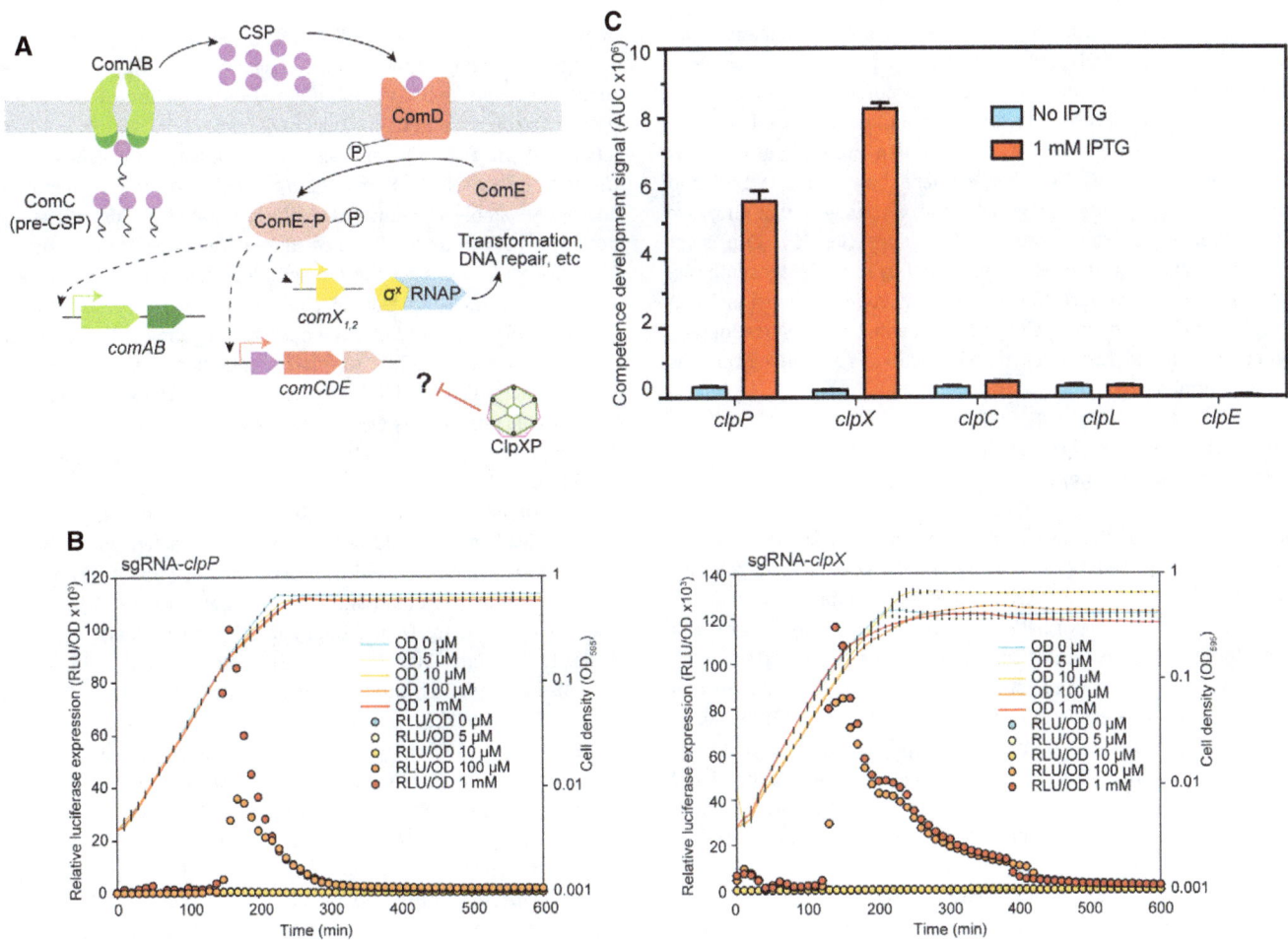

Figure 6. The ATPase ClpX and the ClpP protease repress competence development.

A Regulatory network of the competence pathway. Competence is induced when the *comC*-encoded competence-stimulating peptide (CSP) is recognized, cleaved, and exported by the membrane transporter (ComAB). Accumulation of CSP then stimulates its receptor (membrane-bound histidine-kinase ComD), which subsequently activates ComE by phosphorylation, which in turn activates the expression of the so-called early competence genes. One of them, *comX*, codes for a sigma factor, which is responsible for the activation of over 100 competence genes, including those required for transformation and DNA repair. Here, we show that the ATPase subunit ClpX works together with the protease ClpP, repressing competence, probably by negatively controlling the basal protein level of the competence regulatory proteins, but the exact mechanism is unknown (question mark).

B Repression of *clpP* or *clpX* by CRISPRi triggers competence development. Activation of competence system is reported by the *ssbB_luc* transcriptional fusion. Detection of competence development was performed in C+Y medium at a pH in which natural competence of the wild-type strain is uninduced. IPTG was added to the medium at the beginning at different final concentrations (0, 5 µM, 10 µM, 100 µM, 1 mM). Cell density (OD_{595}) and luciferase activity of the bacterial cultures were measured every 10 min. The values represent averages of three replicates with SEM.

C Influence of repression of *clpP*, *clpX*, *clpC*, *clpL*, and *clpE* on competence development. AUC (area under the curve) of the relative luciferase expression curve in panel (B) (1 mM IPTG and no IPTG) and Fig EV5 was calculated and used to represent the competence development signal. The values represent averages of three replicates with SEM.

while multiple bands with a size smaller than 15 kDa appeared. TA of *S. pneumoniae*, including wall teichoic acid (WTA) and membrane-anchored lipoteichoic acid (LTA), are polymers with identical repeating units (RU) (Fischer *et al*, 1993). Addition of one RU can lead to about a 1.3 kDa increase in molecular weight (Gisch *et al*, 2013). Interestingly, the weight interval between the extra smaller bands from bacterial cells with depleted SPD_1197 or SPD_1198 seemed to match the molecular weight of the RU, suggesting that SPD_1197 and SPD_1198 play a role in TA precursor polymerization. Although repression of the genes associated with peptidoglycan synthesis (*murT*, *gatD* and *pbp2x*) made the 4 main TA bands weaker, the pattern of the TA bands was not changed. Likely, the reduction in the TA of these three strains is due to the reduction in peptidoglycan, which constitutes the anchor for wall TA. Additionally, a CRISPRi strain targeting *tarI* of the *lic1* locus, which is involved in an early step of TA precursor synthesis, was included as a control. Note that *tarI* is cotranscribed with the other four genes of the *lic1* locus, including *tarJ*, *licA*, *licB*, and *licC*. Likely, CRISPRi knockdown of *tarI* will repress transcription of the entire *lic1* locus and thus block the synthesis of TA precursors. In line with this, we observed a reduction in the total amount of teichoic acid chains when *tarI* was repressed by CRISPRi (Appendix Fig S14).

The TA chains of *S. pneumoniae* are thought to be polymerized before they are transported to the outside of the membrane by the flippase TacF (Damjanovic *et al*, 2007), and so far it is not known which protein(s) function(s) as TA polymerase (Denapaite *et al*, 2012). In line with SPD_1198 being the TA polymerase, homology analysis shows that it contains a predicted polymerase domain. The large cytoplasmic part of SPD_1197 may aid in the assembly of the TA biosynthetic machinery by protein–protein interactions (Denapaite *et al*, 2012). Together, we here show that SPD_1197 and SPD_1198 are essential for growth and we suggest that they are responsible for polymerization of TA chains (Fig 5F). Consistent with the nomenclature used for genes involved in TA biosynthesis, we named *spd_1198 tarP* (for teichoic acid ribitol polymerase) and *spd_1197 tarQ* (in operon with *tarP*, sequential alphabetical order). Whether TarP and TarQ interact and function as a complex remains to be determined.

The essential ATPase ClpX and the protease ClpP repress competence development

We wondered whether we could also employ CRISPRi to probe gene regulatory networks in which essential genes play a role. An important pathway in *S. pneumoniae* is development of competence for genetic transformation, which is under the control of a well-studied two-component quorum sensing signaling network (Claverys *et al*, 2009). Several lines of evidence have shown that the highly conserved ATP-dependent Clp protease, ClpP, in association with an ATPase subunit (either ClpC, ClpE, ClpL, or ClpX), is involved in regulation of pneumococcal competence (Charpentier *et al*, 2000; Chastanet *et al*, 2001) (Fig 6A). Identification of the ATPase subunit responsible for ClpP-dependent repression of competence was hampered because of the essentiality, depending on the growth medium and laboratory conditions, of several *clp* mutants including *clpP* and *clpX* (Chastanet *et al*, 2001). To address this issue, we employed CRISPRi and constructed sgRNAs targeting *clpP*, *clpC*, *clpE*, *clpL*, and *clpX*. Competence development was quantified using a *luc* construct, driven by a competence-specific promoter (Slager *et al*, 2014). As shown in Fig 6B, when expression of ClpP or ClpX was repressed by addition of IPTG, competence development was enhanced, while depleting any of the other ATPase subunits (ClpC, ClpE, and ClpL) had no effect on competence (Figs 6C and EV5). This shows that ClpX is the main ATPase subunit responsible for ClpP-dependent repression of competence.

Discussion

Here, we developed an IPTG-inducible CRISPRi system to study essential genes in *S. pneumoniae* (Fig 1). In addition, we adopted a simple and efficient one-step sgRNA engineering strategy using infusion cloning. This approach resulted in ~89% positive sgRNA clones after a single round of transformation, thus enabling high-throughput cloning of sgRNAs.

Growth analysis of the CRISPRi strains targeting the 348 potentially essential genes showed that individual repression of 73% of the targeted genes led to growth phenotypes, using a stringent cutoff for phenotype detection (Figs 2B and C, and EV2). There could be several reasons why CRISPRi knockdown of the remaining 94 genes did not cause a detectable growth phenotype. Tn-seq sometimes incorrectly assigns an essential function to non-essential genes (van

Opijnen *et al*, 2009; van Opijnen & Camilli, 2013). Also, Tn-seq relies on a round of growth on blood agar plates, while our CRISPRi phenotypes were only assayed in liquid C+Y medium. Additionally, we used stringent cutoffs for phenotype definition, which will miss genes with mild growth or lysis phenotypes. Certain genes might also not be repressed well enough by CRISPRi to show a phenotype (in case of stable proteins that only require a few molecules for growth). This can be for instance caused when the sgRNA targets a PAM site far away from the transcription start site, when there is poor access of the sgRNA-dCas9 complex to the target DNA or when there are polar effects within the operon alleviating the essentiality. We can also not exclude a suppressor mutation arising in some of the "No phenotype" CRISPRi strains, as most CRISPRi knockdowns with growth phenotypes eventually grew out to the same final OD and contain a loss-of-function mutation in the coding sequence of *dcas9* (Fig EV4).

Based on analysis of the CRISPRi knockdowns, several previously "hypothetical" genes could be functionally characterized and annotated. For instance, combined with BlastP analysis and determination of *oriC-ter* ratios, we could annotate the pneumococcal primosomal machinery, including DnaA, DnaB, DnaC, DnaD, DnaG, and DnaI (Table 1, Appendix Figs S2 and S12). Note that *spd_2030* (*dnaC*) was mis-annotated as *dnaB* in several databases, such as in NCBI (ProteinID: ABJ54728), KEGG (Entry: SPD_2030), Uniprot (Entry: A0A0H2ZNF7), which may be due to the different naming of primosomal proteins in *E. coli* and *Bacillus subtilis* (Smits *et al*, 2011; Briggs *et al*, 2012). By characterizing CRISPRi-based knockdowns with cell morphology defects, we identified four essential cell wall biosynthesis genes (*murT*, *gatD*, *tarP*, and *tarQ*), which are promising candidates for future development of novel antimicrobials.

This work and other studies highlight that high-throughput phenotyping by CRISPRi is a powerful approach for hypothesis-forming and functional characterization of essential genes (Peters *et al*, 2016). We also show that CRISPRi can be used to unravel gene regulatory networks in which essential genes play a part (Fig 6). While we shed light on the function of just several previously uncharacterized essential genes, the here-described library contains richer information that needs to be further explored. In addition, CRISPRi screens can be used for mechanism of action (MOA) studies with new bioactive compounds. Indeed, CRISPRi was recently successfully employed to show that *B. subtilis* UppS is the molecular target of compound MAC-0170636 (Peters *et al*, 2016). We anticipate that the here-described pneumococcal CRISPRi library can function as a novel drug target discovery platform, can be applied to explore host–microbe interactions, and will provide a useful tool to increase our knowledge concerning pneumococcal cell biology.

Materials and Methods

Strains, growth conditions, and transformation

Oligonucleotides are shown in Dataset EV4 and strains in Appendix Table S1. *Streptococcus pneumoniae* D39 and its derivatives were cultivated in C+Y medium, pH = 6.8 (Slager *et al*, 2014) or Columbia agar with 2.5% sheep blood at 37°C. Transformation of *S. pneumoniae* was performed as previously described (Martin *et al*, 2000), and CSP-1 was used to induce competence.

Transformants were selected on Columbia agar supplemented with 2.5% sheep blood at appropriate concentrations of antibiotics (100 µg/ml spectinomycin, 250 µg/ml kanamycin, 1 µg/ml tetracycline, 40 µg/ml gentamycin, 0.05 µg/ml erythromycin). For construction of depletion strains with the Zn^{2+}-inducible promoter, 0.1 mM $ZnCl_2$/0.01 mM $MnCl_2$ was added to induce the ectopic copy of the target gene (mentioned as 0.1 mM Zn^{2+} for convenience). Working stock of the cells, called "T2 cells", were prepared by growing the cells in C+Y medium to OD_{600} 0.4, and then resuspending the cells with equal volume of fresh medium with 17% glycerol.

Escherichia coli MC1061 was used for subcloning of plasmids, and competent cells were prepared by $CaCl_2$ treatment. The *E. coli* transformants were selected on LB agar with appropriate concentrations of antibiotics (100 µg/ml spectinomycin, 100 µg/ml ampicillin, 50 µg/ml kanamycin).

Construction of an IPTG-inducible CRISPRi system in *S. pneumoniae*

Streptococcus pyogenes dcas9 (*dcas9sp*) was obtained from Addgene (Addgene #44249, Qi *et al*, 2013) and subcloned into plasmid pJWV102 (Veening laboratory collection) with the IPTG-inducible promoter P_{lac} (Sorg, 2016) replacing P_{Zn}, resulting in plasmid pJWV102-P_{lac}-*dcas9sp*. pJWV102-P_{lac}-*dcas9sp* was integrated into the *bgaA* locus in *S. pneumoniae* D39 by transformation. To control P_{lac} expression, a codon-optimized *E. coli lacI* gene driven by the constitutive promoter PF6 was inserted at the *prsA* locus in *S. pneumoniae* D39 (Sorg, 2016), leading to the construction of strain DCI23. DCI23 was used as the host strain for the insertion of gene-specific sgRNAs and enables the CRISPRi system. The DNA fragment encoding the single-guide RNA targeting luciferase (sgRNA*luc*) was ordered as a synthetic DNA gBlock (Integrated DNA Technologies) containing the constitutive P3 promoter (Sorg *et al*, 2015). The sgRNA*luc* sequence is transcribed directly after the +1 of the promoter and contains 19 nucleotides in the base-pairing region, which binds to the non-template (NT) strand of the coding sequence of luciferase, followed by an optimized single-guide RNA (Chen *et al*, 2013) (Fig EV1A). Then, the sgRNA*luc* with P3 promoter was cloned into pPEP1 (Sorg *et al*, 2015) with removing the chloramphenicol resistance marker (pPEPX) leading to the production of plasmid pPEPX-P3-sgRNA*luc*, which integrates into the region between *amiF* and *treR* of *S. pneumoniae* D39. The pPEPX-P3-sgRNA*luc* is used as the template for generation of other sgRNAs by infusion cloning or by the inverse PCR method. The *lacI* gene with gentamycin resistance marker and flanked *prsA* regions was subcloned into pPEPY (Veening laboratory collection), resulting in plasmid pPEPY-PF6-*lacI*. This plasmid can be used to amplify *lacI* and integrate it at the *prsA* locus while selecting for gentamycin resistance. The entire pneumococcal CRISPRi system, consisting of plasmids pJWV102-P_{lac}-*dcas9sp*, pPEPY-PF6-*lacI*, and pPEPX-P3-sgRNA*luc*, is available from Addgene (ID 85588, 85589, and 85590, respectively).

Selection of essential genes

To identify each gene's contribution to fitness for basal level growth, we performed Tn-seq in *S. pneumoniae* D39 essentially as described before (Zomer *et al*, 2012; Burghout *et al*, 2013), but with growing cells in C+Y medium at 37°C. Possibly essential genes were identified using ESSENTIALS (Zomer *et al*, 2012). Based on that, we included all the identified essential genes and added extra essential genes identified in serotype 4 strain TIGR4 (van Opijnen *et al*, 2009; van Opijnen & Camilli, 2012). Note that in the Tn-seq study of 2012, fitness of each gene under 17 *in vitro* and 2 *in vivo* conditions was determined and genes were grouped into different classes (van Opijnen & Camilli, 2012). Finally, 391 genes were selected (Dataset EV1).

Oligonucleotides for the CRISPRi library

The 20-nt guide sequences of the sgRNAs targeting different genes were selected with CRISPR Primer Designer (Yan *et al*, 2015). Briefly, we searched within the coding sequence of each essential gene for a 14-nt specificity region consisting of the 12-nt "seed" region of the sgRNA and GG of the 3-nt PAM (GGN). sgRNAs with more than one binding site within the pneumococcal genome, as determined by a BLAST search, were discarded. Next, we took a total length of 21 nt (including the +1 of the P3 promoter and 20 nt of perfect match to the target) and the full-length sgRNA's secondary structure was predicted using ViennaRNA (Lorenz *et al*, 2011), and the sgRNA sequence was accepted if the dCas9 handle structure was folded correctly (Larson *et al*, 2013). We chose the guide sequences as close as possible to the 5′ end of the coding sequence of the targeted gene (Qi *et al*, 2013). The sequences of the sgRNAs (20 nt) are listed in Dataset EV3.

Cloning of sgRNA

We used infusion cloning instead of inverse PCR recommended by Larson *et al* (2013) because significantly higher cloning efficiencies were obtained with infusion cloning. Two primers, sgRNA_inF_plasmid_linearize_R and sgRNA_inF_plasmid_linearize_F, were designed for linearization of plasmid pPEPX-P3-sgRNA*luc*. These two primers bind directly upstream and downstream of the 19-bp guide sequence for *luc*. To fuse the 20-nt new guide sequence into the linearized vector, two 50-nt complementary primers were designed for each target gene. Each primer contains 15 nt at one end, overlapping with the sequence on the 5′ end of the linearized vector, followed by the 20-nt specific guide sequence for each target gene; and 15 nt overlapping with the sequence on the 3′ end of the linearized vector (Fig EV2A). The two 50-nt complementary primers were annealed in TEN buffer (10 mM Tris, 1 mM EDTA, 100 mM NaCl, pH 8) by heating at 95°C for 5 min and cooling down to room temperature. The annealed product was fused with the linearized vector using the Quick-Fusion Cloning kit (BiMake, Cat. B22612) according to the manufacturer with the exception of using only one half of the recommended volume per reaction. Each reaction was directly used to transform competent *S. pneumoniae* D39 strain DCI23.

Luciferase assay

Streptococcus pneumoniae strains XL28 and XL29 were grown to OD_{600} = 0.4 in 5-ml tubes at 37°C and then diluted 1:100 in fresh C+Y medium with or without 1 mM IPTG. Then, in triplicates, 250-µl diluted bacterial culture was mixed with 50 µl of 6× luciferin solution in C+Y medium (2.7 mg/ml, D-Luciferin sodium salt,

SYNCHEM OHG) in 96-well plates (Polystyrol, white, flat, and clear bottom; Corning). Optical density at 595 nm (OD$_{595}$) and luminescence signal were measured every 10 min for 10 h using a Tecan Infinite F200 Pro microtiter plate reader.

Growth assays

For growth curves of strains of the CRISPRi library, T2 cells were thawed and diluted 1:1,000 into fresh C+Y medium with or without 1 mM IPTG. Then, 300 μl of bacterial culture was added into each well of 96-well plates. OD$_{595}$ was measured every 10 min for 18 h with a Tecan Infinite F200 Pro microtiter plate reader. Specially, for the data shown in Fig 3, Appendix Figs S10 and S11, T2 cells were diluted 1:100 in C+Y medium. For growth assays of the depletion strains with the Zn^{2+}-inducible promoter, T2 cells were thawed and diluted 1:100 into fresh C+Y medium with or without 0.1 mM Zn^{2+}.

Detection of teichoic acids

Sample preparation

T2 cells of *S. pneumoniae* strains were inoculated into fresh C+Y medium with 0.1 mM Zn^{2+} by 1:50 dilution, and then grown to OD$_{600}$ 0.15 at 37°C. Cells were collected at 8,000 *g* for 3 min and resuspended with an equal volume of fresh C+Y medium without Zn^{2+}. Bacterial cultures were diluted 1:10 into C+Y with or without 0.1 mM Zn^{2+} or 1 mM IPTG (for CRISPRi strains) and then incubated at 37°C. When OD$_{600}$ reached 0.3, cells were centrifuged at 8,000 *g* for 3 min. The pellets were washed once with cold TE buffer (10 mM Tris–Cl, pH 7.5; 1 mM EDTA, pH 8.0), and resuspended with 150 μl of TE buffer. Cells were lysed by sonication.

Detection of teichoic acid with phosphoryl choline antibody MAb TEPC-15

Protein concentration of the whole-cell lysate was determined with the DC protein assay kit (Bio-Rad Cat. 500-0111). Whole-cell lysates were mixed with equal volumes of 2× SDS protein loading buffer (100 mM Tris–HCl, pH 6.8; 4% SDS; 0.2% bromophenol blue; 20% glycerol; 10 mM DTT) and boiled at 95°C for 5 min. 2 μg of protein was loaded, followed by SDS–PAGE on a 12% polyacrylamide gel with cathode buffer (0.1 M Tris, 0.1 M tricine, 0.1% SDS) on top of the wells and anode buffer (0.2 M Tris/Cl, pH 9.9) in the bottom. After electrophoresis, samples in the gel were transferred onto a polyvinylidene difluoride (PVDF) membrane as described (Minnen *et al*, 2011). Teichoic acid was detected with anti-PC specific monoclonal antibody TEPC-15 (M1421, Sigma) by 1:1,000 dilution as first antibody, and then with anti-mouse IgG HRP antibody (GE Healthcare UK Limited) with 1:5,000 dilution as second antibody. The blots were developed with ECL prime Western blotting detection reagent (GE Healthcare UK Limited), and the images were obtained with a Bio-Rad imaging system.

Microscopy

To detect the morphological changes after knockdown of the target genes, strains in the CRISPRi library were induced with IPTG and depletion strains were incubated in C+Y medium without Zn^{2+}, stained with DAPI (DNA dye) and Nile red (membrane dye), and then studied by fluorescence microscopy. Specifically, 10 μl of thawed T2 cells was added into 1 ml of fresh C+Y medium, with or without 1 mM IPTG, in a 1.5-ml Eppendorf tube, followed by 2.5 h of incubation at 37°C. After that, 1 μl of 1 mg/ml Nile red was added into the tube and cells were stained for 4 min at room temperature. Then, 1 μl of 1 mg/ml DAPI was added and the mix was incubated for one more minute. Cells were spun down at 8,000 *g* for 2 min, and then, the pellets were suspended with 30 μl of fresh C+Y medium. 0.5 μl of cell suspension was spotted onto a PBS agarose pad on microscope slides. DAPI, Nile red, and phase contrast images were acquired with a Deltavision Elite (GE Healthcare, USA). Microscopy images were analyzed with ImageJ.

For fluorescence microscopy of strains containing zinc-inducible GFP fusions, strains were grown in C+Y medium to OD$_{600}$ 0.1, followed by 10 times dilution in fresh C+Y medium with 0.1 mM Zn^{2+}. After 1 h of incubation, cells were spun down, washed with PBS, and resuspended in 50 μl PBS. 0.5 μl of cell suspension was spotted onto a PBS agarose pad on microscope slides. Visualization of GFP was performed as described previously (Kjos *et al*, 2015).

For electron microscopy, T2 cells of *S. pneumoniae* strains were inoculated into C+Y medium with 0.1 mM Zn^{2+} and incubated at 37°C. When OD$_{600}$ reached 0.15, the bacterial culture was centrifuged at 8,000 *g* for 2 min. The pellets were resuspended into C+Y without Zn^{2+} such that OD$_{600}$ was 0.015, and then, cells were incubated again at 37°C. Bacterial cultures were put on ice to stop growth when OD$_{600}$ reached 0.35. Cells were collected by centrifugation and washed once with distilled water. A small pellet of cells was cryo-fixed in liquid ethane using the sandwich plunge freezing method (Baba, 2008) and freeze-substituted in 1% osmium tetroxide, 0.5% uranyl acetate, and 5% distilled water in acetone using the fast low-temperature dehydration and fixation method (McDonald & Webb, 2011). Cells were infiltrated overnight with Epon 812 (Serva, 21045) and polymerized at 60°C for 48 h. 90-nm-thick sections were cut with a Reichert ultramicrotome and imaged with a Philips CM12 transmission electron microscope running at 90 kV.

Competence assays

The previously described *ssbB_luc* competence reporter system, amplified from strain MK134 (Slager *et al*, 2014), was transformed into the CRISPRi strains (sgRNA*clpP*, sgRNA*clpX*, sgRNA*clpL*, sgRNA*clpE*, sgRNA*clpC*). Luminescence assays for detection of activation of competence system were performed as previously described (Slager *et al*, 2014). IPTG was added into C+Y medium (at a non-permissive pH for competence development) at the beginning of cultivation to different final concentrations.

Acknowledgements

We thank A. Zomer, P. Burghout, and P. Hermans for help with acquiring and analyzing the Tn-seq data. We thank K. Kuipers and M. I. de Jonge for providing the TEPC-15 antibodies and for the TA Western blotting protocol. V. Benes and B. Haase (GeneCore, EMBL, Heidelberg) are thanked for sequencing support. XL is supported by China Scholarship Council (No. 201506210151). ADP was supported by a "Marie Curie IF" grant (call H2020-MSCA-IF-2014, number 657546). KK is supported by a NWO VENI fellowship (563.14.003). MK is supported by a grant from the Research Council of Norway (250976/F20). Work in the Veening laboratory is supported by the EMBO Young Investigator

Program, a VIDI fellowship (864.12.001) from the Netherlands Organization for Scientific Research, Earth and Life Sciences (NWO-ALW), and ERC Starting Grant 337399-PneumoCell.

Author contributions

XL and J-WV designed the study. XL, MK, AD, CG, SPK, JS, and JWV performed experiments and analyzed the data. KK performed electron microscopy. RAS developed the IPTG-inducible system. JS analyzed the RNA-Seq data and performed the CRISPRi growth analysis. XL, JS, MK, AD, CG, J-RZ, and J-WV wrote the manuscript with input from all authors. All authors read and approved the final manuscript.

References

Aprianto R, Slager J, Holsappel S, Veening JW (2016) Time-resolved dual RNA-seq reveals extensive rewiring of lung epithelial and pneumococcal transcriptomes during early infection. *Genome Biol* 17: 198

Avery OT, Macleod CM, McCarty M (1944) Studies on the chemical nature of the substance inducing transformation of pneumococcal types: induction of transformation by a desoxyribonucleic acid fraction isolated from pneumococcus type Iii. *J Exp Med* 79: 137–158

Baba M (2008) Electron microscopy in yeast. *Methods Enzymol* 451: 133–149

Bartual SG, Straume D, Stamsas GA, Munoz IG, Alfonso C, Martinez-Ripoll M, Havarstein LS, Hermoso JA (2014) Structural basis of PcsB-mediated cell separation in *Streptococcus pneumoniae*. *Nat Commun* 5: 3842

Bikard D, Hatoum-Aslan A, Mucida D, Marraffini LA (2012) CRISPR interference can prevent natural transformation and virulence acquisition during *in vivo* bacterial infection. *Cell Host Microbe* 12: 177–186

Bikard D, Jiang W, Samai P, Hochschild A, Zhang F, Marraffini LA (2013) Programmable repression and activation of bacterial gene expression using an engineered CRISPR-Cas system. *Nucleic Acids Res* 41: 7429–7437

Briggs GS, Smits WK, Soultanas P (2012) Chromosomal replication initiation machinery of low-G+C-content Firmicutes. *J Bacteriol* 194: 5162–5170

Brown S, Santa Maria JP Jr, Walker S (2013) Wall teichoic acids of gram-positive bacteria. *Annu Rev Microbiol* 67: 313–336

Burghout P, Zomer A, van der Gaast-de Jongh CE, Janssen-Megens EM, Francoijs KJ, Stunnenberg HG, Hermans PW (2013) *Streptococcus pneumoniae* folate biosynthesis responds to environmental CO2 levels. *J Bacteriol* 195: 1573–1582

Charpentier E, Novak R, Tuomanen E (2000) Regulation of growth inhibition at high temperature, autolysis, transformation and adherence in *Streptococcus pneumoniae* by ClpC. *Mol Microbiol* 37: 717–726

Chastanet A, Prudhomme M, Claverys JP, Msadek T (2001) Regulation of *Streptococcus pneumoniae* clp genes and their role in competence development and stress survival. *J Bacteriol* 183: 7295–7307

Chen B, Gilbert LA, Cimini BA, Schnitzbauer J, Zhang W, Li GW, Park J, Blackburn EH, Weissman JS, Qi LS, Huang B (2013) Dynamic imaging of genomic loci in living human cells by an optimized CRISPR/Cas system. *Cell* 155: 1479–1491

Chewapreecha C, Harris SR, Croucher NJ, Turner C, Marttinen P, Cheng L, Pessia A, Aanensen DM, Mather AE, Page AJ, Salter SJ, Harris D, Nosten F, Goldblatt D, Corander J, Parkhill J, Turner P, Bentley SD (2014) Dense genomic sampling identifies highways of pneumococcal recombination. *Nat Genet* 46: 305–309

Chung KM, Hsu HH, Govindan S, Chang BY (2004) Transcription regulation of ezrA and its effect on cell division of *Bacillus subtilis*. *J Bacteriol* 186: 5926–5932

Claverys JP, Martin B, Polard P (2009) The genetic transformation machinery: composition, localization, and mechanism. *FEMS Microbiol Rev* 33: 643–656

Damjanovic M, Kharat AS, Eberhardt A, Tomasz A, Vollmer W (2007) The essential *tacF* gene is responsible for the choline-dependent growth phenotype of *Streptococcus pneumoniae*. *J Bacteriol* 189: 7105–7111

Denapaite D, Bruckner R, Hakenbeck R, Vollmer W (2012) Biosynthesis of teichoic acids in *Streptococcus pneumoniae* and closely related species: lessons from genomes. *Microbial drug resistance* 18: 344–358

Figueiredo TA, Sobral RG, Ludovice AM, Almeida JM, Bui NK, Vollmer W, de Lencastre H, Tomasz A (2012) Identification of genetic determinants and enzymes involved with the amidation of glutamic acid residues in the peptidoglycan of *Staphylococcus aureus*. *PLoS Pathog* 8: e1002508

Fischer W, Behr T, Hartmann R, Peter-Katalinić J, Egge H (1993) Teichoic acid and lipoteichoic acid of *Streptococcus pneumoniae* possess identical chain structures. *Eur J Biochem* 215: 851–857

Gisch N, Kohler T, Ulmer AJ, Muthing J, Pribyl T, Fischer K, Lindner B, Hammerschmidt S, Zahringer U (2013) Structural reevaluation of *Streptococcus pneumoniae* lipoteichoic acid and new insights into its immunostimulatory potency. *J Biol Chem* 288: 15654–15667

Goranov AI, Breier AM, Merrikh H, Grossman AD (2009) YabA of *Bacillus subtilis* controls DnaA-mediated replication initiation but not the transcriptional response to replication stress. *Mol Microbiol* 74: 454–466

Hammerschmidt S, Wolff S, Hocke A, Rosseau S, Muller E, Rohde M (2005) Illustration of pneumococcal polysaccharide capsule during adherence and invasion of epithelial cells. *Infect Immun* 73: 4653–4667

Henriques AO, Glaser P, Piggot PJ, Moran CP Jr (1998) Control of cell shape and elongation by the *rodA* gene in *Bacillus subtilis*. *Mol Microbiol* 28: 235–247

Ikeda M, Sato T, Wachi M, Jung HK, Ishino F, Kobayashi Y, Matsuhashi M (1989) Structural similarity among *Escherichia coli* FtsW and RodA proteins and *Bacillus subtilis* SpoVE protein, which function in cell division, cell elongation, and spore formation, respectively. *J Bacteriol* 171: 6375–6378

Irwin CR, Farmer A, Willer DO, Evans DH (2012) In-fusion(R) cloning with vaccinia virus DNA polymerase. *Methods Mol Biol* 890: 23–35

Jerga A, Rock CO (2009) Acyl-Acyl carrier protein regulates transcription of fatty acid biosynthetic genes via the FabT repressor in *Streptococcus pneumoniae*. *J Biol Chem* 284: 15364–15368

Johnston C, Martin B, Fichant G, Polard P, Claverys JP (2014) Bacterial transformation: distribution, shared mechanisms and divergent control. *Nat Rev Microbiol* 12: 181–196

de Jong A, van der Meulen S, Kuipers OP, Kok J (2015) T-REx: transcriptome analysis webserver for RNA-seq expression data. *BMC Genom* 16: 663

Kaur R, Casey JR, Pichichero ME (2016) Emerging *Streptococcus pneumoniae* strains colonizing the nasopharynx in children after 13-valent pneumococcal conjugate vaccination in comparison to the 7-valent era, 2006-2015. *Pediatr Infect Dis J* 35: 901–916

Kawai Y, Ogasawara N (2006) *Bacillus subtilis* EzrA and FtsL synergistically regulate FtsZ ring dynamics during cell division. *Microbiology* 152: 1129–1141

Kharat AS, Denapaite D, Gehre F, Bruckner R, Vollmer W, Hakenbeck R, Tomasz A (2008) Different pathways of choline metabolism in two choline-independent strains of *Streptococcus pneumoniae* and their impact on virulence. *J Bacteriol* 190: 5907–5914

Khattar MM, Begg KJ, Donachie WD (1994) Identification of FtsW and characterization of a new *ftsW* division mutant of *Escherichia coli*. *J Bacteriol* 176: 7140–7147

Kim L, McGee L, Tomczyk S, Beall B (2016) Biological and epidemiological features of antibiotic-resistant *Streptococcus pneumoniae* in pre- and post-conjugate vaccine eras: a United States perspective. *Clin Microbiol Rev* 29: 525–552

Kjos M, Aprianto R, Fernandes VE, Andrew PW, van Strijp JA, Nijland R, Veening JW (2015) Bright fluorescent *Streptococcus pneumoniae* for live-cell imaging of host-pathogen interactions. *J Bacteriol* 197: 807–818

Kocaoglu O, Tsui HC, Winkler ME, Carlson EE (2015) Profiling of beta-lactam selectivity for penicillin-binding proteins in *Streptococcus pneumoniae* D39. *Antimicrob Agents Chemother* 59: 3548–3555

Kuipers K, Gallay C, Martinek V, Rohde M, Martinkova M, van der Beek SL, Jong WS, Venselaar H, Zomer A, Bootsma H, Veening JW, de Jonge MI (2016) Highly conserved nucleotide phosphatase essential for membrane lipid homeostasis in *Streptococcus pneumoniae*. *Mol Microbiol* 101: 12–26

Larson MH, Gilbert LA, Wang X, Lim WA, Weissman JS, Qi LS (2013) CRISPR interference (CRISPRi) for sequence-specific control of gene expression. *Nat Protoc* 8: 2180–2196

Li Y, Weinberger DM, Thompson CM, Trzcinski K, Lipsitch M (2013) Surface charge of *Streptococcus pneumoniae* predicts serotype distribution. *Infect Immun* 81: 4519–4524

Lorenz R, Bernhart SH, Honer Zu Siederdissen C, Tafer H, Flamm C, Stadler PF, Hofacker IL (2011) ViennaRNA package 2.0. *Algorithms Mol Biol* 6: 26

Lu YJ, Rock CO (2006) Transcriptional regulation of fatty acid biosynthesis in *Streptococcus pneumoniae*. *Mol Microbiol* 59: 551–566

Maggi S, Massidda O, Luzi G, Fadda D, Paolozzi L, Ghelardini P (2008) Division protein interaction web: identification of a phylogenetically conserved common interactome between *Streptococcus pneumoniae* and *Escherichia coli*. *Microbiology* 154: 3042–3052

Martin B, Prudhomme M, Alloing G, Granadel C, Claverys JP (2000) Cross-regulation of competence pheromone production and export in the early control of transformation in *Streptococcus pneumoniae*. *Mol Microbiol* 38: 867–878

Massidda O, Novakova L, Vollmer W (2013) From models to pathogens: how much have we learned about *Streptococcus pneumoniae* cell division? *Environ Microbiol* 15: 3133–3157

McClure R, Balasubramanian D, Sun Y, Bobrovskyy M, Sumby P, Genco CA, Vanderpool CK, Tjaden B (2013) Computational analysis of bacterial RNA-Seq data. *Nucleic Acids Res* 41: e140

McDonald KL, Webb RI (2011) Freeze substitution in 3 hours or less. *J Microsc* 243: 227–233

Meeske AJ, Riley EP, Robins WP, Uehara T, Mekalanos JJ, Kahne D, Walker S, Kruse AC, Bernhardt TG, Rudner DZ (2016) SEDS proteins are a widespread family of bacterial cell wall polymerases. *Nature* 537: 634–638

Minnen A, Attaiech L, Thon M, Gruber S, Veening JW (2011) SMC is recruited to *oriC* by ParB and promotes chromosome segregation in *Streptococcus pneumoniae*. *Mol Microbiol* 81: 676–688

Mobegi FM, van Hijum SA, Burghout P, Bootsma HJ, de Vries SP, van der Gaast-de Jongh CE, Simonetti E, Langereis JD, Hermans PW, de Jonge MI, Zomer A (2014) From microbial gene essentiality to novel antimicrobial drug targets. *BMC Genom* 15: 958

Mohammadi T, van Dam V, Sijbrandi R, Vernet T, Zapun A, Bouhss A, Diepeveen-de Bruin M, Nguyen-Disteche M, de Kruijff B, Breukink E (2011) Identification of FtsW as a transporter of lipid-linked cell wall precursors across the membrane. *EMBO J* 30: 1425–1432

Morgan C, Rosenkranz HS, Carr HS, Rose HM (1967) Electron microscopy of chloramphenicol-treated *Escherichia coli*. *J Bacteriol* 93: 1987–2002

Morlot C, Noirclerc-Savoye M, Zapun A, Dideberg O, Vernet T (2004) The D, D-carboxypeptidase PBP3 organizes the division process of *Streptococcus*

pneumoniae. *Mol Microbiol* 51: 1641–1648

Munch D, Roemer T, Lee SH, Engeser M, Sahl HG, Schneider T (2012) Identification and *in vitro* analysis of the GatD/MurT enzyme-complex catalyzing lipid II amidation in *Staphylococcus aureus*. *PLoS Pathog* 8: e1002509

Ng WL, Robertson GT, Kazmierczak KM, Zhao J, Gilmour R, Winkler ME (2003) Constitutive expression of PcsB suppresses the requirement for the essential VicR (YycF) response regulator in *Streptococcus pneumoniae* R6. *Mol Microbiol* 50: 1647–1663

Noirclerc-Savoye M, Morlot C, Gerard P, Vernet T, Zapun A (2003) Expression and purification of FtsW and RodA from *Streptococcus pneumoniae*, two membrane proteins involved in cell division and cell growth, respectively. *Protein Expr Purif* 30: 18–25

Noirclerc-Savoye M, Le Gouellec A, Morlot C, Dideberg O, Vernet T, Zapun A (2005) *In vitro* reconstitution of a trimeric complex of DivIB, DivIC and FtsL, and their transient co-localization at the division site in *Streptococcus pneumoniae*. *Mol Microbiol* 55: 413–424

Noirot-Gros MF, Dervyn E, Wu LJ, Mervelet P, Errington J, Ehrlich SD, Noirot P (2002) An expanded view of bacterial DNA replication. *Proc Natl Acad Sci USA* 99: 8342–8347

Ochman H, Gerber AS, Hartl DL (1988) Genetic applications of an inverse polymerase chain reaction. *Genetics* 120: 621–623

van Opijnen T, Bodi KL, Camilli A (2009) Tn-seq: high-throughput parallel sequencing for fitness and genetic interaction studies in microorganisms. *Nat Methods* 6: 767–772

van Opijnen T, Camilli A (2012) A fine scale phenotype-genotype virulence map of a bacterial pathogen. *Genome Res* 22: 2541–2551

van Opijnen T, Camilli A (2013) Transposon insertion sequencing: a new tool for systems-level analysis of microorganisms. *Nat Rev Microbiol* 11: 435–442

Peters JM, Colavin A, Shi H, Czarny TL, Larson MH, Wong S, Hawkins JS, Lu CH, Koo BM, Marta E, Shiver AL, Whitehead EH, Weissman JS, Brown ED, Qi LS, Huang KC, Gross CA (2016) A comprehensive, CRISPR-based functional analysis of essential genes in bacteria. *Cell* 165: 1493–1506

Pinho MG, Kjos M, Veening JW (2013) How to get (a)round: mechanisms controlling growth and division of coccoid bacteria. *Nat Rev Microbiol* 11: 601–614

Prina E, Ranzani OT, Torres A (2015) Community-acquired pneumonia. *Lancet* 386: 1097–1108

Prudhomme M, Berge M, Martin B, Polard P (2016) Pneumococcal competence coordination relies on a cell-contact sensing mechanism. *PLoS Genet* 12: e1006113

Qi LS, Larson MH, Gilbert LA, Doudna JA, Weissman JS, Arkin AP, Lim WA (2013) Repurposing CRISPR as an RNA-guided platform for sequence-specific control of gene expression. *Cell* 152: 1173–1183

Rajagopal M, Walker S (2016) Envelope structures of gram-positive bacteria. *Curr Top Microbiol Immunol* 404: 1–44

Reinscheid DJ, Gottschalk B, Schubert A, Eikmanns BJ, Chhatwal GS (2001) Identification and molecular analysis of PcsB, a protein required for cell wall separation of group B streptococcus. *J Bacteriol* 183: 1175–1183

Roggiani M, Goulian M (2015) Chromosome-membrane interactions in bacteria. *Annu Rev Genet* 49: 115–129

Sham LT, Barendt SM, Kopecky KE, Winkler ME (2011) Essential PcsB putative peptidoglycan hydrolase interacts with the essential FtsXSpn cell division protein in *Streptococcus pneumoniae* D39. *Proc Natl Acad Sci USA* 108: E1061–E1069

Sham LT, Tsui HC, Land AD, Barendt SM, Winkler ME (2012) Recent advances in pneumococcal peptidoglycan biosynthesis suggest new vaccine and antimicrobial targets. *Curr Opin Microbiol* 15: 194−203

Slager J, Kjos M, Attaiech L, Veening JW (2014) Antibiotic-induced replication stress triggers bacterial competence by increasing gene dosage near the origin. *Cell* 157: 395−406

Smits WK, Merrikh H, Bonilla CY, Grossman AD (2011) Primosomal proteins DnaD and DnaB are recruited to chromosomal regions bound by DnaA in *Bacillus subtilis. J Bacteriol* 193: 640−648

Song JH, Ko KS, Lee JY, Baek JY, Oh WS, Yoon HS, Jeong JY, Chun J (2005) Identification of essential genes in *Streptococcus pneumoniae* by allelic replacement mutagenesis. *Mol Cells* 19: 365−374

Sorg RA, Veening JW (2015) Microscale insights into pneumococcal antibiotic mutant selection windows. *Nat Commun* 6: 8773

Sorg RA, Kuipers OP, Veening JW (2015) Gene expression platform for synthetic biology in the human pathogen *Streptococcus pneumoniae. ACS Synth Biol* 4: 228−239

Sorg RA (2016) Engineering approaches to investigate pneumococcal gene expression regulation and antibiotic resistance development. PhD thesis, ISBN 978-90-367-9258-5

Thanassi JA, Hartman-Neumann SL, Dougherty TJ, Dougherty BA, Pucci MJ (2002) Identification of 113 conserved essential genes using a high-throughput gene disruption system in *Streptococcus pneumoniae. Nucleic Acids Res* 30: 3152−3162

Verhagen LM, de Jonge MI, Burghout P, Schraa K, Spagnuolo L, Mennens S, Eleveld MJ, van der Gaast-de Jongh CE, Zomer A, Hermans PW, Bootsma HJ (2014) Genome-wide identification of genes essential for the survival of *Streptococcus pneumoniae* in human saliva. *PLoS ONE* 9: e89541

Vollmer W, Tomasz A (2001) Identification of the teichoic acid phosphorylcholine esterase in *Streptococcus pneumoniae. Mol Microbiol* 39: 1610−1622

Wu KF, Huang J, Zhang YQ, Xu WC, Xu HM, Wang LB, Cao J, Zhang XM, Yin YB (2014) A novel protein, RafX, is important for common cell wall polysaccharide biosynthesis in *Streptococcus pneumoniae*: implications for bacterial virulence. *J Bacteriol* 196: 3324−3334

Yan M, Zhou SR, Xue HW (2015) CRISPR primer designer: design primers for knockout and chromosome imaging CRISPR-Cas system. *J Integr Plant Biol* 57: 613−617

Zapun A, Philippe J, Abrahams KA, Signor L, Roper DI, Breukink E, Vernet T (2013) *In vitro* reconstitution of peptidoglycan assembly from the Gram-positive pathogen *Streptococcus pneumoniae. ACS Chem Biol* 8: 2688−2696

Zhao H, Sun Y, Peters JM, Gross CA, Garner EC, Helmann JD (2016) Depletion of undecaprenyl pyrophosphate phosphatases disrupts cell envelope biogenesis in *Bacillus subtilis. J Bacteriol* 198: 2925−2935

Zomer A, Burghout P, Bootsma HJ, Hermans PW, van Hijum SA (2012) ESSENTIALS: software for rapid analysis of high throughput transposon insertion sequencing data. *PLoS ONE* 7: e43012

Zusman DR, Carbonell A, Haga JY (1973) Nucleoid condensation and cell division in *Escherichia coli* MX74T2 ts52 after inhibition of protein synthesis. *J Bacteriol* 115: 1167−1178

9

A proteomic atlas of insulin signalling reveals tissue-specific mechanisms of longevity assurance

Luke S Tain[1], Robert Sehlke[1,2], Chirag Jain[1], Manopriya Chokkalingam[2], Nagarjuna Nagaraj[3], Paul Essers[1], Mark Rassner[1], Sebastian Grönke[1], Jenny Froelich[1], Christoph Dieterich[4,5], Matthias Mann[3] iD, Nazif Alic[6], Andreas Beyer[2,7,*] iD & Linda Partridge[1,6,**] iD

Abstract

Lowered activity of the insulin/IGF signalling (IIS) network can ameliorate the effects of ageing in laboratory animals and, possibly, humans. Although transcriptome remodelling in long-lived IIS mutants has been extensively documented, the causal mechanisms contributing to extended lifespan, particularly in specific tissues, remain unclear. We have characterized the proteomes of four key insulin-sensitive tissues in a long-lived *Drosophila* IIS mutant and control, and detected 44% of the predicted proteome (6,085 proteins). Expression of ribosome-associated proteins in the fat body was reduced in the mutant, with a corresponding, tissue-specific reduction in translation. Expression of mitochondrial electron transport chain proteins in fat body was increased, leading to increased respiration, which was necessary for IIS-mediated lifespan extension, and alone sufficient to mediate it. Proteasomal subunits showed altered expression in IIS mutant gut, and gut-specific over-expression of the RPN6 proteasomal subunit, was sufficient to increase proteasomal activity and extend lifespan, whilst inhibition of proteasome activity abolished IIS-mediated longevity. Our study thus uncovered strikingly tissue-specific responses of cellular processes to lowered IIS acting in concert to ameliorate ageing.

Keywords ageing; insulin/IGF; mitochondria; proteasome; proteome
Subject Categories Ageing; Post-translational Modifications, Proteolysis & Proteomics; Signal Transduction

Introduction

Reduced activity of the highly evolutionarily conserved insulin/IGF-like signalling (IIS) network in laboratory model organisms can extend lifespan (Fontana *et al*, 2010; Kenyon, 2010), maintain function in older ages (Barzilai *et al*, 2012) and ameliorate pathology associated with multiple age-related diseases (Niccoli & Partridge, 2012; Johnson *et al*, 2015). Furthermore, lowered activity of the IIS network is implicated in human survival to advanced ages (Suh *et al*, 2008; Tazearslan *et al*, 2011; Deelen *et al*, 2013; He *et al*, 2014). Understanding exactly how reducing IIS network activity ameliorates the effects of ageing could hence pave the way to prevention of ageing-related disease in humans.

The IIS network responds to nutrient availability, growth factors and stress signals to regulate multiple processes, including development, growth, metabolism, stress resistance, reproduction and lifespan (Fontana *et al*, 2010; Kenyon, 2010). These highly pleiotropic phenotypes make it difficult to pinpoint the mechanisms that specifically ameliorate ageing and hence to understand whether they can be separated from other, not necessarily desirable, pleiotropic effects.

Long-lived IIS mutants show a major and tissue-specific rearrangement of RNA transcript expression, as a consequence of alteration of the activity of target transcription factors (TFs; Fontana *et al*, 2010; Kenyon, 2010). In *Caenorhabditis elegans* and *Drosophila*, the single Forkhead boxO, FoxO, transcription factor (DAF-16 and dFOXO, respectively) is required for blunted IIS to extend lifespan (Kenyon *et al*, 1993; Slack *et al*, 2011; Yamamoto & Tatar, 2011), with other TFs also required in both organisms (Kenyon, 2010; Tepper *et al*, 2013; Slack *et al*, 2015). In *Drosophila*, dFOXO is required only for longevity and xenobiotic resistance, and not for other phenotypes of reduced IIS (Slack *et al*, 2011). Hence, in *Drosophila*, *dfoxo* dependence triages at least some of the gene expression

1 Max-Planck Institute for Biology of Ageing, Cologne, Germany
2 CECAD Cologne Excellence Cluster on Cellular Stress Responses in Aging Associated Diseases, Cologne, Germany
3 Department of Proteomics and Signal Transduction, Max-Planck-Institute of Biochemistry, Martinsried, Germany
4 Section of Bioinformatics and Systems Cardiology, Department of Internal Medicine III and Klaus Tschira Institute for Integrative Computational Cardiology, University of Heidelberg, Heidelberg, Germany
5 DZHK (German Centre for Cardiovascular Research), Partner site Heidelberg/Mannheim, Heidelberg, Germany
6 Institute of Healthy Ageing, and GEE, UCL, London, UK
7 Center for Molecular Medicine Cologne (CMMC), University of Cologne, Cologne, Germany
   *Corresponding author. E-mail: andreas.beyer@uni-koeln.de
   **Corresponding author. E-mail: partridge@age.mpg.de

changes potentially causal in longevity from several other pheno-types caused by IIS mutants, and facilitates identification of the mechanisms involved.

Profiling of the RNA transcriptome has identified genes and molecular mechanisms that may ameliorate ageing in IIS mutants in *C. elegans* (Murphy *et al*, 2003; Halaschek-Wiener *et al*, 2005; Oh *et al*, 2006; McElwee *et al*, 2007; Ewald *et al*, 2015; Kaletsky *et al*, 2016) and *Drosophila* (Teleman *et al*, 2008; Alic *et al*, 2011). Increased detoxification (Amador-Noguez *et al*, 2007; Selman *et al*, 2009; Ewald *et al*, 2015; Afschar *et al*, 2016) and reduced transla-tion (Selman *et al*, 2009; Afschar *et al*, 2016; Kaletsky *et al*, 2016) are two functional signatures consistently associated with longevity across different model organisms (McElwee *et al*, 2007). However, in general RNA profiling in *C. elegans* has been carried out on whole worms, potentially leaving tissue-specific mechanisms undetected. Furthermore, a wide range of post-transcriptional mechanisms may modify expression of proteins (Barrett *et al*, 2012; Batista & Chang, 2013; Cech & Steitz, 2014; Liu *et al*, 2016; MacInnes, 2016), with recent reports of the correlation between expression levels in the transcriptome and the proteome ranging from 40 to 84% (Wolkow, 2000; Schwanhäusser *et al*, 2011; Li *et al*, 2014).

Tissue-specific modulation of IIS can also extend lifespan, in the worm, *Drosophila* and the mouse (Wolkow, 2000; Blüher *et al*, 2003; Giannakou, 2004; Demontis & Perrimon, 2010; Alic *et al*, 2014). Conditional knockout models in mice, and studies in humans, have revealed that the responses of gene expression to reduced IIS are tissue-specific (Nandi *et al*, 2004; Rask-Madsen & Kahn, 2012). Recently, tissue-specific, global regulation of RNA expression in IIS mutants has been undertaken in the *Drosophila* gut and fat body (Alic *et al*, 2014) and in neurons in *C. elegans* (Alic *et al*, 2014; Kaletsky *et al*, 2016). However, proteomic profiling of individual tissues in the key invertebrate model organisms *Droso-phila* and *C. elegans* has been limited, due to their small size, and the requirement for large quantities of starting material. Initial investigations into the proteome of wild-type *Drosophila* focused on protein identification, using samples of low complexity, and either subcellular or biochemical fractionation to achieve increasing depth (Veraksa *et al*, 2005; Brunner *et al*, 2007; Aradska *et al*, 2015). Recent advances in mass spectrometry, sample preparation tech-niques and data analysis now allow the quantification of near complete proteomes and proteomic expression profiling (Nagaraj *et al*, 2012; Mann *et al*, 2013; Azimifar *et al*, 2014; Kim *et al*, 2014; Deshmukh *et al*, 2015). These new developments have reduced the amount of starting material required for shotgun proteomics and created an opportunity to investigate how attenuated IIS affects the proteome of individual fly tissues and hence to identify candidate mechanisms that ameliorate the effects of ageing.

In this study, we profiled the changes in the tissue-specific proteomes of IIS mutant *Drosophila* and their responses to removal of the key dFOXO transcription factor. We profiled four key insulin-sensitive tissues: the brain, gut, fat body and muscle. The majority of the responses to a systemic reduction in IIS were highly tissue-specific, and 60% of them were not detected in previous transcrip-tional studies. Proteins associated with the ribosome and the mitochondrial electron transport chain were differentially expressed in the fat body, and led to a reduction in translation and increased mitochondrial respiration. The gut showed a proteomic signature of proteasome–ubiquitin-mediated catabolism, and was associated

with elevated proteasomal assembly and activity. Increased respira-tion and proteasome activity in the fat body and gut, respectively, were required for the increased longevity of IIS mutants. Impor-tantly, manipulation of mitochondrial biogenesis in the fat body and of proteasomal activity in the gut could each extend lifespan in wild-type flies. Our tissue-specific, proteomic analysis has thus revealed how individual tissues of an organism can act in concert, through diverse tissue-specific responses, to ameliorate the ageing of the whole organism in response to reduced IIS.

Results

The tissue-specific proteome of wild-type flies

We first characterized the tissue-specific proteomes of control, wild-type flies. We dissected the brain, thorax (containing predominately muscle), intestinal tract with Malpighian tubules (referred to as gut from here on) and abdominal fat body of young (10 days), adult, female flies. We then performed single-shot, label-free proteomics with a Q-Exactive benchtop quadrupole-Orbitrap mass spectrometer (Fig 1A). We achieved 99% confidence of identification at both protein and peptide levels, and in total identified and quantified 96,404 peptides corresponding to 6,085 proteins, representing 44% coverage of the predicted, protein-coding genes (Fig 1B, Dataset EV1).

The four tissues shared a common core of 1,916 proteins (Fig 1C). Gene ontology (GO) enrichment analysis showed that these served a variety of house-keeping functions, and included mitochondrial and ribosomal constituents (see Dataset EV2 for GO analysis). The four tissues also showed marked differences, with 6–26% of the proteins identified as present in a tissue being unique to it (Fig 1C). To further understand the proteomic features that distin-guish individual tissues, we performed principal component (PC) analysis. 74% of the variation between tissues could be captured by the first two PCs. Mapping GO terms onto the proteins contributing to PC1 revealed that proteins with neuronal functions accounted for much of the effect, separating the brain, which had the largest number of unique proteins, from the other three tissues. PC2 detected mainly differences in developmental origins, structure and metabolic activity (Fig 1D and Dataset EV3). Biological replicates from each wild-type tissue clustered tightly together (PCA, Fig 1D), confirming reproducibility (Fig EV1).

Reduced IIS remodels tissue-specific proteomes

We next identified the effect on the proteomes of down-regulation of IIS. Ablation of the median neurosecretory cells (mNSCs) in the fly brain, which secrete three of the seven *Drosophila* insulin-like peptides (Dilps) into circulation, causes a robust extension of life-span (Broughton *et al*, 2010). We compared protein expression levels in mNSC-ablated (*InsP3-Gal4/UAS-rpr*) flies to those of wild-type controls (*w^{Dah}*). In total, expression of 2,372 proteins was significantly altered upon reduced IIS (10% FDR, Fig 2 and Dataset EV4), 982 of which showed absolute fold changes larger than two in at least one tissue. 60% of these changes were not previously identi-fied at the RNA transcript level in whole IIS mutant flies (McElwee *et al*, 2007; Alic *et al*, 2011). Thus substantial, new information was gained by tissue-specific proteome analysis.

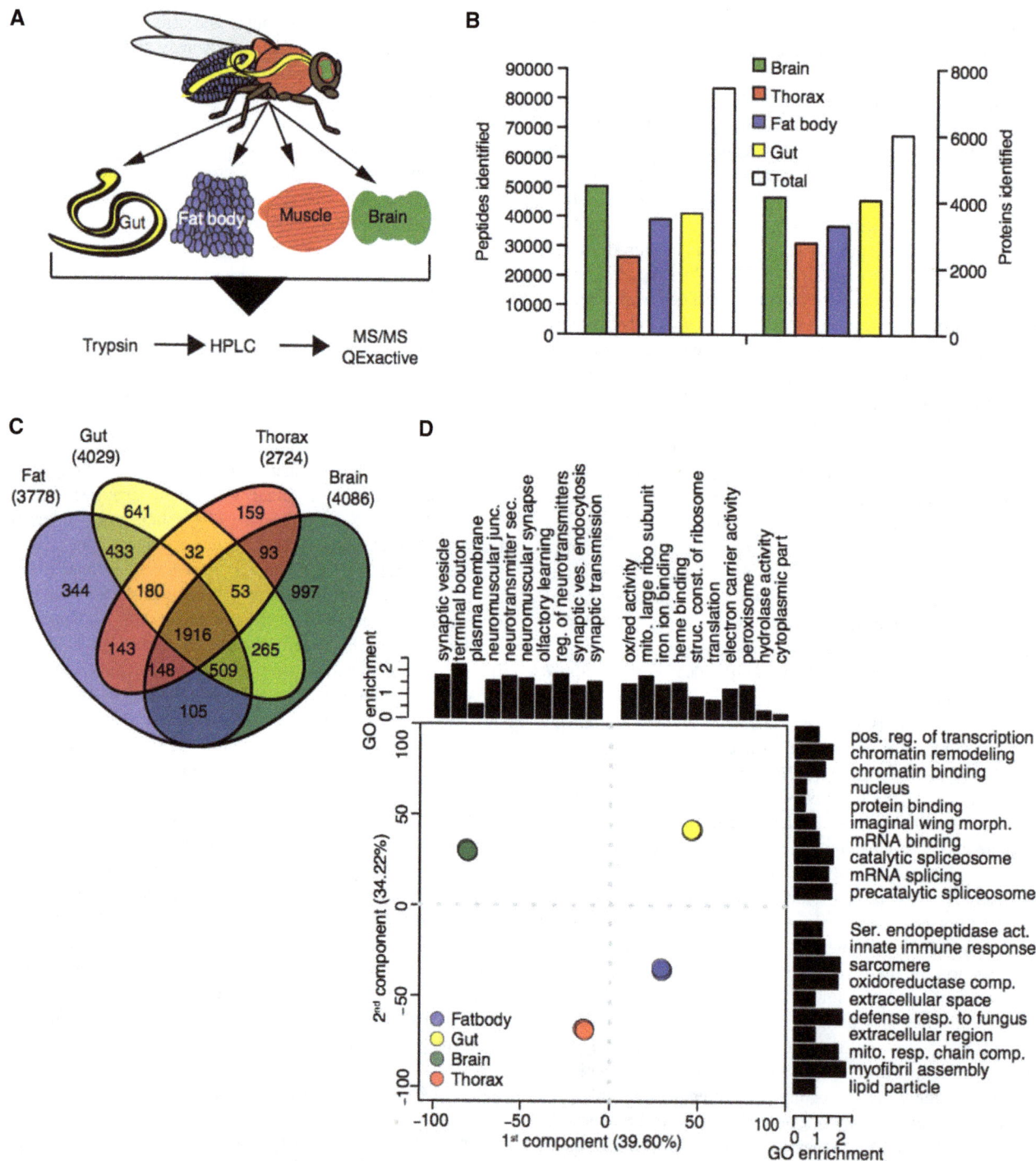

Figure 1. *Drosophila* **tissues contain common and tissue-specific proteomes.**

A Schematic of experimental design.

B Tissue-specific and total number of detected peptides and proteins.

C Venn diagram of proteins detected in wild-type fly tissues.

D Principal component analysis of the four wild-type tissues GO terms enriched in the 5% tails of contribution along each dimension. Replicates show little variation and overlay each other.

The brain and gut showed the highest numbers of differentially regulated proteins (1,312 and 1,062, respectively), whilst the fat body (256) and thorax (119) showed comparatively few. Changes in expression were almost entirely tissue-specific (85%), with only two proteins, β-tubulin (60D) and l(1)G0230, showing changes in all four (Dataset EV4). Only 13% of the proteins that were

Tissue-specific differential regulation upon
reduced IIS

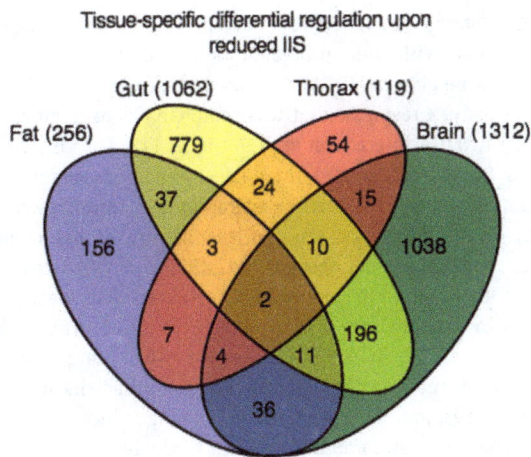

Figure 2. Proteome response to reduced IIS is highly tissue-specific.
Venn diagram of differentially expressed proteins upon reduced IIS, comparing tissues from control (w^{Dah}) and mNSC-ablated flies. Total number of differentially expressed, tissue-specific proteins is shown in parentheses.

set of proteins that are most likely *dfoxo*-independent, but ablation responsive, we identified those for which the fold difference in responsiveness to mNSC ablation in the wild-type and *dfoxo*-null background was significantly less than the minimum fold change of *dfoxo*-dependent proteins in the same tissue. Additionally, we included the proteins that showed an exaggerated responsiveness to the ablation in the *dfoxo*-null background. In this way, we defined a set of 196 proteins as *dfoxo*-independent, with high confidence. Indeed, 52% of these showed significant differences in expression even between the $dfoxo^{\Delta/\Delta}$ and $dfoxo^{\Delta/\Delta}$ *Insp3-Gal4/UAS-rpr* (Fig EV2).

To further increase the power of our analysis of *dfoxo* dependency, we employed network propagation, which combines differential expression with physical interactors of individual proteins to identify altered subnetworks in each tissue (Vanunu *et al*, 2010). Differentially expressed *dfoxo*-dependent or *dfoxo*-independent proteins were mapped independently onto the *Drosophila* protein–protein interaction network (Murali *et al*, 2011), and the associated *P*-value of each protein was negative logarithmic transformed and propagated to adjacent interacting proteins. We then clustered the *dfoxo*-dependent and *dfoxo*-independent responses to reduced IIS in each tissue and identified functional categories of these clusters with GO enrichment analysis (Dataset EV5).

We focused on clusters whose regulation was predicted as *dfoxo*-dependent within the network propagation analysis, since these should include processes causally linked to longevity. *dfoxo*-dependent responses in the brain were enriched for mitochondrial electron transport chain, mRNA splicing and nucleosome components (Fig 3), whilst in the gut they were enriched for proteasome and ubiquitin-mediated protein catabolism (Fig 3). Mitochondrial electron transport chain was also enriched within *dfoxo*-dependent responses in the fat body, as were ribosomal constituents and proteins involved in nucleosome assembly. Amongst the functional groups identified by network propagation, we further focused on three strong candidates for mediating the longevity of IIS mutant flies. First, we chose the ribosome, which we identified as an almost uniquely fat body-specific response. Second, we focused on the mitochondrial electron transport chain, the only response identified as common to multiple tissues, the brain, gut and fat body. Finally, we examined proteasome–ubiquitin protein catabolism as the most enriched cluster in the gut.

differentially regulated in a single tissue were also detected as solely expressed in that tissue, and thus, the tissue-specific nature of the differential regulation by IIS was mainly due to tissue-specific responses rather than tissue-specific expression patterns (Dataset EV4).

Identifying tissue-specific, differential protein expression potentially causal in extension of lifespan

To identify candidate proteins that could be causal in the increased longevity of mNSC-ablated flies, we classified them according to their requirement for *dfoxo* for their differential regulation. We took the set of proteins that changed expression with ablation of mNSCs in wild-type flies and asked whether their response to mNSC ablation was abrogated in a *dfoxo*-null background. Accordingly, we profiled the tissue-specific proteomes of mNSC-ablated flies lacking *dfoxo* (*Insp3-Gal4/UAS-rpr; dfoxo$^{\Delta/\Delta}$*) and the corresponding *dfoxo$^{\Delta/\Delta}$* controls. We constructed a linear model using all four genotypes and identified proteins whose expression was significantly affected by the interaction between ablation of mNSCs and *dfoxo* deletion. We excluded from this set any proteins whose response to reduced IIS was exaggerated, as opposed to abrogated, in the *dfoxo*-null background. We thus identified 361 proteins that required *dfoxo* for their change in expression in IIS mutant flies (Fig EV2). To assess whether those 361 proteins were more likely to be direct or indirect targets of *dfoxo*, we searched for predicted *dfoxo*-binding motifs within 1 kb of their transcriptional start sites using MEME (Bailey *et al*, 2009). 45% of the 361 proteins came from genes with *dfoxo*-binding motifs (Dataset EV4) and may therefore be directly regulated by dFOXO. However, the remaining 55% of those 361 proteins came from genes lacking *dfoxo*-binding motifs which suggest that although their expression was *dfoxo*-dependent, they were not directly regulated transcriptionally by dFOXO.

The proteins whose expression showed no significant interaction between response to mNSC ablation and the presence of *dfoxo* could have been truly *dfoxo*-independent or false negatives. To define a

Reduced IIS alters fat body-specific, *dfoxo*-dependent protein expression to regulate translation

We next determined if the tissue-specific, *dfoxo*-dependent proteomic responses led to tissue-specific functional changes. Our proteomic and bioinformatics analysis identified a fat body-specific, *dfoxo*-dependent reduction in ribosome-associated proteins in response to reduced IIS (Fig 3). To determine whether this signature was reflected in altered physiology, we quantified tissue-specific translation, using ^{35}S incorporation to detect *de novo* protein synthesis. Individual tissues were dissected and incubated in medium containing ^{35}S-labelled methionine and cysteine. We separated samples by SDS–PAGE and quantified *de novo* labelled proteins (Fig EV3). No significant change was detected in head, thorax or gut (Fig EV3), but there was significantly less incorporation of labelled amino acids into proteins in the fat body of the IIS mutants,

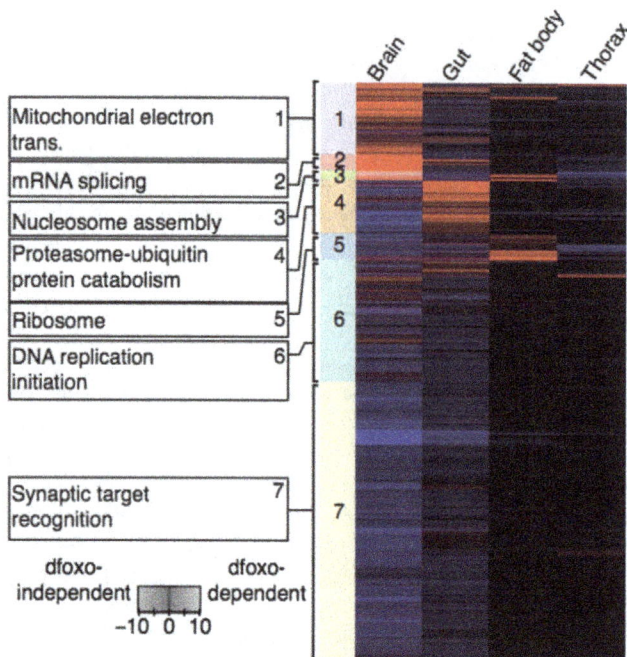

Figure 3. Hierarchical clustering and GO enrichment analysis of dFOXO-dependent and dFOXO-independent IIS-mediated regulation of the proteome.

Tissue-specific heatmap of significantly regulated *dfoxo*-dependent (red) and *dfoxo*-independent (blue) proteins in response to reduced IIS. Coloured side bars represent network propagated score clustering and associated most significantly enriched GO terms.

and that reduction was dependent on *dfoxo* (two-way ANOVA $P < 0.05$; Fig EV3). This finding suggests that IIS-mediated, *dfoxo*-dependent reduction in translation in the fat body may contribute to longevity.

Reduced IIS tissue-specifically and *dfoxo*-dependently regulates respiration

Changes in mitochondrial function have been linked to ageing in several model organisms (Bratic & Larsson, 2013). Paradoxically, however, both reduced mitochondrial respiratory chain protein levels (Copeland *et al*, 2009) and increased mitochondrial biogenesis (Rera *et al*, 2011) can extend lifespan. Our analysis highlighted a tissue-specific regulation of mitochondrial proteins that was *dfoxo*-dependent in the brain and fat body and independent of *dfoxo* in the gut (Fig 3). In total, we detected 60 proteins belonging to the annotation mitochondrial electron transport chain, with 6–35% of them in different tissues showing significantly altered levels upon IIS reduction (Fig 4A). They were coordinately up-regulated in the fat body, with no clear coordinated change in other organs (Fig 4A). Using Western blotting, we directly confirmed that NDUFS3, a complex I subunit, was up-regulated in a *dfoxo*-dependent manner in the fat body (Fig 4B).

To determine the functional significance of the observed changes in protein levels, we provided substrates for the individual complexes of the electron transport chain to measure respiration in actual respiratory state (PGMP3), the potential maximum respiratory state (Vmax) and role of complex I (rotenone sensitive). In agreement with our proteomic and bioinformatic analysis, endogenous respiration (PGMP3) was up-regulated by reduced IIS in the fat body, a response that was completely dependent on *dfoxo* (two-way ANOVA interaction term $P < 0.01$; Fig 4C). This was also the case for the Vmax and rotenone-sensitive respiratory states, suggesting that the increased respiration was due to increased capacity of the respiratory chain, at least in part through complex I (Fig 4C). In contrast, endogenous (PGMP3) and Vmax respiratory states in the gut were reduced in mNSC-ablated flies and, as predicted by our analysis of the proteome, this occurred independently of *dfoxo* (Fig 4C). Respiration in the head and thorax was unchanged by reduced IIS (Fig 4C). We confirmed the increase in endogenous respiration of the fat body, and the reduced endogenous respiration of the gut, in an independent model of reduced IIS, long-lived *dilp2-3,5* mutant flies (Gronke *et al*, 2010; Fig 4D).

To further understand the mechanistic basis for increased respiration in the fat bodies of mNSC-ablated flies, we examined mitochondrial biogenesis. We first examined mitochondrial DNA (mtDNA) levels in the fat body, in the presence and absence of *dfoxo*, and found that they were increased in the MNC-ablated flies in a *dfoxo*-dependent manner (two-way ANOVA interaction term $P = < 0.01$; Figs 4E and EV4A). Mitochondrial biogenesis is driven by nuclear transcription factors, including nuclear respiratory factors 1 and 2 (nrf; Evans & Scarpulla, 1990; Virbasius *et al*, 1993), which are co-activated by PGC-1 to regulate the expression of mitochondrial proteins (Puigserver & Spiegelman, 2003; Scarpulla, 2008; Tiefenböck *et al*, 2010). *Drosophila* has single homologs of PGC-1 and nrf-2 (Gershman *et al*, 2007), named *spargel* and *delg*, respectively (Tiefenböck *et al*, 2010), and loss of either reduces transcription of nuclear encoded mitochondrial genes (Tiefenböck *et al*, 2010). *Spargel* and *delg* are both expressed at low levels in adult *Drosophila* fat body (Graveley *et al*, 2011), and we did not detect them, or their regulation in our analysis (Dataset EV1). However, as with many TFs and cofactors, the activity of Spargel and delg could be regulated on the post-translational level (Li *et al*, 2007). We tested whether *spargel* or *delg* were required for the increased respiration in the fat bodies of IIS mutant flies. To achieve fat body-specific knockdown of *spargel* and *delg*, we performed this analysis in *dilp2-3,5* mutants. We found that increased respiration in the fat body of the IIS mutant was entirely dependent on *Spargel* and *delg* expression in the fat body (two-way ANOVA interaction term $P = < 0.001$; Fig EV4B and C). Furthermore, knockdown of expression of *Spargel* or *delg* in the fat body of *dilp2-3,5* mutants reduced the extent of their increased longevity, but did not affect the lifespan of wild-type controls (Figs 4F, and EV4D and E). Experimentally increased expression of *Spargel* in the adult fat body, using the GeneSwitch driver S106-GS, was sufficient to increase respiration (Fig EV4F) and lifespan (Figs 4G and EV4G), whilst over-expression in the gut alone was not (Fig EV4H).

Overall, our data reveal that lowered systemic IIS coopts different mechanisms to regulate respiration in different tissues. Reduced IIS decreased respiration in the gut independently of *dfoxo*, whilst simultaneously increasing respiration in the fat body in a *dfoxo-* and *spargel/delg*-dependent manner. Furthermore, increased mitochondrial biogenesis, and thus respiration, in the fat body is both necessary and sufficient to extend lifespan.

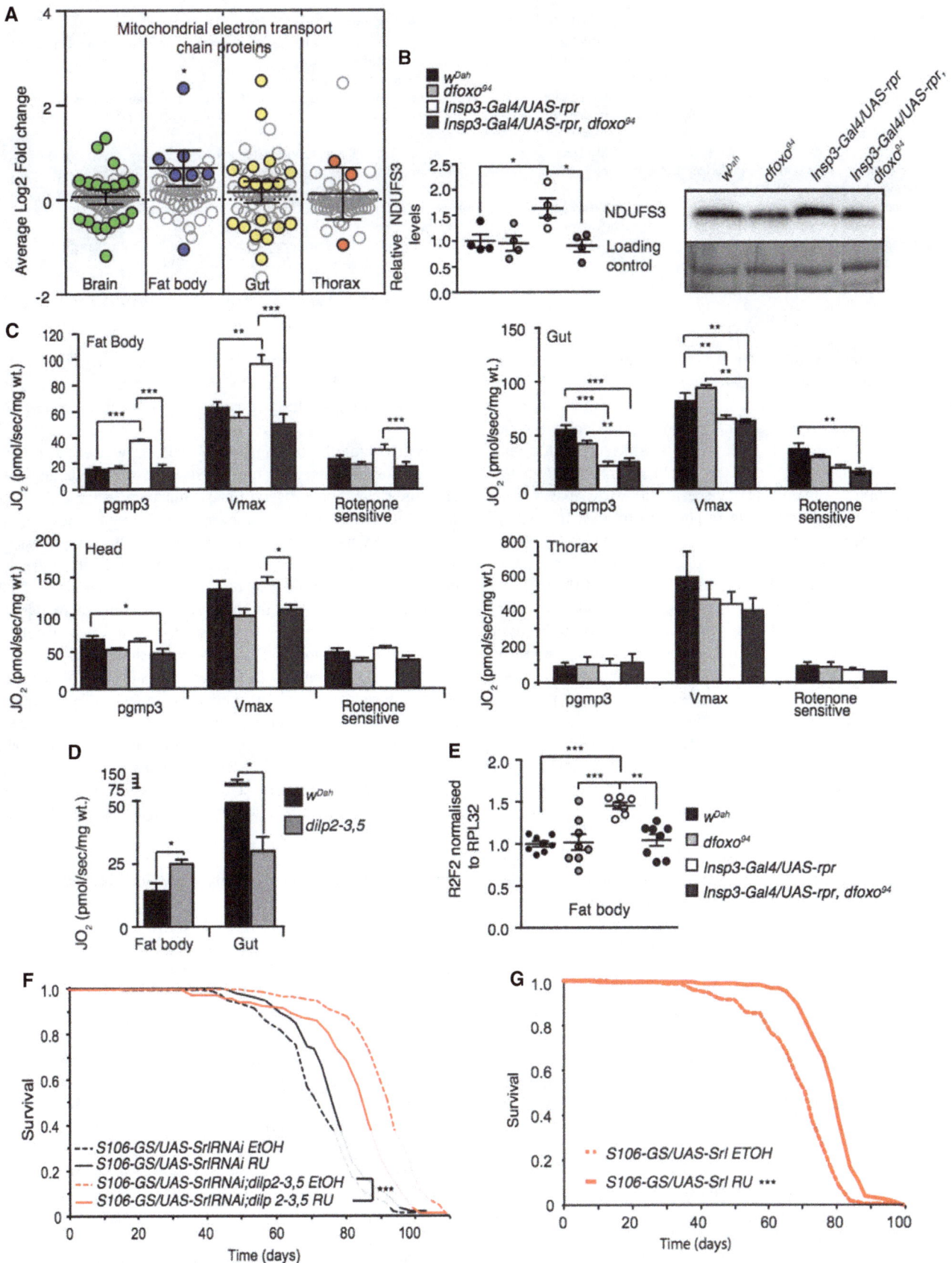

Figure 4.

◀ **Figure 4. Reduced IIS differentially regulates tissue-specific mitochondrial function and biogenesis.**

A Average log2 fold change of IIS-regulated proteins associated with the functional term mitochondrial electron transport chain. One-sample *t*-test shows directional significance. Grey circles show all detected proteins, and coloured circles show significantly regulated proteins.

B Western blot analysis of fat body NDUFS3, and stain free loading control (*n* = 4), significance determined by *t*-test.

C Fat body, gut, head and thorax oxygen consumption normalized to dry wt. in 10 days old control, *dfoxo*$^{\Delta94}$ mutants, mNSC ablation, mNSC-ablated flies lacking *dfoxo*. Oxygen consumption was assessed by using substrates entering the level of complex I (PGMP3), complex I+II once uncoupled by CCP (Vmax) and rotenone-sensitive complex I+II + rotenone (*n* = 5).

D Respiratory measurements of an independent IIS mutant (*dilp2-3,5*) compared to controls (*w*^{Dah}) (*n* = 5), significance determined by *t*-test.

E Relative mtDNA levels compared to nuclear DNA in fat body (*n* = 8). Relative mtDNA levels in the gut, head and thorax are shown in Fig EV4A. Significance determined by one-way ANOVA.

F Lifespan analysis of *dilp2-3,5* mutant and control flies with reduced expression of *Spargel* in the fat body (*S106-GS/UAS-SrlRNAi*) induced by RU or none induced (*n* = 150/genotype/treatment). Lifespan analysis of genetic controls shown in Fig EV4D. Statistical significance was determined by Log Rank test.

G Lifespan analysis of flies over-expressing *Spargel*, in the adult fat body (*n* = 150/genotype/treatment). Genetic control lifespan shown in Fig EV4G. Statistical significance was determined by Log Rank test.

Data information: Bars indicate mean ± SEM (*$P < 0.05$; **$P < 0.01$; ***$P < 0.001$; *n* = 5).

Reduced IIS alters the *dfoxo*-dependent gut proteome, increasing proteostasis to maintain gut health

The proteasome is a major site of protein degradation in the cell (Lecker *et al*, 2006), and its dysfunction is associated with ageing and disease (Vilchez *et al*, 2014). Our proteomic and bioinformatics analysis identified a prominent, gut-specific, *dfoxo*-dependent regulation of proteasomal proteins in response to reduced IIS (Fig 3). The 26S proteasome consists of 33 proteins, 14 of which (7-α & 7-β subunits) constitute the 20S proteasome core, whilst 19 constitute the 19S regulatory particle (Tomko & Hochstrasser, 2013). In total, we detected 94% of the proteasomal subunits, and 36% were differentially regulated specifically in the gut in response to reduced IIS (Fig 5A). However, we did not detect a consistent pattern or direction of regulation (4 up- and 8 down-regulated, Fig 5A). Proteasomal subunit Rpt6R showed the greatest degree of regulation, increasing 2.6-fold in the gut. We confirmed this change and its *dfoxo* dependency, by Western blot analysis of the guts of mNSC-ablated flies compared to controls (Fig 5B). Rpt6, a 19S proteasomal subunit, plays a key role, along with Rpn6, in regulating the assembly and activity of the 26S proteasome holoenzyme (Park *et al*, 2011; Pathare *et al*, 2012; Sokolova *et al*, 2015). Hence, up-regulation of this subunit could indicate an up-regulation of proteasomal assembly. Indeed, the guts of mNSC-ablated flies showed a threefold increase in assembly of the 26S proteasome compared to controls (Fig 5C), as measured using an in-gel assay (Vernace *et al*, 2007). These data suggest that reduced IIS results in increased proteasomal assembly in the gut, possibly through increased levels of Rpt6.

We next determined whether the increased proteasomal assembly in the gut affected proteasomal function. To measure proteasome activity, isolated tissue homogenates were incubated with a fluorogenic (AMC, 7-Amino-4-methylcoumarin) proteasome substrate (Z-Leu-Leu-Glu-AMC), allowing the quantification of peptidylglutamyl-peptide hydrolysing (caspase-like) activity of the proteasome. As predicted by our bioinformatic analysis (Fig 3), no differences in activity were detected in the fat body, head or thorax in response to reduced IIS (Fig 5D). However, in agreement with our assembly data, reduced IIS resulted in a highly significant, *dfoxo*-dependent, threefold increase in proteasomal activity in the gut (two-way ANOVA interaction term $P < 0.001$; Fig 5D). To establish the generality of this finding, we measured proteasomal activity in long-lived *dilp2-3,5* mutants (Gronke *et al*, 2010) and found

that, again, proteasome activity was increased in gut but not fat body (Fig 5E). Thus, increased proteasome activity in the gut in response to reduced IIS is a candidate mechanism for increased longevity.

To determine the functional relevance of gut-specific increased proteasomal activity, we quantified the level of ubiquitinated proteins. Ubiquitination serves many functions in the cell (Grabbe *et al*, 2011). Poly-ubiquitination, specifically on Lysine 48 (K48), however, targets the protein for degradation by the proteasome, and accumulation of poly-ubiquitinated proteins, often seen with increasing age, indicates loss of proteostasis (Tonoki *et al*, 2009). Based on our tissue-specific proteasomal activity assay (Fig 5D), we hypothesized that K48-ubiquitinated proteins levels would not differ in the head, thorax or fat body as a result of altered IIS activity, but would be reduced in the gut. We performed Western blot analysis against K48 poly-ubiquitinated proteins (Fig 5F) and, in agreement with our hypothesis, we detected a gut-specific, *dfoxo*-dependent reduction in K48 poly-ubiquitinated proteins in response to reduced IIS (Fig 5F). Therefore, increased proteasomal assembly/activity, and thus clearance of K48 poly-ubiquitinated proteins, may underlie IIS-mediated longevity through enhanced proteome maintenance.

Increased gut proteasomal activity maintains gut integrity and extends lifespan

We next determined the role of increased proteasomal activity in the extended lifespan of IIS mutants. We orally administered bortezomib, a potent inhibitor of the proteasome, to flies with reduced IIS and controls, for the duration of their lifespan. At a concentration of 2 μM, bortezomib did not reduce the lifespan of wild-type flies (Fig 6A), but significantly reduced the lifespan-extension of mNSC-ablated flies (Fig 6A) and, independently, of *dilp2-3,5* mutants (Fig EV5A). Using Cox proportional hazards, we confirmed that treatment with bortezomib had a significantly different effect on survival in the two models of reduced IIS flies compared to their wild-type controls ($P < 0.01$). Higher concentrations of bortezomib (3 μM) completely abolished the difference in lifespan between wild-type and ablated flies, but also reduced the survival of wild-type flies (Fig EV5B).

To determine the functional consequence of increased proteasome activity in the gut, we quantified age-associated changes in gut integrity in control and long-lived flies. Feeding non-absorbable

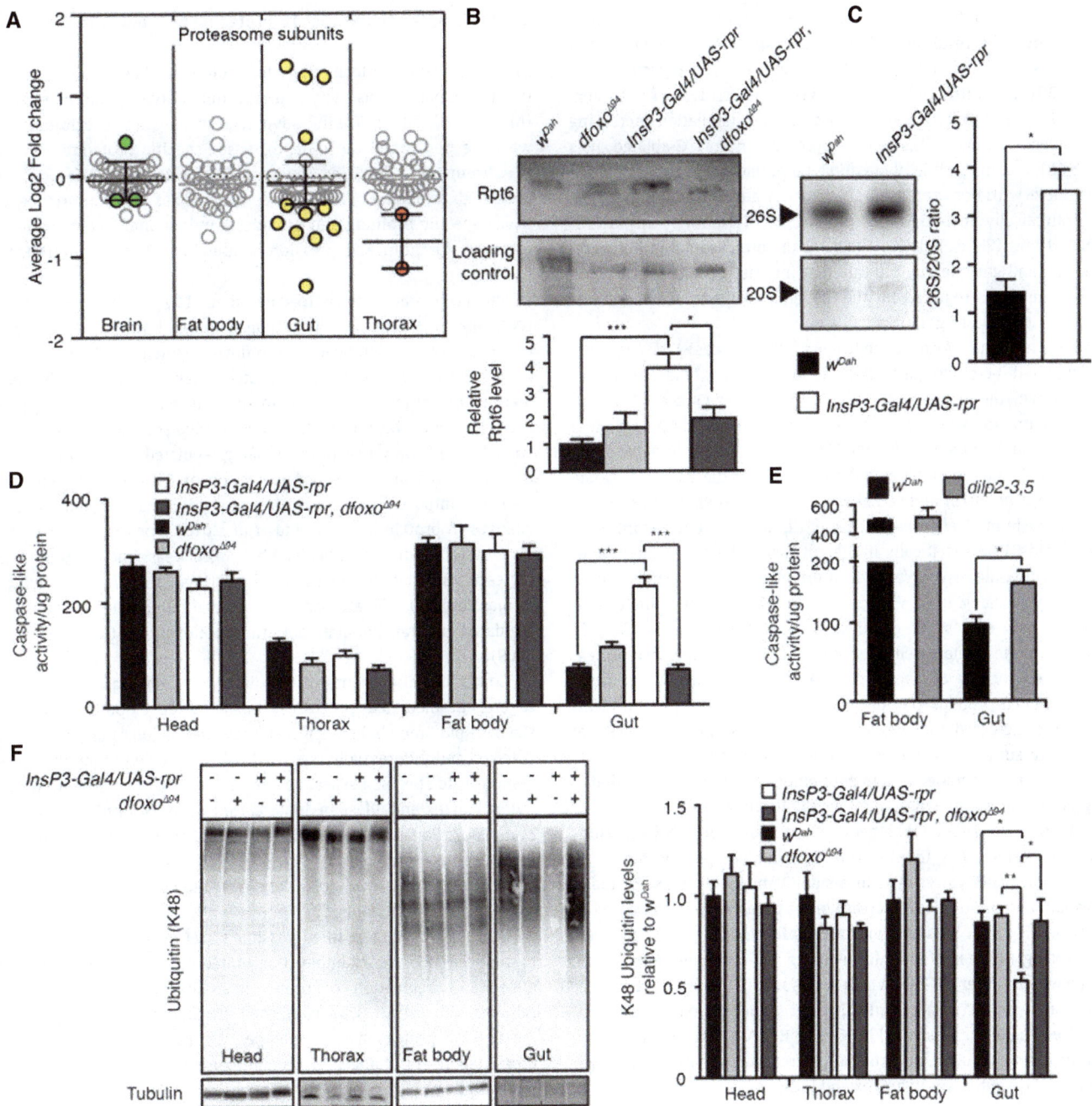

Figure 5. IIS-*dfoxo*-dependent regulation of proteasome activity in the gut enhances proteome maintenance.

A   Tissue-specific average log2 fold change of IIS-regulated proteins associated to the proteasome.

B   Western blot and quantification analysis of Rpt6 in wild-type (w^{Dah}) ($n = 8$), $dfoxo^{\Delta94}$ mutants ($n = 3$), mNSC-ablated flies (*InsP3-Gal4/UAS-rpr*) ($n = 7$) and mNSC-ablated flies lacking *dfoxo* (*InsP3-Gal4/UAS-rpr, dfoxo$^{\Delta94}$*) ($n = 4$), normalized to protein loading and significance determined by one-way ANOVA.

C   Assessment of proteasome assembly by in-gel caspase-like (LLE) proteasome activity in mNSC-ablated fly gut and control flies (w^{Dah}). Quantification shows the activity ratio 26S/20S ($n = 5$), significance determined by t-test.

D   Tissue-specific caspase-like activity was assessed using fluorogenic substrate (LLE-AMC, Enzo Life Science) in wild-type (w^{Dah}; $n =$ head 14, fat body 5, gut 8, thorax 8), $dfoxo^{\Delta94}$ mutants ($n =$ head 9, fat body 5, gut 7, thorax 8), mNSC-ablated flies (*InsP3-Gal4/UAS-rpr*; $n =$ head 14, fat body 5, gut 7, thorax 9) and mNSC-ablated flies lacking *dfoxo* (*InsP3-Gal4/UAS-rpr, dfoxo$^{\Delta94}$*; $n =$ head 7, fat body 5, gut 9, thorax 9), significance determined by two-way ANOVA and *post hoc* pairwise tests.

E   Proteasomal caspase-like activity of an independent IIS mutant (*dilp2-3,5*, grey) compared to controls (w^{Dah}) (gut $n = 9$, fat body $n = 7$). Statistical significance was determined by t-test.

F   Representative Western blot showing tissue-specific levels of K48-linked poly-ubiquitin and their quantification (head, thorax and fat body Westerns performed in quadruplicate). For gut samples wild-type (w^{Dah}) and $dfoxo^{\Delta94}$ mutants $n = 11$, *InsP3-Gal4/UAS-rpr* $n = 12$ and *InsP3-Gal4/UAS-rpr, dfoxo$^{\Delta94}$* $n = 10$, significance determined by two-way ANOVA and *post hoc* pairwise tests.

Data information: *$P < 0.05$, **$P < 0.01$, ***$P < 0.001$. All error bars show SEM.

blue food dye to flies allows direct assessment of gut integrity, with loss of integrity resulting in blue dye leaking into the hemolymph giving "smurf" flies (Rera *et al*, 2012). As previously reported (Rera *et al*, 2012), control flies show an age-associated loss of gut integrity, with the proportion of "smurfs" in the population increasing with age; however, this was suppressed in mNSC-ablated flies (Fig 6B), and independently in *dilp2-3,5* mutants (Fig EV5C). Whilst treatment with bortezomib (2 μM) did not affect aged control flies, it dramatically increased the proportion of "smurfs" in mNSC-ablated flies (Fig 6C). These effects of the proteasomal inhibitor on lifespan and gut integrity indicate that increased proteasomal activity in the gut is required for the lifespan extension of IIS mutants because it increases gut integrity.

Together, the *dfoxo* dependence of the increased proteasomal activity and increased gut integrity of mNSC-ablated flies, as well as the known role of proteostasis in ageing (Lopez-Otin *et al*, 2013), prompted us to experimentally test whether increasing proteasomal assembly and activity in the adult gut was sufficient to extend lifespan. RPN6 aids in stabilizing the assembly of the 26S proteasome (Pathare *et al*, 2012) and can increase lifespan when over-expressed ubiquitously in *C. elegans* (Vilchez *et al*, 2012). We therefore over-expressed RPN6 specifically in the gut, using the constitutive, mid-gut-specific Gal4 driver *Np1*, which drives expression exclusively in enterocytes (Jiang *et al*, 2009; Alic *et al*, 2014). Constitutive, gut-specific over-expression of RPN6 was sufficient to significantly increase proteasome assembly in the gut (Fig 6D), increase proteasomal activity (Fig 6E) and induce a 26% reduction in K48 poly-ubiquitinated proteins (Fig 6F), suggesting increased protein clearance and increased gut integrity (Fig 6G). Thus, increased levels of RPN6 are sufficient to increase proteasome assembly and activity, and increased clearance of K48 poly-ubiquitinated proteins, enhancing proteome maintenance, and thus the health of the gut.

We next determined whether over-expression of RPN6 was sufficient to extend the fly lifespan. Constitutive, gut-specific over-expression of RPN6 resulted in a significant 5% increase in mean lifespan, and up to a 19% increase in maximal lifespan compared to controls (Fig 6H). We independently confirmed this finding by over-expressing RPN6 in the adult gut using the inducible, gut-specific GeneSwitch driver *TIGS-2*, which was sufficient to increase proteasome activity, increase gut integrity and induce a small, but significant, extension of lifespan (Figs 6I and EV6A–C). Hence, increased gut proteome maintenance, through increased proteasome function, can increase gut integrity and extend lifespan.

Discussion

Reduced activity of the evolutionarily conserved insulin/IGF-like signalling network extends longevity and reduces age-associated pathologies, extending health and vitality (Fontana *et al*, 2010). Several studies have examined transcriptome remodelling in response to reduced IIS, to uncover the genes and molecular mechanisms underlying the longevity phenotype. However, the specific combination of expression changes and physiological processes that are required to prolong healthy lifespan remains elusive. Here, we have quantified the tissue-specific proteome of *Drosophila* in response to reduced IIS, using genetic manipulations that allowed us to focus on expression profiles associated with longevity.

Establishing the tissue-specific proteome of *Drosophila*

In recent years, the technology for proteomics has begun to reach the accessibility, sensitivity and reproducibility of genomic studies (Mann *et al*, 2013). Taking advantage of advancing technologies, we have performed an in-depth, tissue-specific proteomic screen in *Drosophila*, detecting over 6,000 proteins, ~40% of the predicted proteome. This dataset emphasizes the importance of tissue-specific profiling and provides a unique and freely available resource for future expression studies in *Drosophila* (Dataset EV1).

Our proteomic analysis revealed that ~15% of the predicted fly proteome is remodelled in response to reduced IIS, with 2,372 proteins showing significant up- or down-regulation. The RNA transcripts encoding 60% of these proteins have not previously been identified as differentially regulated by reduced IIS in whole fly profiles. This may be due to the highly tissue-specific nature of the changes, with only two proteins being regulated in all tissues. Differences between patterns of expression of RNA and protein may also be important, due to post-transcriptional regulation and changes in protein stability (Liu *et al*, 2016). For example, expression of GCN4 in nutritionally challenged yeast increases due to increased translation, not transcription (Ingolia *et al*, 2009), whilst in flies subject to dietary restriction several mitochondrial genes are regulated by translational, not transcriptional, control (Zid *et al*, 2009).

Confirming the quality of this dataset, proteins associated with several IIS-mediated processes, including development, growth, sleep, lipid metabolism, translation and adult lifespan, were detected and differentially regulated. We were able to pinpoint these responses to specific tissues. For example, global reduction in translation can extend lifespan in *C. elegans* and *Drosophila* (Hansen *et al*, 2007; Pan *et al*, 2007; Wang *et al*, 2014) and is an evolutionarily conserved response to reduced IIS (McElwee *et al*, 2007; Stout *et al*, 2013; Essers *et al*, 2016), and our study showed this response to be specific to the fat body.

Our dataset also identified possible novel mediators of responses to reduced IIS. For example, our proteomic analysis suggested a gut-specific regulation of proteasomal function in response to reduced IIS, which was confirmed by finding a corresponding gut-specific proteasomal phenotype. We have also characterized *dfoxo*-dependent changes in protein expression in IIS mutant flies, separating those changes associated with IIS-mediated longevity from those changes associated to other IIS-mediated phenotypes. Furthermore, we determined which IIS responsive protein-coding genes contain predicted dFOXO-binding motifs, identifying possible direct and indirect targets of dFOXO. Additionally, we examined transcriptional changes in several regulated candidate genes associated with mitochondria and the proteasome (Fig EV7A–C). Some candidate genes were regulated in a *dfoxo*-dependent manner, consistent with a direct regulation by dFOXO; however, many did not reflect the changes seen at the protein level, suggesting indirect regulation by dFOXO, possibly though post-transcriptional and/or post-translational regulation (Fig EV7A–C). Importantly, we have thus identified longevity-associated changes in the proteome of IIS mutant flies and the processes that they regulate, and we have experimentally demonstrated the role of some of these processes in extension of lifespan.

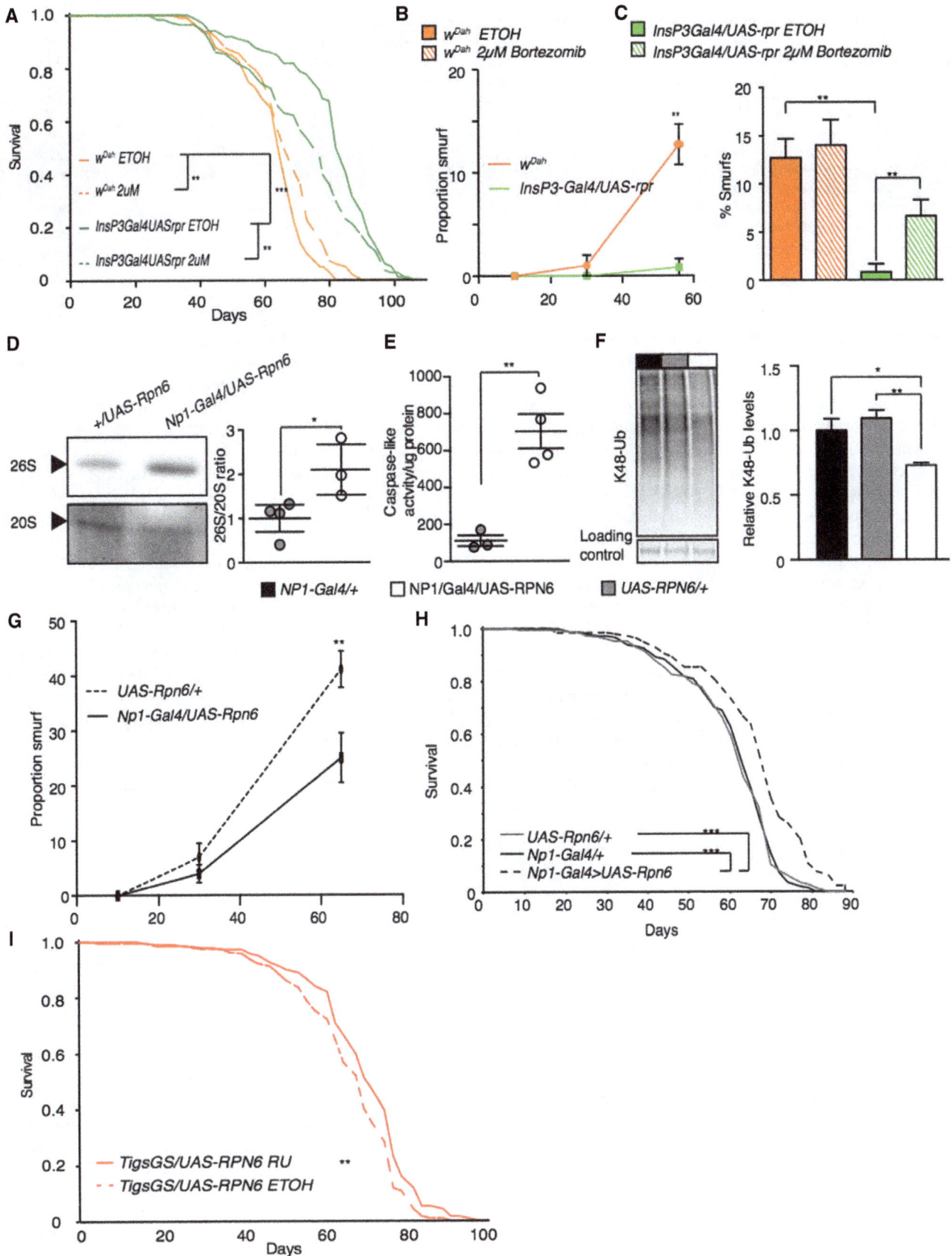

Figure 6.

Figure 6. Proteasome activity is necessary for IIS-mediated longevity, and sufficient to extend lifespan.

A Lifespan analysis of wild-type flies and mNSC-ablated flies treated with 2 µM proteasome inhibitor (bortezomib) or vehicle (EtOH) ($n = 100$).

B, C Assessment of age-related gut integrity through "Smurf" assay at either different ages (10, 30, 65 days, $n = 100$), significance determined by t-test, or in aged (55 days, $n = 90$) flies in the presence of 2 µM proteasome inhibitor (bortezomib), two-way ANOVA.

D Assessment of proteasome assembly by in-gel caspase-like (LLE) proteasome activity in NP1-Gal4/UAS-RPN6 ($n = 3$) flies compared to +/UAS-RPN6 control flies ($n = 4$), significance determined by t-test. Quantification shows the activity ratio 26S/20S ($n > 4$).

E Gut proteasome activity (caspase-like) of NP1-Gal4/UAS-RPN6 ($n = 4$) flies compared to +/UAS-RPN6 control flies ($n = 3$), significance determined by t-test.

F Representative Western blot showing gut-specific levels of K48-linked poly-ubiquitin and their quantification ($n = 6$), significance determined by one-way ANOVA.

G Assessment of age-related gut integrity through "Smurf" assay at difference ages (10, 30, 65 days) in NP1-Gal4/UAS-RPN6 flies and +/UAS-RPN6 control flies ($n > 100$), significance determined by t-test.

H, I Lifespan analysis of flies with gut-specific over-expression of UAS-RPN6 using a constitutive gut-specific driver (Np1-Gal4) or in the adult using Tigs-Gal4-GS in the presence of RU486 (200 µm) or vehicle (EtOH) ($n = 150$).

Data information: *$P < 0.05$, **$P < 0.01$, ***$P < 0.001$. All error bars show SEM. Statistical significance between lifespan analyses (A, H, I) was determined by Log Rank test.

Tissue-specific regulation of mitochondrial number and respiration as a regulator of the IIS longevity response

Mitochondrial respiration and the production of reactive oxygen species (ROS) have been proposed as a mechanism of ageing (Bratic & Larsson, 2013). According to this theory, ROS are produced as a by-product of mitochondrial respiration and on interaction with cellular macromolecules they induce damage (Harman, 1956). However, the link between mitochondria and ageing is not simple, and the validity of this theory has been questioned (Bratic & Larsson, 2013).

Seemingly in agreement with reduced ROS being beneficial, reducing mitochondrial respiration extends lifespan in several organisms (Dillin et al, 2002; Rea et al, 2007; Copeland et al, 2009). In contrast, many pro-longevity interventions are associated with increased activation of mitochondrial respiration (Evans & Scarpulla, 1990; Virbasius et al, 1993; Ristow & Schmeisser, 2011), including models of dietary/caloric restriction (Mootha et al, 2003; Baker et al, 2006) and IIS, as found here and in C. elegans (Zarse et al, 2012). Declining mitochondrial number and function is associated with ageing (Yen et al, 1989; Shigenaga et al, 1994; Hebert et al, 2015). Thus, increasing or maintaining mitochondrial biogenesis with age may be beneficial.

We found increased, fat body-specific respiration that was causal in the longevity of IIS mutants. Furthermore, we and others (Rera et al, 2011), have shown that increasing mitochondrial biogenesis, extends lifespan, although we found the effect to be confined to the fat body, whilst Rera et al (2011) found an effect in both the fat body and the gut, the discrepancy possibly due to the use of different fly stains. We also found that the increased fat body-specific respiration and longevity of IIS mutant flies are dependent on PGC1-α in the fat body, indicating that PGC1-α is a downstream mediator of IIS, not only for growth (Tiefenböck et al, 2010; Mukherjee & Duttaroy, 2013), but also, as found here, longevity.

How increased respiration can, directly or indirectly, extend lifespan remains unclear. Growing evidence suggests that ROS, produced by mitochondrial respiration, is increased by pro-longevity interventions and may act as signalling molecules to activate the expression of pro-longevity genes (Ristow & Schmeisser, 2011). On the other hand, inhibition of target of rapamycin (TOR) in yeast increases respiration, but reduces ROS production, due to increased electron transport chain subunit expression and increased efficiency of electron transport, preventing a protracted electron transit time and ROS formation (Bonawitz et al, 2007). Furthermore, in flies, increased mitochondrial biogenesis protects against high fat diets, preventing lipid accumulation and heart dysfunction (Diop et al, 2015). To determine the exact mechanisms involved, analysis of tissue-specific determination of ROS production and electron transport chain complex status under impaired IIS and other pro-longevity interventions will be needed.

Interestingly, although not associated with longevity, we also found that respiration in the gut is decreased under reduced IIS, independently of dfoxo activity. Our analysis suggested that this response was associated with down-regulation of Complex III protein subunits. When Complex III subunits are ubiquitously reduced throughout development and adulthood, flies become considerably longer lived and produce fewer offspring (Copeland et al, 2009). However, adult-specific expression fails to extend lifespan, yet flies remain less fecund (Copeland et al, 2009). Since the down-regulation of respiration we observed on reducing IIS in the gut was not associated with increased lifespan, and occurred independently of dfoxo, the reduction in gut respiration may underlie the reduced fecundity of IIS mutants.

Tissue-specific proteostasis as a regulator of the IIS longevity response

Gradual loss of proteostasis is considered a hallmark of ageing (Lopez-Otin et al, 2013). We found that, under reduced IIS, Drosophila tissue specifically regulate proteostasis, an evolutionarily conserved response that correlates with longevity. In flies, the age-associated decline of proteostasis has been linked to declining proteasome function (Tsakiri et al, 2013), whilst maintaining proteasome function is associated with longevity (Chondrogianni et al, 2000; Pérez et al, 2009; Vilchez et al, 2012; Ungvari et al, 2013). In yeast, C. elegans, and Drosophila, ubiquitous over-expression of proteasomal subunits can extend lifespan, possibly through increased maintenance of the cellular proteome (Chen et al, 2006; Tonoki et al, 2009; Kruegel et al, 2011; Vilchez et al, 2012). Increasing the maintenance of the cellular proteome and extending life span can also be achieved by reducing translation (Hansen et al, 2007; Pan et al, 2007; Wang et al, 2014). We found in Drosophila that increased proteasome activity and decreased translation correlated with longevity. However, these responses were highly tissue-specific. Only the fat body of Drosophila responded to reduced IIS by decreasing translation, and only the gut responded by increasing proteasome activity. We were able to recapitulate increased proteasome activity by gut-specific over-expression of

RPN6, which was sufficient to extend lifespan in wild-type flies. Importantly, we could also show that increased proteasomal activity is required for the longevity of IIS mutants. Understanding how manipulation of one, or relatively few, protein subunits affects the function of a large protein complex is important for translating functional studies towards amelioration of the effects of human ageing. Whilst each subunit plays a role, only some subunits, that is the catalytic subunits, are essential for proteasome activity. However, Rpt3, Rpt6 and RPN6 play an equally vital role, those of opening the catalytic core and assembly maintaining structural integrity of the proteasome (Murata *et al*, 2009; Tian *et al*, 2011; Pathare *et al*, 2012; Sokolova *et al*, 2015). RPN6 over-expression also influences lifespan (Vilchez *et al*, 2012), and we suggest through increased 26S proteasomal assembly, which recapitulates the phenotype seen in IIS mutant flies. It will therefore be important to determine whether similar regulation occurs in the tissues of long-lived mammalian IIS models.

Tissue-tissue communication and rebuilding the fly

Tissue-specific reduction in IIS activity can improve function in that specific tissue, and can also act at a distance to improve the function of other tissues through systemic effects (Zhang *et al*, 2013; García-Cáceres *et al*, 2016; Kaletsky *et al*, 2016). In these cases, tissue-to-tissue communication, involving FoxO and other key regulators, promotes longevity (Murphy *et al*, 2007; Bai *et al*, 2012; Zhang *et al*, 2013; Alic *et al*, 2014). The fly genome encodes ~300 proteins predicted to be targeted for secretion. We found 35% of these proteins to be regulated in response to reduced IIS, but very few dependent upon *dfoxo*, and thus associated with longevity. Future investigations to directly assay the proteomic changes in circulating hemolymph will help to determine which proteins may be involved in communication between tissues.

A system-level approach to analyse the tissue-specific proteome response to reduced IIS, followed by functional genetic and molecular analysis, has allowed the identification of IIS-mediated, tissue-specific pro-longevity processes. Despite their highly specific responses to lowered IIS, individual tissues each contributed to the increase in lifespan in the IIS mutant flies. As well as its effect on ageing, IIS co-ordinates tissue-specific processes to regulate early life traits such as development, growth and reproduction in response to nutrition and other cues. The level of IIS appears to have evolved to optimize these early life fitness traits, but evidently this level is too high for optimal health at later ages not normally subject to natural selection, resulting in pro-longevity effects of reduced IIS.

Here, we focused on two pro-longevity effects, gut-specific proteasome assembly/activity and fat body-specific respiration/mitochondrial biogenesis. Increasing proteasome activity in the gut was sufficient to extend lifespan by 11%, and elevating mitochondrial respiration in the fat body can extend lifespan to the same extent (11%). However, reducing IIS extends lifespan to a greater extent. For example, mNSCs-ablated flies are ~30% longer lived than controls. Thus, it is tempting to suggest that increasing both proteasome activity and mitochondrial respiration in the same fly could lead to an additive increase in lifespan, and that manipulation of other processes identified here, such as reduced translation in the fat body, may extend lifespan even further (Fig 7), a hypothesis that should be tested in future experiments.

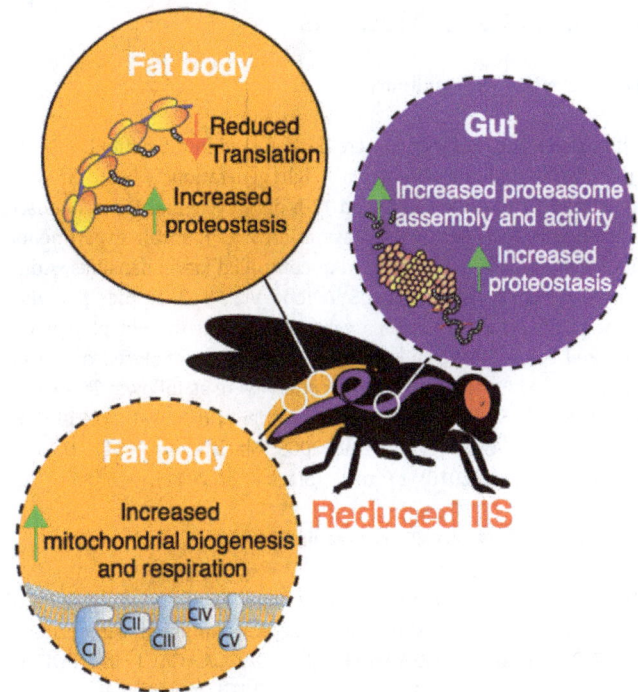

Figure 7. Model of tissue-specific responses to reduced IIS that may mediate longevity.
Colours denote different tissues (fat body, orange; gut, purple). Dashed lines show tissue-specific responses that are both necessary and sufficient to extend lifespan.

In summary, tissue-specific proteomic analysis of the responses to reduced IIS in *Drosophila* revealed that proteins from 15% of *Drosophila*'s protein-coding genes were regulated in response to reduced IIS across the four tissues examined here. Surprisingly, our study detected considerably more regulated proteins than our previous RNA transcript analyses in IIS mutants, highlighting the importance of tissue-specific, and proteome, profiling to understand the effects of IIS. Here we focused specifically on those underlying responses to reduced IIS that are causal for longevity. Reducing translation, specifically in the fat body, is associated with longevity and not other IIS phenotypes, and tissue-specific respiration was also increased through elevated mitochondrial biogenesis. Elevated mitochondrial biogenesis and respiration in the fat body were required for IIS-mediated longevity, and directly increasing mitochondrial biogenesis in the fat body extended life span. Expression of proteasomal subunits was modulated in the gut of the fly by reduced IIS. These changes resulted in increased proteasome assembly and activity, increased proteome maintenance and increased gut integrity with age. Inhibition of these changes abolished IIS-mediated longevity, and directly increasing gut-specific proteasome function extended life span. Importantly, we showed that these responses are pro-longevity and largely, tissue-specific. Due to the level of conservation of these processes, our study suggests the development of tissue-targeted pharmacological agents via homing peptides (Svensen *et al*, 2012) or chemical hybridization (Finan *et al*, 2016) to inhibit translation, activate the proteasome or increase respiration may be beneficial in prolonging longevity and health in mammals.

Materials and Methods

Fly stocks and fly husbandry

All mutants and transgenes were backcrossed into a white Dahomey (w^{Dah}) wild-type strain for at least eight generations. Fly stocks were kept at 25°C on a 12-h light and 12-h dark cycle and fed a standard sugar/yeast/agar diet (SYA; Bass et al, 2007). In all experiments, once-mated females were reared at controlled larval densities. Adult flies were aged (10 days) in SYA food vials (10–25 flies per vial) prior to analysis. Dissections were carried out in cold phosphate-buffered saline (PBS) and tissues either directly analysed or frozen on dry ice. Fly strains used in this study were as follows: Tigs-Gene-switch (Tigs-GS) (Buch et al, 2008; Poirier et al, 2008), UAS-RPN6, UAS-rpr, InsP3-Gal4 (Buch et al, 2008; Slack et al, 2011), UAS-Srl (Tiefenböck et al, 2010), $dfoxo^{\Delta 94}$ (Slack et al, 2011).

Generation of UAS-RPN6 transgenic flies

To generate a transgenic fly stock for conditional over-expression of Drosophila Rpn6, the Rpn6 ORF was PCR amplified with primers SOL710 (ATGAATTCGCAAGATGGCCGGAGCAACAC) and SOL711 (ATGGTACCTTACGACAGCTTCTTAGCCTTC) and cDNA LD18931 as template (Drosophila Genomics Resource Center) and subsequently cloned into the pUAST attB vector using the EcoRI and KpnI restriction sites of the primers. Transgenic flies were generated using the φC31 and attP/attB integration system (Bischof et al, 2007), and transgenes were inserted into the attP40 landing-site.

Peptide generation

Fly tissues (50/sample) from six biological replicates were lysed in pre-heated (95°C) 6M guanidine hydrochloride, 10 mM TCEP, 40 mM CAA, 100 mM Tris pH 8,5 lysis buffer. Following shaking at 18,407 g (95°C), tissues were sonicated for five cycles (Bioruptor plus). Lysis buffer was then diluted 11-fold in digestion buffer (25 mM Tris 8.5 pH, 10% acetyl nitride) and vortexed prior to a trypsin gold (Promega) overnight (37°C) digest (1 μg trypsin/50 μg protein). Samples were sonicated again for five cycles, prior to a further trypsin gold (Promega) overnight (37°C) digest (0.5 μg trypsin/50 μg protein) for 4 h (37°C) with gentle agitation. Samples were then placed in a SpeedVac (5 min, 37°C) to remove acetyl nitride. Peptides were desalted using SDB.XC Stage Tips (Rappsilber et al, 2003). Peptides were then eluted using (80% acetyl nitride, 0.1% formic acid) and placed in a SpeedVac (55 min, 29°C) to remove acetyl nitride, and quantified via Nanodrop.

MS/MS

Peptides were loaded on a 50-cm column with 75 μm inner diameter, packed in-house with 1.8-μm C18 particles (Dr Maisch GmbH, Germany). Reversed phase chromatography was performed using the Thermo EASY-nLC 1000 with a binary buffer system consisting of 0.1% formic acid (buffer A) and 80% acetonitrile in 0.1% formic acid (buffer B). The peptides were separated by a segmented gradient of 0 to 20 to 40% buffer B in 0 to 85 to 140 min for a 160-min gradient run with a flow rate of 350 nl/min. The Q-Exactive was operated in the data-dependent mode with survey scans acquired at a resolution of 120,000; the resolution of the MS/MS scans was set to 15,000. Up to the 20 most abundant isotope patterns with charge ≥ 2 and < 7 from the survey scan were selected with an isolation window of 1.5 Th and fragmented by HCD (20) with normalized collision energies of 27. The maximum ion injection times for the survey scan and the MS/MS scans were 50 and 100 ms, respectively, and the AGC target value for the MS and MS/MS scan modes was set to 1E6 and 1E5, respectively. The MS AGC underfill ratio was set to 20% or higher. Repeat sequencing of peptides was kept to a minimum by dynamic exclusion of the sequenced peptides for 45 s.

Protein identification and quantification

The .raw data files were analysed. Protein identification was carried out using MaxQuant (Cox & Mann, 2008) version 1.5.0.4 using the integrated Andromeda search engine (Cox et al, 2011). The data were searched against the canonical and isoform, Swiss-Prot and TrEMBL, Uniprot sequences corresponding to Drosophila melanogaster (20,987 entries). The database was automatically complemented with sequences of contaminating proteins by MaxQuant. For peptide identification, cysteine carbamidomethylation was set as "fixed" and methionine oxidation and protein N-terminal acetylation as "variable" modification. The in-silico digestion parameter of the search engine was set to "Trypsin/P", allowing for cleavage after lysine and arginine, also when followed by proline. The minimum number of peptides and razor peptides for protein identification was 1; the minimum number of unique peptides was 0. Protein and peptide identification was performed with FDR of 0.01. The "second peptide" option was on, allowing for the identification of co-fragmented peptides. In order to transfer identifications to non-sequenced peptides in the separate analyses, the option "Match between runs" was turned on using a "Match time window" of 0.5 min and "Alignment time window" of 20 min. Protein and peptide identifications were performed within, not across, tissue groups. Label-free quantification (LFQ) and normalization was done using MaxQuant (Cox et al, 2014). The default parameters for LFQ were used, except that the "LFQ min. ratio count" parameter was set to 1. Unique plus razor peptides were used for protein quantification. LFQ analysis was done separately on each tissue.

Perseus informatics analysis

The results of the LFQ analyses were loaded into the Perseus statistical framework (http://www.perseus-framework.org/) version 1.4.1.2. Protein contaminants, reverse database identifications and proteins "Only identified by site" were removed, and LFQ intensity values were log2 transformed. After categorical annotation into four categories based on genotype ($dfoxo^{\Delta 94}/w^{Dah}$) and ablation of mNSCs (yes/no), the data were filtered in order to contain a minimum of four valid values in at least one category. The remaining missing values were replaced, separately for each column, from normal distribution using width of 0.3 and down shift of 1.8. Based on quality control analysis, seven of the 90 samples (three fat body samples, three thorax samples and one brain sample) were excluded from the subsequent analysis due to technical failure of the chromatography. With those exceptions, the data were of high quality and reproducible between replicates. Information on Genotype and tissue-specific average LFQ values and associated SEM values are included in Dataset EV1.

Bioinformatics

PCA analysis was carried out using the FactoMineR R package (Lê et al, 2008). The LIMMA package (Bioconductor suite, R) was used for differential expression analysis (Schwämmle et al, 2013). Multiplicity-corrected P-values were calculated using the Benjamini-Hochberg procedure and significance determined using an adjusted P-value cut-off of 0.1.

Experimental groups were specified as follows: (A) w^{Dah}, (B) $dfoxo^{\Delta94}$ mutant, (C) InsP3-Gal4/UAS-rpr and (D) InsP3-Gal4/UAS-rpr, $dfoxo^{\Delta94}$. Hypotheses were tested as linear combinations against the null hypothesis of no change: ablation-induced changes in the wild-type background (C vs. A), ablation-induced changes in the $dfoxo^{\Delta94}$ background (D vs. B) and their interaction term (C–A vs. D–B). The subset of proteins differentially expressed in C vs. A was further subdivided into two categories: Proteins that changed in response to mNSC ablation in a dfoxo-dependent manner, and those that behaved similarly regardless of genetic background were dfoxo-independent: Foxo-dependent: Differential in C vs. A and the interaction term. If a protein was regulated in the same direction in C vs. A and D vs. B, it was required to show a stronger response in the first contrast. Foxo-independent: Differential in C vs. A. Protein expression levels were required to be equivalent with regard to differential expression in the interaction term. In any given tissue, a protein was identified as equivalent if the 95% CI of its log2 fold change fell within an interval ($[-t; t]$). The parameter t was determined as minimum log-fold change of any significantly differentially expressed protein detected in the interaction term of each respective tissue: brain ($t = 0.072$), fat body ($t = 0.177$), gut ($t = 0.155$) and thorax ($t = 0.216$). Additionally, this subset included proteins found to be significant in the interaction term if they showed a stronger response to mNSC ablation in equal direction to that in the $dfoxo^{\Delta94}$ background compared to the wild-type background.

Binding site identification

We used FIMO, MEME suite (Bailey et al, 2009) to identify genes whose transcription start site was within 1,000 bp of a dFOXO binding motif, using the default P-value threshold ($1e^{-4}$). Sequences corresponding to genes of measured proteins were extracted from the BDGP6 reference genome, whilst the dFOXO binding motif was taken from Fly Factor Survey (Zhu et al, 2011). We then mapped FIMO hits to proteins whose expression changed under reduced IIS conditions or those that were identified as dfoxo-dependent and used a hypergeometric test to evaluate the significance of the overlap. Selection and background were limited to the set of genes corresponding to measured proteins.

Gene ontology term enrichment

topGO was used for GO term enrichment analysis and gene annotation performed using the Bioconductor annotation package org. dm.e.g.db. To identify enriched GO terms, the one-sided elim Fisher procedure was used ($\alpha \leq 0.05$; Alexa et al, 2006). The enrichment score of a GO term is defined as log2 (#Detected significant genes/#Expected significant genes). For individual tissue-specific contrasts, the subsets of differentially expressed proteins were tested for term enrichment against a tissue-specific reference background. Cell plots show the most specific significantly enriched categories with a minimum of five significant associated proteins. To functionally characterize the individual tissues of wild-type flies, GO enrichment analysis was carried out for proteins detected exclusively in the respective tissue, against the background of all detected proteins. For PCA map functional annotation (Fig 1D), the top 5% and lowest 5% quantiles of proteins according to their contributions to each dimension were tested for GO term enrichment, against the background of all detected proteins. The plot shows the 10 most significant terms with a minimum of 10 significant proteins, for each direction of the two dimensions.

Network propagation

Network propagation (Vanunu et al, 2010) was done using the D. melanogaster PPI network DroID (Murali et al, 2011) filtered for high confidence edges (40% confidence, $n = 34,866$). The network was converted to an adjacency matrix and normalized with the Laplacian transformation (using the graph.laplacian function in the igraph R package). Differentially regulated proteins were classified into dfoxo-dependent and dfoxo-independent (see Bioinformatics section above). For the analysis of dfoxo-dependent proteins, the P-values of proteins not belonging to this group were excluded (set to 1). Likewise, for the evaluation of the dfoxo-independent set, P-values of proteins not detected as such were excluded. Finally, the P-values were −log2 transformed and mapped to the network. After mapping, the transformed P-values on the individual nodes were diffused to their adjacent nodes using the spreading coefficient of 0.8 (corresponds to the percentage of sharing to neighbours). This spreading changed the scores of all proteins in the network, and it was iteratively repeated until protein scores do not change anymore.

We noticed that the network topology (i.e. the network structure) induces a bias: certain regions of the network tend to accumulate high scores during the propagation even if there is no actual signal. In order to correct for this bias, we computed a node-specific topology bias in the following way: we started with equal scores on each node (e.g. each node has a score of 1). Then, we perform the propagation as described above. The resulting scores reflect the bias and are subtracted from the node scores after propagating the actual node scores. Subsequently, the corrected propagated scores were clustered based on their Euclidean distances (using the hclust function in R) allowing for the identification of enriched clusters for each tissue. For visualizing the enriched clusters from dfoxo-dependent and dfoxo-independent proteins on the same heatmap, the propagated profiles were subtracted (dfoxo-dependent smoothed scores and dfoxo-independent smoothed scores). GO enrichment analysis of individual clusters was done using the topGO R package defining all proteins that are part of the network as background.

^{35}S-methionine/^{35}S-cysteine incorporation assay

Incorporation assays were performed as previously described. Briefly, 5–8 tissues were dissected and collected in DMEM

(#41965-047, Gibco). For labelling, DMEM was replaced with methionine and cysteine free DMEM (#21-013-24, Gibco), supplemented with ^{35}S-labelled methionine and cysteine (#NEG772, Perkin-Elmer) and incubated in uncapped Eppendorf tubes on a shaking platform at room temperature (60 min). Samples were then washed in ice cold PBS and lysed in RIPA buffer (150 mM sodium chloride, 1.0% NP-40, 0.5% sodium deoxycholate, 0.1% SDS, 50 mM Tris, pH 8.0). Lysates were cleared by centrifugation at 15,871 g, 4°C for 10 min. Protein was precipitated by adding 1 volume of 20% TCA, incubating for 15 min on ice and centrifugation at 15,871 g, 4°C for 15 min. The pellet was washed twice in acetone and resuspended in 4 M guanidine HCl. Half the sample was added to 10 ml of scintillation fluid (Ultima Gold, Perkin-Elmer) and counted for 5 min per sample in a scintillation counter (Perkin-Elmer). The remaining sample was used to determine protein content using BCA (Pierce), following the manufacturers protocol. Scintillation counts were then normalized to total protein content prior to statistical analysis.

Respiratory rate measurements

Flies (10 days) were dissected in PBS and transferred to respiratory buffer (120 mM sucrose, 50 mM KCl, 20 mM Tris–HCl, 4 mM KH$_2$PO$_4$, 2 mM MgCl$_2$, 1 mM EGTA, 0.01% digitonin, 0.05% BSA, pH 7.2). Oxygen consumption was measured using an oxygraph chamber (OROBOROS) at 25°C. Complex I-dependent respiration was assessed using the substrates proline (10 mM), pyruvate (10 mM), malate (5 mM) and glutamate (5 mM), along with ADP (1.25 mM). The respiration was uncoupled by the addition of CCCP (0.3 μM). Maximum flux was then measured by adding complex II substrates succinate (10 mM) and glycerol-3-phosphate (5 mM). Rotenone-sensitive flux was measured in the presence of rotenone (3 μM). Finally, dry weights of tissues were determined to normalize the oxygen consumption.

20s proteasome activity assay

Proteasome caspase-like activity was measured as previously described (Vernace et al, 2007). Briefly, tissues were dissected in PBS and homogenized in 25 mM Tris (pH 7.5), and debris cleared by centrifugation. Protein content was measured, and 15 μg used to measure activity by incubating with 12.5 μM Z-Leu-Leu-Glu-AMC (Enzo Life Sciences) at a total volume of 200 μl. AMC fluorescence was measured at 360 nM excitation and 460 nM emission using a spectrofluorimeter (Tecan). Free AMC was used as a standard every 2 min for 30 min.

In-gel proteasome assembly/activity assay

In-gel activity/assembly assay was performed as previously described (Vernace et al, 2007). Briefly, individual tissues were dissected in PBS and homogenized in proteasome buffer (50 mM Tris–HCl, pH 7.4, 5 mM MgCl$_2$, 1 mM ATP, 1 mM DTT and 10% glycerol) on ice, centrifuged at 16,000 g (4°C, 10 min) and equal amounts of proteins loaded on a Bio-Rad TGX 7.5% precast gel. Gels were incubated in proteasome buffer containing 0.4 mM Z-Leu-Leu-Glu-AMC (Enzo Life Sciences) for 15 min (37°C). Proteasome bands were visualized with UV light (360 nm) on a ChemiDoc station

(Bio-Rad). Differential activity was estimate by densitometry using ImageJ.

Western blotting

Western blots were carried out on protein extracts of individual dissected tissues. Proteins were quantified using BCA (Pierce), and equal amounts loaded on Any-KD pre-stained SDS–PAGE gels (Bio-Rad) and blotted according to standard protocols. Antibody dilutions were as follows: K48-poly Ubiquitin (Cell Signaling Technology D9D5 #8081; 1:1,000), Rpt6 (Santa Cruz biotechnology 9E3 #65752; 1:100), NDUFS3 (Abcam 17D95 #14711; 1:1,000). Appropriate secondary antibodies conjugated to horseradish peroxidase were used at a dilution of 1:10,000.

Life span analysis

Once-mated flies, reared at standard densities, were transferred to vials (10-25/vial). Flies were transferred to fresh vials three times a week, and deaths scored on transferal. Standard SYA food (Bass et al, 2007) was used throughout but, where needed, RU486, EtOH (vehicle control) was added to the food at a final concentration of 200 μM, or for bortezomib at 1–3 μm.

Gut integrity assay ("Smurf" assay)

Flies were grown and aged on standard SYA food until 48 h prior to the assay and then switched to SYA food containing 2.5% (w/v) Brilliant Blue FCF (AppliChem). Flies were then examined for "smurfing" and were recorded as "smurf" or not "smurf", as previously described (Rera et al, 2012; Vizcaíno et al, 2014), except flies were kept on dyed food for 48 h.

Quantitative real-time PCR

Total RNA was extracted using Trizol (Invitrogen Corp.) according to the manufacturer's instructions, including a DNase treatment. cDNA was prepared using SuperScript III first-strand synthesis kit (Invitrogen Corp.). Quantitative real-time PCR was performed in a 7900HT real-time PCR system (Applied Biosystems). Relative expression (fold induction) was calculated using the $\Delta\Delta C_T$ method and Rpl32 or Actin5c as a normalization control.

Statistical analysis

Statistical analysis was performed using Graphpad Prism and JMP. Individual statistical tests are mentioned in appropriate figure legends. Lifespan assays were recorded using Excel and survival was analysed using log rank test.

Acknowledgements

Stocks from the Bloomington Drosophila Stock Center (NIH P40OD018537) were used in this study. We would also like to acknowledge Ilian Atanassov and the Proteomic facility at the Max Planck Institute for Biology of Ageing for help during MS analysis. We acknowledge funding from the Max Planck Society (to LT, RS, CJ, MR, SG, JF, NN, MM, LP), a Wellcome Trust Strategic Award (WT098565/Z/12/Z) (to LP), the European Union's Framework Programme for Research and Innovation Horizon (2014–2020) under the Marie Sklodowska-Curie Grant Agreement no. 655623 (to PE), a Bundesministerium für Bildung und Forschung Grant SyBACol 0315893A-B (to AB, CD, MC, LT, LP). We also acknowledge funding form Biotechnology and Biological Sciences Research Council (NA: BB/M029093/1) (to NA), the Medical Research Council (NA: MR/L018802/1) (to NA) and the Royal Society (NA: RG140694) (to NA). CD was supported by the Klaus Tschira Stiftung gGmbH [00.219b.2013]. The research leading to these results has received funding from the European Research Council under the European Union's Seventh Framework Programme (FP7/2007-2013)/ERC grant agreement number 268739 (to LP).

Author contributions

LST, LP designed the experiments. LST, CJ, NN, PE, MR, SG, and JF performed the experiments. LST, RS, MC, NN, CD, MM, NA, AB, and LP designed the data analysis. LST, RS, and MC performed the analysis. LST, NA, and LP wrote the manuscript.

References

Afschar S, Toivonen JM, Hoffmann JM, Tain LS, Wieser D, Finlayson AJ, Driege Y, Alic N, Emran S, Stinn J, Froehlich J, Piper MD, Partridge L (2016) Nuclear hormone receptor DHR96 mediates the resistance to xenobiotics but not the increased lifespan of insulin-mutant *Drosophila*. *Proc Natl Acad Sci USA* 113: 1321–1326

Alexa A, Rahnenführer J, Lengauer T (2006) Improved scoring of functional groups from gene expression data by decorrelating GO graph structure. *Bioinformatics* 22: 1600–1607

Alic N, Andrews TD, Giannakou ME, Papatheodorou I, Slack C, Hoddinott MP, Cocheme HM, Schuster EF, Thornton JM, Partridge L (2011) Genome-wide dFOXO targets and topology of the transcriptomic response to stress and insulin signalling. *Mol Syst Biol* 7: 502–518

Alic N, Tullet JM, Niccoli T, Broughton S, Hoddinott MP, Slack C, Gems D, Partridge L (2014) Cell-nonautonomous effects of dFOXO/DAF-16 in aging. *Cell Rep* 6: 608–616

Amador-Noguez D, Dean A, Huang W, Setchell K, Moore D, Darlington G (2007) Alterations in xenobiotic metabolism in the long-lived Little mice. *Aging Cell* 6: 453–470

Aradska J, Bulat T, Sialana FJ, Birner-Gruenberger R, Erich B, Lubec G (2015) Gel-free mass spectrometry analysis of *Drosophila melanogaster* heads. *Proteomics* 15: 3356–3360

Azimifar SB, Nagaraj N, Cox J, Mann M (2014) Cell-type-resolved quantitative proteomics of murine liver. *Cell Metab* 20: 1076–1087

Bai H, Kang P, Tatar M (2012) *Drosophila* insulin-like peptide-6 (dilp6) expression from fat body extends lifespan and represses secretion of *Drosophila* insulin-like peptide-2 from the brain. *Aging Cell* 11: 978–985

Bailey TL, Boden M, Buske FA, Frith M, Grant CE, Clementi L, Ren J, Li WW, Noble WS (2009) MEME SUITE: tools for motif discovery and searching. *Nucleic Acids Res* 37: W202–W208

Baker DJ, Betik AC, Krause DJ, Hepple RT (2006) No decline in skeletal muscle oxidative capacity with aging in long-term calorically restricted rats: effects are independent of mitochondrial DNA integrity. *J Gerontol A Biol Sci Med Sci* 61: 675–684

Barrett LW, Fletcher S, Wilton SD (2012) Regulation of eukaryotic gene expression by the untranslated gene regions and other non-coding elements. *Cell Mol Life Sci* 69: 3613–3634

Barzilai N, Huffman DM, Muzumdar RH, Bartke A (2012) The critical role of metabolic pathways in aging. *Diabetes* 61: 1315–1322

Bass TM, Grandison RC, Wong R, Martinez P, Partridge L, Piper MDW (2007) Optimization of dietary restriction protocols in *Drosophila*. *J Gerontol A Biol Sci Med Sci* 62: 1071–1081

Batista PJ, Chang HY (2013) Long noncoding RNAs: cellular address codes in development and disease. *Cell* 152: 1298–1307

Bischof J, Maeda RK, Hediger M, Karch F, Basler K (2007) An optimized transgenesis system for *Drosophila* using germ-line-specific phiC31 integrases. *Proc Natl Acad Sci USA* 104: 3312–3317

Blüher M, Kahn BB, Kahn CR (2003) Extended longevity in mice lacking the insulin receptor in adipose tissue. *Science* 299: 572–574

Bonawitz ND, Chatenay-Lapointe M, Pan Y, Shadel GS (2007) Reduced TOR signaling extends chronological life span via increased respiration and upregulation of mitochondrial gene expression. *Cell Metab* 5: 265–277

Bratic A, Larsson N-G (2013) The role of mitochondria in aging. *J Clin Invest* 123: 951–957

Broughton SJ, Slack C, Alic N, Metaxakis A, Bass TM, Driege Y, Partridge L (2010) DILP-producing median neurosecretory cells in the *Drosophila* brain mediate the response of lifespan to nutrition. *Aging Cell* 9: 336–346

Brunner E, Ahrens CH, Mohanty S, Baetschmann H, Loevenich S, Potthast F, Deutsch EW, Panse C, de Lichtenberg U, Rinner O, Lee H, Pedrioli PGA, Malmstrom J, Koehler K, Schrimpf S, Krijgsveld J, Kregenow F, Heck AJR, Hafen E, Schlapbach R et al (2007) A high-quality catalog of the *Drosophila melanogaster* proteome. *Nat Biotechnol* 25: 576–583

Buch S, Melcher C, Bauer M, Katzenberger J, Pankratz MJ (2008) Opposing effects of dietary protein and sugar regulate a transcriptional target of *Drosophila* insulin-like peptide signaling. *Cell Metab* 7: 321–332

Cech TR, Steitz JA (2014) The noncoding RNA revolution-trashing old rules to forge new ones. *Cell* 157: 77–94

Chen Q, Thorpe J, Dohmen JR, Li F, Keller JN (2006) Ump1 extends yeast lifespan and enhances viability during oxidative stress: central role for the proteasome? *Free Radic Biol Med* 40: 120–126

Chondrogianni N, Petropoulos I, Franceschi C, Friguet B, Gonos ES (2000) Fibroblast cultures from healthy centenarians have an active proteasome. *Exp Gerontol* 35: 721–728

Copeland JM, Cho J, Lo T Jr, Hur JH, Bahadorani S, Arabyan T, Rabie J, Soh J, Walker DW (2009) A mitochondrial ATP synthase subunit interacts with TOR signaling to modulate protein homeostasis and lifespan in *Drosophila*. *Curr Biol* 19: 1591–1598

Cox J, Mann M (2008) MaxQuant enables high peptide identification rates, individualized p.p.b.-range mass accuracies and proteome-wide protein quantification. *Nat Biotechnol* 26: 1367–1372

Cox J, Neuhauser N, Michalski A, Scheltema RA, Olsen JV, Mann M (2011) Andromeda: a peptide search engine integrated into the MaxQuant environment. *J Proteome Res* 10: 1794–1805

Cox J, Hein MY, Luber CA, Paron I, Nagaraj N, Mann M (2014) Accurate proteome-wide label-free quantification by delayed normalization and maximal peptide ratio extraction, termed MaxLFQ. *Mol Cell Proteomics* 13: 2513–2526

Deelen J, Uh H-W, Monajemi R, van Heemst D, Thijssen PE, Böhringer S, van den Akker EB, de Craen AJM, Rivadeneira F, Uitterlinden AG, Westendorp RGJ, Goeman JJ, Slagboom PE, Houwing-Duistermaat JJ, Beekman M (2013) Gene set analysis of GWAS data for human longevity highlights the relevance of the insulin/IGF-1 signaling and telomere maintenance pathways. *Age* 35: 235–249

Demontis F, Perrimon N (2010) FOXO/4E-BP signaling in *Drosophila* muscles regulates organism-wide proteostasis during aging. *Cell* 143: 813–825

Deshmukh AS, Murgia M, Nagaraj N, Treebak JT, Cox J, Mann M (2015) Deep proteomics of mouse skeletal muscle enables quantitation of protein isoforms, metabolic pathways, and transcription factors. *Mol Cell Proteomics* 14: 841–853

Dillin A, Hsu A-L, Arantes-Oliveira N, Lehrer-Graiwer J, Hsin H, Fraser AG, Kamath RS, Ahringer J, Kenyon C (2002) Rates of behavior and aging specified by mitochondrial function during development. *Science* 298: 2398–2401

Diop SB, Bisharat-Kernizan J, Birse RT, Oldham S, Ocorr K, Bodmer R (2015) PGC-1/spargel counteracts high-fat-diet-induced obesity and cardiac lipotoxicity downstream of TOR and brummer ATGL lipase. *Cell Rep* 10: 1572–1584

Essers P, Tain LS, Nespital T, Goncalves J, Froehlich J, Partridge L (2016) Reduced insulin/insulin-like growth factor signaling decreases translation in *Drosophila* and mice. *Sci Rep* 6: 30290

Evans MJ, Scarpulla RC (1990) NRF-1: a trans-activator of nuclear-encoded respiratory genes in animal cells. *Genes Dev* 4: 1023–1034

Ewald CY, Landis JN, Porter Abate J, Murphy CT, Blackwell TK (2015) Dauer-independent insulin/IGF-1-signalling implicates collagen remodelling in longevity. *Nature* 519: 97–101

Finan B, Clemmensen C, Zhu Z, Stemmer K, Gauthier K, Müller L, De Angelis M, Moreth K, Neff F, Perez-Tilve D, Fischer K, Lutter D, Sánchez-Garrido MA, Liu P, Tuckermann J, Malehmir M, Healy ME, Weber A, Heikenwalder M, Jastroch M et al (2016) Chemical hybridization of glucagon and thyroid hormone optimizes therapeutic impact for metabolic disease. *Cell* 167: 843–850.e14

Fontana L, Partridge L, Longo VD (2010) Extending healthy life span-from yeast to humans. *Science* 328: 321–326

García-Cáceres C, Quarta C, Varela L, Gao Y, Gruber T, Legutko B, Jastroch M, Johansson P, Ninkovic J, Yi C-X, Le Thuc O, Szigeti-Buck K, Cai W, Meyer CW, Pfluger PT, Fernandez AM, Luquet S, Woods SC, Torres-Alemán I, Kahn CR et al (2016) Astrocytic insulin signaling couples brain glucose uptake with nutrient availability. *Cell* 166: 867–880

Gershman B, Puig O, Hang L, Peitzsch RM, Tatar M, Garofalo RS (2007) High-resolution dynamics of the transcriptional response to nutrition in *Drosophila*: a key role for dFOXO. *Physiol Genomics* 29: 24–34

Giannakou ME (2004) Long-lived *Drosophila* with overexpressed dFOXO in adult fat body. *Science* 305: 361

Grabbe C, Husnjak K, Dikic I (2011) The spatial and temporal organization of ubiquitin networks. *Nat Rev Mol Cell Biol* 12: 295–307

Graveley BR, May G, Brooks AN, Carlson JW, Cherbas L, Davis CA, Duff M, Eads B, Landolin J, Sandler J, Wan KH, Andrews J, Brenner SE, Cherbas P, Gingeras TR, Hoskins R, Kaufman T, Celniker SE (2011) The developmental transcriptome of *Drosophila melanogaster*. *Nature* 471: 473–479

Gronke S, Clarke DF, Broughton S, Andrews TD, Partridge L (2010) Molecular evolution and functional characterization of *Drosophila* insulin-like peptides. *PLoS Genet* 6: e1000857

Halaschek-Wiener J, Khattra JS, McKay S, Pouzyrev A, Stott JM, Yang GS, Holt RA, Jones SJM, Marra MA, Brooks-Wilson AR, Riddle DL (2005) Analysis of long-lived *C. elegans* daf-2 mutants using serial analysis of gene expression. *Genome Res* 15: 603–615

Hansen M, Taubert S, Crawford D, Libina N, Lee S-J, Kenyon C (2007) Lifespan extension by conditions that inhibit translation in *Caenorhabditis elegans*. *Aging Cell* 6: 95–110

Harman D (1956) Aging: a theory based on free radical and radiation chemistry. *J Gerontol* 11: 298–300

He Y-H, Lu X, Yang L-Q, Xu L-Y, Kong Q-P (2014) Association of the insulin-like growth factor binding protein 3 (IGFBP-3) polymorphism with longevity in Chinese nonagenarians and centenarians. *Aging (Albany NY)* 6: 944–956

Hebert SL, Marquet-de Rougé P, Lanza IR, McCrady-Spitzer SK, Levine JA,

Middha S, Carter RE, Klaus KA, Therneau TM, Highsmith EW, Nair KS (2015) Mitochondrial aging and physical decline: insights from three generations of women. *J Gerontol A Biol Sci Med Sci* 70: 1409–1417

Ingolia NT, Ghaemmaghami S, Newman JRS, Weissman JS (2009) Genome-wide analysis *in vivo* of translation with nucleotide resolution using ribosome profiling. *Science* 324: 218–223

Jiang H, Patel PH, Kohlmaier A, Grenley MO, McEwen DG, Edgar BA (2009) Cytokine/Jak/Stat signaling mediates regeneration and homeostasis in the *Drosophila* midgut. *Cell* 137: 1343–1355

Johnson SC, Dong X, Vijg J, Suh Y (2015) Genetic evidence for common pathways in human age-related diseases. *Aging Cell* 14: 809–817

Kaletsky R, Lakhina V, Arey R, Williams A, Landis J, Ashraf J, Murphy CT (2016) The *C. elegans* adult neuronal IIS/FOXO transcriptome reveals adult phenotype regulators. *Nature* 529: 92–96

Kenyon C, Chang J, Gensch E, Rudner A, Tabtiang R (1993) A *C. elegans* mutant that lives twice as long as wild-type. *Nature* 366: 461–464

Kenyon CJ (2010) The genetics of ageing. *Nature* 464: 504–512

Kim M-S, Pinto SM, Getnet D, Nirujogi RS, Manda SS, Chaerkady R, Madugundu AK, Kelkar DS, Isserlin R, Jain S, Thomas JK, Muthusamy B, Leal-Rojas P, Kumar P, Sahasrabuddhe NA, Balakrishnan L, Advani J, George B, Renuse S, Selvan LDN et al (2014) A draft map of the human proteome. *Nature* 509: 575–581

Kruegel U, Robison B, Dange T, Kahlert G, Delaney JR, Kotireddy S, Tsuchiya M, Tsuchiyama S, Murakami CJ, Schleit J, Sutphin G, Carr D, Tar K, Dittmar G, Kaeberlein M, Kennedy BK, Schmidt M (2011) Elevated proteasome capacity extends replicative lifespan in *Saccharomyces cerevisiae*. *PLoS Genet* 7: e1002253

Lê S, Josse J, Husson F (2008) FactoMineR: an R package for multivariate analysis. *J Stat Softw* 25: 1–18

Lecker SH, Goldberg AL, Mitch WE (2006) Protein degradation by the ubiquitin-proteasome pathway in normal and disease states. *J Am Soc Nephrol* 17: 1807–1819

Li X, Monks B, Ge Q, Birnbaum MJ (2007) Akt/PKB regulates hepatic metabolism by directly inhibiting PGC-1α transcription coactivator. *Nature* 447: 1012–1016

Li JJ, Bickel PJ, Biggin MD (2014) System wide analyses have underestimated protein abundances and the importance of transcription in mammals. *PeerJ* 2: e270

Liu Y, Beyer A, Aebersold R (2016) On the dependency of cellular protein levels on mRNA abundance. *Cell* 165: 535–550

Lopez-Otin C, Blasco MA, Partridge L, Serrano M, Kroemer G (2013) The hallmarks of aging. *Cell* 153: 1194–1217

MacInnes AW (2016) The role of the ribosome in the regulation of longevity and lifespan extension. *Wiley Interdiscip Rev RNA* 7: 198–212

Mann M, Kulak NA, Nagaraj N, Cox J (2013) The coming age of complete, accurate, and ubiquitous proteomes. *Mol Cell* 49: 583–590

McElwee JJ, Schuster E, Blanc E, Piper MD, Thomas JH, Patel DS, Selman C, Withers DJ, Thornton JM, Partridge L, Gems D (2007) Evolutionary conservation of regulated longevity assurance mechanisms. *Genome Biol* 8: R132

Mootha VK, Lindgren CM, Eriksson K-F, Subramanian A, Sihag S, Lehar J, Puigserver P, Carlsson E, Ridderstråle M, Laurila E, Houstis N, Daly MJ, Patterson N, Mesirov JP, Golub TR, Tamayo P, Spiegelman B, Lander ES, Hirschhorn JN, Altshuler D et al (2003) PGC-1alpha-responsive genes involved in oxidative phosphorylation are coordinately downregulated in human diabetes. *Nat Genet* 34: 267–273

Mukherjee S, Duttaroy A (2013) Spargel/dPGC-1 is a new downstream effector in the insulin-TOR signaling pathway in *Drosophila*. *Genetics* 195: 433–441

Murali T, Pacifico S, Yu J, Guest S, Roberts GG, Finley RL (2011) DroID 2011: a comprehensive, integrated resource for protein, transcription factor, RNA and gene interactions for *Drosophila*. *Nucleic Acids Res* 39: D736–D743

Murata S, Yashiroda H, Tanaka K (2009) Molecular mechanisms of proteasome assembly. *Nat Rev Mol Cell Biol* 10: 104–115

Murphy CT, McCarroll SA, Bargmann CI, Fraser A, Kamath RS, Ahringer J, Li H, Kenyon C (2003) Genes that act downstream of DAF-16 to influence the lifespan of *Caenorhabditis elegans*. *Nature* 424: 277–283

Murphy CT, Lee S-J, Kenyon C (2007) Tissue entrainment by feedback regulation of insulin gene expression in the endoderm of *Caenorhabditis elegans*. *Proc Natl Acad Sci USA* 104: 19046–19050

Nagaraj N, Kulak NA, Cox J, Neuhauser N, Mayr K, Hoerning O, Vorm O, Mann M (2012) System-wide perturbation analysis with nearly complete coverage of the yeast proteome by single-shot ultra HPLC runs on a bench top Orbitrap. *Mol Cell Proteomics* 11: M111.013722

Nandi A, Kitamura Y, Kahn CR, Accili D (2004) Mouse models of insulin resistance. *Physiol Rev* 84: 623–647

Niccoli T, Partridge L (2012) Ageing as a risk factor for disease review. *Curr Biol* 22: R741–R752

Oh SW, Mukhopadhyay A, Dixit BL, Raha T, Green MR, Tissenbaum HA (2006) Identification of direct DAF-16 targets controlling longevity, metabolism and diapause by chromatin immunoprecipitation. *Nat Genet* 38: 251–257

Pan KZ, Palter JE, Rogers AN, Olsen A, Chen D, Lithgow GJ, Kapahi P (2007) Inhibition of mRNA translation extends lifespan in *Caenorhabditis elegans*. *Aging Cell* 6: 111–119

Park S, Kim W, Tian G, Gygi SP, Finley D (2011) Structural defects in the regulatory particle-core particle interface of the proteasome induce a novel proteasome stress response. *J Biol Chem* 286: 36652–36666

Pathare GR, Nagy I, Bohn S, Unverdorben P, Hubert A, Körner R, Nickell S, Lasker K, Sali A, Tamura T, Nishioka T, Förster F, Baumeister W, Bracher A (2012) The proteasomal subunit Rpn6 is a molecular clamp holding the core and regulatory subcomplexes together. *Proc Natl Acad Sci USA* 109: 149–154

Pérez VI, Buffenstein R, Masamsetti V, Leonard S, Salmon AB, Mele J, Andziak B, Yang T, Edrey Y, Friguet B, Ward W, Richardson A, Chaudhuri A (2009) Protein stability and resistance to oxidative stress are determinants of longevity in the longest-living rodent, the naked mole-rat. *Proc Natl Acad Sci USA* 106: 3059–3064

Poirier L, Shane A, Zheng J, Seroude L (2008) Characterization of the *Drosophila* gene-switch system in aging studies: a cautionary tale. *Aging Cell* 7: 758–770

Puigserver P, Spiegelman BM (2003) Peroxisome proliferator-activated receptor-gamma coactivator 1 alpha (PGC-1 alpha): transcriptional coactivator and metabolic regulator. *Endocr Rev* 24: 78–90

Rappsilber J, Ishihama Y, Mann M (2003) Stop and go extraction tips for matrix-assisted laser desorption/ionization, nanoelectrospray, and LC/MS sample pretreatment in proteomics. *Anal Chem* 75: 663–670

Rask-Madsen C, Kahn CR (2012) Tissue-specific insulin signaling, metabolic syndrome, and cardiovascular disease. *Arterioscler Thromb Vasc Biol* 32: 2052–2059

Rea SL, Ventura N, Johnson TE (2007) Relationship between mitochondrial electron transport chain dysfunction, development, and life extension in *Caenorhabditis elegans*. *PLoS Biol* 5: e259

Rera M, Bahadorani S, Cho J, Koehler CL, Ulgherait M, Hur JH, Ansari WS, Lo T, Jones DL, Walker DW (2011) Modulation of longevity and tissue homeostasis by the *Drosophila* PGC-1 homolog. *Cell Metab* 14: 623–634

Rera M, Clark RI, Walker DW (2012) Intestinal barrier dysfunction links metabolic and inflammatory markers of aging to death in *Drosophila*. *Proc Natl Acad Sci USA* 109: 21528–21533

Ristow M, Schmeisser S (2011) Extending life span by increasing oxidative stress. *Free Radic Biol Med* 51: 327–336

Scarpulla RC (2008) Transcriptional paradigms in mammalian mitochondrial biogenesis and function. *Physiol Rev* 88: 611–638

Schwämmle V, León IR, Jensen ON (2013) Assessment and improvement of statistical tools for comparative proteomics analysis of sparse data sets with few experimental replicates. *J Proteome Res* 12: 3874–3883

Schwanhäusser B, Busse D, Li N, Dittmar G, Schuchhardt J, Wolf J, Chen W, Selbach M (2011) Global quantification of mammalian gene expression control. *Nature* 473: 337–342

Selman C, Tullet JM, Wieser D, Irvine E, Lingard SJ, Choudhury AI, Claret M, Al-Qassab H, Carmignac D, Ramadani F, Woods A, Robinson IC, Schuster E, Batterham RL, Kozma SC, Thomas G, Carling D, Okkenhaug K, Thornton JM, Partridge L et al (2009) Ribosomal protein S6 kinase 1 signaling regulates mammalian life span. *Science* 326: 140–144

Shigenaga MK, Hagen TM, Ames BN (1994) Oxidative damage and mitochondrial decay in aging. *Proc Natl Acad Sci USA* 91: 10771–10778

Slack C, Giannakou ME, Foley A, Goss M, Partridge L (2011) dFOXO-independent effects of reduced insulin-like signaling in *Drosophila*. *Aging Cell* 10: 735–748

Slack C, Alic N, Foley A, Cabecinha M, Hoddinott MP, Partridge L (2015) The Ras-Erk-ETS-signaling pathway is a drug target for longevity. *Cell* 162: 72–83

Sokolova V, Li F, Polovin G, Park S (2015) Proteasome activation is mediated via a functional switch of the Rpt6 C-terminal tail following chaperone-dependent assembly. *Sci Rep* 5: 14909

Stout GJ, Stigter ECA, Essers PB, Mulder KW, Kolkman A, Snijders DS, van den Broek NJF, Betist MC, Korswagen HC, MacInnes AW, Brenkman AB (2013) Insulin/IGF-1-mediated longevity is marked by reduced protein metabolism. *Mol Syst Biol* 9: 679

Suh Y, Atzmon G, Cho M-O, Hwang D, Liu B, Leahy DJ, Barzilai N, Cohen P (2008) Functionally significant insulin-like growth factor I receptor mutations in centenarians. *Proc Natl Acad Sci USA* 105: 3438–3442

Svensen N, Walton JGA, Bradley M (2012) Peptides for cell-selective drug delivery. *Trends Pharmacol Sci* 33: 186–192

Tazearslan C, Huang J, Barzilai N, Suh Y (2011) Impaired IGF1R signaling in cells expressing longevity-associated human IGF1R alleles. *Aging Cell* 10: 551–554

Teleman AA, Hietakangas V, Sayadian AC, Cohen SM (2008) Nutritional control of protein biosynthetic capacity by insulin via Myc in *Drosophila*. *Cell Metab* 7: 21–32

Tepper RG, Ashraf J, Kaletsky R, Kleemann G, Murphy CT, Bussemaker HJ (2013) PQM-1 complements DAF-16 as a key transcriptional regulator of DAF-2-mediated development and longevity. *Cell* 154: 676–690

Tian G, Park S, Lee MJ, Huck B, McAllister F, Hill CP, Gygi SP, Finley D (2011) An asymmetric interface between the regulatory and core particles of the proteasome. *Nat Struct Mol Biol* 18: 1259–1267

Tiefenböck SK, Baltzer C, Egli NA, Frei C (2010) The *Drosophila* PGC-1 homologue Spargel coordinates mitochondrial activity to insulin signalling. *EMBO J* 29: 171–183

Tomko RJ, Hochstrasser M (2013) Molecular architecture and assembly of the eukaryotic proteasome. *Annu Rev Biochem* 82: 415–445

Tonoki A, Kuranaga E, Tomioka T, Hamazaki J, Murata S, Tanaka K, Miura M (2009) Genetic evidence linking age-dependent attenuation of the 26S proteasome with the aging process. *Mol Cell Biol* 29: 1095–1106

Tsakiri EN, Sykiotis GP, Papassideri IS, Gorgoulis VG, Bohmann D, Trougakos IP (2013) Differential regulation of proteasome functionality in reproductive vs. somatic tissues of *Drosophila* during aging or oxidative stress. *FASEB J* 27: 2407–2420

Ungvari Z, Csiszar A, Sosnowska D, Philipp EE, Campbell CM, McQuary PR, Chow TT, Coelho M, Didier ES, Gelino S, Holmbeck MA, Kim I, Levy E, Sonntag WE, Whitby PW, Austad SN, Ridgway I (2013) Testing predictions of the oxidative stress hypothesis of aging using a novel invertebrate model of longevity: the giant clam (*Tridacna derasa*). *J Gerontol A Biol Sci Med Sci* 68: 359–367

Vanunu O, Magger O, Ruppin E, Shlomi T, Sharan R (2010) Associating genes and protein complexes with disease via network propagation. *PLoS Comput Biol* 6: e1000641

Veraksa A, Bauer A, Artavanis-Tsakonas S (2005) Analyzing protein complexes in *Drosophila* with tandem affinity purification-mass spectrometry. *Dev Dyn* 232: 827–834

Vernace VA, Arnaud L, Schmidt-Glenewinkel T, Figueiredo-Pereira ME (2007) Aging perturbs 26S proteasome assembly in *Drosophila melanogaster*. *FASEB J* 21: 2672–2682

Vilchez D, Morantte I, Liu Z, Douglas PM, Merkwirth C, Rodrigues APC, Manning G, Dillin A (2012) RPN-6 determines *C. elegans* longevity under proteotoxic stress conditions. *Nature* 489: 263–268

Vilchez D, Saez I, Dillin A (2014) The role of protein clearance mechanisms in organismal ageing and age-related diseases. *Nat Commun* 5: 1–13

Virbasius JV, Virbasius CA, Scarpulla RC (1993) Identity of GABP with NRF-2, a multisubunit activator of cytochrome oxidase expression, reveals a cellular role for an ETS domain activator of viral promoters. *Genes Dev* 7: 380–392

Vizcaíno JA, Deutsch EW, Wang R, Csordas A, Reisinger F, Ríos D, Dianes JA, Sun Z, Farrah T, Bandeira N, Binz P-A, Xenarios I, Eisenacher M, Mayer G, Gatto L, Campos A, Chalkley RJ, Kraus H-J, Albar JP, Martinez-Bartolomé S et al (2014) ProteomeXchange provides globally coordinated proteomics data submission and dissemination. *Nat Biotechnol* 32: 223–226

Vizcaíno JA, Csordas A, Del-Toro N, Dianes JA, Griss J, Lavidas I, Mayer G, Perez-Riverol Y, Reisinger F, Ternent T, Xu Q-W, Wang R, Hermjakob H (2016) 2016 update of the PRIDE database and its related tools. *Nucleic Acids Res* 44: 11033

Wang D, Cui Y, Jiang Z, Xie W (2014) Knockdown expression of eukaryotic initiation factor 5 C-terminal domain containing protein extends lifespan in *Drosophila melanogaster*. *Biochem Biophys Res Commun* 446: 465–469

Wolkow CA (2000) Regulation of *C. elegans* life-span by insulinlike signaling in the nervous system. *Science* 290: 147–150

Yamamoto R, Tatar M (2011) Insulin receptor substrate chico acts with the transcription factor FOXO to extend *Drosophila* lifespan. *Aging Cell* 10: 729–732

Yen TC, Chen YS, King KL, Yeh SH, Wei YH (1989) Liver mitochondrial respiratory functions decline with age. *Biochem Biophys Res Commun* 165: 944–1003

Zarse K, Schmeisser S, Groth M, Priebe S, Beuster G, Kuhlow D, Guthke R, Platzer M, Kahn CR, Ristow M (2012) Impaired insulin/IGF1 signaling extends life span by promoting mitochondrial L-proline catabolism to induce a transient ROS signal. *Cell Metab* 15: 451–465

Zhang P, Judy M, Lee S-J, Kenyon C (2013) Direct and indirect gene regulation by a life-extending FOXO protein in *C. elegans*: roles for GATA factors and lipid gene regulators. *Cell Metab* 17: 85–100

Zhu LJ, Christensen RG, Kazemian M, Hull CJ, Enuameh MS, Basciotta MD, Brasefield JA, Zhu C, Asriyan Y, Lapointe DS, Sinha S, Wolfe SA, Brodsky MH (2011) FlyFactorSurvey: a database of *Drosophila* transcription factor binding specificities determined using the bacterial one-hybrid system. *Nucleic Acids Res* 39: D111–D117

Zid BM, Rogers AN, Katewa SD, Vargas MA, Kolipinski MC, Lu TA, Benzer S, Kapahi P (2009) 4E-BP extends lifespan upon dietary restriction by enhancing mitochondrial activity in *Drosophila*. *Cell* 139: 149 160

10

Adaptive resistance of melanoma cells to RAF inhibition via reversible induction of a slowly dividing de-differentiated state

Mohammad Fallahi-Sichani[1],* (iD), Verena Becker[1], Benjamin Izar[2,3], Gregory J Baker[1], Jia-Ren Lin[4], Sarah A Boswell[1] (iD), Parin Shah[2] (iD), Asaf Rotem[2] (iD), Levi A Garraway[2,3,5] & Peter K Sorger[1,4,5],** (iD)

Abstract

Treatment of *BRAF*-mutant melanomas with MAP kinase pathway inhibitors is paradigmatic of the promise of precision cancer therapy but also highlights problems with drug resistance that limit patient benefit. We use live-cell imaging, single-cell analysis, and molecular profiling to show that exposure of tumor cells to RAF/ MEK inhibitors elicits a heterogeneous response in which some cells die, some arrest, and the remainder adapt to drug. Drug-adapted cells up-regulate markers of the neural crest (e.g., NGFR), a melanocyte precursor, and grow slowly. This phenotype is transiently stable, reverting to the drug-naïve state within 9 days of drug withdrawal. Transcriptional profiling of cell lines and human tumors implicates a c-Jun/ECM/FAK/Src cascade in de-differentiation in about one-third of cell lines studied; drug-induced changes in c-Jun and NGFR levels are also observed in xenograft and human tumors. Drugs targeting the c-Jun/ECM/FAK/Src cascade as well as BET bromodomain inhibitors increase the maximum effect (E_{max}) of RAF/MEK kinase inhibitors by promoting cell killing. Thus, analysis of reversible drug resistance at a single-cell level identifies signaling pathways and inhibitory drugs missed by assays that focus on cell populations.

Keywords adaptive and reversible drug resistance; *BRAF^V600E* melanomas; de-differentiated NGFR^High state; RAF and MEK inhibitors
Subject Categories Molecular Biology of Disease; Quantitative Biology & Dynamical Systems; Signal Transduction

Introduction

Small-molecule inhibitors of MAP kinases (MAPK), such as RAF inhibitors (e.g., vemurafenib and dabrafenib), MEK inhibitors (e.g., selumetinib and trametinib), or their combination, benefit a majority of melanoma patients whose tumors carry activating V600E/K mutations in the *BRAF* oncogene, but they commonly fail to cure disease due to acquired resistance. Acquired resistance has been shown to involve a diversity of oncogenic mutations in components of the MAPK pathway (Nazarian *et al*, 2010; Poulikakos *et al*, 2011; Wagle *et al*, 2011, 2014; Villanueva *et al*, 2013; Long *et al*, 2014; Van Allen *et al*, 2014; Moriceau *et al*, 2015) or parallel signaling networks such as the PI3K/AKT kinase cascade (Shi *et al*, 2014a,b). In some cases, however, the emergence of drug-resistant clones cannot be fully explained by known genetic mechanisms (Hugo *et al*, 2015). It is thought that genetically distinct, fully drug-resistant clones arise from tumor cells that survive the initial phases of therapy due to drug adaptation (or tolerance) (Emmons *et al*, 2016). Reversible (non-genetic) drug adaptation can be reproduced in cultured cells, and combination therapies that block adaptive mechanisms *in vitro* have shown promise in improving rates and durability of response (Lito *et al*, 2013). Thus, better understanding of mechanisms involved in drug adaptation is likely to improve the effectiveness of melanoma therapy by delaying or controlling acquired resistance.

Adaptation to RAF inhibitors involves cell-autonomous changes such as up-regulation or rewiring of mitogenic signaling cascades as well as non-cell-autonomous changes in the microenvironment such as paracrine signaling from stromal cells (Gopal *et al*, 2010; Lito *et al*, 2012; Abel *et al*, 2013; Hirata *et al*, 2015; Obenauf *et al*, 2015). Understanding these mechanisms is made more complex by variability in adaptive responses from one tumor cell line to the next (Fallahi-Sichani *et al*, 2015). Differences in early adaptive signaling (involving the PI3K/AKT, JNK/c-Jun, and NF-κB networks) exist even among *BRAF^V600E* cell lines with comparably high sensitivity to brief (3–4 days of) vemurafenib treatment (Fallahi-Sichani *et al*, 2015).

1 Department of Systems Biology, Program in Therapeutic Sciences, Harvard Medical School, Boston, MA, USA
2 Department of Medical Oncology, Dana–Farber Cancer Institute, Boston, MA, USA
3 Broad Institute of Harvard and MIT, Cambridge, MA, USA
4 HMS LINCS Center and Laboratory of Systems Pharmacology, Harvard Medical School, Boston, MA, USA
5 Ludwig Center at Harvard, Harvard Medical School, Boston, MA, USA
 *Corresponding author. E-mail: mohammad_fallahisichani@hms.harvard.edu
 **Corresponding author. E-mail: peter_sorger@hms.harvard.edu

There is growing evidence that a reversible drug-tolerant state associated with chromatin modifications (e.g., enhanced histone demethylase activity) can be induced in cancer cells following drug exposure (Sharma *et al*, 2010). In the case of *BRAF*-mutant melanomas, a comparable vemurafenib-tolerant state has been associated with changes in the expression of differentiation markers, including: MITF, a key regulator of melanocyte lineage; NGFR (the low affinity nerve growth factor receptor, also known as p75NTR or CD271), a neural crest marker; and receptor tyrosine kinases such as AXL, EGFR, and PDGFRβ (Johannessen *et al*, 2013; Konieczkowski *et al*, 2014; Muller *et al*, 2014; Sun *et al*, 2014; Ravindran Menon *et al*, 2015; Smith *et al*, 2016; Tirosh *et al*, 2016).

Despite a wealth of data on signaling networks involved in drug adaptation, most of our knowledge comes from studying bulk tumor cell populations. This makes it hard to determine whether proteins involved in adaptation are weakly active in all cells or highly active in a subset of cells. In addition, the phenotypic consequences of drug adaptation (e.g., the emergence of slowly proliferating cells) have primarily been studied using fixed time assays following 1–2 weeks of drug exposure, when drug-adapted cells exhibit high activity in multiple pro-growth signaling cascades (Ravindran Menon *et al*, 2015). It therefore remains unclear how the initial responses to drug relate to subsequent phenotypes such as cell death or adaptation. Continuous-time, single-cell assays are required to tease out these aspects of drug adaptation.

In this paper, we monitor the responses of $BRAF^{V600E}$ melanoma cells to vemurafenib in real time using live-cell imaging and then analyze the resulting cell states using molecular and phenotypic profiling. We find that vemurafenib-treated $BRAF^{V600E}$ cells exhibit a range of fates over the first 3–4 days of drug exposure; a subset of cells undergoes apoptosis, a second subset remains arrested in the G0/G1 phase of the cell cycle, and a third subset enters a slowly cycling drug-resistant state. The slowly cycling resistant state is maintained when cells are grown in the presence of drug, but it is reversible upon 9 days of outgrowth in medium lacking drug, resulting in the regeneration of a population of cells exhibiting the three behaviors of drug-naïve cells. We find that adaptive resistance is associated with de-differentiation along the melanocyte lineage and up-regulation of neural crest markers such as NGFR. These changes can also be detected in naïve and drug-treated patient-matched human tumors by RNA profiling and histopathology. We identify kinase inhibitors and epigenome modifiers (e.g., BET inhibitors) that appear to block acquisition of the slowly cycling NGFR^{High} state in cell lines and in a $BRAF^{V600E}$ melanoma xenograft model and thereby increase sensitivity to vemurafenib. The data and methods used in this paper are freely available and formatted to interchange standards established by the NIH LINCS project (http://www.lincsproject.org/) to promote reuse and enhance reproducibility.

Results

Live-cell imaging and single-cell analysis uncover a slowly cycling drug-resistant state involved in adaptation to RAF inhibitors

To study the dynamics of $BRAF^{V600E}$ inhibition in melanoma cells, we performed live-cell imaging on two vemurafenib-sensitive cell lines at concentrations near the IC$_{50}$ for cell killing (COLO858 and

MMACSF; IC$_{50}$ ~0.1–0.5 μM; we subsequently expanded the analysis to additional lines, as described below). The cells expressed a dual cell cycle reporter (Tyson *et al*, 2012), comprising (i) mCherry-geminin, a protein that is absent during G0/G1, accumulates during S/G2/M, and disappears rapidly late in M phase concurrent with cytokinesis and birth of daughter cells (Sakaue-Sawano *et al*, 2008), and (ii) H2B-Venus, which labels chromatin, allowing mitotic chromosome condensation and disintegration of nuclei during apoptosis to be scored morphologically (Fig 1A and B). Within 24 h of exposure to 1 μM vemurafenib, COLO858 and MMACSF cells were observed to accumulate in the G0/G1 phase of the cell cycle (with low mCherry-geminin levels and decondensed chromatin; Fig 1C and Movie EV1). Between t = 24 and 48 h, ~50% of COLO858 and ~40% of MMACSF cells underwent apoptosis and other cells in the population stayed arrested in G0/G1 (Fig 1C–E).

Subsequently, however, the fates of the two cell lines diverged: MMACSF cells remained arrested, but ~20% of COLO858 cells re-entered S phase between 48 and 84 h (highlighted in pink in Fig 1B and C). Single-cell traces of these adapted cells showed that they underwent division every ~65 h, as compared to doubling time of ~24 h for DMSO-treated COLO858 cells; the additional cell division time was spent in G0/G1 (Fig 1F and Appendix Fig S1). Thus, COLO858 cells exhibited a mixed response to vemurafenib within the first 84 h of exposure to 1 μM vemurafenib with ~60% of cells undergoing apoptosis (primarily between 24 and 48 h of drug exposure), ~20% arresting in the G0/G1 phase of the cell cycle, and ~20% exiting arrest and entering a slowly dividing state.

Exposure of COLO858 cells to 1 μM vemurafenib for an additional ~4 days (for a total of ~8 days) revealed that cells continued to divide slowly in the presence of drug. Approximately half of the cells tracked from day 4 to 8 did not divide at all and the other half divided only once, yielding an estimated doubling time of 48–96 h (Fig 1G). The ratio of cycling versus non-cycling cells in the population appeared to be ~1:1 throughout this period. Thus, slow proliferation and occasional generation of non-dividing cells represent a durable state for COLO858 cells exposed to vemurafenib.

Generation of dividing adapted cells is not explained by MAPK pathway re-activation

Incomplete inhibition or re-activation of the MAPK pathway has been identified as a major cause of adaptive resistance to vemurafenib (Lito *et al*, 2012). This arises because MAPK signaling is a key regulator of proliferation in melanoma cells. We observed that exposure of COLO858 and MMACSF cells to vemurafenib or vemurafenib plus trametinib (a potent and selective inhibitor of MEK kinase) inhibited p-ERK$^{T202/Y204}$ levels by up to ~15-fold relative to untreated cells; p-ERK$^{T202/Y204}$ is a marker of MAPK pathway activity that can be scored by single-cell imaging. Levels of p-ERK$^{T202/Y204}$ in drug-treated cells did not significantly change during 24–72 h of treatment suggesting no recovery of MAPK signaling within this period (Fig 2A and B, and Appendix Fig S2A and B). We conclude that COLO858 cells that re-enter the cell cycle do not re-activate the MAPK pathway. To investigate this further, we co-stained cells for p-ERK$^{T202/Y204}$ and the proliferation markers p-Rb$^{S807/811}$ and Ki-67. These epitopes are present at high levels during S/G2/M, but absent in G0/G1 phases of the cell cycle. No significant difference in p-ERK$^{T202/Y204}$ levels was observed between

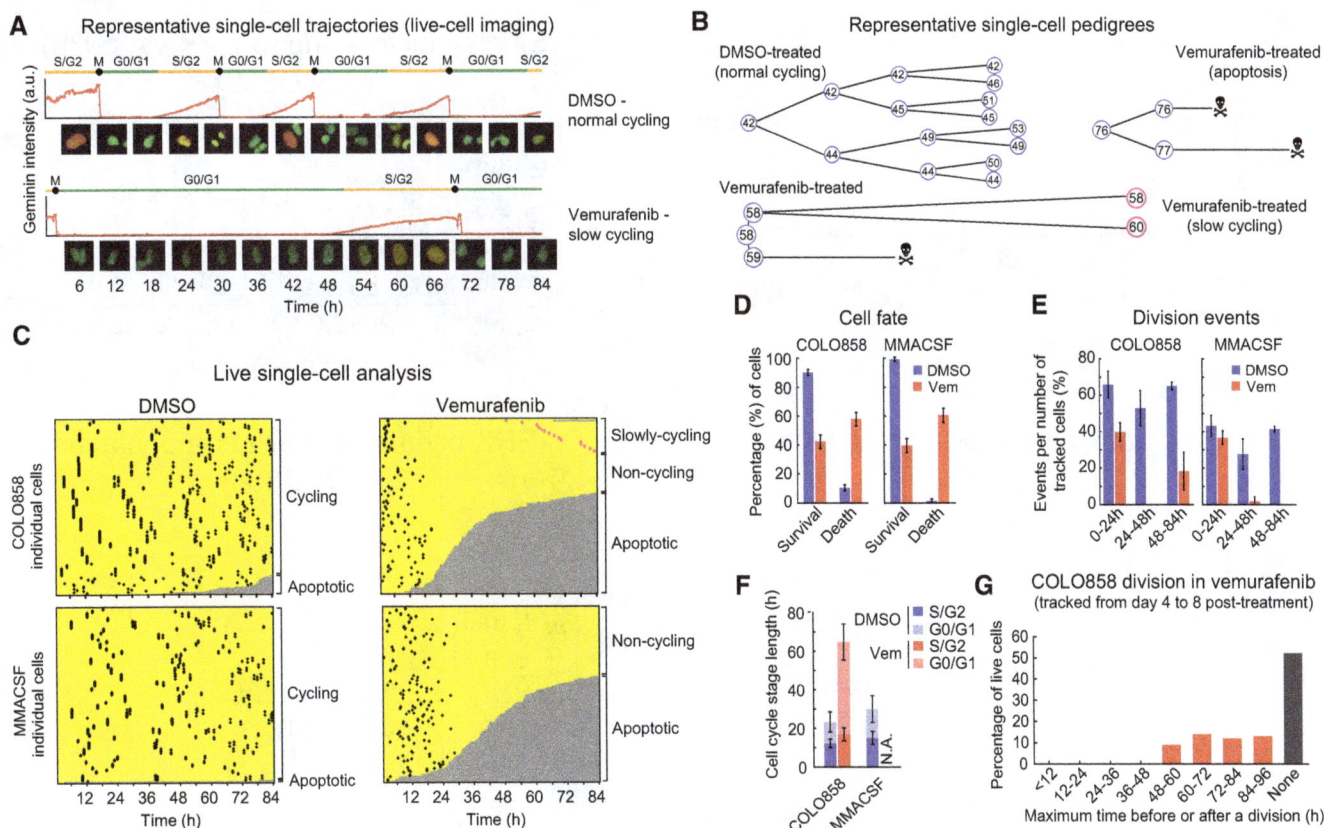

Figure 1.   Live-cell imaging uncovers a slowly cycling drug-resistant state involved in adaptation to RAF inhibition.

Time-lapse imaging of COLO858 and MMACSF cells stably expressing H2B-Venus and mCherry-geminin exposed to 1 μM vemurafenib or DMSO for 4–8 days.

A, B   Representative images and cell cycle phases (A) and representative maps of cell lineage (B) are depicted for COLO858 under DMSO and vemurafenib conditions.

C   Single-cell analysis of division and death events. Horizontal axes represent single-cell tracks with time. Division events are displayed as black or pink (in the case of slowly cycling cells) dots. Transition from yellow to gray indicates cell death.

D   Percentage of surviving and dead cells among cells tracked for 84 h.

E   Percentage of division events among cells tracked during indicated time intervals.

F   Length of different cell cycle phases (G0/G1 and S/G2) in cells tracked for 84 h. No data are reported for MMACSF-vemurafenib because all cells stopped dividing ~24 h after treatment and no single cell divided more than once.

G   Division times for COLO858 cells tracked between days 4 and 8 post-treatment with 1 μM vemurafenib. Minimum doubling times were estimated for 100 individual cells by identifying the longest time interval before or after which a cell divides.

Data information: Data in (D–F) are presented as mean ± SD using 3–4 groups of cells imaged from multiple wells (see Materials and Methods).

---

drug-treated COLO858 cells in G0/G1 (cells that scored as p-Rb^{Low} or Ki-67^{Low}) and cells that had re-entered cell division and were in S/G2/M (cells that scored as p-Rb^{High} or Ki-67^{High}) (Fig 2C and D, and Appendix Fig S2C). We conclude that the re-entry of a subset of vemurafenib-treated COLO858 cells into the cell cycle does not require detectable re-activation of the MAPK pathway, suggesting an adaptation that makes MAPK signaling less essential for proliferation.

Adaptation reduces drug maximal effect and the incremental benefit of raising the drug dose

To better understand how the presence of slowly dividing, drug-adapted cells influences vemurafenib responsiveness at a population level, as conventionally assayed, we performed sequential drug treatment with RAF and MEK inhibitors and then counted surviving cells (Fig 3A). First, COLO858 and MMACSF cells were exposed for

24 h to a dose of vemurafenib below the IC$_{50}$ (0.01–0.32 μM) with the goal of inducing adaptation but minimizing cell death. 1 μM vemurafenib was then added without changing media, and cells were grown for a further 72 h prior to counting the number of viable and apoptotic cells. This protocol resulted in a final total drug concentration of 1.01–1.32 μM, which is above the IC$_{50}$; treatment of cells with DMSO served as a control. We found that pre-treatment of COLO858 cells with vemurafenib reduced cell killing by a second bolus of drug in a dose-dependent manner (Fig 3B): A maximal 2.5-fold increase in the number of surviving cells was observed following 0.1 μM drug pre-treatment (relative to DMSO pre-treated cells). No such effect was observed in MMACSF cells, in which cell viability fell monotonically with increasing total drug concentration. When the concentration of vemurafenib in the second bolus of drug was varied across a range of 0–5 μM, as a means of analyzing dose–response relationships, we observed a significant reduction in E$_{max}$ (the fraction of cells arrested or killed at maximum drug dose) and a

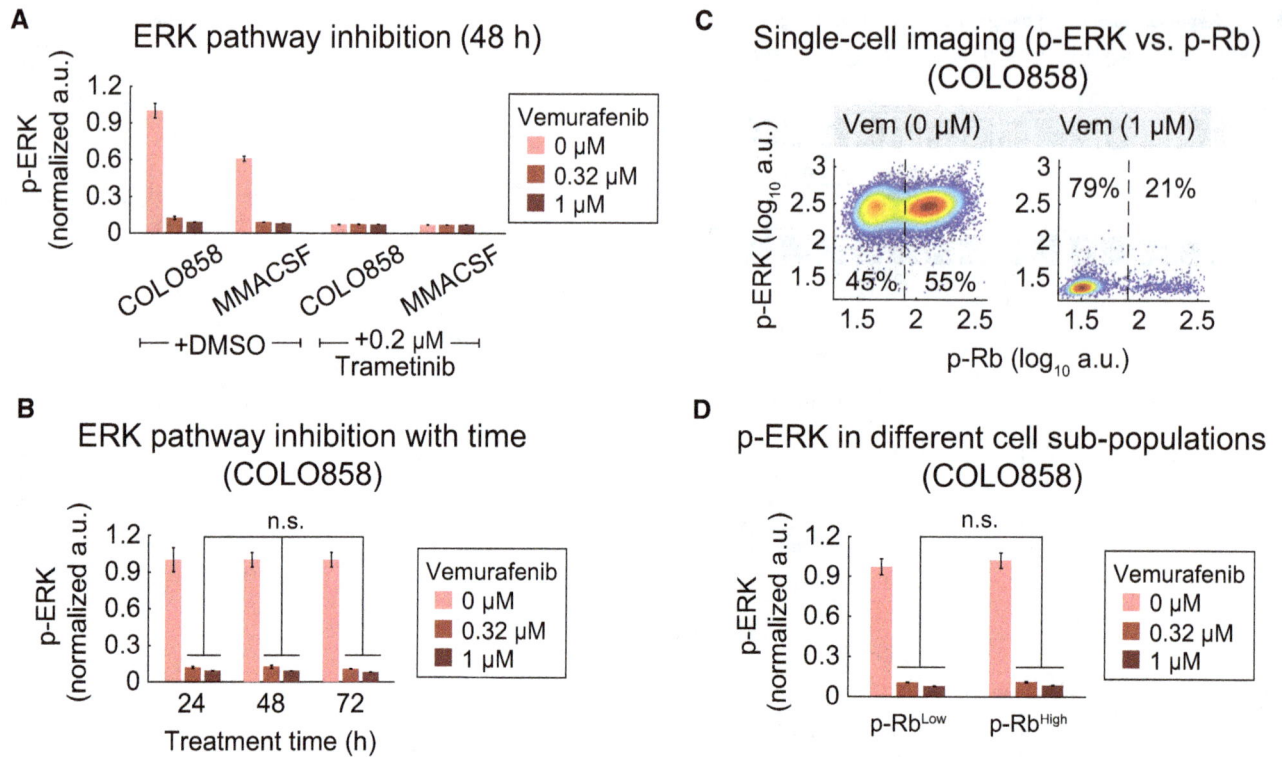

Figure 2. Drug adaptation is not explained by MAPK pathway re-activation.

A   p-ERK$^{T202/Y204}$ levels as measured in duplicate by immunofluorescence in COLO858 and MMACSF cells treated for 48 h with vemurafenib in combination with DMSO or trametinib at indicated doses.

B   p-ERK$^{T202/Y204}$ variation with time (24, 48, 72 h) in COLO858 cells treated in duplicate with vemurafenib at indicated doses.

C   Covariate single-cell analysis of p-Rb$^{S807/811}$ versus p-ERK in COLO858 cells 72 h after exposure to indicated doses of vemurafenib. Vertical dashed lines were used to gate p-Rb^{High} versus p-Rb^{Low} cells.

D   Mean p-ERK levels in p-Rb^{High} versus p-Rb^{Low} subpopulations of COLO858 cells treated in duplicate with indicated doses of vemurafenib for 72 h.

Data information: Data in (A, B, D) are presented as mean ± SD and are normalized to DMSO-treated COLO858 cells at each time point. Statistical significance was determined by two-way ANOVA.

decrease in Hill slope for COLO858 but not MMACSF cells (Fig EV1A and B). Sequential dosing experiments were also repeated using cells pre-treated with a 1:1 molar ratio of vemurafenib plus trametinib. Pre-treatment of COLO858 with RAF/MEK inhibitor combination also led to subsequent resistance (Figs 3B and EV1C and D). We conclude that pre-treatment of COLO858 with sublethal doses of RAF inhibitor or a RAF/MEK inhibitor combination has a significant effect on subsequent drug response primarily by lowering maximal effect (E_{max}) and reducing the incremental effect of rising drug concentration (Hill slope).

Drug-adapted, slowly cycling cells up-regulate genes associated with a de-differentiated NGFR^{High} state

To identify genes associated with acquisition of the slowly cycling, vemurafenib-adapted state, we performed RNA sequencing (RNA-seq) on COLO858 cells exposed to drug for 24 and 48 h; drug-treated MMACSF cells served as a control. Genes differentially expressed in COLO858 or MMACSF cells relative to DMSO-treated controls were selected based on a statistical cutoff of $q < 0.01$. Among these genes, we focused on the subset differing in degree of enrichment by twofold or more between the two cell lines (see

Materials and Methods). Genes enriched in vemurafenib-treated COLO858 cells relative to MMACSF cells comprised 479 up- and 646 down-regulated genes at 24 h and 853 up- and 713 down-regulated genes at 48 h (Fig 4A and Dataset EV1). The top GO terms included neural differentiation, neurogenesis, and cytoskeleton regulation (Fig 4B and Dataset EV2): Genes involved in these processes were enriched in COLO858 cells and reduced or unchanged in MMACSF cells. For example, mRNA for NGFR (UniProtKB: P08138), a neural crest marker, increased ~15-fold in vemurafenib-treated COLO858 cells, representing one of the highest fold-changes in the dataset ($q = 3 \times 10^{-4}$); in contrast, NGFR fell ~fourfold in MMACSF cells ($q = 5 \times 10^{-4}$). It has previously been show that cells expressing NGFR represent an intermediate in the process by which melanocytes differentiate from neural crest cells (Mica *et al*, 2013) and NGFR is used clinically as a histopathological marker to distinguish desmoplastic melanomas, tumors that are negative for conventional melanocytic markers, from other skin neoplasms (Lazova *et al*, 2010). In addition to promoting NGFR expression in COLO858 cells, vemurafenib exposure led to up-regulation of neurogenesis genes such as *S100B*, *CNTN6*, *L1CAM*, *FYN*, *MAP2*, and *NCAM1*, further evidence that cells acquire a less differentiated, more neural crest-like state (Fig 4A). The expression of genes involved in cell cycle

A

Sequential drug treatments

24 h pre-treatment with
low dose of drug or vehicle

⬇

Subsequent treatment at
multiple doses for 72 h

DMSO

Vemurafenib

DMSO

Vemurafenib
+ Trametinib (1:1)

Vemurafenib
(0-5 µM)

Vemurafenib
(0-5 µM)

Vem + Tram
(0-5 µM, 1:1)

Vem + Tram
(0-5 µM, 1:1)

B

Measuring adaptation to pre-treatment by viability assays

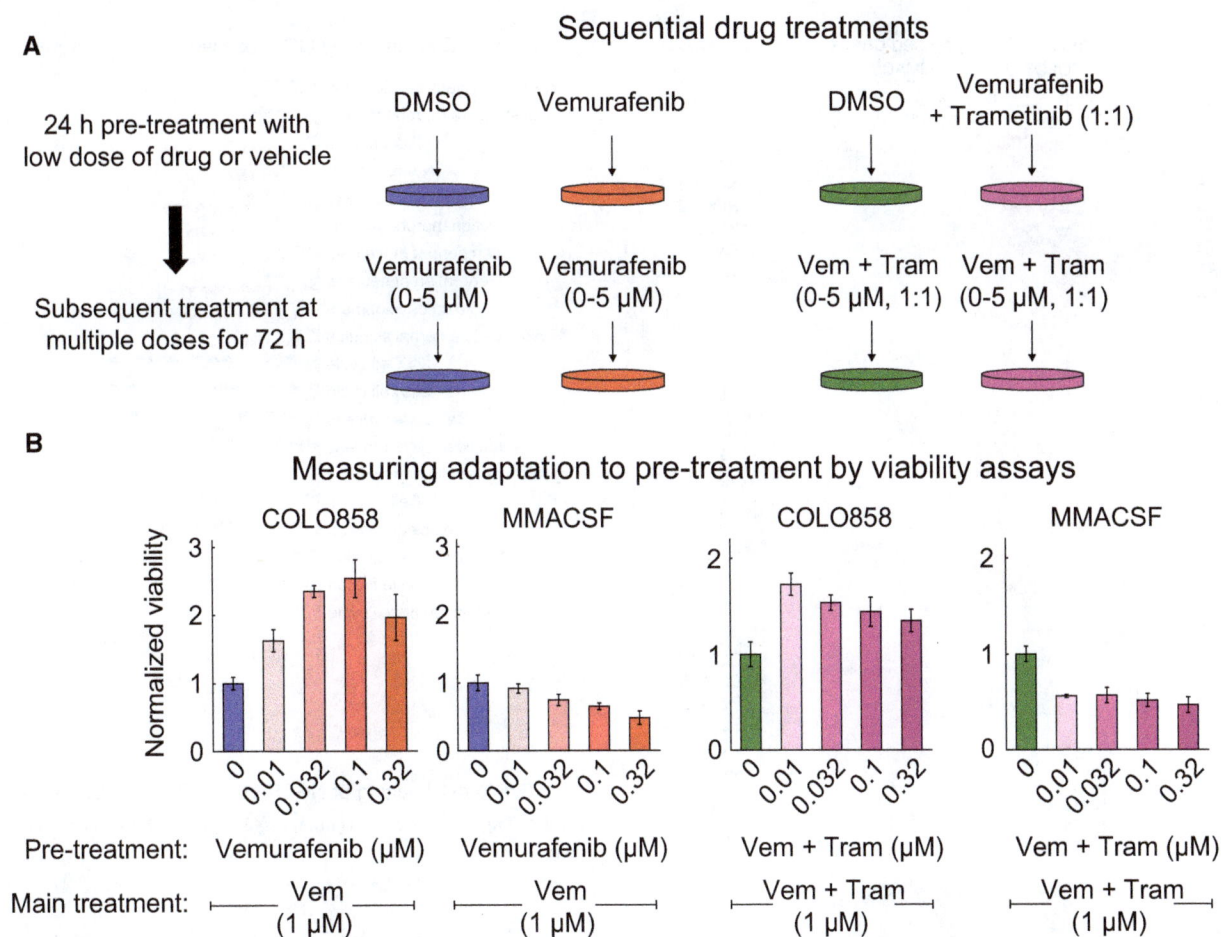

Pre-treatment: Vemurafenib (µM)   Vemurafenib (µM)   Vem + Tram (µM)   Vem + Tram (µM)

Main treatment: Vem (1 µM)   Vem (1 µM)   Vem + Tram (1 µM)   Vem + Tram (1 µM)

Figure 3. Sequential drug treatments reveal adaptive resistance to RAF and MEK inhibitors.

A Schematic outline of a two-stage drug treatment experiment to measure the impact of drug adaptation on the response of COLO858 and MMACSF cells to RAF/MEK inhibitors.

B Viability of cells pre-treated with DMSO, multiple doses of vemurafenib, or vemurafenib plus trametinib, and then treated with an additional 1 µM vemurafenib or vemurafenib plus trametinib (without change of media). Cell viability measured in four replicates was normalized to the viability of cells pre-treated with DMSO. Data are presented as mean ± SD.

progression also changed upon drug exposure: In both COLO858 and MMACSF cells, cell cycle genes were down-regulated by 24 h, but in COLO858, they rose again to their original levels by 48 h (Fig 4A and B), consistent with live-cell imaging data showing that COLO858 cells transiently arrest and then re-enter the cell cycle.

To follow changes in NGFR protein levels in control and drug-treated cells, we co-stained for NGFR and the proliferation marker Ki-67; NGFR protein levels were low in drug-naïve COLO858 cells and increased up to ~sevenfold by 48 h and ~25-fold by 72 h of vemurafenib exposure, consistent with mRNA data. In contrast, NGFR levels fell in MMACSF cells following 48–72 h in drug (Fig 4C). Time-course studies of COLO858 cells helped to reveal how drug-adapted NGFR^{High} cells arose. Within 24 h of vemurafenib treatment, the population of cells shifted from a largely (> 80%) proliferative Ki-67^{High}/NGFR^{Low} state to a non-mitotic Ki-67^{Low}/NGFR^{Low} state (Fig 4D). Non-mitotic cells then up-regulated NGFR, acquiring a Ki-67^{Low}/NGFR^{High} state by t = 48 h, after which they gradually re-entered the cell cycle. By t = 72 h, > 90% of cells in the population were NGFR^{High}. Among these NGFR^{High} cells, ~40%

eventually became Ki-67^{High}, showing that they had begun to proliferate. We conclude that a subset of cells exposed to vemurafenib transiently exits the cell cycle and induces an adaptive response that makes them drug-resistant and NGFR^{High}; cells subsequently re-enter the cell cycle and proliferate slowly.

Vemurafenib-induced de-differentiation of cells and adaptive resistance are reversible upon drug removal

To determine whether the NGFR^{High}, drug-adapted state is reversible, we exposed COLO858 cells to vemurafenib (at 0.32 µM) for 48 h and then isolated NGFR^{High} and NGFR^{Low} cells by fluorescence-activated cell sorting (FACS). These cell populations differed in NGFR levels ~fourfold on average (Fig 5A and Appendix Fig S3). FACS-sorted cells were allowed to grow in fresh medium, and samples were fixed every 24 h and the levels of NGFR and Ki-67 expression measured by immunofluorescence (Fig 5B). We observed that NGFR levels progressively increased in the NGFR^{Low} pool and fell in the NGFR^{High} pool, so that by day 9 average receptor

A Differentially expressed genes

COLO858   MMACSF

Time: 24 h   48 h   24 h   48 h

Replicate: (I) (II)   (I) (II)   (I) (II)   (I) (II)

CLDN1, TUBA1A, SPARC, PLAT, JUN, TGFBR2, ACTA2

FYN, NCAM1, ROCK2, BCL2, FN1, TPM1

MET, CALD1, SNAI2, L1CAM, MMP15, S100B, MAP2, TGFB2, MYL6, COL3A1, NGFR, CNTN6

$\log_2$ fold change in (FPKM+1)

CCND1, CDC25A, POLA2, E2F8, CCNA2, CDC25C, MCM2, MCM3, CDT1, GMNN, POLA1, E2F2, E2F7

CCNE2, CDC20, BUB1, AURKA, CCNB1, CCNB2, CDK1, PLK1, E2F1, CDC6, AURKB

B Enrichment by GO processes

Nervous system development
Anatomical structure morphogenesis
Locomotion
Neurogenesis
Cell development
Neuron projection morphogenesis
Cellular component movement
Generation of neurons
Axon development
Response to external stimulus
Cell cycle
Mitotic cell cycle
Cell cycle process
Mitotic cell cycle process
Single-organism organelle organization
Cell division
Nuclear division
Mitotic nuclear division
Organelle fission
Regulation of cell cycle

24 h only
48 h only
Both 24 and 48 h

$-\log_{10}$ (FDR)

C NGFR protein expression

COLO858   MMACSF

NGFR (a.u.)

Vemurafenib (μM)

Treatment time: 48 h, 72 h

D Single-cell imaging of NGFR vs. Ki-67 in COLO858 cells

No drug   24 h Vem (1 μM)   48 h Vem (1 μM)   72 h Vem (1 μM)

NGFR ($\log_{10}$ a.u.)

Ki-67 ($\log_{10}$ a.u.)

<1%   1%   48%   56%
<1%   1%   11%   35%
84%   47%   8%   3%

NGFR / Ki-67

Change of cell state with time

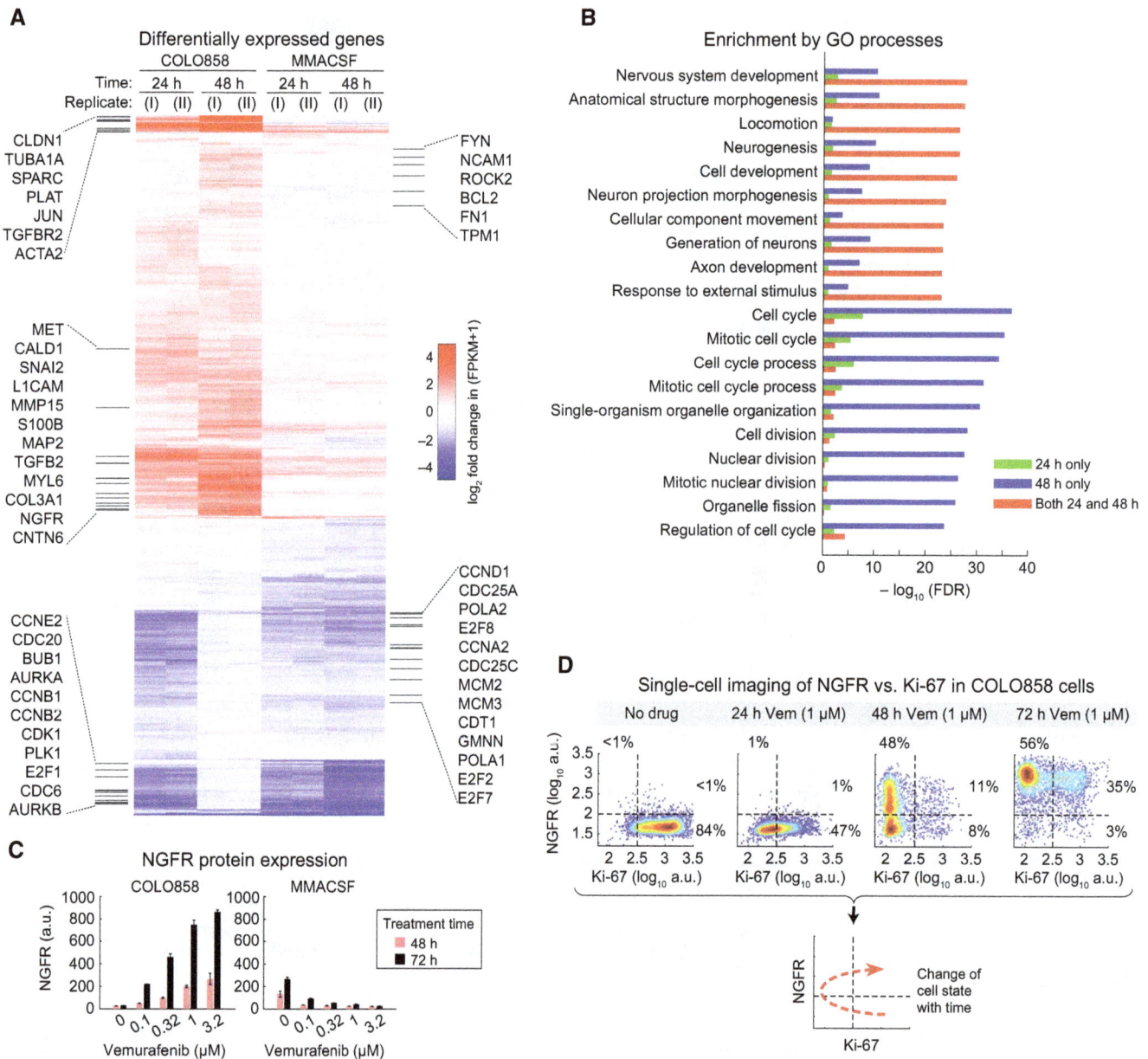

Figure 4. Drug resistance is associated with de-differentiation of cells to a slowly cycling NGFR^High phenotype.

A Differentially up-regulated genes in COLO858 relative to MMACSF cells treated with 0.2 μM vemurafenib for 24 and 48 h ($\log_2$ (ratio) ≥ 1). Selected genes involved in neurogenesis, neural differentiation and myelination (red), cell adhesion, ECM remodeling and epithelial–mesenchymal transition (brown), and cell cycle regulation (blue) are highlighted.

B Top Gene Ontology (GO) biological processes differentially regulated between COLO858 and MMACSF cells.

C NGFR protein levels measured in duplicate by immunofluorescence in COLO858 and MMACSF cells treated with indicated doses of vemurafenib for 48 or 72 h. Data are presented as mean ± SD.

D Covariate single-cell analysis of Ki-67 versus NGFR in COLO858 cells 24–72 h after exposure to 1 μM vemurafenib or DMSO.

expression levels were indistinguishable in the two pools of cells. Return of NGFR levels to pre-treatment levels was accompanied by an increase in Ki-67 staining showing that rapid proliferation had resumed.

To measure vemurafenib sensitivity, NGFR^High and NGFR^Low cells were exposed to drug, 2 days after sorting (the time required for cells to completely re-adhere), and drug response was measured using a conventional 3-day assay and analyzed using a recently developed "growth rate inhibition" (GR) metric that corrects for differences in cell proliferation rates (Hafner *et al*, 2016). We observed that NGFR^High cells were significantly less sensitive to vemurafenib in comparison with NGFR^Low cells ($P = 6 \times 10^{-5}$) (Fig 5C). In contrast, when cells from NGFR^High and NGFR^Low pools were allowed to grow for 9 days in the absence of drug, responsiveness to vemurafenib was indistinguishable (Fig 5D). Moreover, when proliferation rates were scored in freshly isolated NGFR^High cells (over a 3-day period), the average cell doubling time was ~32 h as compared to ~18 h for cells in the NGFR^{Low} pool. (Fig 5E). Finally, when cells that had undergone one cycle of vemurafenib-induced NGFR up-regulation were allowed to reset to the pre-treatment state by outgrowth in the absence of drug and then re-exposed to vemurafenib for 48 h, NGFR was up-regulated to the same degree as in drug-naïve cells (Fig 5F).

These experiments demonstrate that the subset of vemurafenib-treated COLO858 cells able to acquire a slowly dividing NGFR^{High}

phenotype is more drug-resistant than the subset of cells in the same initial population that remains NGFR^{Low}. As expected, the magnitude of the difference in drug resistance and growth rate observed in studies of FACS-sorted cells was smaller than in live-cell imaging experiments. This is because analysis of sorted cells involves waiting for cells to re-adhere in the absence of drug; during this period, the adapted phenotype relaxes to the pre-treatment state. In contrast, in live-cell studies, cells are continuously exposed to drug and the adapted phenotype is maintained. Overall, we

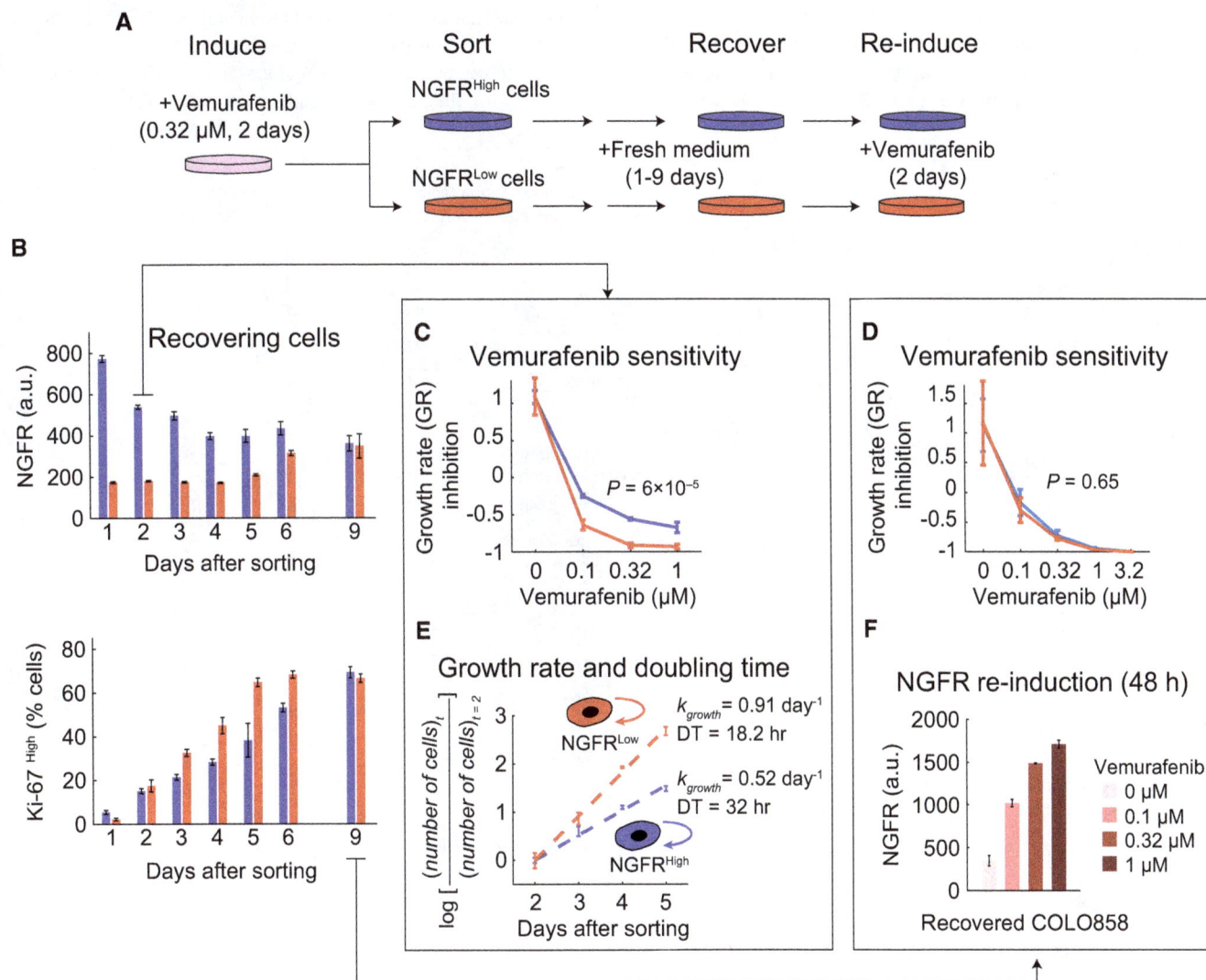

Figure 5. Vemurafenib-induced de-differentiation of cells and adaptive resistance are reversible upon drug removal.

A   Schematic outline of an experiment involving induction of the slowly cycling NGFR^{High} state in COLO858 cells following 48-h treatment with 0.32 µM vemurafenib, sorting cells to obtain NGFR^{Low} and NGFR^{High} subpopulations, recovering each cell subpopulation in fresh growth medium for 1–9 days, and re-inducing recovered cells with vemurafenib.

B   NGFR and Ki-67 protein levels measured by immunofluorescence in cells grown for 9 days in fresh medium (n = 4).

C, D   Growth rate (GR) inhibition assay performed on FACS-sorted NGFR^{High} and NGFR^{Low} pools of cells after 2 (C) or 9 (D) days of outgrowth in fresh medium. Measurements were performed in 4 (C) or 6 (D) replicates.

E   Growth rate and doubling time measurements in 4 replicates in FACS-sorted NGFR^{High} and NGFR^{Low} cells during 2–5 days of outgrowth in fresh medium.

F   NGFR levels measured in duplicate by immunofluorescence in COLO858 cells recovered after 9 days of outgrowth in fresh media and subsequently re-exposed for 48 h to four doses of vemurafenib.

Data information: Data in (B–F) are presented as mean ± SD. Statistical significance was determined by two-way ANOVA.

conclude that the vemurafenib-induced, slowly cycling, NGFR[High] state is transiently stable allowing NGFR[High] and NGFR[Low] cells to inter-convert on a time scale of about a week in culture. Such behavior is inconsistent with a genetic difference between the two populations of cells, but similar to the transiently heritable cell-to-cell variability previously shown to play a role in cellular response to pro-apoptotic ligands (Flusberg *et al*, 2013) and other small-molecule drugs (Cohen *et al*, 2008; Sharma *et al*, 2010).

Induction of an NGFR[High] state involves extracellular matrix (ECM) components, focal adhesion, and the AP1 transcription factor c-Jun

To identify biochemical pathways involved in NGFR up-regulation, we performed pathway enrichment analysis on genes differentially regulated in vemurafenib-treated COLO858 and MMACSF cells. Cell adhesion, ECM remodeling, and epithelial-to-mesenchymal transition (EMT) were among the top enriched pathways (Fig EV2 and Dataset EV2). Genes up-regulated in vemurafenib-treated COLO858 cells included the ECM components thrombospondin-1 (THBS1; TSP-1; UniProtKB: P07996), an adhesive glycoprotein that mediates cell–cell and cell–ECM interactions, the laminin subunits LAMA1 and LAMC1 (UniProtKB: P25391 and P11047), CCN signaling protein NOV (Perbal, 2004) (UniProtKB: P48745), and several integrin family receptors (Fig 6A and B). Gene set enrichment analysis (GSEA) showed that similar molecules and pathways accompany increased NGFR expression in 25 *BRAF*[V600E] melanoma cell lines found in the Cancer Cell Line Encyclopedia (CCLE) and 128 *BRAF*[V600E] melanoma biopsies in The Cancer Genome Atlas (TCGA) (Fig 6C).

To identify potential transcriptional regulators of genes up-regulated in the NGFR[High] state, we used DAVID (http://

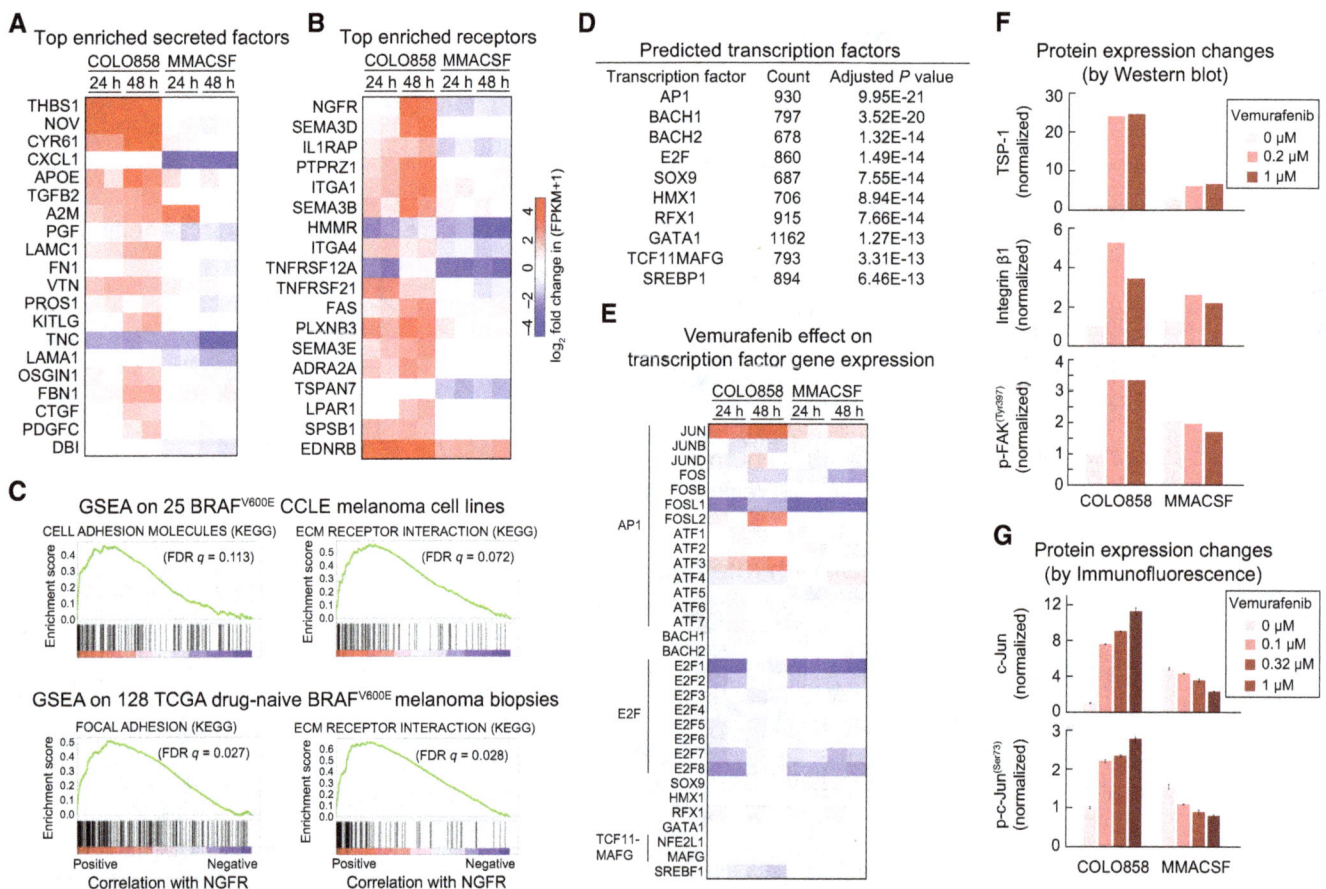

Figure 6. The NGFR[High] state involves extracellular matrix (ECM) components, focal adhesion, and the AP1 transcription factor c-Jun.

A, B   Top differentially regulated genes encoding secreted proteins (A) and cell surface receptors (B) between COLO858 and MMACSF cells.

C   Ranked GSEA plots of top KEGG pathways significantly correlated with NGFR expression in 25 *BRAF*[V600E] melanoma cell lines from the CCLE (top) and tumor biopsies of 128 *BRAF*[V600E] melanoma patients in TCGA (bottom).

D, E   A list of transcription factor candidates predicted (by DAVID; see Materials and Methods) to regulate differentially expressed genes between vemurafenib-treated COLO858 and MMACSF cells (D), and the corresponding transcription factor gene expression levels in these cells (E).

F   Quantified Western blot measurements (see Materials and Methods) for thrombospondin-1 (THBS1; TSP-1), integrin β1, and p-FAK[Y397] in COLO858 and MMACSF cells treated for 48 h with indicated doses of vemurafenib. Data are first normalized to HSP90α/β levels in each cell line at each treatment condition and then to DMSO-treated COLO858 cells.

G   c-Jun and p-c-Jun[S73] changes as measured in duplicate by immunofluorescence in COLO858 and MMACSF cells treated for 48 h with indicated doses of vemurafenib. Data are normalized to DMSO-treated COLO858 cells.

Data information: Data in (F, G) are presented as mean ± SD.

david.abcc.ncifcrf.gov) (Fig 6D) and then examined expression levels for the top 10 transcription factor candidates (Fig 6E). DAVID identified the AP1 family of transcription factors as the top candidates for regulators of the adapted state in COLO858 cells ($P \approx 10^{-20}$) (Fig 6D). Moreover, *JUN*—which encodes the AP1 transcription factor c-Jun—was the most differentially enriched candidate: In COLO858, it was up-regulated ~ninefold within 24 h and ~23-fold within 48 h of exposure to vemurafenib, but changed at most ~twofold in MMACSF cells within 48 h of treatment (Fig 6E). When we focused DAVID and differential expression analysis specifically on receptors and secreted/ECM proteins, AP1 factors and *JUN* were again predicted to be key differential regulators of vemurafenib response in COLO858 and MMACSF cells (Fig EV3A).

To investigate the involvement of ECM proteins and receptors in drug adaptation, we performed Western blotting on extracts from COLO858 and MMACSF cells, focusing on the subset of genes for which antibodies are available: TSP-1, integrin β1 (a subunit of a cell adhesion receptor that binds to TSP-1), and an activating phosphorylation site on the focal adhesion kinase (p-FAK^{Y397}). Following 48 h in 0.2 or 1 μM vemurafenib, p-FAK levels (normalized to the levels of HSP90α/β) increased ~3.5-fold in COLO858 cells relative to DMSO-treated cells, whereas they fell slightly (~25%) in MMACSF cells (Figs 6F and EV3B). TSP-1 levels (also normalized to HSP90α/β levels) increased by ~25-fold in COLO858 cells but only ~twofold in MMACSF cells. Integrin β1 was induced > fivefold in COLO858 cells, but only ~twofold in MMACSF cells. Both c-Jun and p-c-Jun^{S73} (the active state of the protein) were substantially elevated by vemurafenib treatment (up to ~12-fold and ~threefold, respectively) in COLO858 cells but down-regulated (by ~50%) in MMACSF cells (Fig 6G). Thus, differences detected at the level of mRNA were reflected in the levels and activities of the corresponding proteins.

To obtain functional data on proteins implicated in the drug-adapted state, we depleted *JUN* or *PTK2* (the FAK gene) in COLO858 cells by siRNA. Depletion of either gene significantly reduced vemurafenib-induced NGFR up-regulation (by ~70%) and increased sensitivity to vemurafenib as compared to cells transfected with control siRNA (Fig 7A). NGFR knockdown, however, did not reduce cell viability nor did exposure of cells to NGF, an NGFR ligand (Fig EV3C). Thus, NGFR appears to be a marker of the vemurafenib-resistant cell state rather than a regulator of drug resistance. In contrast, siRNA experiments directly implicate c-Jun and FAK in drug adaptation.

Concurrent inhibition of RAF/MEK signaling and a putative c-Jun/FAK/Src cascade overcomes vemurafenib resistance in NGFR^{High} cells

To begin to identify signaling proteins involved in acquisition of an NGFR^{High}, drug-adapted state, we exposed COLO858 cells to vemurafenib in combination with a range of small-molecule kinase inhibitors, including defactinib and PF562271, two compounds that target FAK; JNK-IN-8, a selective inhibitor of c-Jun N-terminal kinases (JNK); and dasatinib and saracatinib, two inhibitors of Src family non-receptor tyrosine kinases that function downstream of FAK (and other receptor tyrosine kinases); see Fig 7 for details on drug dosing and nominal target information. When COLO858 cells

were treated with these drugs and vemurafenib at 0.32 or 1 μM for 48 h, induction of NGFR was reduced from threefold to fivefold to less than 1.5-fold (and in some cases to levels below those of drug-naïve cells; Fig 7B). In contrast, exposing COLO858 cells to trametinib plus vemurafenib enhanced NGFR induction as compared to vemurafenib alone (Fig 7B). Thus, kinases targeted by the approved drug dasatinib (Sprycel®), investigational drugs defactinib, and saracatinib as well as tool compounds PF562271 and JNK-IN-8 are involved in vemurafenib-induced NGFR up-regulation. Super-induction of NGFR by vemurafenib plus trametinib shows that MAPK signaling is a negative regulator of this process (Appendix Fig S4A).

To better understand the effects of kinase inhibitors on vemurafenib adaptation, we measured NGFR and Ki-67 in single cells by imaging. Assays were performed across doses, and data were z-scored and visualized as a 2D landscape (Fig 7C). This showed that the effects of JNK, FAK, and Src inhibitors in vemurafenib-treated cells (green arrow) were orthogonal to the effects of trametinib (red arrow): Whereas the fraction of NGFR^{High} cells rose as the fraction of Ki-67^{High} cells fell with trametinib, with JNK, FAK, or Src inhibitors both NGFR and Ki-67 were suppressed (Fig 7C). Co-drugging COLO858 cells with trametinib reduced p-ERK$^{T202/Y204}$ levels, but co-drugging with JNK, FAK, or Src inhibitors had no effect on p-ERK$^{T202/Y204}$ levels relative to vemurafenib alone (Fig EV4). We conclude that whereas trametinib acts to enhance both the therapeutic effect of vemurafenib (i.e., inhibition of proliferation) and its counter-therapeutic effect (i.e., induction of the NGFR^{High} state), JNK, FAK, or Src inhibitors have orthogonal activities that reduce drug adaptation.

The biological significance of these findings was confirmed by dose–response studies showing that combining vemurafenib with JNK, FAK, or Src inhibitors increased killing of COLO858 but not MMACSF cells (Fig 7D and Appendix Fig S4B). Moreover, the effectiveness of JNK, FAK, or Src inhibitors rose when MAPK signaling was more fully inhibited by a combination of vemurafenib and trametinib (Fig 7D; right panel). The primary effect of combining MAPK inhibitors with drugs such as dasatinib or saracatinib was an increase in E_{max} and a reduction in the fraction of surviving cells. In Fig 7D, log–log drug dose–response plots highlight that co-drugging primarily affected the 1–10% of $BRAF^{V600E}$ cells that survived MAPK inhibitors; such differences were much less obvious when using log-linear dose–response plots conventionally used to score interaction. We conclude that drugs targeting kinases that lie within pathways identified by gene sent enrichment analysis as up-regulated in vemurafenib-treated COLO858 cells substantially increase cell killing.

A screen identifies chromatin modifications as additional contributors to the NGFR^{High} state

The acquisition of transiently heritable states in drug-treated melanoma cells is reminiscent of reversible drug tolerance previously shown to involve chromatin modifications sensitive to HDAC inhibition (Sharma *et al*, 2010; Ravindran Menon *et al*, 2015). We therefore performed a focused screen with a library of small-molecule inhibitors of epigenome-modifying enzymes to identify those that reduced NGFR induction by vemurafenib. COLO858 cells were exposed for 48 h to a sublethal dose of

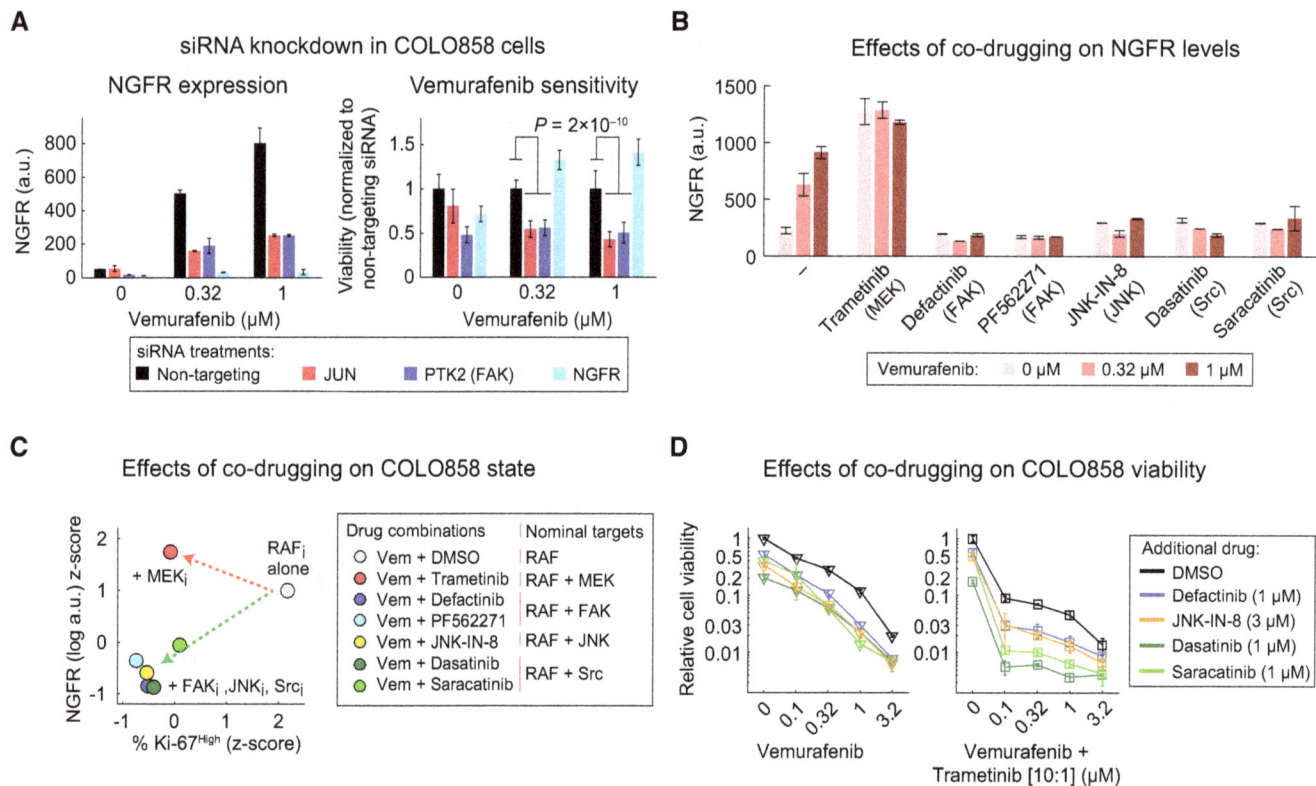

Figure 7. Concurrent inhibition of RAF/MEK signaling and the c-Jun/FAK/Src cascade blocks the NGFR^{High} state and increases cell killing.

A NGFR levels as measured by immunofluorescence (left panel) and relative cell viability (right panel) in COLO858 cells following treatment in duplicate with indicated doses of vemurafenib in the presence of siRNAs targeting JUN, PTK2, and NGFR for 72 h. Viability data for each siRNA condition at each dose of vemurafenib were normalized to cells treated with the same dose and the non-targeting siRNA.

B NGFR protein levels measured by immunofluorescence in duplicate in COLO858 cells treated for 48 h with indicated doses of vemurafenib, in combination with DMSO, MEK inhibitor trametinib (0.6 μM), FAK inhibitors defactinib (3 μM) and PF562271 (3 μM), JNK inhibitor JNK-IN-8 (3 μM), or Src inhibitors dasatinib (3 μM) and saracatinib (3 μM).

C Pairwise comparison between drug combination-induced changes in NGFR and Ki-67 in COLO858 cells treated for 48 h with vemurafenib at 0.32 and 1 μM in combination with DMSO or two doses of trametinib (0.2, 0.6 μM), defactinib (1, 3 μM), PF562271 (1, 3 μM), dasatinib (1, 3 μM), saracatinib (1, 3 μM), and JNK-IN-8 (1, 3 μM). NGFR and Ki-67 levels were measured by immunofluorescence. For each signal, data were averaged across two replicates, two doses of vemurafenib, and two doses of the second drug, log-transformed, and z-score-scaled across seven different drug combinations.

D Relative viability of COLO858 cells treated for 72 h with vemurafenib or vemurafenib plus trametinib (10:1 dose ratio) in combination with DMSO, JNK-IN-8, dasatinib, saracatinib, and defactinib at indicated doses. Viability data were measured in three replicates and normalized to DMSO-treated controls.

Data information: Data in (A, B, D) are presented as mean ± SD. Statistical significance was determined by two-way ANOVA.

vemurafenib (0.32 μM) in combination with one of 41 inhibitors of HDAC, BET, and other chromatin-targeting compounds at three doses (see Materials and Methods for a list of compounds); NGFR levels were measured by imaging (Fig 8A). Three BET bromodomain inhibitors, (+)-JQ1, I-BET, and I-BET151, were found to consistently suppress NGFR up-regulation when combined with vemurafenib in technical and biological replicates (Fig 8A and Appendix Fig S5A). All three compounds also reduced Ki-67 levels (Appendix Fig S5B) and increased killing of COLO858 cells when combined with vemurafenib, as measured by 3-day viability assays and evaluation of drug E_{max} (Fig 8B; left panel). For example, when applied to COLO858 cells, 0.32 μM JQ1 slowed down cell division but was not measurably cytotoxic (Movie EV2) and no more apoptosis was detected than in a DMSO-only control (Fig EV5). However, a combination of 0.32 μM JQ1 and 1 μM vemurafenib increased apoptosis to

> 90% of cells (1 μM vemurafenib alone induced 40% apoptosis under these conditions; Fig EV5). Thus, JQ1 and vemurafenib are synergistic in cell killing by conventional Loewe criteria. JQ1 and the other BET inhibitors were even more effective in promoting cell killing when vemurafenib and trametinib were used in combination (Fig 8B; right panel). Moreover, on a plot of Ki-67 versus NGFR levels, the effects of BET inhibitors (green arrow) were orthogonal to those of trametinib (red arrow), a property shared with JNK, FAK, and Src inhibitors (Fig 8C). BET inhibitors reduced c-Jun up-regulation induced by vemurafenib to a significant degree (by an average of ~50%; $P < 2 \times 10^{-8}$) but did not fully block it (Fig 8D). From these data, we conclude that chromatin modifications are likely to be involved in the acquisition of the transiently heritable, NGFR^{High}, vemurafenib-resistant cell state and that multiple BET inhibitors can block this effect.

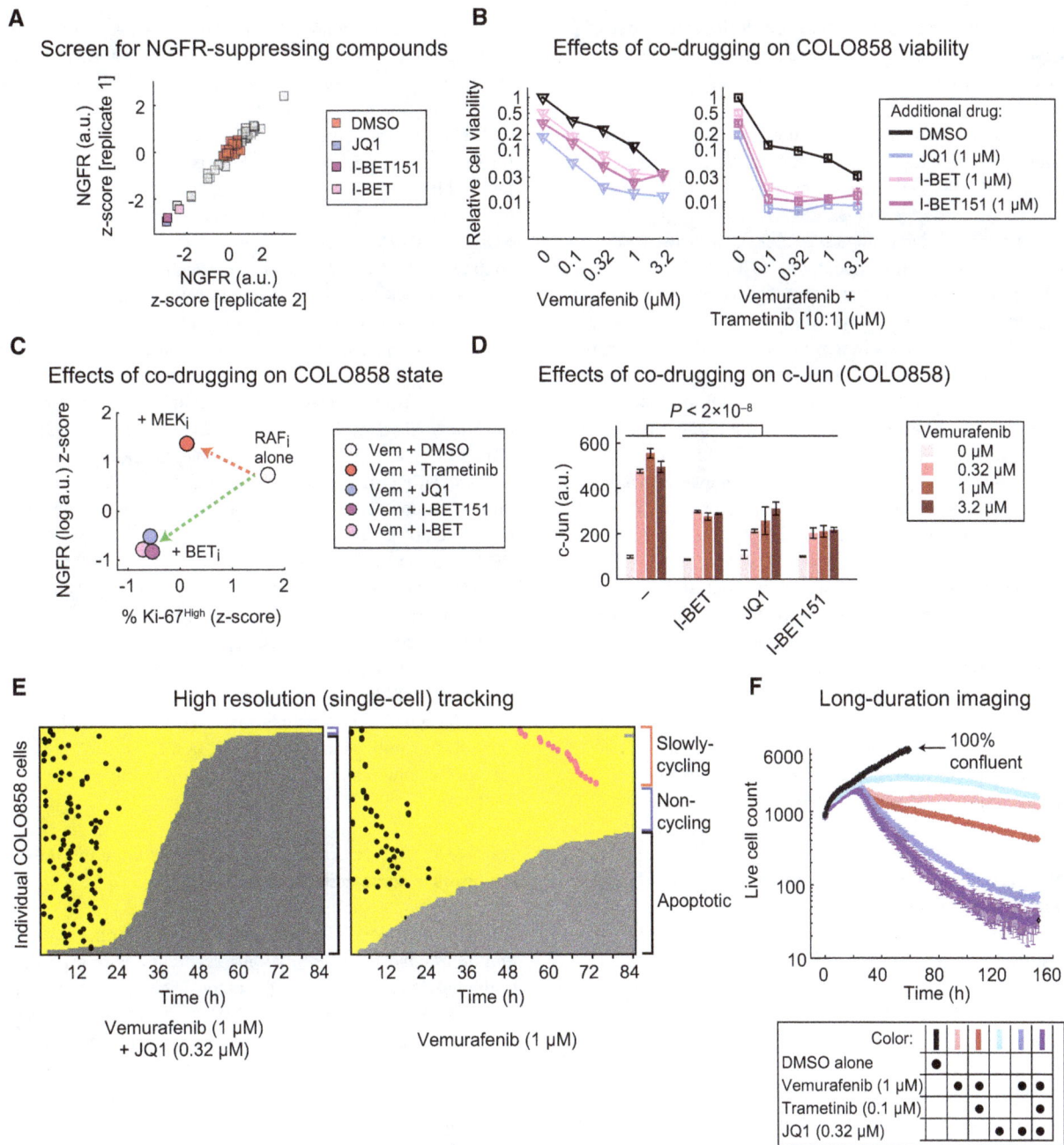

Figure 8. BET inhibitors suppress the slowly cycling NGFR^High state and effectively reduce the cancer cell population with time.

A   COLO858 cells were treated for 48 h in duplicate with vemurafenib (at 0.32 μM) in combination with DMSO or three doses (0.11, 0.53, and 2.67 μM) of each of 41 compounds in a chromatin-targeting library. NGFR protein levels were measured by immunofluorescence, averaged across three doses of each compound, and z-scored.

B   Relative viability of COLO858 cells treated for 72 h with vemurafenib or vemurafenib plus trametinib (10:1 dose ratio) in combination with DMSO, (+)-JQ1, I-BET, and I-BET151 at indicated doses. Viability data were measured in three replicates and normalized to DMSO-treated controls.

C   Pairwise comparison between drug-induced changes in NGFR and Ki-67 in COLO858 cells treated with vemurafenib at 0.32, 1, and 3.2 μM in combination with DMSO or trametinib (0.2 μM), I-BET (1 μM), I-BET151 (1 μM), and (+)-JQ1 (1 μM) for 48 h. Data for each drug combination were averaged across two replicates and three doses of vemurafenib, log-transformed, and z-score-scaled.

D   c-Jun protein levels measured by immunofluorescence in duplicate in COLO858 cells treated for 48 h with indicated doses of vemurafenib, in combination with DMSO, I-BET (1 μM), (+)-JQ1 (1 μM), and I-BET151 (1 μM).

E   Single-cell analysis of division and death events following live-cell imaging of COLO858 cells treated with 1 μM vemurafenib in combination with DMSO or (+)-JQ1 (0.32 μM) for 84 h. Data are presented as described in Figure 1.

F   Time-lapse analysis of COLO858 cells treated in three replicates for ~1 week with different drug combinations at indicated doses. Data for DMSO-treated cells are shown until day 3, the time at which cells reach ~100% confluency.

Data information: Data in (B, D, F) are presented as mean ± SD.

NGFR-suppressing drug combinations block the emergence of slowly cycling cells and effectively reduce the cancer cell population with time

To link the activities of drugs that inhibit induction of NGFR by vemurafenib to the kinetics of cell killing, we performed live-cell imaging of COLO858 cells in the presence of 1 μM vemurafenib alone or in combination with 0.32 μM JQ1. Co-drugging eliminated the emergence of slowly cycling cells (pink) and increased the fraction of cells undergoing apoptosis to > 95% (Fig 8E and Appendix Fig S5C and D, and Movie EV3). We also monitored cell growth every 45 min for ~7 days using a live-cell microscope (an IncuCyte® Live Cell Analysis System) that is placed in an incubator and results in minimal perturbation of growth conditions. In vemurafenib-treated cells (Fig 8F; pink line), both cell division and cell death were observed between 0 and 30 h after which cell number was nearly constant. Co-drugging with trametinib increased cell killing (red), but by the end of one week, the number of viable cells was still ~50% of the initial number ($t = 0$). Exposure of cells to JQ1 alone induced cytostasis with little cell killing (light blue), whereas the combination of vemurafenib plus JQ1 was highly cytotoxic, resulting in continuous cell killing throughout the 7-day assay period (blue); the triple combination of vemurafenib, trametinib, and JQ1 was even more effective (purple). Live-cell analysis of COLO858 cells exposed to combinations of vemurafenib, trametinib, and the FAK inhibitor defactinib yielded comparable findings (Appendix Fig S5C–F). These data show that a drug identified on the basis of its ability to block acquisition of an NGFR^{High} state also blocks the emergence of slowly growing, vemurafenib-adapted cells and, as a consequence, causes a sustained increase in the rate of cell killing.

JNK, FAK, Src, and BET inhibitors overcome the NGFR^{High} state in additional BRAF$^{V600E/D}$ melanoma lines

To investigate the generality of the biology described above, we analyzed seven additional BRAF$^{V600E/D}$ melanoma cell lines. In two of these lines (A375 and WM115), NGFR levels were high in the absence of vemurafenib but increased modestly in a dose-dependent manner following 48-h exposure to 0.1–1 μM vemurafenib (Fig 9A and Appendix Fig S6A–C). Concomitantly, these lines exhibited up to ~sevenfold dose-dependent increase in c-Jun levels (Appendix Fig S6D). The five other lines we examined exhibited no detectable increase in NGFR or c-Jun levels upon exposure to vemurafenib. These findings are consistent with previous data showing that

adaptation to vemurafenib is heterogeneous across cell lines (Fallahi-Sichani et al, 2015), but overall, a statistically significant and positive correlation was observed between vemurafenib-induced c-Jun and NGFR levels (Pearson's $\rho = 0.86$, $P = 0.001$) (Fig 9B).

When we measured the levels of TSP-1, integrin β1, and p-FAK^{Y397} in A375 and WM115 cells, we observed vemurafenib-induced increases in expression and/or high basal levels, in contrast to low basal levels and an absence of induction in drug-treated NGFR^{Low} MZ7MEL cells (Appendix Fig S6E and Fig EV3B). In common with COLO858 cells, co-drugging A375 and WM115 cell lines with JNK-IN-8, dasatinib, saracatinib, defactinib, and either vemurafenib or vemurafenib plus trametinib increased cell killing (and reduced E_{max}), but co-drugging had no significant effect on killing of MZ7MEL cells (Fig 9C and Appendix Fig S6F). When we repeated a focused screen for epigenome-targeting compounds in A375 and WM115 cells, we identified the same three BET inhibitors JQ1, I-BET, and I-BET151 as capable of blocking vemurafenib-induced NGFR up-regulation (Appendix Fig S7A–C). All three of these compounds enhanced cell killing when combined with vemurafenib or vemurafenib plus trametinib (Fig 9D). On a plot of NGFR versus Ki-67 levels, the effects of co-drugging A375 or WM115 cells with vemurafenib and inhibitors of BET proteins, JNK, FAK, or Src were orthogonal to those of co-drugging with trametinib, in all cases reducing the fraction of Ki-67^{High} and NGFR^{High} cells relative to vemurafenib alone but without further reducing p-ERK levels (Fig 9E and F, and Appendix Fig S7D and E). From these data, we conclude that even though basal NGFR levels vary significantly among COLO858, A375, and WM115 cells, all three lines exhibit similar drug adaptation in the presence of MAPK inhibitors.

The NGFR^{High} state is associated with resistance to MAPK inhibitors in some melanoma patients

When tumor biopsies from drug-naïve melanoma patients were immunostained for NGFR, we observed variability from one tumor to the next and, within a single tumor, from one region to the next: NGFR^{High}/MITF^{Low} and NGFR^{Low}/MITF^{High} domains were present in 4/11 samples and the former stained less strongly for Ki-67 (Fig 10A and Appendix Fig S8). We obtained biopsies from a patient prior to the onset of therapy, 2 weeks after initiation of therapy with dabrafenib plus trametinib and subsequent to relapse and then measured NGFR, Ki-67, and c-Jun levels by immunostaining. Relative to the pre-treatment biopsy, the on-treatment biopsy exhibited a reduction in the fraction of Ki-67^{High} cells from ~23% to ~4%,

Figure 9. JNK, FAK, Src, and BET inhibitors overcome the NGFR^{High} drug-resistant state in additional BRAF$^{V600E/D}$ melanoma lines.

A   NGFR protein levels measured in duplicate by immunofluorescence in seven BRAF$^{V600E/D}$ cell lines treated with vemurafenib at indicated doses for 48 h.

B   Correlation between vemurafenib-induced changes in c-Jun and NGFR protein levels across nine BRAF$^{V600E/D}$ melanoma cell lines. Cells were treated with five doses of vemurafenib (0, 0.1, 0.32, 1, and 3.2 μM) for 48 h. c-Jun and NGFR protein levels measured by immunofluorescence at each condition were averaged across two replicates and normalized to DMSO-treated controls. The area under the dose–response curve (AUC) for the two measurements (c-Jun and NGFR) was calculated, z-score-scaled across nine cell lines, and their pairwise Pearson's correlation was reported.

C, D   Relative viability of A375 and WM115 cells treated in 3 replicates for 72 h with vemurafenib or vemurafenib plus trametinib (10:1 dose ratio) in combination with indicated kinase inhibitors (C) or BET inhibitors (D).

E, F   Pairwise comparison between NGFR and Ki-67 levels in A375 and WM115 cells treated with vemurafenib in combination with indicated kinase inhibitors (E) or BET inhibitors (F). Drug doses, time points, and data normalization are similar to Figs 7C and 8C.

Data information: Data in (A, C, D) are presented as mean ± SD.

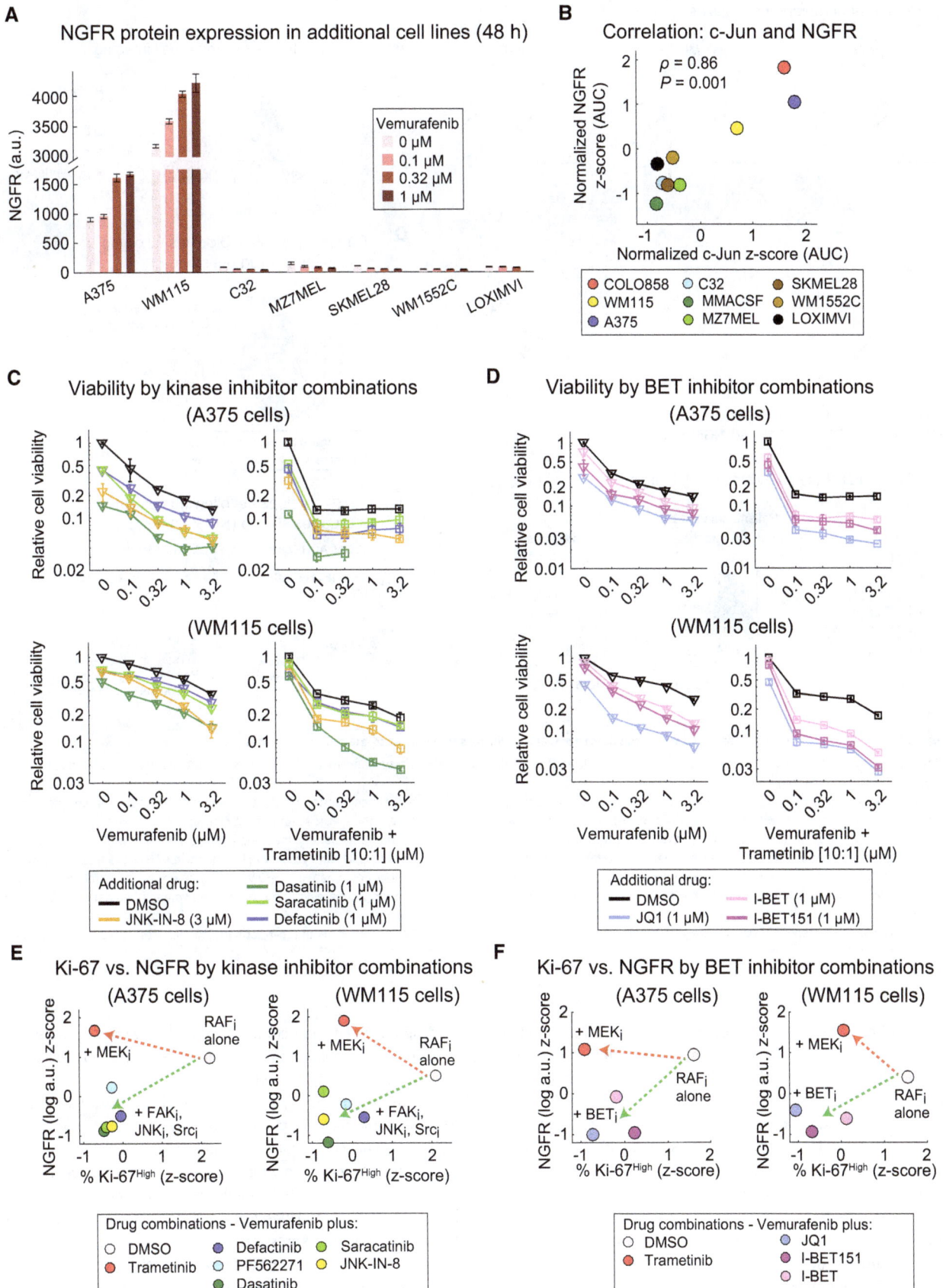

Figure 9.

A Pre-treatment melanoma biopsies

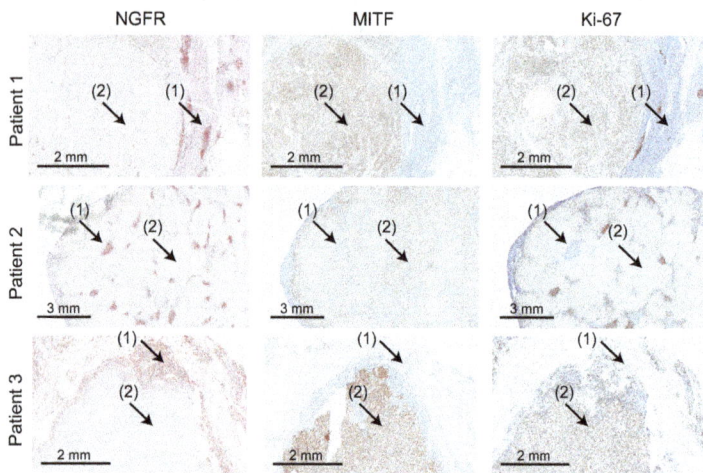

Tumor Domains (by antigen levels):
(1): NGFR^{High}/MITF^{Low}/Ki-67^{Low} (2): NGFR^{Low}/MITF^{High}/Ki-67^{High}

B   Ki-67 vs NGFR levels by single-cell imaging

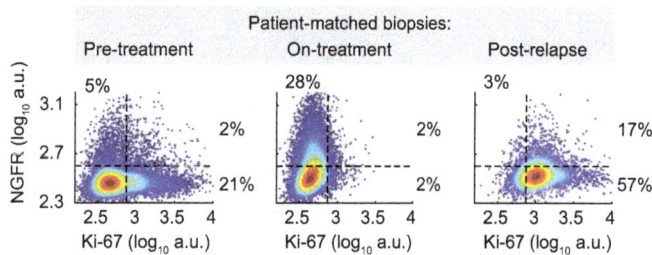

C   c-Jun levels by single-cell imaging

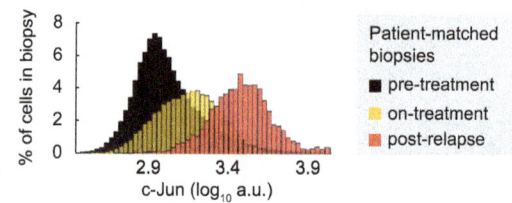

D   Transcript levels in biopsies following acquistion of drug resistance

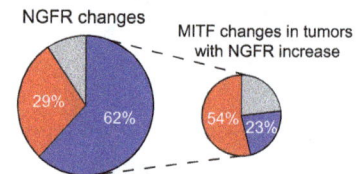

E   GSEA on biopsies following acquistion of drug resistance (KEGG pathways)

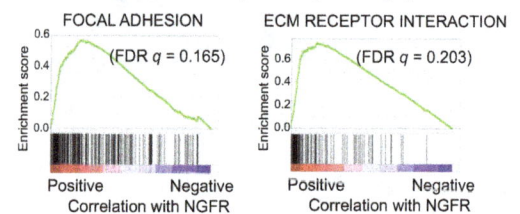

Figure 10.   The NGFR^{High} state is associated with resistance to MAPK inhibitors in a subset of melanoma patients.

A   Immunohistochemical analysis of vemurafenib-naïve tumors from three melanoma patients stained for NGFR, MITF, and Ki-67 (see Materials and Methods for patient clinical information).

B   Covariate single-cell analysis of Ki-67 versus NGFR measured by immunofluorescence in pre-treatment, on-treatment (with dabrafenib and trametinib combination for 2 weeks), and post-relapse tumor biopsies of a *BRAF*-mutant melanoma patient (see Materials and Methods for patient clinical information).

C   Cell population histograms representing c-Jun variations measured by immunofluorescence in the same patient-matched biopsies as shown in (B).

D   NGFR gene expression changes in 21 matched pairs of pre-treatment and post-resistance tumor biopsies analyzed by RNA sequencing. MITF changes are shown for tumors with a post-resistance NGFR increase (increase = log$_2$ (fold-change) > 0.5, decrease = log$_2$ (fold-change) < −0.5, no change = |log$_2$ (fold-change)| ≤ 0.5). Gene expression data from patients treated with RAF inhibitor, MEK inhibitor, or their combination were analyzed by combining two published datasets (Sun *et al*, 2014; Hugo *et al*, 2015).

E   Ranked GSEA plots of top KEGG pathways significantly correlated with NGFR expression in 18 matched pairs of pre-treatment and post-resistance tumor biopsies (Hugo *et al*, 2015).

consistent with the anticipated effects of dabrafenib/trametinib therapy, but the fraction of NGFR^{High} cells increased from ~7% to ~30% (Fig 10B). The distribution of signal intensities across single cells suggested that these changes primarily involved a switch from a Ki-67^{High}/NGFR^{Low} state to a Ki-67^{Low}/NGFR^{High} state following initiation of therapy. This change was associated with an increase in c-Jun expression (Fig 10C; compare yellow and black distributions). Following relapse, the fraction of Ki-67^{High} cells increased dramatically (to ~74%) reflecting re-acquisition of proliferative potential and ~20% of these cells were NGFR^{High}. Relapse was also accompanied by a dramatic increase in c-Jun levels (Fig 10C; red distribution). Although the number of samples in this study is low, the data are consistent with a heterogeneous distribution of NGFR^{High}/Ki-67^{Low} and NGFR^{Low}/Ki-67^{High} domains in melanoma tumors, and

a drug-mediated induction of NGFR^{High}/Ki-67^{Low} state that is concomitant with c-Jun up-regulation, a situation reminiscent of our observations in cultured cells.

To investigate changes in NGFR levels across a larger cohort of *BRAF*-mutant melanoma patients, we analyzed two published RNA-seq datasets involving matched samples from 21 tumors pre-treatment and following emergence of resistance to different combinations of RAF/MEK inhibitors (Sun *et al*, 2014; Hugo *et al*, 2015). In 62% of biopsies from patients with acquired drug resistance, NGFR gene expression increased as compared to pre-treatment levels (Fig 10D). MITF expression fell in only 50% of these biopsies, suggesting that therapy-induced NGFR up-regulation and MITF down-regulation do not necessarily occur concomitantly. GSEA of these tumors identified ECM–receptor interactions and focal adhesion among the top enriched KEGG pathways correlated with NGFR gene expression levels, consistent with data obtained in cell lines (Fig 10E, and Datasets EV3 and EV4). We conclude that melanomas exhibit variability in differentiation status pre- and post-treatment but that acquisition of an NGFR^{High} state is associated with resistance to RAF/MEK-targeted therapy in about half of melanomas examined.

JQ1 suppresses induction of an NGFR^{High} state in $BRAF^{V600E}$ melanoma xenografts

To test whether the NGFR^{High} phenotype can be blocked *in vivo* by drugs identified as effective in cell lines, we analyzed A375 cells grown as xenografts in nude mice. A375 cells are among the most widely used xenograft models for *BRAF*-mutant melanoma. Mice were exposed for 5 days to RAF inhibitor dabrafenib (at a 25 mg/kg dose) alone or in combination with JQ1 (at a 50 mg/kg dose). Four xenograft tumors per condition were excised, fixed, sectioned, and then co-stained for NGFR and Ki-67. Analysis of staining intensity at a single-cell level revealed heterogeneity from one region of tumor to the next and a reciprocal relationship between regions of the tumor that were NGFR^{High} and Ki-67^{High} (Fig 11A), a pattern similar to what was observed in human tumors. Treatment of animals with dabrafenib plus JQ1 significantly reduced the fraction of NGFR^{High} cells as compared to dabrafenib alone (or vehicle-treated controls) and the combination also reduced the fraction of Ki-67^{High} cells relative to dabrafenib or vehicle (Fig 11B). These data mimic two key aspects of what we observed in cultured A375 cells (which have high NGFR levels in the basal state): First, JQ1 can reduce NGFR levels, and second, JQ1 and dabrafenib can combine to reduce the fraction of proliferating Ki-67^{High} cells. Moreover, as these experiments were being conducted, a study was published on tumor burden in mice engrafted with A375 tumors. It showed that JQ1 and vemurafenib have synergistic effects of tumor shrinkage (Paoluzzi *et al*, 2016). Together, these findings establish that the effects of co-drugging with JQ1 and MAPK inhibitors observed in cell culture can also be obtained in xenograft models. This sets the stage for large-scale pre-clinical evaluation of drugs such as BET bromodomain inhibitors as a means of blocking drug adaptation and increasing

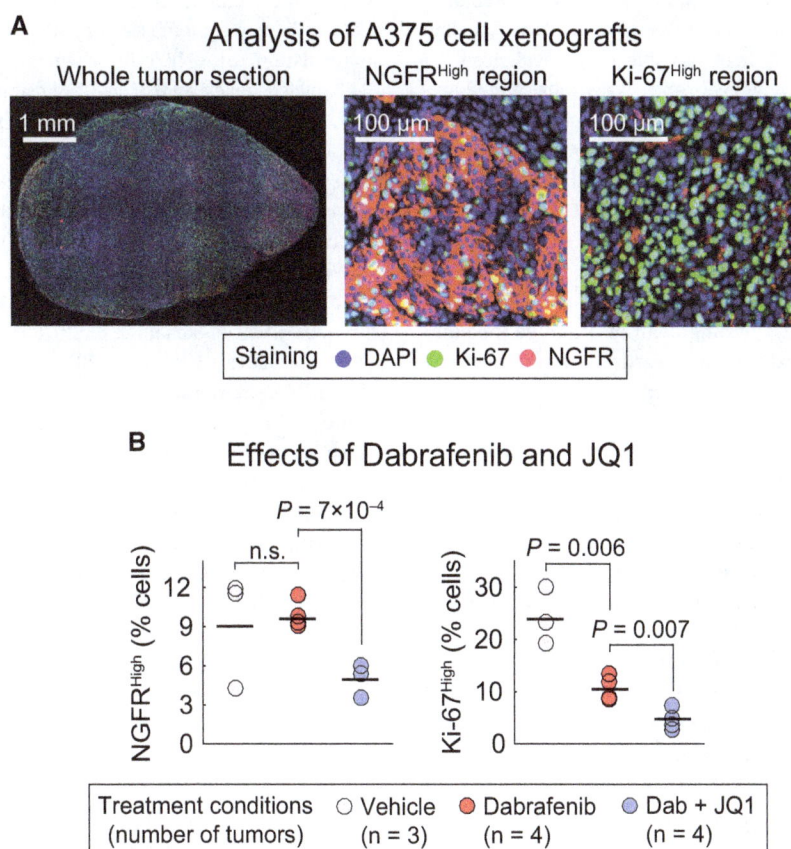

A Analysis of A375 cell xenografts

Whole tumor section   NGFR^{High} region   Ki-67^{High} region

Staining ● DAPI ● Ki-67 ● NGFR

B Effects of Dabrafenib and JQ1

Treatment conditions   ○ Vehicle   ● Dabrafenib   ● Dab + JQ1
(number of tumors)   (n = 3)   (n = 4)   (n = 4)

Figure 11. The NGFR^{High} phenotype can be suppressed by JQ1 in $BRAF^{V600E}$ melanoma xenografts.

A Immunofluorescence analysis of A375 melanoma xenograft tumors co-stained for Ki-67 and NGFR proteins. Selected images from a whole tumor section as well as NGFR^{High}/Ki-67^{Low} and NGFR^{Low}/Ki-67^{High} regions of a vehicle-treated tumor are shown to highlight the spatial and cell-to-cell heterogeneity in Ki-67 and NGFR protein expression.

B Percentage of Ki-67^{High} and NGFR^{High} cells in tumors treated for 5 days with dabrafenib (25 mg/kg) only, dabrafenib (25 mg/kg) in combination with JQ1 (50 mg/kg), or vehicle. Number of tumors (mice) analyzed per condition is shown. Solid horizontal lines represent the mean of measurements. Up to 50,000 individual cells per tumor were analyzed for NGFR and Ki-67 intensities. Statistical significance was determined using two-tailed two-sample *t*-test.

cell killing by MAPK inhibitors in a subset of *BRAF*-mutant melanomas.

Discussion

In this paper, we use time-lapse, live-cell imaging, and single-cell analysis to show that *BRAF*-mutant melanoma cells exhibit time-variable and heterogeneous phenotypes when exposed to MAPK pathway inhibitors such as vemurafenib, dabrafenib, and trametinib near the IC_{50} for cell killing. Cells initially undergo growth arrest, consistent with the known requirement for MAPK activity in proliferation. Apoptosis peaks between 48 and 72 h and typically kills 40–60% of cells, while other cells enter a G0/G1 arrest. In a subset of lines, a subpopulation of cells overcomes drug-mediated cell cycle arrest and begins to divide threefold to fourfold more slowly than drug-naïve cells. Such adapted cells exhibit elevated neural crest markers including NGFR and neurogenesis genes, suggestive of drug-induced de-differentiation and consistent with previous studies associating increased NGFR levels or loss of melanocyte differentiation markers (e.g., MITF) with resistance to MAPK pathway inhibitors (Konieczkowski *et al*, 2014; Muller *et al*, 2014; Ravindran Menon *et al*, 2015). In culture, the generation of slowly cycling $NGFR^{High}$ cells reduces drug maximal effect, as evidenced by short-term (3-day) viability assays and week-long time-lapse imaging. Slowly cycling, drug-adapted cells are likely to contribute to residual disease and eventual emergence of genetically distinct drug-resistant clones (Frick *et al*, 2015; Hata *et al*, 2016).

Reversible drug resistance

The slowly cycling $NFGR^{High}$ state induced by vemurafenib is only transiently stable: After 9 days of outgrowth in drug-free medium, such cells reset to their initial state as measured by restoration of vemurafenib sensitivity, increased proliferation rate, and reduced expression of NGFR. Such metastable, bidirectional changes in cell state are inconsistent with selection of pre-existing genetic variants but are more durable than transiently heritable differences generated by stochastic fluctuation in protein levels (Cohen *et al*, 2008; Gascoigne & Taylor, 2008; Flusberg *et al*, 2013). Instead, the phenomenon is reminiscent of drug-tolerant persisters (DTPs), which constitute < 1% of drug-naïve cell populations, become enriched following exposure to high concentrations of anti-cancer drugs (> 100-fold above IC_{50} values) for > 9 days, and have hyperactive IGF-1R signaling (Sharma *et al*, 2010). Like $NGFR^{High}$ melanoma cells, DTPs are sensitive to some kinase inhibitors and to inhibitors of epigenome-modifying enzymes, HDACs in the case of DTPs, and BET inhibitors in the case of vemurafenib-adapted melanoma cells. However, time-lapse imaging shows that melanoma cells responding to vemurafenib induce a slowly dividing drug-adapted state more rapidly and in a larger fraction of cells (> 20% of cells by 3 days near the vemurafenib IC_{50}) than has been observed for DTPs. We speculate that differences between these phenomena originate primarily from differences in the strength and timing of imposed selective pressures; whereas acutely high doses of drug lead to selection of a small percentage of intrinsically, highly insensitive cells (i.e., DTPs) (Sharma *et al*, 2010; Roesch *et al*, 2013), lower and more realistic drug doses provide a larger fraction of cells with

sufficient time to induce an adaptive mechanism and become drug insensitive; once induced, this resistance appears to protect cells from higher doses of drug (Ravindran Menon *et al*, 2015).

Our data add to a growing body of research suggesting that tumor cells can reversibly undergo dynamic changes that create subpopulations of cells with different proliferative potentials and sensitivity to apoptosis. Stochastic fluctuation in protein levels (Spencer *et al*, 2009), DTPs, and $NGFR^{High}$ melanoma cells represent three distinguishable but related mechanisms of achieving a state of reversible drug resistance. Such cells are thought to be the basis of residual disease and to provide a pool for further genetic or epigenetic changes that eventually induce the growth of drug-resistant clones (Hata *et al*, 2016).

A speculative pathway for reversible drug resistance in melanoma

Based on data from drug-adapted cells in culture, the efficacy of co-drugging these cells with kinase and BET bromodomain inhibitors and analysis of gene expression profiles in human melanoma biopsies, we propose a speculative model for the adaptive resistance to RAF/MEK inhibition characterized in this paper. Exposure of *BRAF*-mutant melanoma cells to MAPK inhibitors initially induces up-regulation of JNK/c-Jun signaling, a known regulator of EMT-related genes and of cell adhesion and ECM molecules (Liu *et al*, 2015; Ramsdale *et al*, 2015). Up-regulation of cell adhesion proteins is accompanied by activation of FAK and downstream Src kinases, causing cells to acquire a distinct epigenetic state and become more neural crest-like. Such cells divide slowly and have a reduced requirement for ERK signaling.

We and others have recently reported that JNK and c-Jun are activated in a subset of melanomas exposed to MAPK inhibitors (Delmas *et al*, 2015; Fallahi-Sichani *et al*, 2015; Ramsdale *et al*, 2015; Riesenberg *et al*, 2015; Titz *et al*, 2016). Our current study links this phenomenon to transiently heritable (reversible) de-differentiation by a subset of cells in the population and to high expression of NGFR, which has previously been observed in human tumors. We also identify compounds other than JNK inhibitors able to block drug adaptation and increase cell killing. Studies on an as-yet limited number of biopsies show that NGFR expression is induced by MAPK inhibitors in human tumors, concomitant with a reduction in cell proliferation; this effect is highly heterogeneous across a single human tumor and also across a xenograft, the former representing a genetically heterogeneous sample and the latter a more homogenous one. In a human tumor analyzed prior to therapy, on therapy and following progression, we find that c-Jun levels increase upon initial MAPK inhibition and rise further when tumors become drug-resistant. Thus, it seems plausible that mechanisms identified in cultured cells are also operative in real tumors. It is important to note, however, that JNK/c-Jun-dependent adaptation marked by an $NGFR^{High}$ state, as described here, appears to occur in only a subset (about one-third) of cell lines studied. Other mechanisms are presumably operative in other cell lines.

Inhibitors of adaptive drug resistance

By targeted screening, we identify two classes of compounds with the potential to block vemurafenib-induced de-differentiation (as marked by elevated NGFR expression): (i) small-molecule kinase inhibitors against components of the postulated c-Jun/FAK/Src cascade and (ii) epigenetic modifiers, including BET bromodomain inhibitors presumed to block the de-differentiation program. Combining vemurafenib with JNK, FAK, or Src kinase inhibitors, or with BET inhibitors suppresses acquisition of the NGFR[High] phenotype, prevents the emergence of slowly cycling drug-resistant cells, and enhances cell killing. In the case of BET inhibitor JQ1, we also show that co-drugging suppresses the NGFR[High] state in BRAF-mutant melanoma xenografts treated with dabrafenib. The primary effect of co-drugging on cultured cells is on maximum effect (E_{max}) and involves reducing viable cell number (in a 3-day assay) from 1% to 10% of the initial population to 0.01–0.1%. In our opinion, such an effect would be missed by most protocols used to screen for drug combinations.

The molecular effects of RAF/MEK and JNK/FAK/Src/BET inhibitors appear to be orthogonal, with the former suppressing MAPK signaling and the latter suppressing the consequent emergence of de-differentiated, adapted cells. Moreover, experiments with vemurafenib and trametinib show that the greater the extent of MAPK inhibition, the greater the extent of adaptation. Thus, inhibitors of adaptation such as dasatinib (Sprycel®) might be expected to combine with MAPK inhibition in therapeutically beneficial ways. Inhibiting Src family kinases has previously been reported to overcome resistance to RAF inhibitors (Girotti et al, 2013), although this was attributed to a role for Src downstream of RTKs rather than FAK. Vemurafenib has also been shown to activate melanoma-associated stromal fibroblasts, increasing ECM production and elevating integrin/FAK/Src signaling to promote vemurafenib resistance in nearby melanoma cells (Hirata et al, 2015). All three of these mechanisms could be involved at the same time, perhaps to different extents in different settings.

Current understanding of biomarkers for vemurafenib-induced de-differentiation in melanomas remains incomplete. For example, the switch of melanoma cells in culture to a drug-resistant NGFR[High] phenotype is not associated with a reduction in MITF levels. Moreover, only about half of NGFR[High] post-resistance biopsies exhibited a reduction in MITF levels, suggesting that therapy-induced NGFR up-regulation and MITF down-regulation are not necessarily concomitant. Low MITF expression in melanomas has previously been linked to increased expression of RTKs such as AXL, EGFR, and PDGFRβ, which activate immediate–early signaling, causing resistance to RAF/MEK inhibitors (Muller et al, 2014). However, the NGFR[High] phenotype we observe is not associated with RTK up-regulation as judged by mRNA expression. These findings raise the question whether we and others are probing different aspects of a unified adaptive mechanism common to all melanomas or whether adaptation is fundamentally different in genetically distinct tumor cells. Answering this question at a single-cell level may help identify novel therapies and biomarkers that have been missed by experiments that focus on bulk tumor cell killing.

Materials and Methods

Cell culture

Melanoma cell lines used in this study were obtained from the Massachusetts General Hospital Cancer Center with the following

primary sources: COLO858 (ECACC), A375, C32, WM115, SKMEL28, and WM1552C (ATCC), LOXIMV1 (DCTD Tumor Repository, National Cancer Institute), MMACSF (RIKEN BioResource Center), and MZ7MEL (Johannes Gutenberg University Mainz). C32, MMACSF, SKMEL28, and WM115 cell lines were grown in DMEM/F12 (Invitrogen) supplemented with 5% fetal bovine serum (FBS) and 1% sodium pyruvate (Invitrogen). COLO858, LOXIMVI, MZ7MEL, and WM1552C cell lines were grown in RMPI 1640 (Corning cellgro) supplemented with 5% FBS and 1% sodium pyruvate (Invitrogen). A375 cells were grown in DMEM with 4.5 g/l glucose, L-glutamine, and sodium pyruvate (Corning cellgro), supplemented with 5% FBS. We added penicillin (50 U/ml) and streptomycin (50 μg/ml) to all growth media.

Reagents and antibodies

Chemical inhibitors from the following sources were dissolved in dimethyl sulfoxide (DMSO) as 10 mM stock solution and used in treatments: vemurafenib (MedChem Express), trametinib (GSK1120212), defactinib, PF562271, pictilisib (GDC0941), palbociclib (PD0332991), and AZD8055 (all from Selleck Chemicals), JNK-IN-8 (EMD Millipore), dasatinib, saracatinib, (+)-JQ1, I-BET, I-BET151, and belinostat (all from Haoyuan Chemexpress). The following primary antibodies with specified animal sources and catalogue numbers were used in specified dilution ratios in immunofluorescence analysis of cells and tissues: p-S6[S240/244] rabbit monoclonal antibody (mAb) (clone D68F8, Cat# 5364), 1:800; p-ERK[T202/Y204] rabbit mAb (clone D13.14.4E, Cat# 4370), 1:800; c-Jun rabbit mAb (clone 60A8, Cat# 9165), 1:800; p-c-Jun[S73] rabbit mAb (clone D47G9, Cat# 3270), 1:800; Ki-67 mouse mAb (clone 8D5, Cat# 9449), 1:400; c-Jun mouse mAb (clone L70B11, Cat# 2315), 1:200; p75NTR (NGFR) rabbit mAb (clone D4B3, Cat# 8238), 1:1,600 (for staining cultured cells) or 1:200 (for staining tissue sections); all from Cell Signaling Technology, and p-Rb[S807/811] goat polyclonal antibody (Cat# sc-16670), 1:400, from Santa Cruz Biotechnology. The following antibodies were diluted 1:1,000 and used in Western blots: p-FAK[Y397] rabbit mAb (clone D20B1, Cat# 8556), FAK rabbit mAb (clone D2R2E, Cat# 13009), integrin β1 rabbit mAb (clone D2E5, Cat# 9699), p75NTR (NGFR) rabbit mAb (clone D4B3, Cat# 8238), β-actin rabbit mAb (clone D6A8, Cat# 8457), all from Cell Signaling Technology; HSP90α/β rabbit polyclonal antibody (Cat# sc-7947) from Santa Cruz Biotechnology; and thrombospondin rabbit polyclonal antibody (Cat# ab85762) from Abcam.

Human tumor specimens

Under IRB-approved protocols, freshly procured and discarded melanoma tumor specimens were formalin-fixed, paraffin-embedded, sectioned, and stained with H&E for histopathological evaluation or immunofluorescence staining. Clinical history of the drug-naive patients (age, sex, BRAF mutation, sequencing method, treatment history) is as follows: patient 1 (74, male, wild-type, whole-exome sequencing, no prior treatment), patient 2 (58, male, BRAF[V600E], targeted sequencing, no prior treatment), patient 3 (86, female, BRAF[V600E], whole-exome sequencing, no prior treatment), and patient 4 (65, male, BRAF[V600E], targeted sequencing, interferon). In the case of patient-matched biopsies from pre-treatment, on-treatment (for 2 weeks), and post-relapse tumors, biopsies were collected from a male patient with metastatic *BRAF*-mutant melanoma, treated with dabrafenib and trametinib combination.

Immunohistochemistry of human tumor specimens

Tumor sections were deparaffinized, and heat-induced epitope retrieval (HIER) was performed on the unit using EDTA for 20 min at 90°C. Sections were incubated for 30 min with primary antibodies including Ki-67 rabbit monoclonal antibody (clone SP6, Cat# VP-RM04) from Vector Laboratories, p75NTR/NGFR rabbit polyclonal antibody (Cat# 119-11668) from RayBiotech, and MITF mouse monoclonal antibody (clone D5, Cat# MA5-14154) from Thermo Scientific, and were then completed with the Leica Refine detection kit (secondary antibody, the DAB chromogen, and the hematoxylin counterstain).

Live-cell reporter constructs

To generate cells expressing fluorescently tagged geminin and H2B, we used the pPB-CAG.EBNXN/pCMV-hyPBase transposase vector system (Allan Bradley, Sanger Institute). First, a pPB-CAG vector containing a multiple cloning site (pPB-CAG-MCS) was generated by annealed oligo cloning of the following primers into the EcoRI and SalI restriction sites of pPB-CAG-EKAREV (Albeck *et al*, 2013) containing a puromycin selection cassette: 5'-aattcggatcccatatgca cgtgctcgagg-3' and 5'-tcgacctcgagcacgtgcatatgggatccg-3'. Next, intermediate pPB-CAG constructs were generated for ERK-KTR-mTurquoise2, H2B-Venus, and mCherry-geminin performing Gibson Assembly (New England Biolabs) at the EcoRI and SalI restriction sites and using the following templates and primers: ERK-KTR (Regot *et al*, 2014) with 5'-tctcatcattttggcaaagaattcggcatgaagggccga aagcct-3' and 5'-ctcaccatactagtggatgggaattgaaag-3' and mTurquoise2 (Goedhart *et al*, 2012) with 5'-ccactagtatggtgagcaagggcgag-3' and 5'-ca cacattccacagggtcgacttacttgtacagctcgtccatg-3'; H2B (Nam & Benezra, 2009) with 5'-tctcatcattttggcaaagaattcggcatgcctgaaccctctaagtctgc-3' and 5'-ctcaccatggtggcgaccggtggatc-3' and Venus with 5'-tcgccaccatgg tgagcaagggcgag-3' and 5'-cacacattccacagggtcgacttatttgtacaattcgtccatc ccc-3'; mCherry with 5'-tctcatcattttggcaaagaattcggcatggtgagcaag ggcgag-3' and 5'-ggatatcccttgtacagctcgtccatgc-3' and geminin (Sakaue-Sawano *et al*, 2008) with 5'-ctgtacaagggatatccatcacactggc-3' and 5'-cacacattccacagggtcgacttacagcgcctttctccg-3'. These intermediate constructs were used as templates for a final round of Gibson cloning to generate pPB-CAG-ERK-KTR-mTurquoise2-P2A-H2B-Venus-P2A-mCherry-geminin in which the DNA coding for three live-cell reporters is separated by self-cleaving P2A sites: ERK-KTR-mTurquoise2 with 5' ctgtctcatcattttggcaaag-3' and 5'-cacgtcgcca gcctgcttaagcaggctgaagttagtagctccgcttcccttgtacagctcgtccatg-3', H2B-Venus with 5'-ttcagcctgcttaagcaggctggcgacgtggaggagaaccccgggccta tgcctgaaccctctaag-3' and 5'-gacatcccccgcttgtttcaataacgaaaaattcgtcg cgcccgagcctttgtacaattcgtccatc-3', mCherry-geminin with 5'-ttttcgtt attgaaacaagcgggggatgtcgaagaaatccgggcccgatggtgagcaagggcgag-3' and 5'-ctgacacacattccacagggtcgacttacagcgcctttctccgttttttc-3'. Plasmid DNA was provided by Marcus Covert (ERK-KTR), Joachim Goedhart (mTurquoise2), Robert Benezra (H2B-mCherry, Addgene plasmid # 20972), Atsushi Miyawaki (geminin), and Allan Bradley (pPB-CAG.EBNXN and pCMV-hyPBase). Positive clones were confirmed by sequencing. To create stable cell lines, cells were co-transfected with the pPB-CAG triple reporter plasmid and pCMV-hyPBase using

FuGene HD (Promega) and transiently selected with puromycin. To enrich for cells stably expressing the live-cell reporter at comparable levels, cells were subjected twice to FACS. Reporter and parental cell lines were confirmed to grow at comparable rates for different vemurafenib concentrations over 72 h of treatment.

Live single-cell imaging and analysis

Cell lines stably expressing the live-cell reporter were seeded into Costar 96-well black clear-bottom tissue culture plates (Corning 3603) in 200 μl full growth medium without phenol red at a density of 4,500 cells per well for COLO858 or 4,000 cells per well for MMACSF; cells were counted using a Cellometer Auto T4 Cell Viability Counter (Nexcelom Bioscience). To facilitate cell tracking, COLO858 reporter cells treated with DMSO were mixed with an equal amount of parental cells, maintaining an overall cell density of 4,500 cells/well. The next day, cells were treated with DMSO or 1 μM vemurafenib, or with 1 μM vemurafenib in combination with DMSO, 3 μM defactinib, 1 μM dasatinib, or 0.32 or 1 μM (+)-JQ1 using an Hewlett-Packard (HP) D300 Digital Dispenser. Within 45–80 min after drug treatment, image acquisition was started using a Nikon Ti motorized inverted microscope with a 10× Plan Fluor 0.30 NA Ph1 objective lens and the Perfect Focus System for continuous maintenance of focus. Plates were placed into an OkoLab cage microscope incubator set to 37°C, 5% CO_2, and 90% humidity to enable stable environmental conditions throughout the experiment. Images were acquired every 6 min for the indicated times with a Hamamatsu ORCA ER cooled CCD camera controlled with Meta-Morph 7 software, using a 2 × 2 binning. For illumination, the Lumencor Spectra-X light engine in combination with a CFP/YFP/mCherry beam splitter (Chroma ID No. 032357) was used. H2B-Venus fluorescence was collected with a 508/24 excitation and a 540/21 emission filter at 200 ms exposure, and mCherry-geminin fluorescence was collected with a 575/22 excitation and a 632/60 emission filter at 400 ms exposure.

Individual cells from up to 10 wells per condition were analyzed. Cell positions and cell death/division events were manually tracked using a custom MATLAB-based script provided by Jose Reyes, Kyle W. Karhohs, and Galit Lahav (Harvard Medical School). Using H2B, a total of 150–217 cells were manually tracked and cell division and death events were recorded. To derive statistical mean and variance, cells from multiple wells were pooled together to generate three or four groups of wells containing ~50–70 cells. Data from 3 to 4 groups of cells were then used to report the mean ± SD. For extracting the geminin signal, the mean intensity of the centroid dilated by 12 pixels was calculated after using a rolling ball background subtraction. To determine the onset of S/G2, a moving average with a window of 40 frames was calculated for the geminin signal and in general an average value above a threshold of 2.0 (COLO858) or 1.5 (MMACSF) was determined as the start of the S/G2 cell cycle stage.

To measure cell division times following longer periods of vemurafenib treatment (i.e., ~8 days), COLO858 cells were initially exposed to 1 μM vemurafenib for 2 days, medium was then changed (to remove apoptotic cells), and cells were treated with 1 μM vemurafenib for an additional 2 days before they were imaged for ~4 days. Minimum doubling times were estimated for 100 individual cells tracked between days 4 and 8 post-treatment by identifying the longest time interval before or after which a cell divides.

Long-term time-lapse live-cell analysis using IncuCyte

COLO858 cells expressing H2B-mVenus were imaged every 45 min for ~1 week after treatment in three replicates with indicated drugs at indicated concentrations with a 4× objective using IncuCyte ZOOM live-cell imager (Essen Bioscience). Dead cells were identified by staining with IncuCyte CytoTox Red Reagent (Essen Bioscience, Cat# 4632). Time-lapse live-cell analysis (following exclusion of dead cells) was performed using ImageJ software.

Apoptosis, cell viability, and growth rate inhibition assays

For 72–96 h viability, apoptosis, or growth rate inhibition assays, cells were seeded in 3–6 replicates at 2,500–5,000 cells per well in 96-well plates (Corning 3603) in 180 μl of full growth media; cells were counted using a Cellometer Auto T4 Cell Viability Counter (Nexcelom Bioscience). Cells were treated the next day using a Hewlett-Packard (HP) D300 Digital Dispenser with compounds at reported doses. To score viability and apoptosis, we used a dye-based imaging assay: The cell-permeable DNA dye Hoechst 33342 was used to mark nuclei, and DEVD-NucView488 caspase-3 substrate was used to mark apoptosis, as previously described in detail (Fallahi-Sichani et al, 2015). Imaging was performed using a 10× objective using a PerkinElmer Operetta High Content Imaging System. Eleven sites were imaged in each well. Image segmentation and analysis were performed using Acapella software (PerkinElmer). The nuclear segmentation with Hoechst 33342 was used to identify individual nuclei and to count cells. To score apoptotic cells, bright spots were detected by dividing NucView488 channel nuclear intensity by the nucleus area and spots brighter than a separating threshold were scored as apoptotic. Relative viability was calculated by subtracting the number of apoptotic cells from the total number of cells to achieve viable cell count at each condition that was normalized to a DMSO-treated control. To compare drug effect on different cell populations that grow at different rates (e.g., FACS-sorted NGFR^{High} and NGFR^{Low} cells), growth rate (GR) inhibition was calculated by normalizing viable cell count data to the growth rate of untreated cells as described previously (Hafner et al, 2016). Data were analyzed using MATLAB 2014b software.

RNA extraction, library construction, and RNA-seq analysis

COLO858 and MMACSF cells were seeded in 10-cm plates in two replicates, treated the next day with either DMSO or 0.2 μM vemurafenib for 24 and 48 h. At the time of harvest for RNA, cells were washed once with PBS and then lysed in the dish with RLT buffer (Qiagen). Samples were immediately processed with Qiashredder and RNeasy kits (Qiagen) and frozen until further use. 10 μg of RNA was DNAse-treated and cleaned up with RNeasy MinElute kit (Qiagen). RNA quality was assessed by Bioanalyzer (Agilent), and all samples had RINs of 9.0 or higher. RNA-seq libraries were constructed using Illumina's TruSeq-stranded mRNA library prep kit and protocol with minor modifications. Briefly, 1 μg of RNA was mixed with 2 μl of a 1:100 dilution of ERCC spike in control Mix2

(Life Technologies) before mRNA purification. Elution and fragmentation was done for 6 min at 94°C. cDNA was synthesized and cleaned up with Ampure beads (Agencourt). Fragments were end-modified and adaptors ligated before another Ampure bead cleanup. Final library amplification was done at 13 cycles and again cleaned up with Ampure beads. The resulting libraries were roughly 380 bp in length as assessed by Bioanalyzer. RAN-seq was performed at the Harvard University Sequencing Facility (FAS Division of Science) on Illumina HiSeq 2000 machines using the standard single-read (1 × 50 bp) protocol. The reads were mapped to the human genome (build hg19) using Tophat (Kim et al, 2013) with default settings, and differential expression analysis was performed using Cuffdiff (Trapnell et al, 2010) running on the web-based Galaxy platform (https://usegalaxy.org/). To identify differentially regulated genes between COLO858 and MMACSF cells, we first selected genes whose expression at 24 or 48 h following treatment changed relative to a DMSO-treated control ($q < 0.01$) in at least one of the cell lines; we then identified those genes with FPKM ≥ 1, that changed in abundance by more than twofold between the two cell lines ($|\log_2 (ratio)| \geq 1$), where "ratio" represents treatment versus DMSO fold-change in COLO858 divided by the treatment versus DMSO fold-change in MMACSF cells. Differentially regulated genes were processed using Metacore (Genego, Inc) software available online (http://portal.genego.com/). Ranked biological processes and pathways were generated using "analyze single experiment" feature with default settings.

Hierarchical clustering

Unsupervised hierarchical clustering of expression levels of differentially regulated genes between COLO858 and MMACSF cell lines was carried out using MATLAB 2014b, using the Chebyshev distance as the metric. (FPKM + 1) values of vemurafenib-treated conditions were normalized to those of DMSO-treated controls and $\log_2$-transformed prior to clustering.

Bioinformatics analysis

Gene expression data analysis and heat-map visualization were performed using MATLAB 2014b software. Differentially expressed genes with transcription factor activity and genes associated with cell surface receptors and secreted peptides/proteins were identified using Advanced Search 2.0 feature of the online Genego database (https://portal.genego.com/). Enriched transcriptional regulators for the list of differentially expressed genes were predicted using the Database for Annotation, Visualization and Integrated Discovery (DAVID) v6.7 (http://david.ncifcrf.gov/) (Huang da et al, 2009a,b), and they were compared to the gene expression levels of transcription factors 24–48 h after vemurafenib treatment.

Gene set enrichment analysis

Gene set enrichment analysis (GSEA) (Mootha et al, 2003; Subramanian et al, 2005) was performed using GSEA v2.2.0 software with 1,000 phenotype permutations. Gene Ontology (GO) biological processes (c5.bp.v5.0.symbols.gmt), and Kyoto Encyclopedia of Genes and Genomes (KEGG) pathway (c2.cp.kegg.v5.0.symbols.gmt) gene sets were obtained from http://www.broadinstitute.

org/gsea/downloads.jsp and used in GSEA. To identify biological processes and pathways most correlated with NGFR expression, we performed GSEA on RNA-seq data of tumors from 128 $BRAF^{V600E}$ melanoma patients included in TCGA (Cancer Genome Atlas Network, 2015), and microarray data of 25 $BRAF^{V600E}$ melanoma cell lines in the CCLE (Barretina et al, 2012), and by selecting NGFR expression levels as the "phenotype" and "Pearson" as the metric for ranking genes. A detailed description of GSEA methodology and interpretation is provided at http://www.broadinstitute.org/gsea/d oc/GSEAUserGuideFrame.html. Briefly, enrichment score indicates "the degree to which a gene set is overrepresented at the top or bottom of a ranked list of genes". The false discovery rate (FDR q-value) is "the estimated probability that a gene set with a given enrichment score represents a false positive finding". "In general, given the lack of coherence in most expression datasets and the relatively small number of gene sets being analyzed, an FDR cutoff of 25% is appropriate".

Fluorescence-activated cell sorting

COLO858 cells were seeded in 15-cm plates, treated the next day with 0.32 µM vemurafenib for 48 h. The cell monolayer was incubated at 37°C with trypsin 0.05% (Gibco) for 1 min, lifted from the plate with a cell scraper and re-suspended in PBS with 2% FBS. The cell suspension was washed in PBS with 2% FBS twice. Cells were counted and assessed for viability by trypan blue exclusion test on a Vi-CELL Cell Viability Counter instrument (Beckman Coulter). The cell suspension was incubated with a PE-conjugated NGFR monoclonal antibody (clone ME20.4-1.H4, Miltenyi Biotec) and with calcein-AM viability marker (Life Technologies) per manufacturers' recommendations. After incubation, the cell suspension was washed twice in PBS with 2% FBS. Cells were sorted on a BD FACSAria II (BD Biosciences) instrument. Unstained and calcein-AM and NGFR-PE single-color controls were used to set appropriate gates. Cells high for calcein were gated, and ~1 million NGFR^{High} or NGFR^{Low} cells were sorted into 15-ml conical tubes (Falcon) that were prepared with 5 ml RMPI 1640 (Corning cellgro) supplemented with 5% FBS and 1% sodium pyruvate (Invitrogen) and plated using culture conditions described above. FACS-sorted NGFR^{High} and NGFR^{Low} cells were counted using a Countess II FL Automated Cell Counter (Life Technologies) and cultured for 9 days in fresh growth media in 96-well plates or treated with vemurafenib for immunofluorescence and growth rate inhibition assays.

In vivo xenograft studies

All animal experiments were conducted in accordance with procedures approved by the Institutional Animal Care and Use Committee (IACUC) at Harvard Medical School. Six-week-old NU/J mice (Jackson Laboratory, Stock# 002019) were transiently anesthetized using 5% vaporized isoflurane and injected subcutaneously in the right flank with 2.5×10^6 A375 human melanoma cells suspended in 200 µl of growth factor-reduced Matrigel (Corning 356230) in PBS (1:1). Tumor xenografts were allowed to grow until the mean volume across the tumors reached 150 mm^3 as measured by digital calipers. Mice were then randomly assigned to treatment (once-daily) for 5 days with one of the following drug combinations: Group 1, 200 µl dabrafenib (25 mg/kg) via oral gavage (OG) plus

500 µl JQ1 (50 mg/kg) via intraperitoneal (IP) injection; Group 2, 200 µl of dabrafenib (25 mg/kg) via OG plus 500 µl of IP vehicle control; Group 3, OG and IP vehicle controls given at equivalent volumes. Dabrafenib was diluted in 0.5% hydroxypropylmethylcellulose and 0.2% Tween-80 in pH 8.0 distilled water, and JQ1 was diluted in 5% dextrose in water. After 5 days, mice were transcardially perfused with oxygenated and heparinized Tyrode's solution; this allowed for simultaneous euthanasia and exsanguination. Flank xenografts were then surgically removed and fixed in 4% paraformaldehyde (PFA) in PBS and stored at 4°C for 48 h. All fixed tumors from a given treatment group were uniformly paraffin-embedded into a single paraffin block holder (i.e., 1 block holder per treatment group) and sectioned at 5 µm thickness: This allowed each microscope slide bearing a single 5-µm paraffin section to contain a representative sample of all the tumors from a given treatment group.

Immunofluorescence staining, quantitation, and analysis

Immunofluorescence assays for cultured cells were performed using cells seeded in 96-well plates (Corning 3603) and then treated the next day using Hewlett-Packard (HP) D300 Digital Dispenser with compounds at reported doses for indicated times in 2–3 replicates. Cells were fixed in 4% paraformaldehyde for 20 min at room temperature and washed with PBS with 0.1% Tween-20 (Sigma-Aldrich) (PBS-T), permeabilized in methanol for 10 min at room temperature, rewashed with PBS-T, and blocked in Odyssey blocking buffer (OBB) for 1 h at room temperature. Cells were incubated overnight at 4°C with primary antibodies in OBB. Cells were then stained with rabbit, mouse, and goat secondary antibodies from Molecular Probes (Invitrogen) labeled with Alexa Fluor 647 (Cat# A31573), Alexa Fluor 488 (Cat# A21202), and Alexa Fluor 568 (Cat# A11057). Cells were washed once in PBS-T, once in PBS, and were then incubated in 250 ng/ml Hoechst 33342 and 1:800 Whole Cell Stain (blue; Thermo Scientific) solution for 20 min. Cells were washed twice with PBS and imaged with a 10× objective using a PerkinElmer Operetta High Content Imaging System. 9–11 sites were imaged in each well. Image segmentation, analysis, and signal intensity quantitation were performed using Acapella software (PerkinElmer). Population-average and single-cell data were analyzed using MATLAB 2014b software. Single-cell density scatter plots were generated using signal intensities for individual cells.

For immunofluorescence analysis of xenograft and human tumor sections, processing and staining steps were performed with a customized protocol using a BOND RX automated immunohistochemistry processor (Leica Biosystems). Briefly, tissue slides were baked at 60°C for 30 min, de-waxed using Bond Dewax Solution (Leica Biosystems, Cat# AR9222), antigen retrieved for 20 min using BOND Epitope Retrieval solution 1 (Leica Biosystems, Cat# AR9961), and blocked with OBB for 30 min. Slides were incubated with secondary antibodies (Invitrogen): goat anti-mouse Alexa 647 (Cat# A21236) and goat anti-rabbit Alexa 555 (Cat# A21428) and scanned with a 10× objective using CyteFinder (RareCyte Inc.) to obtain pre-staining images for registering background due to non-specific binding of secondary antibodies (e.g., in necrotic regions of tumors). Slides were then incubated with a fluorophore inactivating solution containing 4.5% H$_2$O$_2$ and 20 mM NaOH in PBS for 2 h to quench the background fluorescence (Lin *et al*, 2015). Slides were then incubated at 4°C overnight with primary antibodies in OBB, washed with PBS, and stained with the same secondary antibodies described above for 2 h at room temperature. Slides were re-scanned with a 10× objective using CyteFinder. Images were stitched together to afford a single aggregate immunofluorescence image of the whole area of all tumor sections per drug treatment group. Fluorescence images were flat-field corrected and single-cell fluorescence intensities were quantified using ImageJ software. Briefly, the Hoechst channel was used to generate single-cell masks and ROIs. For Ki-67 nuclear staining, the masks were applied and the mean/integrated pixel intensity per cell was obtained. For NGFR staining, the nuclear masks were then converted to ring-shaped ROIs (2-pixel wide), and the ROIs were applied to quantify NGFR signal intensities per cell. We then used pre-staining images (generated from slides stained with secondary antibodies only) to exclude high-background regions of the tumor (with fluorescence intensities higher than a separating threshold) from further analysis. Removal of high-background regions, and analysis of population-average and single-cell data were performed using MATLAB 2014b software.

Western blots and quantitation

To prepare protein lysates, cells were seeded in 10-cm plates and treated the next day with either DMSO or two doses of vemurafenib (0.2, 0.32, or 1 µM) for 48 h at 37°C in full growth media. Cells were transferred to ice, washed with ice-cold PBS, and lysed with a 1% NP-40 lysis buffer (1% NP-40, 150 mM NaCl, 20 mM Tris pH 7.5, 10 mM NaF, 1 mM EDTA pH 8.0, 1 mM $ZnCl_2$ pH 4.0, 1 mM $MgCl_2$, 1 mM Na_3VO_4, 10% glycerol) supplemented with complete mini/EDTA-free protease inhibitors (Roche, Cat# 11836170001). Protein concentration of cleared lysates was measured using a BCA assay kit (Thermo Scientific, Cat# 23225). Lysates were adjusted to equal protein concentrations for each cell line, 4× NuPage LDS sample buffer (Invitrogen) supplemented with 50 mM DTT was added, and samples were heated for 10 min at 70°C and loaded on Novex 3–8% Tris–Acetate gels (Invitrogen). Western blots were performed using the iBlot Gel Transfer Stacks PVDF system (Invitrogen). After blocking with OBB (Licor), membranes were incubated with primary antibodies diluted 1:1,000 in OBB. As secondary antibody, donkey anti-rabbit IgG coupled to IRDye 800CW (Licor, Cat# 926-32213) was diluted 1:15,000 in OBB including 0.01% SDS. Membranes were scanned on an Odyssey CLx scanner (Licor) with 700 and 800 nm channels set to automatic intensity at a 169 µm resolution. Protein levels were quantified with the Image Studio 4.0.21 software (Licor) using the built-in manual analysis tool with a median local background correction. All intensity values were corrected using the respective HSP90α/β or β-actin levels and then normalized to the DMSO control sample for COLO858. Western blots were performed in three independent replicates, and a representative blot with quantification is shown.

siRNA transfection

siRNAs against *JUN*, *PTK2*, and *NGFR* genes and a non-targeting control were from Dharmacon. COLO858 cells seeded in 96-well plates were transfected using transfection reagent DharmaFECT 2

(Dharmacon) in antibiotic-free growth medium. Cells were treated the next day with vemurafenib at indicated doses for 48–72 h followed by fixation for immunofluorescence staining.

Compound screening using a chromatin-targeting library

A375, COLO858, and WM115 cells were plated in two replicates at 4,500, 7,000, and 8,000 cells per well, respectively; cells were counted using a Countess II FL Automated Cell Counter (Life Technologies). Cells were treated the next day with either DMSO or three different doses (0.11, 0.53, and 2.67 µM) of each of 41 compounds in the Harvard Medical School Library of Integrated Network-based Cellular Signature (HMS-LINCS) IV chromatin-targeting library (see https://lincs.hms.harvard.edu/db/libraries/LINCS-IV/ for the list of compounds) using previously prepared 384-well dilution plates and a Seiko pin transfer robot system. Immediately after, A375 and WM115 cells were then treated with 1 µM vemurafenib and COLO858 cells were treated with 0.32 µM vemurafenib using an HP D300 Digital Dispenser. Forty-eight hours after treatment, cells were fixed and analyzed for NGFR expression using immunofluorescence microscopy, as described earlier. To identify candidate compounds with potential for suppressing NGFR expression, the NGFR measurements were averaged across the three doses of chromatin-targeting compounds and z-scored. The data were scatter-plotted between the two replicates. Compounds whose NGFR levels were consistently among the five lowest z-scores in all the three cell lines were selected for further analysis.

Statistical analysis

All data (with error bars) are presented as mean ± SD using indicated numbers of replicates. Statistical significance for cell-based experiments performed at different drug or growth factor doses, time points, and replicates was determined based on two-way analysis of variance (ANOVA). Correlation between measurements was evaluated based on pairwise Pearson's correlation coefficient. Statistical significance for xenograft experiments was determined using the two-tailed two-sample *t*-test. Significance was set at $P < 0.05$. Statistical analyses were performed using MATLAB 2014b software.

Acknowledgements

This work was funded by NIH grants U54-HL127365, CA139980, and GM107618 (to PKS), and NCI grant K99CA194163 (to MF-S), a Merck Fellowship of the Life Sciences Research Foundation (to MF-S), grants from the Adelson Medical Research Foundation and the Melanoma Research Alliance (to LAG), the Ludwig Center at Harvard (to PKS and LAG), and a DFCI Wong Family Award (to BI). We thank the Nikon Imaging Center and ICCB-Longwood Screening Facility at HMS, the BWH Pathology Core and J Reyes, SH Davis, KW Karhohs, G Lahav, G Berriz, J Muhlich, E Williams, D Wrobel, R Benezra, A Bradley, MW Covert, J Goedhart, A Miyawaki, C Shamu, NS Gray, C Yoon, LN Kwong, GM Murphy, and C Lian for reagents and assistance.

Author contributions

MF-S designed the project and performed cell-based experiments and computational/statistical analysis. VB performed cell-based experiments. GJB performed mouse experiments. BI analyzed clinical samples. JRL analyzed clinical and xenograft samples. SAB assisted with gene expression analysis and PS and AR with cell culture. LAG and PKS supervised the research. All authors wrote and reviewed the manuscript.

References

Abel EV, Basile KJ, Kugel CH III, Witkiewicz AK, Le K, Amaravadi RK, Karakousis GC, Xu X, Xu W, Schuchter LM, Lee JB, Ertel A, Fortina P, Aplin AE (2013) Melanoma adapts to RAF/MEK inhibitors through FOXD3-mediated upregulation of ERBB3. *J Clin Invest* 123: 2155–2168

Albeck JG, Mills GB, Brugge JS (2013) Frequency-modulated pulses of ERK activity transmit quantitative proliferation signals. *Mol Cell* 49: 249–261

Barretina J, Caponigro G, Stransky N, Venkatesan K, Margolin AA, Kim S, Wilson CJ, Lehar J, Kryukov GV, Sonkin D, Reddy A, Liu M, Murray L, Berger MF, Monahan JE, Morais P, Meltzer J, Korejwa A, Jane-Valbuena J, Mapa FA et al (2012) The Cancer Cell Line Encyclopedia enables predictive modelling of anticancer drug sensitivity. *Nature* 483: 603–607

Cancer Genome Atlas Network (2015) Genomic classification of cutaneous melanoma. *Cell* 161: 1681–1696

Cohen AA, Geva-Zatorsky N, Eden E, Frenkel-Morgenstern M, Issaeva I, Sigal A, Milo R, Cohen-Saidon C, Liron Y, Kam Z, Cohen L, Danon T, Perzov N, Alon U (2008) Dynamic proteomics of individual cancer cells in response to a drug. *Science* 322: 1511–1516

Delmas A, Cherier J, Pohorecka M, Medale-Giamarchi C, Meyer N, Casanova A, Sordet O, Lamant L, Savina A, Pradines A, Favre G (2015) The c-Jun/RHOB/AKT pathway confers resistance of BRAF-mutant melanoma cells to MAPK inhibitors. *Oncotarget* 6: 15250–15264

Emmons MF, Faiao-Flores F, Smalley KS (2016) The role of phenotypic plasticity in the escape of cancer cells from targeted therapy. *Biochem Pharmacol* 122: 1–9

Fallahi-Sichani M, Moerke NJ, Niepel M, Zhang T, Gray NS, Sorger PK (2015) Systematic analysis of BRAF(V600E) melanomas reveals a role for JNK/c-Jun pathway in adaptive resistance to drug-induced apoptosis. *Mol Syst Biol* 11: 797

Flusberg DA, Roux J, Spencer SL, Sorger PK (2013) Cells surviving fractional killing by TRAIL exhibit transient but sustainable resistance and inflammatory phenotypes. *Mol Biol Cell* 24: 2186–2200

Frick PL, Paudel BB, Tyson DR, Quaranta V (2015) Quantifying heterogeneity and dynamics of clonal fitness in response to perturbation. *J Cell Physiol* 230: 1403–1412

Gascoigne KE, Taylor SS (2008) Cancer cells display profound intra- and interline variation following prolonged exposure to antimitotic drugs. *Cancer Cell* 14: 111–122

Girotti MR, Pedersen M, Sanchez-Laorden B, Viros A, Turajlic S, Niculescu-Duvaz D, Zambon A, Sinclair J, Hayes A, Gore M, Lorigan P, Springer C, Larkin J, Jorgensen C, Marais R (2013) Inhibiting EGF receptor or SRC family kinase signaling overcomes BRAF inhibitor resistance in melanoma. *Cancer Discov* 3: 158–167

Goedhart J, von Stetten D, Noirclerc-Savoye M, Lelimousin M, Joosen L, Hink MA, van Weeren L, Gadella TW Jr, Royant A (2012) Structure-guided evolution of cyan fluorescent proteins towards a quantum yield of 93%. *Nat Commun* 3: 751

Gopal YN, Deng W, Woodman SE, Komurov K, Ram P, Smith PD, Davies MA (2010) Basal and treatment-induced activation of AKT mediates resistance to cell death by AZD6244 (ARRY-142886) in Braf-mutant human cutaneous melanoma cells. *Cancer Res* 70: 8736–8747

Hafner M, Niepel M, Chung M, Sorger PK (2016) Growth rate inhibition metrics correct for confounders in measuring sensitivity to cancer drugs. *Nat Methods* 13: 521–527

Hata AN, Niederst MJ, Archibald HL, Gomez-Caraballo M, Siddiqui FM, Mulvey HE, Maruvka YE, Ji F, Bhang HE, Krishnamurthy Radhakrishna V, Siravegna G, Hu H, Raoof S, Lockerman E, Kalsy A, Lee D, Keating CL, Ruddy DA, Damon LJ, Crystal AS et al (2016) Tumor cells can follow distinct evolutionary paths to become resistant to epidermal growth factor receptor inhibition. *Nat Med* 22: 262–269

Hirata E, Girotti MR, Viros A, Hooper S, Spencer-Dene B, Matsuda M, Larkin J, Marais R, Sahai E (2015) Intravital imaging reveals how BRAF inhibition generates drug-tolerant microenvironments with high integrin beta1/FAK signaling. *Cancer Cell* 27: 574–588

Huang da W, Sherman BT, Lempicki RA (2009a) Systematic and integrative analysis of large gene lists using DAVID bioinformatics resources. *Nat Protoc* 4: 44–57

Huang da W, Sherman BT, Lempicki RA (2009b) Bioinformatics enrichment tools: paths toward the comprehensive functional analysis of large gene lists. *Nucleic Acids Res* 37: 1–13

Hugo W, Shi H, Sun L, Piva M, Song C, Kong X, Moriceau G, Hong A, Dahlman KB, Johnson DB, Sosman JA, Ribas A, Lo RS (2015) Non-genomic and immune evolution of melanoma acquiring MAPKi resistance. *Cell* 162: 1271–1285

Johannessen CM, Johnson LA, Piccioni F, Townes A, Frederick DT, Donahue MK, Narayan R, Flaherty KT, Wargo JA, Root DE, Garraway LA (2013) A melanocyte lineage program confers resistance to MAP kinase pathway inhibition. *Nature* 504: 138–142

Kim D, Pertea G, Trapnell C, Pimentel H, Kelley R, Salzberg SL (2013) TopHat2: accurate alignment of transcriptomes in the presence of insertions, deletions and gene fusions. *Genome Biol* 14: R36

Konieczkowski DJ, Johannessen CM, Abudayyeh O, Kim JW, Cooper ZA, Piris A, Frederick DT, Barzily-Rokni M, Straussman R, Haq R, Fisher DE, Mesirov JP, Hahn WC, Flaherty KT, Wargo JA, Tamayo P, Garraway LA (2014) A melanoma cell state distinction influences sensitivity to MAPK pathway inhibitors. *Cancer Discov* 4: 816–827

Lazova R, Tantcheva-Poor I, Sigal AC (2010) P75 nerve growth factor receptor staining is superior to S100 in identifying spindle cell and desmoplastic melanoma. *J Am Acad Dermatol* 63: 852–858

Lin JR, Fallahi-Sichani M, Sorger PK (2015) Highly multiplexed imaging of single cells using a high-throughput cyclic immunofluorescence method. *Nat Commun* 6: 8390

Lito P, Pratilas CA, Joseph EW, Tadi M, Halilovic E, Zubrowski M, Huang A, Wong WL, Callahan MK, Merghoub T, Wolchok JD, de Stanchina E, Chandarlapaty S, Poulikakos PI, Fagin JA, Rosen N (2012) Relief of profound feedback inhibition of mitogenic signaling by RAF inhibitors attenuates their activity in BRAFV600E melanomas. *Cancer Cell* 22: 668–682

Lito P, Rosen N, Solit DB (2013) Tumor adaptation and resistance to RAF inhibitors. *Nat Med* 19: 1401–1409

Liu J, Han Q, Peng T, Peng M, Wei B, Li D, Wang X, Yu S, Yang J, Cao S, Huang K, Hutchins AP, Liu H, Kuang J, Zhou Z, Chen J, Wu H, Guo L, Chen Y, Chen Y et al (2015) The oncogene c-Jun impedes somatic cell reprogramming. *Nat Cell Biol* 17: 856–867

Long GV, Stroyakovskiy D, Gogas H, Levchenko E, de Braud F, Larkin J, Garbe C, Jouary T, Hauschild A, Grob JJ, Chiarion Sileni V, Lebbe C, Mandala M, Millward M, Arance A, Bondarenko I, Haanen JB, Hansson J, Utikal J, Ferraresi V et al (2014) Combined BRAF and MEK inhibition versus BRAF inhibition alone in melanoma. *N Engl J Med* 371: 1877–1888

Mica Y, Lee G, Chambers SM, Tomishima MJ, Studer L (2013) Modeling neural crest induction, melanocyte specification, and disease-related pigmentation defects in hESCs and patient-specific iPSCs. *Cell Rep* 3: 1140–1152

Mootha VK, Lindgren CM, Eriksson KF, Subramanian A, Sihag S, Lehar J, Puigserver P, Carlsson E, Ridderstrale M, Laurila E, Houstis N, Daly MJ, Patterson N, Mesirov JP, Golub TR, Tamayo P, Spiegelman B, Lander ES, Hirschhorn JN, Altshuler D et al (2003) PGC-1alpha-responsive genes involved in oxidative phosphorylation are coordinately downregulated in human diabetes. *Nat Genet* 34: 267–273

Moriceau G, Hugo W, Hong A, Shi H, Kong X, Yu CC, Koya RC, Samatar AA, Khanlou N, Braun J, Ruchalski K, Seifert H, Larkin J, Dahlman KB, Johnson DB, Algazi A, Sosman JA, Ribas A, Lo RS (2015) Tunable-combinatorial mechanisms of acquired resistance limit the efficacy of BRAF/MEK cotargeting but result in melanoma drug addiction. *Cancer Cell* 27: 240–256

Muller J, Krijgsman O, Tsoi J, Robert L, Hugo W, Song C, Kong X, Possik PA, Cornelissen-Steijger PD, Foppen MH, Kemper K, Goding CR, McDermott U, Blank C, Haanen J, Graeber TG, Ribas A, Lo RS, Peeper DS (2014) Low MITF/AXL ratio predicts early resistance to multiple targeted drugs in melanoma. *Nat Commun* 5: 5712

Nam HS, Benezra R (2009) High levels of Id1 expression define B1 type adult neural stem cells. *Cell Stem Cell* 5: 515–526

Nazarian R, Shi H, Wang Q, Kong X, Koya RC, Lee H, Chen Z, Lee MK, Attar N, Sazegar H, Chodon T, Nelson SF, McArthur G, Sosman JA, Ribas A, Lo RS (2010) Melanomas acquire resistance to B-RAF(V600E) inhibition by RTK or N-RAS upregulation. *Nature* 468: 973–977

Obenauf AC, Zou Y, Ji AL, Vanharanta S, Shu W, Shi H, Kong X, Bosenberg MC, Wiesner T, Rosen N, Lo RS, Massague J (2015) Therapy-induced tumour secretomes promote resistance and tumour progression. *Nature* 520: 368–372

Paoluzzi L, Hanniford D, Sokolova E, Osman I, Darvishian F, Wang J, Bradner JE, Hernando E (2016) BET and BRAF inhibitors act synergistically against BRAF-mutant melanoma. *Cancer Med* 5: 1183–1193

Perbal B (2004) CCN proteins: multifunctional signalling regulators. *Lancet* 363: 62–64

Poulikakos PI, Persaud Y, Janakiraman M, Kong X, Ng C, Moriceau G, Shi H, Atefi M, Titz B, Gabay MT, Salton M, Dahlman KB, Tadi M, Wargo JA, Flaherty KT, Kelley MC, Misteli T, Chapman PB, Sosman JA, Graeber TG et al (2011) RAF inhibitor resistance is mediated by dimerization of aberrantly spliced BRAF(V600E). *Nature* 480: 387–390

Ramsdale R, Jorissen RN, Li FZ, Al-Obaidi S, Ward T, Sheppard KE, Bukczynska PE, Young RJ, Boyle SE, Shackleton M, Bollag G, Long GV, Tulchinsky E, Rizos H, Pearson RB, McArthur GA, Dhillon AS, Ferrao PT (2015) The transcription cofactor c-JUN mediates phenotype switching and BRAF inhibitor resistance in melanoma. *Sci Signal* 8: ra82

Ravindran Menon D, Das S, Krepler C, Vultur A, Rinner B, Schauer S, Kashofer K, Wagner K, Zhang G, Bonyadi Rad E, Haass NK, Soyer HP, Gabrielli B, Somasundaram R, Hoefler G, Herlyn M, Schaider H (2015) A stress-induced early innate response causes multidrug tolerance in melanoma. *Oncogene* 34: 4448–4459

Regot S, Hughey JJ, Bajar BT, Carrasco S, Covert MW (2014) High-sensitivity measurements of multiple kinase activities in live single cells. *Cell* 157: 1724–1734

Riesenberg S, Groetchen A, Siddaway R, Bald T, Reinhardt J, Smorra D, Kohlmeyer J, Renn M, Phung B, Aymans P, Schmidt T, Hornung V, Davidson I, Goding CR, Jonsson G, Landsberg J, Tuting T, Holzel M (2015) MITF and c-Jun antagonism interconnects melanoma dedifferentiation with pro-inflammatory cytokine responsiveness and myeloid cell recruitment. *Nat Commun* 6: 8755

Roesch A, Vultur A, Bogeski I, Wang H, Zimmermann KM, Speicher D, Korbel C, Laschke MW, Gimotty PA, Philipp SE, Krause E, Patzold S, Villanueva J, Krepler C, Fukunaga-Kalabis M, Hoth M, Bastian BC, Vogt T, Herlyn M (2013) Overcoming intrinsic multidrug resistance in melanoma by blocking the mitochondrial respiratory chain of slow-cycling JARID1B(high) cells. *Cancer Cell* 23: 811–825

Sakaue-Sawano A, Kurokawa H, Morimura T, Hanyu A, Hama H, Osawa H, Kashiwagi S, Fukami K, Miyata T, Miyoshi H, Imamura T, Ogawa M, Masai H, Miyawaki A (2008) Visualizing spatiotemporal dynamics of multicellular cell-cycle progression. *Cell* 132: 487–498

Sharma SV, Lee DY, Li B, Quinlan MP, Takahashi F, Maheswaran S, McDermott U, Azizian N, Zou L, Fischbach MA, Wong KK, Brandstetter K, Wittner B, Ramaswamy S, Classon M, Settleman J (2010) A chromatin-mediated reversible drug-tolerant state in cancer cell subpopulations. *Cell* 141: 69–80

Shi H, Hong A, Kong X, Koya RC, Song C, Moriceau G, Hugo W, Yu CC, Ng C, Chodon T, Scolyer RA, Kefford RF, Ribas A, Long GV, Lo RS (2014a) A novel AKT1 mutant amplifies an adaptive melanoma response to BRAF inhibition. *Cancer Discov* 4: 69–79

Shi H, Hugo W, Kong X, Hong A, Koya RC, Moriceau G, Chodon T, Guo R, Johnson DB, Dahlman KB, Kelley MC, Kefford RF, Chmielowski B, Glaspy JA, Sosman JA, van Baren N, Long GV, Ribas A, Lo RS (2014b) Acquired resistance and clonal evolution in melanoma during BRAF inhibitor therapy. *Cancer Discov* 4: 80–93

Smith MP, Brunton H, Rowling EJ, Ferguson J, Arozarena I, Miskolczi Z, Lee JL, Girotti MR, Marais R, Levesque MP, Dummer R, Frederick DT, Flaherty KT, Cooper ZA, Wargo JA, Wellbrock C (2016) Inhibiting drivers of non-mutational drug tolerance is a salvage strategy for targeted melanoma therapy. *Cancer Cell* 29: 270–284

Spencer SL, Gaudet S, Albeck JG, Burke JM, Sorger PK (2009) Non-genetic origins of cell-to-cell variability in TRAIL-induced apoptosis. *Nature* 459: 428–432

Subramanian A, Tamayo P, Mootha VK, Mukherjee S, Ebert BL, Gillette MA, Paulovich A, Pomeroy SL, Golub TR, Lander ES, Mesirov JP (2005) Gene set enrichment analysis: a knowledge-based approach for interpreting genome-wide expression profiles. *Proc Natl Acad Sci USA* 102: 15545–15550

Sun C, Wang L, Huang S, Heynen GJ, Prahallad A, Robert C, Haanen J, Blank C, Wesseling J, Willems SM, Zecchin D, Hobor S, Bajpe PK, Lieftink C, Mateus C, Vagner S, Grernrum W, Hofland I, Schlicker A, Wessels LF et al (2014) Reversible and adaptive resistance to BRAF(V600E) inhibition in melanoma. *Nature* 508: 118–122

Tirosh I, Izar B, Prakadan SM, Wadsworth MH II, Treacy D, Trombetta JJ, Rotem A, Rodman C, Lian C, Murphy G, Fallahi-Sichani M, Dutton-Regester K, Lin JR, Cohen O, Shah P, Lu D, Genshaft AS, Hughes TK, Ziegler CG, Kazer SW et al (2016) Dissecting the multicellular ecosystem of metastatic melanoma by single-cell RNA-seq. *Science* 352: 189–196

Titz B, Lomova A, Le A, Hugo W, Kong X, Ten Hoeve J, Friedman M, Shi H, Moriceau G, Song C, Hong A, Atefi M, Li R, Komisopoulou E, Ribas A, Lo RS, Graeber TG (2016) JUN dependency in distinct early and late BRAF inhibition adaptation states of melanoma. *Cell Discov* 2: 16028

Trapnell C, Williams BA, Pertea G, Mortazavi A, Kwan G, van Baren MJ, Salzberg SL, Wold BJ, Pachter L (2010) Transcript assembly and quantification by RNA-Seq reveals unannotated transcripts and isoform switching during cell differentiation. *Nat Biotechnol* 28: 511–515

Tyson DR, Garbett SP, Frick PL, Quaranta V (2012) Fractional proliferation: a method to deconvolve cell population dynamics from single-cell data. *Nat Methods* 9: 923–928

Van Allen EM, Wagle N, Sucker A, Treacy DJ, Johannessen CM, Goetz EM, Place CS, Taylor-Weiner A, Whittaker S, Kryukov GV, Hodis E, Rosenberg M, McKenna A, Cibulskis K, Farlow D, Zimmer L, Hillen U, Gutzmer R, Goldinger SM, Ugurel S *et al* (2014) The genetic landscape of clinical resistance to RAF inhibition in metastatic melanoma. *Cancer Discov* 4: 94–109

Villanueva J, Infante JR, Krepler C, Reyes-Uribe P, Samanta M, Chen HY, Li B, Swoboda RK, Wilson M, Vultur A, Fukunaba-Kalabis M, Wubbenhorst B, Chen TY, Liu Q, Sproesser K, DeMarini DJ, Gilmer TM, Martin AM, Marmorstein R, Schultz DC *et al* (2013) Concurrent MEK2 mutation and BRAF amplification confer resistance to BRAF and MEK inhibitors in melanoma. *Cell Rep* 4: 1090–1099

Wagle N, Emery C, Berger MF, Davis MJ, Sawyer A, Pochanard P, Kehoe SM, Johannessen CM, Macconaill LE, Hahn WC, Meyerson M, Garraway LA (2011) Dissecting therapeutic resistance to RAF inhibition in melanoma by tumor genomic profiling. *J Clin Oncol* 29: 3085–3096

Wagle N, Van Allen EM, Treacy DJ, Frederick DT, Cooper ZA, Taylor-Weiner A, Rosenberg M, Goetz EM, Sullivan RJ, Farlow DN, Friedrich DC, Anderka K, Perrin D, Johannessen CM, McKenna A, Cibulskis K, Kryukov G, Hodis E, Lawrence DP, Fisher S *et al* (2014) MAP kinase pathway alterations in BRAF-mutant melanoma patients with acquired resistance to combined RAF/MEK inhibition. *Cancer Discov* 4: 61–68

11

A Notch positive feedback in the intestinal stem cell niche is essential for stem cell self-renewal

Kai-Yuan Chen[1,2,†] ⓘ, Tara Srinivasan[3,†], Kuei-Ling Tung[4,†], Julio M Belmonte[5] ⓘ, Lihua Wang[2,4], Preetish Kadur Lakshminarasimha Murthy[6], Jiahn Choi[2], Nikolai Rakhilin[1,2], Sarah King[3], Anastasia Kristine Varanko[4], Mavee Witherspoon[6], Nozomi Nishimura[3] ⓘ, James A Glazier[5], Steven M Lipkin[7], Pengcheng Bu[2,8,*] ⓘ & Xiling Shen[1,2,3,**] ⓘ

Abstract

The intestinal epithelium is the fastest regenerative tissue in the body, fueled by fast-cycling stem cells. The number and identity of these dividing and migrating stem cells are maintained by a mosaic pattern at the base of the crypt. How the underlying regulatory scheme manages this dynamic stem cell niche is not entirely clear. We stimulated intestinal organoids with Notch ligands and inhibitors and discovered that intestinal stem cells employ a positive feedback mechanism via direct Notch binding to the second intron of the Notch1 gene. Inactivation of the positive feedback by CRISPR/Cas9 mutation of the binding sequence alters the mosaic stem cell niche pattern and hinders regeneration in organoids. Dynamical system analysis and agent-based multiscale stochastic modeling suggest that the positive feedback enhances the robustness of Notch-mediated niche patterning. This study highlights the importance of feedback mechanisms in spatiotemporal control of the stem cell niche.

Keywords gene editing; intestine stem cell; Notch signaling; organoid; positive feedback
Subject Categories Quantitative Biology & Dynamical Systems; Signal Transduction; Stem Cells

Introduction

The stem cell niche provides a spatial environment that regulates stem cell self-renewal and differentiation (Lander *et al*, 2012). One example is at the base of the intestinal crypt, where self-renewing LGR5[+] crypt base columnar (CBC) cells and lysozyme-secreting Paneth cells form a mosaic pattern in which each Paneth cell is separated from others by LGR5[+] cells (Barker *et al*, 2007; Sato *et al*, 2011a). As proliferative intestinal stem cells (ISCs), CBCs mostly divide symmetrically, compete with each other in a neutral drift process, and regenerate the intestinal epithelium in 3–5 days (Lopez-Garcia *et al*, 2010; Snippert *et al*, 2010). The stem cell niche is capable of recovering from radiation or chemical damages to restore tissue homeostasis (Buczacki *et al*, 2013; Metcalfe *et al*, 2014).

Regulation of the niche is a concerted effort involving various signaling pathways. Paneth cells provide niche factors including epidermal growth factor (EGF), Wnt ligands (WNT3A), Notch ligands, and bone morphogenetic protein (BMP) inhibitor Noggin to support CBC self-renewal, while pericryptal stromal cells underneath the niche also supply additional Wnt ligands (WNT2B) (Barker, 2014). Among the pathways, juxtacrine Notch signaling pathway is often linked to developmental patterning (Artavanis-Tsakonas *et al*, 1999; Kopan & Ilagan, 2009). Notch signaling is mediated through direct cell-to-cell contact of membrane-bound Notch ligands on one cell and transmembrane Notch receptors on adjacent cells. The extracellular domain of Notch receptors binds Notch ligands, which activates receptor cleavage that releases the Notch receptor intracellular domain (NICD) to translocate to the nucleus. NICD interacts with the DNA-binding protein RBPJk to activate expression of downstream genes, such as the HES family

1 School of Electrical and Computer Engineering, Cornell University, Ithaca, NY, USA
2 Department of Biomedical Engineering, Duke University, Durham, NC, USA
3 Department of Biomedical Engineering, Cornell University, Ithaca, NY, USA
4 Department of Biological and Environmental Engineering, Cornell University, Ithaca, NY, USA
5 Biocomplexity Institute and Department of Physics, Indiana University, Bloomington, IN, USA
6 School of Mechanical Aerospace Engineering, Cornell University, Ithaca, NY, USA
7 Departments of Medicine, Genetic Medicine and Surgery, Weill Cornell Medical College, New York, NY, USA
8 Key Laboratory of RNA Biology, Key Laboratory of Protein and Peptide Pharmaceutical, Institute of Biophysics, Chinese Academy of Sciences, Beijing, China
*Corresponding author. E-mail: bupc@ibp.ac.cn
**Corresponding author. E-mail: xs37@duke.edu
†These authors contributed equally to this work

transcription factors. Notch signaling is essential for intestinal stem cell self-renewal and crypt homeostasis (Fre *et al*, 2005; van der Flier & Clevers, 2009). Among Notch receptors, inhibition of both Notch1 and Notch2 completely depletes proliferative stem/progenitor cells in the intestinal epithelium (Riccio *et al*, 2008). Inhibition of Notch1 alone is sufficient to cause a defective intestinal phenotype, while inhibition of Notch2 alone causes no significant phenotype (Wu *et al*, 2010). Notch3 and Notch4 are not expressed in the intestinal epithelium (Fre *et al*, 2011). Among Notch ligands, DLL1 and DLL4 are essential and function redundantly, and inactivation of both causes loss of stem and progenitor cells; in contrast, JAG1 is not essential (Pellegrinet *et al*, 2011).

In this study, we characterized the response of Notch signaling components in LGR5[+] CBCs from intestinal organoids and identified a direct Notch positive feedback loop. Perturbation to the positive feedback by CRISPR/CAS9 mutation of the binding sequence significantly reduced the number of CBCs in the stem cell niche. Computational modeling suggests that the positive feedback may contribute to robustness of the system when proliferation rates are high.

Results

Notch lateral inhibition and positive feedback

We characterized Notch signaling response in CBCs and Paneth cells (Fig 1A) using the *in vitro* intestinal organoid system (Sato *et al*, 2009), from which LGR5-EGFP[+] CBCs and CD24[+] Paneth cells were isolated using an established protocol (Sato *et al*, 2011a). Immunofluorescence (IF) confirmed that the sorted CD24[+] Paneth cells express lysozyme (Fig EV1A). RT–qPCR on purified CBCs and Paneth cells confirmed that Notch receptors (Notch1, Notch2) and signaling effectors (Hes1, Hes5) are enriched in CBCs, while Notch ligands (Dll1, Dll4, Jag1) and the secretory lineage regulator, Atoh1 (Yang *et al*, 2001), are enriched in Paneth cells, largely consistent with previous microarray measurements (Sato *et al*, 2011a; Fig 1B). Inhibition of Notch receptor cleavage by the γ-secretase inhibitor DAPT reduced the number of CBCs and increased the number of Paneth cells, whereas Notch activation by recombinant ligand JAG1 embedded in Matrigel (Sato *et al*, 2009; Van Landeghem *et al*, 2012; VanDussen *et al*, 2012; Yamamura *et al*, 2014; Mahapatro *et al*, 2016; Srinivasan *et al*, 2016) or EDTA (Rand *et al*, 2000) increased the number of CBCs and decreased the number of Paneth cells

(Fig 1C). Inhibition of Notch by DAPT up-regulated ligand expression, indicating that active Notch signaling suppresses ligand expression (Fig 1D and E). This is consistent with a lateral inhibition (LI) mechanism previously reported in several developmental systems, where ligands on a "sender" cell (in this case, Paneth cell) activate receptors on a "receiver" cell (in this case, CBC), which, in turn, suppresses ligand expression in the receiver cell (Collier *et al*, 1996). This intercellular feedback scheme causes bifurcation between adjacent cells, resulting in two opposite Notch signaling states (Fig 1F).

Additionally, Notch activation by recombinant JAG1 embedded in Matrigel (Sato *et al*, 2009; Van Landeghem *et al*, 2012; VanDussen *et al*, 2012; Yamamura *et al*, 2014; Mahapatro *et al*, 2016; Srinivasan *et al*, 2016) or EDTA (Rand *et al*, 2000) significantly increased receptor (Notch1/2) expression, while DAPT significantly reduced receptor expression (Fig 1D and E). This suggests the existence of a positive feedback loop, where activated Notch receptors up-regulate their own expression (Fig 1F).

NICD directly activates Notch1 transcription

Although both Notch1 and Notch2 form positive autoregulation, Notch1 has a stronger response than Notch2 (Fig 1D and E). This is consistent with previous reports showing that Notch1 and Notch2 are somewhat functionally redundant, but Notch1 is more critical to stem cell self-renewal and crypt homeostasis, while Notch2 is dispensable (Wu et al, 2010). We performed lineage tracing using tamoxifen-inducible Notch1[CreER] × ROSA26[tdTomato] transgenic mouse reporter strains (Fre *et al*, 2011; Oh *et al*, 2013). After induction, labeled Notch1[+] cells showed a similar pattern that largely overlaps with CBCs in the niche (Fig 2A). From day 1 to day 3, marked progeny of Notch1[+] cells expanded out of the niche and overtook the trans-amplifying (TA) progenitor compartments; by day 30, the marked clones of the original Notch1[+] cells replaced the entire epithelium (Figs 2B and EV1B). These lineage tracing experiments confirmed that Notch1 is active in CBCs, which is consistent with previous findings (Wu et al, 2010; Fre *et al*, 2011; Pellegrinet *et al*, 2011; Oh *et al*, 2013).

We analyzed the LICR ChIP-Seq dataset of mouse small intestinal cells from ENCODE using the UCSC genome browser (Consortium, 2012) to investigate regulation of Notch1 and Notch2 transcription. The second intron region of the Notch1 gene is highly enriched with enhancer histone marks H3K4me1 and H3K27ac, while no such

---

Figure 1. Notch levels in niche cells.

A Left: Cross-sectional view of murine intestinal crypt bottoms with co-immunofluorescence (co-IF) showing intermingled LGR5-EGFP[+] (green) CBCs and lysozyme[+] (LYZ, red) Paneth cells. Scale bar: 50 μm. Right: Schematic illustration of a niche pattern in both longitudinal and cross-sectional views of a crypt.

B RT–qPCR quantification of Notch signaling components in CBC and Paneth cell populations. The experiment was performed in triplicate and presented mean ± SEM (***$P \leq 0.001$, **$P \leq 0.01$, *$P \leq 0.05$; Student's *t*-test).

C Representative FACS plots of organoids treated with DMSO, JAG1 (embedded in Matrigel), EDTA or DAPT for 48 h, including gated analysis to isolate CD24[high]/SSC[high] Paneth cells and LGR5-EGFP[+] CBCs according to an established protocol (Sato *et al*, 2011c).

D RT–qPCR quantification of Notch signaling components in CBCs and Paneth cell populations after organoids were treated with Matrigel-embedded JAG1 (top), EDTA (middle), or DAPT (bottom). The experiments were performed in triplicate and presented mean ± SEM (**$P \leq 0.01$, *$P \leq 0.05$; Student's *t*-test).

E Western blot analysis of Notch signaling components from conditions described in (D). Actin was used as a loading control.

F Schematic illustration of lateral inhibition and positive feedback between neighboring cells. Transparent colors and dotted lines represent low expression/activity levels.

Source data are available online for this figure.

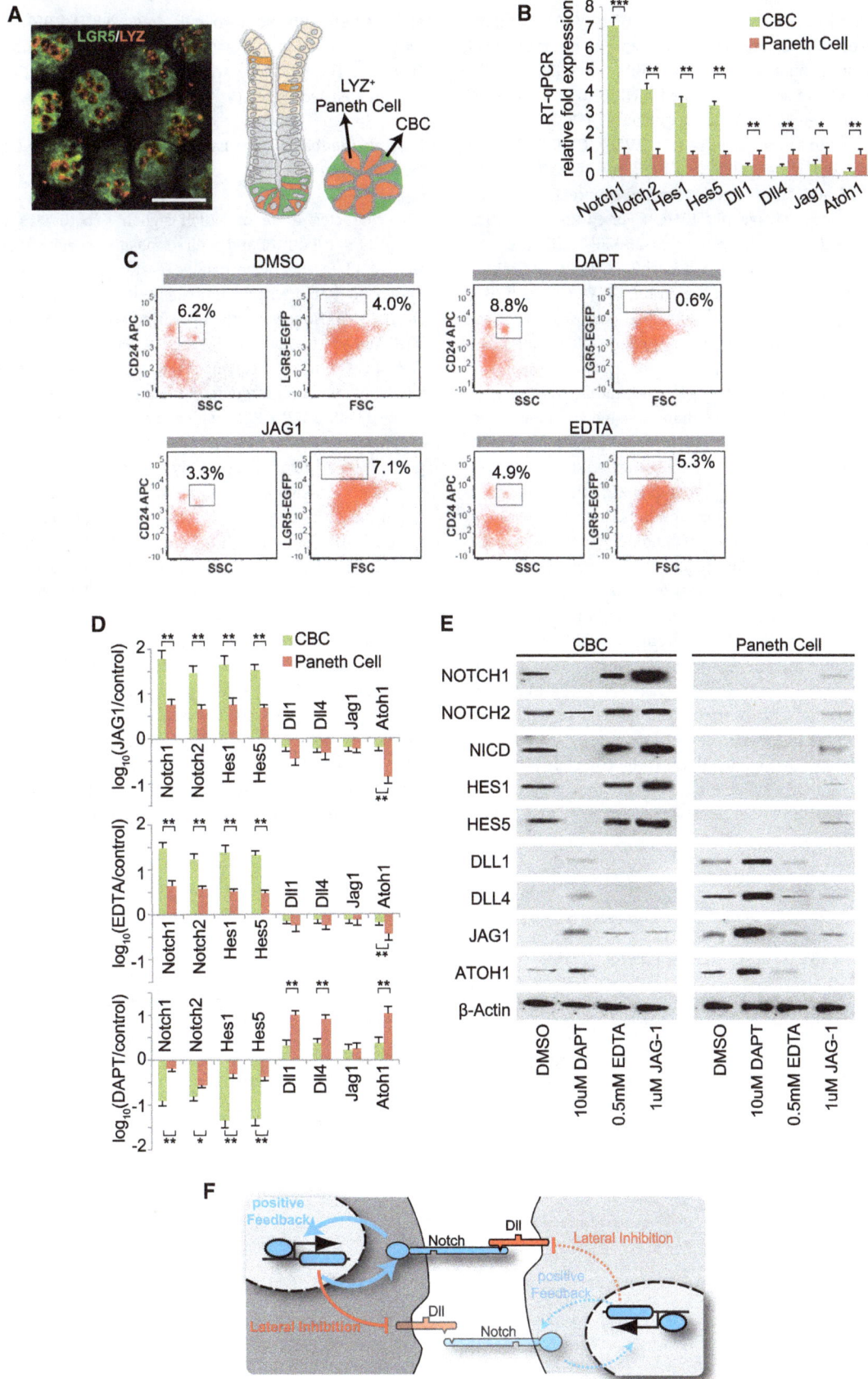

Figure 1.

regions were found in the Notch2 sequence (Fig EV2A). Computational analysis of this region with MotifMap (Wang *et al*, 2014) predicted a putative binding motif for RBPJk, the DNA-binding protein that forms an effector complex with NICD to activate Notch signaling. A unique eight base pair sequence (TTCCCACG, Chr2: 26,349,981–26,349,988) was identified (Fig EV2B). ChIP-PCR shows that NICD binds to this sequence in CBCs, and the binding was enhanced by JAG1 activation of receptors and suppressed by DAPT inhibition of receptor cleavage (Fig EV2C). For further validation, we crossed a LGR5-EGFP strain with a Rosa26-YFP-NICD strain (Oh *et al*, 2013) to generate a tamoxifen-inducible LGR5-EGFP-CreERT2 × Rosa26-YFP-NICD (NICD-OE) mouse strain. ChIP-PCR analysis of tamoxifen-induced NICD-expressing intestinal cells from NICD-OE mice also showed elevated NICD binding compared to uninduced control (Fig EV2D).

To further validate this enhancer sequence motif, we performed a luciferase reporter assay with the enhancer sequence cloned to the pGL4.27 [luc2P/minP] luciferase reporter vector containing a minimal promoter (Fig 2C). We compared the wild-type binding sequence with three mutated sequences (Fig 2C): (i) partial mutation of 3 nts in the binding sequence and 3nts of adjacent flanking region, (ii) partial deletion of 2nts of binding sequence, and (iii) mutation of the entire 8 nt binding sequence. The luciferase reporter vectors were transfected into intestine cells directly isolated from the mouse intestine with a pRL-SV40 control vector containing no binding sequence. The luciferase signal from the wild-type enhancer sequence was significantly higher than those from the mutated sequences or the control vector (Fig 2C). Jag1 and DAPT treatments also elicited stronger responses from the wild-type sequence than the mutated sequences or the control vector (Fig 2C).

We then performed a pull-down assay to confirm interaction between NICD and the identified binding motif. Oligonucleotides of the wild-type and mutated enhancer sequences were synthesized and labeled with biotin to pull down NICD/RBPJk complex from mouse intestinal crypt lysates, which was validated by Western blot (Fig 2D). The wild-type sequence has stronger NICD binding than the mutated sequences.

Positive feedback is critical to self-renewal, niche homeostasis, and recovery

Our characterization of Notch signaling pathways in niche cells suggests that both LI and a direct positive feedback are active. The role of LI in the niche can be easily rationalized, because LI is known to regulate developmental patterns (Collier *et al*, 1996; Kim *et al*, 2014). However, it is unclear whether the Notch positive feedback has any function. CRISPR-Cas9 vectors were designed to target the NICD binding sequence (Fig 2E, Table EV1). CRISPR/Cas9 vectors with specific guide RNAs (gRNAs) were transfected into single LGR5-EGFP CBCs, which were subsequently propagated as organoids. After puromycin selection, individual colonies were picked and sequenced separately to confirm CRISPR/Cas9 editing in the cells. Sequencing results indicate the presence of indels in the target NICD binding region formed through non-homologous end joining (NHEJ) (Fig EV2E). The mutated binding motif significantly reduced NICD binding compared to the empty vector (EV) control in CBCs sorted from organoids treated with DMSO (control), JAG1, or DAPT, according to ChIP-qPCR (Fig EV2F), which was consistent with the outcomes of the pull-down assay (Fig 2D). The mutations also significantly decreased Notch1 transcript levels measured by RT–qPCR (Fig EV2G) and NICD levels measured by Western blot (Fig EV2H). Expression levels of Notch signaling components (Notch1, Notch2, Hes1, Hes5) and LGR5 all decreased in CRISPR/Cas9-targeted cells with the mutated binding motif (Fig EV2I). Taken together, the data suggest that, when Notch receptors are activated, the resulting NICD/RBPJk complex bind to the Notch1

Figure 2. Notch1 positive feedback.

A   Representative image indicating Notch1 expression (red) in intestinal crypt bottoms of tamoxifen-induced Notch1^{CreER} × Rosa26^{tdTomato} mice. Scale bar: 50 µm.

B   Representative images of intestinal crypts showing progeny of Notch1$^+$ cells 1 day (left) and 3 days (right) after tamoxifen induction in Notch1^{CreER} × Rosa26^{tdTomato} mice. Dotted lines label the boundary of cells. Scale bar: 20 µm.

C   Luciferase reporter assay of NICD binding sequences. Left: Luciferase reporter vector map with the wild-type and three mutated NICD binding sequences cloned into the enhancer site. Blue represents wild-type sequence, and red represents mutated sequences. Right: Luciferase activity in the four sequences with normalization to a pRL-SV40 control vector. Jag1 or DAPT was added to stimulate or suppress Notch signaling for 48 h. The experiment was performed in replicates (*n* = 4) and presented mean + SEM (*$P \leq 0.05$, **$P \leq 0.01$; Student's *t*-test compares other conditions to WT sequence in normal condition separately).

D   Western blot following DNA pull-down showing NICD-DNA interaction. DNA pull-down was performed using mouse intestine crypt lysates with biotin-labeled oligonucleotide duplex of wild-type or mutated sequences containing the putative NICD/RBPJk binding site, followed by Western blotting to validate NICD binding on the pull-down sequences. Actin was used for input control.

E   Design of gRNAs for CRISPR/Cas9 mutagenesis to target the putative NICD/RBPJk binding motif on mouse Notch1 sequence.

F   Single LGR5-EGFP$^+$ CBCs were transfected with either an empty vector (control) or CRISPR/Cas9 gRNAs. Shown are representative brightfield images over 15 days and co-IF images indicating LGR5-EGFP (green) and LYZ (red) expression with DAPI nuclear staining. Scale bar represents 100 µm in low magnification and 25 µm in high magnification images, respectively.

G   Single LGR5-EGFP CBCs were transfected with either an empty vector (control) or CRISPR/Cas9 gRNAs. Left: Colony-forming efficiency measured after 5 days. Quantitative analysis calculated from 1,000 cells/replicate. The experiment was performed in triplicate and presented mean ± SEM (**$P \leq 0.01$; Student's *t*-test). Right: Quantitative comparison of organoid diameters after 15 days. The experiment was performed in triplicate and presented mean ± SEM (**$P \leq 0.01$; Student's *t*-test).

H   Single LGR5-EGFP ISCs were transfected with either an empty vector (control) or CRISPR/Cas9 gRNAs. Ratio of LGR5-EGFP$^+$ CBCs/LYZ$^+$ Paneth cells as determined by FACS analysis after 15 days. The experiment was performed in triplicate and presented mean ± SEM (**$P \leq 0.01$; Student's *t*-test).

I   Single empty vector control or CRISPR/Cas9-positive feedback knockout (PF KO) LGR5-EGFP$^+$ CBCs were transfected with an RBPJk-dsRed reporter construct and grown into organoids, which were subsequently treated with DMSO, DAPT, or JAG1 for 48 h. Left: Representative FACS plots for RBPJk-dsRED and LGR5-EGFP expression indicating a gated double positive fraction for each condition. Right: Mean fluorescence intensity (MFI) of RBPJk-dsRed expression of the entire cell population. The experiment was performed in triplicate and presented mean ± SEM (**$P \leq 0.01$; Student's *t*-test).

Source data are available online for this figure.

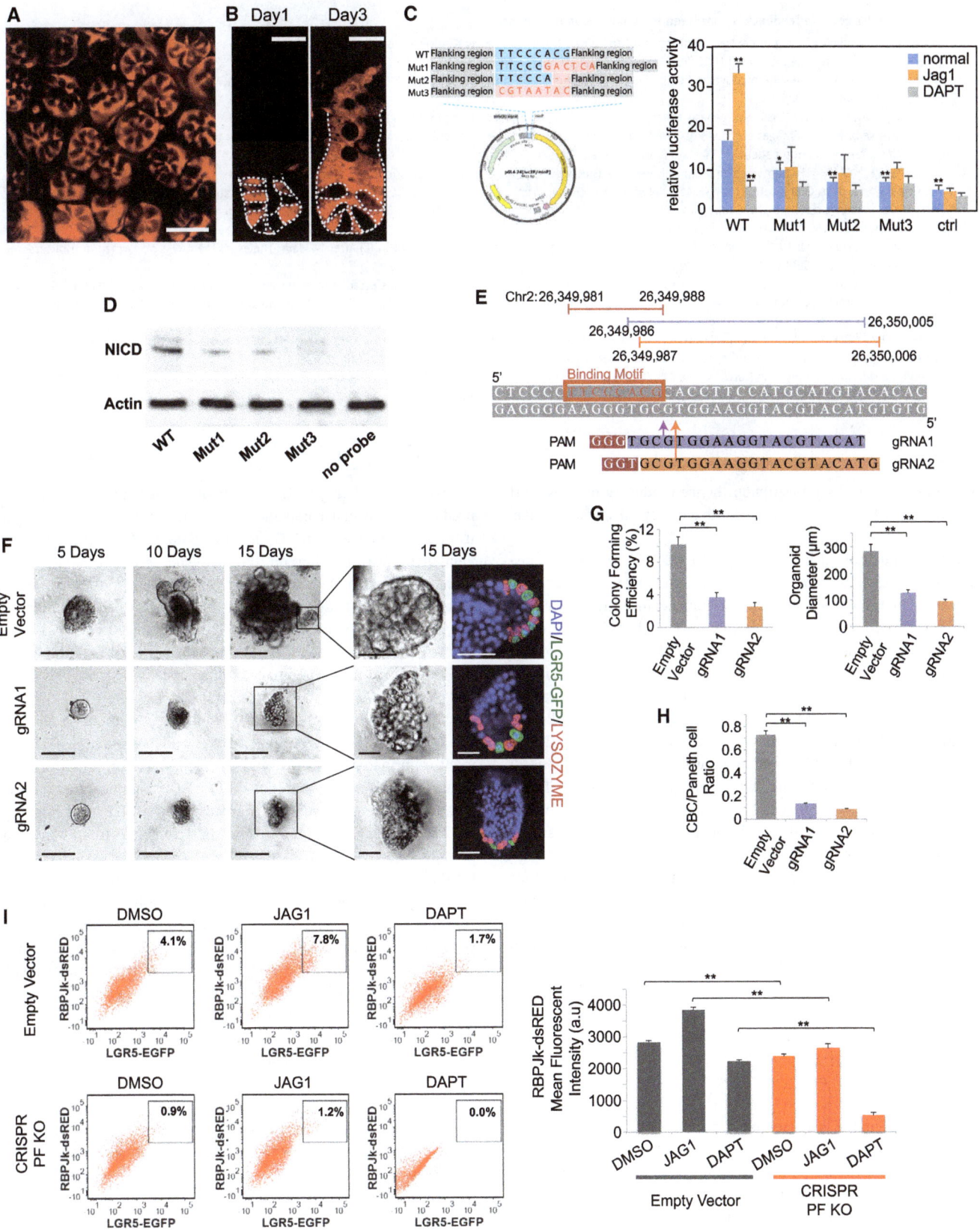

Figure 2.

Figure 3. Notch1 positive feedback is conserved in human colon organoids.

A Top: Sequence and chromatogram of NICD binding motif to human Notch1 following ChIP-PCR from EPHB2^{high}OLFM4^{high} colon stem cells. Bottom: Human Notch1 gene. Red line indicates the location of NICD/RBPJk binding motif on human Notch1.

B Agarose gel analysis of ChIP-PCR products indicating active binding of NICD to the motif in Notch1 sequence in EPHB2^{high}OLFM4^{high} colon stem cells treated with DMSO, DAPT, or JAG1.

C Design of gRNAs for CRISPR/Cas9 mutagenesis to target the putative NICD/RBPJk binding motif on human Notch1.

D Representative brightfield images of organoids derived from single EPHB2^{high}OLFM4^{high} colon stem cells transfected with either an empty vector control or CRISPR/Cas9 gRNAs after 7 days (top panel) and 14 days (bottom panel). Scale bar represents 50 μm.

E Single EPHB2^{high}OLFM4^{high} human colon stem cells were transfected with either an empty vector control or CRISPR/Cas9 gRNAs. Left: Colony-forming efficiency measured after 7 days. Quantitative analysis calculated from 1,000 cells/replicate. The experiment was performed in triplicate and presented mean ± SEM (**$P \leq 0.01$; Student's t-test). Right: Quantitative comparison of organoid diameters after 14 days for each condition. The experiment was performed in triplicate and presented mean ± SEM (**$P \leq 0.01$; Student's t-test).

F Percentage of EPHB2^{high}OLFM4^{high} stem cells based on FACS analysis for each condition described in (E) after 14 days. The experiment was performed in triplicate and presented mean ± SEM (**$P \leq 0.01$; Student's t-test).

G RT–PCR measurements indicating NOTCH1 expression in EPHB2^{high}OLFM4^{high} colon stem cells transfected with either an empty vector control or CRISPR/Cas9 gRNAs and subsequently treated with DMSO, DAPT or Matrigel-embedded JAG1. The experiment was performed in triplicate and presented mean ± SEM (**$P \leq 0.01$; Student's t-test).

H Single EPHB2^{high}OLFM4^{high} colon stem cells were transfected with either an empty vector control or CRISPR/Cas9 gRNAs. Shown is Western blot analysis for NICD expression in sorted EPHB2^{high}OLFM4^{high} colon stem cells from each condition. Actin was used as a loading control.

I RT–PCR measurements indicating Notch1/2, Hes1/5, Olfm4, and Lgr5 expression in EPHB2^{high}OLFM4^{high} colon stem cells for each condition described in (H). The experiment was performed in triplicate and presented mean ± SEM (**$P \leq 0.01$; Student's t-test).

gene and enhances its transcription, hence producing more Notch1 receptors and forming a positive feedback loop in intestinal stem cells.

CRISPR mutation of the binding motif (PF KO) reduced colony-forming efficiency and growth rate of intestinal organoids markedly (Fig 2F and G). Furthermore, the mutation significantly reduced the number of CBCs and the ratio of CBC to Paneth cell in the niche (Figs 2F and H, and EV2J). Next, to understand how this positive feedback influences Notch signaling and cell fate, we transfected sorted LGR5-EGFP$^+$ CBCs with an RBPJk-dsRED reporter as an indicator of Notch activity and grew them into organoids, followed by FACS analysis. In PF KO organoids, the Notch^{high}/LGR5^{high} (dsRed^{high}/GFP^{high}) CBC population was significantly reduced, responded less to JAG1, and was completed depleted by DAPT compared to the empty vector control (Fig 2I). Therefore, the positive feedback amplifies Notch signaling, and maintains CBC self-renewal and the mosaic niche pattern.

To test whether additional Notch activation can compensate for the Notch1 positive feedback function, CRISPR/Cas9 vectors with specific guide RNAs (gRNAs) were transfected into tamoxifen-inducible NICD-OE intestine cells derived from a LGR5-EGFP-CreERT2 × Rosa26-YFP-NICD strain (Oh et al, 2013; Fig EV3). NICD overexpression enhanced colony-forming efficiencies and organoid sizes in both empty vector control and CRISPR mutated organoids (Fig EV3A–C). NICD overexpression largely restored the colony formation efficiencies of mutated organoid to wild-type (empty vector) levels, but not quite their sizes.

The Notch1 PF motif is conserved in human colon organoids

Like their mouse counterparts, human intestinal and colon epithelia also contain LGR5$^+$ cells and are highly regenerative. A similar computational analysis of the human genome identified an analogous NICD/RBPJk binding region (TTCCCACG, Chr9: 139,425,108–139,425,115) located on the second intron of the human Notch1 sequence (Fig 3A), which also showed high enrichment of H3K4me1 and H3K27ac enhancer chromatin marks in several

human cell lines (Fig EV4A). We then derived human colon organoids using normal colon tissue in surgically resected specimens from colorectal cancer (CRC) patients (Sato et al, 2011b). ChIP-PCR validated NICD binding on the predicted sequence (Fig 3B) in human colon stem cells marked by EPHB2^{high}OLFM4^{high} expression (Jung et al, 2011). Consistent with mouse CBCs, NICD binding to the motif was suppressed by DAPT and enhanced by JAG1 treatment. We then designed CRISPR-Cas9 vectors to mutate the NICD/RBPJk binding motif in human colon stem cells (Figs 3C and EV4B, and Table EV2). ChIP-qPCR validated that the CRISPR/Cas9-mediated mutation reduced NICD binding to the motif in all three conditions (DMSO, JAG1, and DAPT), and prevented JAG1 treatment from increasing NICD binding (Fig EV4C). Suppression of the Notch1 PF by the mutations (PF KO) significantly reduced organoid-forming efficiency, size of organoids, and the percentage of EPHB2^{high}OLFM4^{high} colon stem cells compared to empty vector control (Figs 3D–F and EV4D). The epithelial cell identity of the EPHB2^{high}OLFM4^{high} cells was validated by their EpCAM expression (Fig EV4D). Without the signal-amplifying Notch1 PF, colon stem cells had lower Notch1 transcript levels (Fig 3G). Mutated colon stem cells also had lower expression levels of NICD, other Notch signaling components, and human colon stem cell markers (LGR5 and OLFM4) (Fig 3H and I). In summary, the Notch1 PF promotes Notch signaling and self-renewal in human colon stem cells.

Notch1 PF enhances robustness in dynamic stem cell niche models

The niche maintains a mosaic Notch^{high} (CBC) and Notch^{low} (Paneth) cell pattern for proper homeostasis. It seems that lateral inhibition (LI) alone should be sufficient for generating such patterns as it did for other developmental patterning (Collier et al, 1996). The positive feedback increases Notch signaling levels in CBCs, but, from a "control" perspective, LI could hypothetically achieve the same objective by simply increasing Notch1 transcription/translation rates. To understand whether the addition of the Notch positive feedback to LI (PFLI) offer any regulatory advantages

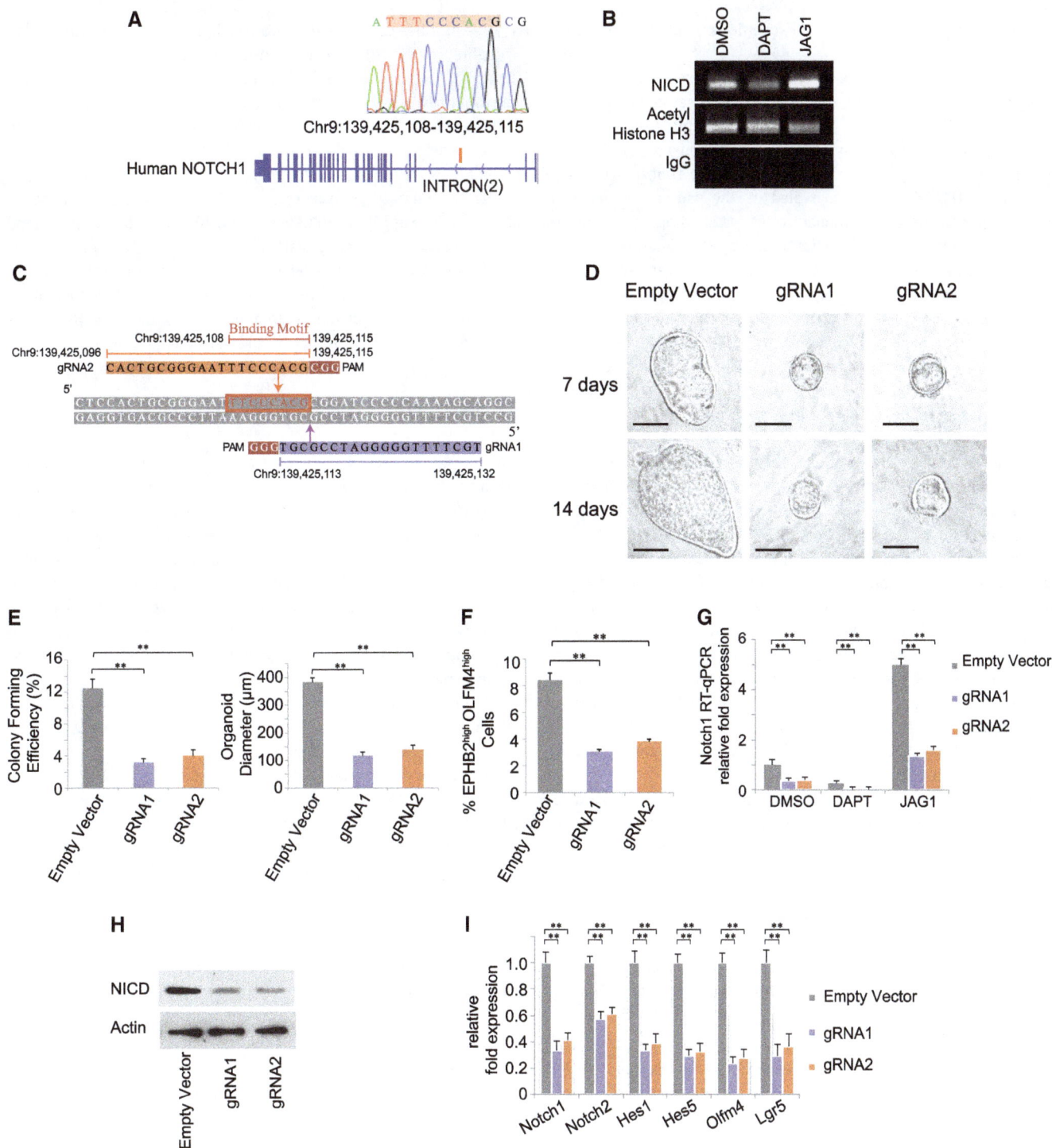

Figure 3.

that cannot be achieved by the LI motif alone, we constructed single-, pair-, and multicell mathematical models to analyze LI with and without PF (Materials and Methods, Table EV3).

We first analyzed Notch response to external ligands in a single cell as an input–output function. In LI, external ligands activate receptors and Notch signaling, which in turn suppresses internal ligand expression. Hence, increasing levels of external ligands leads to a monotonic decrease of internal ligands (Figs 4A and EV5A). On the other hand, PFLI causes bifurcation and generates a more switch-like response with hysteresis (Figs 4A and EV5B).

Pair-cell analysis suggests that both LI and PFLI could achieve intercellular bistability, with two neighboring cells settling in opposite (Notch^{high} vs. Notch^{low}) states. Nevertheless, PFLI is much more robust in generating bistability than LI alone and is less dependent on cooperativity (Hill coefficient) of the reactions (Figs 4B and EV5C and D).

Next, we used the maximum Lyapunov exponent (MLE) to analyze the stability of patterning in multicellular systems (Sprinzak *et al*, 2011). PFLI is able to maintain stable patterns over a much wider parameter range than LI, especially when cooperativity of reaction is low, suggesting that PFLI is a more robust patterning mechanism (Fig 4C). We then scaled up the pair-cell model to a multicellular model with stationary cells surrounding each other to explore Notch patterning dynamics. In steady state analysis, both LI and PFLI can generate stable mosaic patterns with varying levels of Notch signaling and ligands (Fig EV6A). However, dynamic simulations from an initial homogeneous state suggest that PFLI reaches the steady state pattern much faster than LI by speeding up divergence of individual cell signaling states (Fig 4D). Taken together, the analyses suggest that the positive feedback increases robustness, stability, and speed of Notch-mediated patterning.

These properties are not necessarily important for relatively stationary patterns with low cellular turnovers. However, they could be important during regeneration when cell divisions and migration perturb the pattern. We therefore constructed a multiscale, agent-based stochastic model using CompuCell3D (Swat *et al*, 2012). The model takes into consideration the three-dimensional structure, cell growth, division, migration, and cell–cell physical contact (Fig 4E, see Materials and Methods). Notch signaling is only modeled at the base of the crypt, while cells above the niche are simply pushed upwards with no specific assumptions made about their properties. This model does not attempt to capture every aspect of the crypt or the exact movements or lineages of individual cells, which involve many signaling pathways and types of cell–cell interaction. Rather, it is solely designed to test how cell division and migration would affect PFLI- vs. LI-mediated Notch patterning.

As expected, PFLI generates bimodality in cells with regard to Notch signaling (NICD) levels (Fig EV6B). When the strength of the PF is reduced, the ratio of NICD^{high} to NICD^{low} cells as well as NICD levels in the NICD^{high} population decrease (Fig EV6B), consistent with our experimental observation that CRISPR/Cas9-mutated PF KO organoids have lower CBC/Paneth cell ratio, and those CBCs have weaker signals (Figs 2F and G, and EV2G–J).

However, can the Notch signaling pattern be maintained by LI alone if we simply change its parameters to increase Notch signaling levels? In other words, is the PF's role limited to maximizing Notch signaling levels, or is PFLI an inherently different control scheme from LI? To address this hypothetical question, we readjusted the maximum Notch transcription rate in the LI model, so that LI and PFLI have equivalent Notch signaling levels. Indeed, both LI and PFLI can generate mosaic Notch signaling patterns when cell proliferation rate is slow (Fig 4E). We then gradually increased the proliferation rate, which leads to increased rates of cell division, migration, and anoikis (Fig EV6C). PFLI is still able to maintain bimodality and binary patterns, whereas LI starts to show less bimodality (more cells with intermediate Notch levels) and more blurred pattern (Fig 4E). Therefore, PFLI is an inherently more robust control motif than LI when the pattern regulation is dynamic and has to cope with increasing rate of cell division, turnover, and movement.

Discussion

The intestinal stem cell niche controls regeneration and homeostasis of the tissue. Here, we report a direct positive feedback, in which NICD cleaved from activated receptors directly bind to the second intron of Notch1 gene to enhance its expression. This positive feedback is active in mouse intestinal and human colon epithelial cells, and its silencing by CRISPR/Cas9 mutation reduced CBC/Paneth cell ratio and limited self-renewal. Dynamical systems analysis and multiscale stochastic simulation further suggest that the PFLI architecture has an inherent advantage over LI with regard to robustness if the signaling pattern is dynamic.

Biological systems such as the stem cell niche are usually robust (Savageau, 1971; Barkai & Leibler, 1997; Alon *et al*, 1999; Stelling *et al*, 2004; Kitano, 2007; Shen *et al*, 2008). They work most of the time, capable of accommodating different conditions and recovering from mistakes and damages. Control theory would predict that they rely on additional mechanisms such as feedback to enhance their regulation. In fact, recent reports showed that Wnt protein works as short-range signals in the intestinal stem cell niche and there is a positive feedback involving Ascl2 and Wnt signaling in intestinal stem cells (Schuijers *et al*, 2015; Farin *et al*, 2016). From an individual cell perspective, positive feedback loops translate gradients into binary decision states, as shown by the single-cell analysis. From a multicellular signaling pattern perspective, such positive feedbacks

---

Figure 4. Computational analysis of Notch patterning with lateral inhibition and positive feedback.

A   Dynamic analysis of the single-cell Notch signaling model. Internal Dll (D$_i$) vs. external Dll (D$_{ext}$) protein levels are plotted. Lateral inhibition (LI) exhibits monostable behavior (top panel) while Notch positive feedback + LI (PFLI) exhibits bifurcation (bottom panel) in response to external Notch activation. The schematic illustrations on the left of each panel show LI and PFLI in single cells with external Dll (D$_{ext}$) inducing Notch signaling activation and following regulation on Notch (N$_i$) Dll (D$_i$).

B   Phase portraits of the pair-cell Notch signaling model. LI (top panel) requires higher cooperativity (Hill coefficient, h) than PFLI (bottom panel) to generate bistability. Lines: nullclines; solid dots: stable steady states; hollow dots: unstable steady states. N$_i$, N$_j$ and D$_i$, D$_j$ refer to Notch and Dll levels in cells i and j, respectively.

C   Multicell maximum Lyapunov exponents (MLE) analysis of LI-only (top) or PFLI (bottom) circuits spanning parameters of production rates of Notch and Dll with varying degrees of cooperativity (H). Yellow regions (positive MLE values) represent states with patterning, and green regions (negative MLE values) represent states without patterning. PFLI generates patterns over a broader parameter range than LI.

D   Multicellular simulation of Notch signaling models showing DLL levels from initially homogeneous unstable steady states to heterogeneous stable steady states (top panel). Middle-left panel: Representative simulation of LI. Middle-right panel: Representative simulation of PFLI. Bottom panel shows the change rates of DLLs levels during patterning with varying relative strength of positive feedback. Red: Signaling dynamics of cells reaching high steady states. Blue: Signaling dynamics of cells reaching low steady states. Gray: Patterning transition period from homogeneous steady states to heterogeneous steady state.

E   Analysis of a stochastic crypt model integrated with Notch signaling simulation. Left: Structure of the crypt model. Right: Representative violin plots indicating NICD level dynamics at crypt bottom with varying turnover rates. Shown also are corresponding representative simulated patterns of NICD levels in crypt bottoms. PFLI shows stronger NICD bimodality and binary patterns than LI when turnover rates become higher.

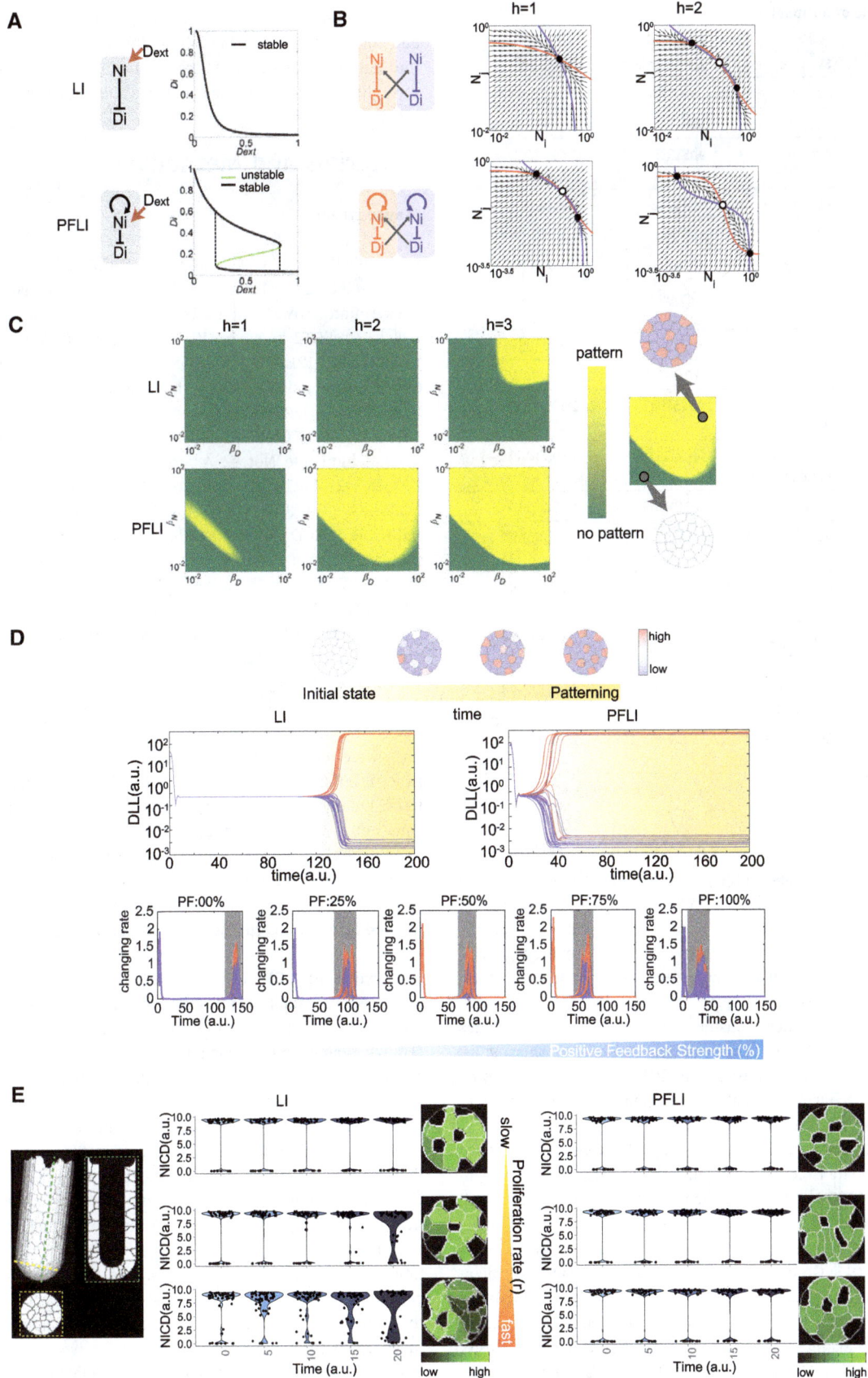

Figure 4.

Table 1. Table of antibodies.

Primary antibody	Supplier	Catalog number	Dilution[a]
anti-ATOH1	Abcam	ab137534	1:1,000 (WB)
anti-β-actin	Abcam	ab6276	1:4,000 (WB)
Anti-CD24 (APC)	Abcam	ab51535	1:500 (FC)
Anti-DLL1	Abcam	ab85346	1:500 (WB)
Anti-DLL4	Abcam	ab7280	1:1,000 (WB)
Anti-EPHB2	R&D Systems	AF467	1:1,000 (FC)
Anti-EpCAM (FITC)	Abcam	ab8666	1:500 (FC)
Anti-GFP	Abcam	ab5450	1:200 (IF)
Anti-HES1	Abcam	ab108937	1:1,000 (WB)
Anti-HES5	Santa Cruz Biotechnology	sc-25395	1:500 (WB)
Anti-JAG1	Santa Cruz Biotechnology	sc-6011	1:500 (WB)
Anti-lysozyme	Abcam	ab108508	1:100 (IF)
Anti-human NICD	R&D Systems	AF3647	1:200 (ChIP)
Anti-mouse NICD	Cell Signaling	4147	1:200 (ChIP)
Anti-NOTCH1	Santa Cruz Biotechnology	sc-9170	1:1,000 (WB)
Anti-NOTCH2	Santa Cruz Biotechnology	sc-32346	1:1,000 (WB)
Anti-OLFM4	Sino Biological Inc.	11639-MM12-50	1:1,000 (FC)

[a]Application: IF, immunofluorescence; WB, Western blotting; FC, flow cytometry.

can enhance the robustness of the patterns, making sure that signaling states are sustained and not easily disrupted by the change of surrounding cells. Therefore, the Notch1 and Ascl2/Wnt positive feedback loops (and potentially many others) may reflect a general regulatory scheme for such systems. Other mechanisms such as cis-inhibition between Notch receptors and ligands (Sprinzak et al, 2010) may further enhance robustness. An approach that combines experimental methods and computational models (Collier et al, 1996; Johnston et al, 2007; Buske et al, 2011) may help us not only identify such mechanisms but also understand their contribution to regulation of the intestinal stem cell niche.

In this study, recombinant JAG1 ligands were embedded in Matrigel to activate Notch signaling in intestinal organoid cells as an alternative to EDTA-mediated Notch signaling activation. Notch ligands are thought to activate receptors via mechanical pulling—hence hybridization of ligands to a surface seems to be required (Wang & Ha, 2013; Gordon et al, 2015). However, ligand hybridization to a 2D surface does not work for 3D cell culture. On the other hand, recombinant ligands such as Dll1, Dll4 (Hicks et al, 2002; Lefort et al, 2006; Mohtashami et al, 2010; Xu et al, 2012; Castel et al, 2013), and JAG1 (Sato et al, 2009; Van Landeghem et al, 2012; VanDussen et al, 2012; Yamamura et al, 2014; Mahapatro et al, 2016; Srinivasan et al, 2016) have been reported to activate

Notch signaling in various cell types, including intestinal organoids and tumor spheroids. One possibility is that ligands embedded in the Matrigel provide the necessary pulling force to activate receptors, although the exact mechanism is still not clear.

Materials and Methods

Mouse strains

LGR5-EGFP mice on a mixed 129/C57BL/6 background and Rosa26-CAG-LSL-tdTomato-WPRE mice on a mixed 129/C57BL/6 background were purchased from The Jackson laboratory. Notch1-CreERT2 knock-in (KI) mice, Notch2-CreERT2 KI mice, and Rosa26-NICD-IRES-YFP KI mice were a generous give from Dr. Spyros Artavanis-Tsakona's laboratory at Harvard University. Subsequently, we generated an inducible Notch1 reporter mouse strain (Notch1-CreERT2 KI × Rosa26-CAG-LSL-tdTomato-WPRE) and an inducible Notch2 reporter mouse strain (Notch2-CreERT2 KI × Rosa26-CAG-LSL-tdTomato-WPRE). We also generated a LGR5-EGFP-CreERT2 × Rosa26-NICD-IRES-YFP KI mouse strain for inducible NICD overexpression (NICD-OE). Genotyping was performed using the following PCR primer pairs for Notch1 (forward: ATAGGAACTTCAAAATGTCGCG; reverse: CACACTTC CAGCGTCTTTGG), Notch2 (forward: ATAGGAACTTCAAAATGT CGCG; reverse: CCCAACGGTGCCAAAAGAGC), and NICD (forward: CTTCACACCCCTCATG ATTGC; reverse: GCAATCGGTCCATGTGA TCC). The thermocyling profile used for PCR amplification is described as follows: 95°C (5 min)/[95°C (30 s), 60°C (30 s), 72°C (60 s)] for 35 cycles/72°C (10 min). Notch1 and Notch2 reporter mice were induced with 75 mg/kg tamoxifen by i.p. injection. The LGR5-EGFP-CreERT2 × Rosa26-NICD-IRES-YFP KI mouse strain was treated daily with 75 mg/kg tamoxifen (i.p. injection) for eight consecutive days to induce Cre enzyme activity and NICD-OE phenotype. All animal experiments were approved by the Cornell Center for Animal Resources and Education (CARE).

Murine intestinal crypt isolation and organoid culture

Eight-week-old LGR5-EGFP or LGR5-EGFP/NICE-OE mice were sacrificed to establish intestinal organoid culture. Briefly, small intestines were harvested, flushed with Ca^{2+}/Mg^{2+}-free PBS to remove debris, and opened up longitudinally to expose luminal surface. A glass coverslip was then gently applied to scrape off villi, and the tissue was cut into 2–3 mm fragments. Intestinal tissues were then washed again with cold PBS and incubated with 2.0 mM EDTA for 45 min on a rocking platform at 4°C. EDTA solution was then decanted without disturbing settled intestinal fragments and replaced with cold PBS. In order to release intestinal crypts in solution, a 10-ml pipette was used to vigorously agitate tissues. The supernatant was collected, and this process was repeated several times to harvest multiple fractions. The crypt fractions were then centrifuged at 84 g for 5 min. Based on microscopic examination, the appropriate enriched crypt fractions were pooled and centrifuged again to obtain a crypt-containing pellet. Advanced DMEM/F12 (Life Technologies) containing Glutamax (Life Technologies) was used to resuspend the cell pellet and subsequently a 40-μm filter was used to purify crypts. Next, single-cell dissociation was achieved by incubating purified crypt solution at 37°C with 0.8 KU/ml DNase (Sigma), 10 µM ROCK pathway inhibitor, Y-27632 (Sigma), and 1 mg/ml trypsin-EDTA (Invitrogen) for 30 min. Single cells were then passed again though a 40-µm filter and resuspended in cold PBS with 0.5% BSA for FACS analysis to collect LGR5-EGFP$^+$ intestinal stem cells (ISCs), which are also called crypt base columnar (CBC) cells.

Single LGR5-EGFP$^+$ CBCs were suspended in Matrigel (BD Biosciences) at a concentration of 1,000 cells or crypts/ml, and 50 µl Matrigel drops were seeded per well on pre-warmed 24-well plates. Matrigel polymerization occurred at 37°C for 10 min and was followed by the addition of complete media to each well. ISC media included the following: Advanced DMEM/F12 supplemented with Glutamax, 10 mM HEPES (Life Technologies), N2 (Life Technologies), B27 without vitamin A (Life Technologies), and 1 µM N-acetylcysteine (Sigma). Growth factors were freshly prepared each passage in an ISC media solution containing 50 ng/ml EGF (Life Technologies), 100 ng/ml Noggin (Peprotech), and 10% R-spondin1-conditioned media (generated in house). The addition of growth factors occurred every 2 days, and the media were fully replaced every 4 days. Organoids were passaged once per week at a ratio of 1:4 by removing organoids from Matrigel with ice-cold PBS. Next, organoids were incubated on ice for 10 min followed by mechanical disruption, centrifugation, and resuspension in fresh Matrigel.

For *in vitro* studies, organoids derived from single LGR5-EGFP ISCs were treated with one of the following: DMSO or 10 µM DAPT (EMD Millipore) added to the media for 48 h (Sikandar *et al*, 2010), or 1uM JAG-1 (AnaSpec) embedded in Matrigel for 48 h (Takeda *et al*, 2011). EDTA was added to ISC media (for a final concentration of 0.5 mM EDTA) to treat organoids for 4 h. Subsequently, organoids were harvested and analyzed by FACS to isolate LGR5-EGFP cells and Paneth cells for RT–PCR or protein analysis. FACS was conducted using a Beckman Coulter flow cytometer with a 40-µm filter. Data analysis was performed using FlowJo software to gate populations according to 7-AAD viability, and forward and side scattering. Cutoff thresholds were provided by unstained ISCs and single stained ISCs when using multiple fluorochromes in order to achieve appropriate compensation.

CRISPR/Cas9 genomic editing

The procedure for CRISPR/Cas9-mediated transfection in mouse ISC organoids has been previously described (Schwank *et al*, 2013). Briefly, guide RNA (gRNA) sequences were designed by Optimized CRISPR Design tool (http://crispr.mit.edu/), and CRISPR/Cas9 plasmids including gRNA sequences were purchased from GenScript. For murine experiments, gRNAs targeting the NICD binding motif on Notch1 included the following: gRNA1: (TACATGCATGGAAGGTGCGT) or gRNA2: (GTACATGCATGGAAGGTGCG) and were cloned into a pGS-gRNA-Cas9-Puro vector backbone. A pGS-CAS9-PURO only vector (no gRNA) was used as a control. Single sorted LGR5-EGFP$^+$ ISCs were transfected using Lipofectamine-2000 (Life Technologies) according to the manufacturer's instructions. Briefly, 4uL Lipofectamine-2000 (diluted in 50 µl Opti-MEM) and 2 µg of CRISPR/Cas9 plasmids (diluted in 50 µl Opti-MEM) were mixed 1:1: and incubated for 5 min at room temperature. Lipofectamine/DNA complexes were then added to single LGR5-EGFP$^+$

ISCs (50 µl/well) in a 24-well plate, which was subsequently centrifuged for 1 h and incubated at 37°C for 4 h. ISCs were then resuspended in Matrigel and overlaid with ISC media (as prepared above) and supplemented with Y-27632 for 48 h. Next, transfected ISCs were selected in media (without R-spondin) containing 300 ng/µl puromycin for 72 h. Selection media were then replaced with ISC media, and organoids were monitored for 15 days followed by FACS analysis or co-immunofluorescence. Individual CRISPR/Cas9-mutated organoid clones were also harvested and lysed for DNA extraction using a QIAmp DNA Mini kit (Qiagen: 51304) according to the manufacturer's instructions. Subsequently, the NICD binding site on mouse Notch1 was amplified by PCR using the following primers (forward: AGAAGAGAAGACAGGAGAAGGA and reverse: GAAGCCACTGACTTTCCTAGAG) and analyzed by Sanger sequencing to visualize mutations. CRISPR/Cas9-mutated organoids derived from single transfected LGR5-EGFP ISCs were also treated for 48 h with DAPT or JAG-1 (as described earlier) before harvesting cells for FACS to isolate LGR5-EGFP$^+$ cells for RT–PCR analysis.

In order to study Notch signaling dynamics, a RBPJk-dsRed reporter on a pGA981-6 vector backbone (Addgene #47683) was transfected into single wild-type or CRISPR/Cas9-mutated sorted LGR5-EGFP ISCs using Lipofectamine-2000 according to the protocol described above. ISCs were then treated for 48 h with 10 µM DAPT or 1 µM JAG-1 and analyzed by microscopy and flow cytometry for LGR5-EGFP and RBPJk-dsRed expression.

Isolation of single cells from human colonic tissue

Approval for this research protocol was obtained from IRB committees at Weill Cornell Medical College and NY Presbyterian Hospital. Patients undergoing colorectal surgery provided written informed consent for use of human tissues. Material was derived from proximal colonic tissue during surgical biopsies. The procedure for isolation of colonic stem cells and organoid culture is previously described (Jung *et al*, 2011). Briefly, colonic specimens were collected and incubated in Advanced DMEM/F12 supplemented with gentamycin (Life Technologies) and fungizone (Life Technologies). Extraneous muscular and sub-mucosal layers were removed from colonic mucosa. The tissue was cut into 1 cm fragments and incubated with 8 mM EDTA for 1 h on a rocking platform at 37°C followed by a 45-min incubation at 4°C. The supernatant was replaced with Advanced DMEM/F12 supplemented with Glutamax, HEPES, and 5% FBS. Vigorous shaking released crypts, which were collected in several fractions. Crypt fractions were then centrifuged (37 *g*, 5 min) and visualized by microscopy to determine which enriched fractions to combine. Subsequently, pooled crypt fractions were centrifuged and resuspended in Advanced DMEM/F12 supplemented with Glutamax, HEPES, N-2, B-27 without vitamin A, 1 mM N-acetyl-L-cysteine, nicotinamide (Sigma), 10 µM Y-27632, 2.5 µM PGE2 (Sigma), 0.5 mg/ml Dispase (BD Biosciences), and 0.8 KU/ml DNase I. Cells were then incubated for 15 min at 37°C followed by mechanical disruption and passage of cell solution through 40-µm filter to obtain a single-cell suspension.

Human colon organoid culture

Single human colon cells were stained with EPHB2 (conjugated to PE), OLFM4 (conjugated to APC), EpCAM-FITC, and 7-AAD according to standard protocols and were suspended in cold PBS with 0.5% BSA for FACS analysis. FlowJo software was used to gate populations according to 7-AAD viability, and forward and side scattering. Cutoff thresholds were provided by unstained ISCs and single stained ISCs when using multiple fluorochromes in order to achieve appropriate compensation. EPHB2^{high}OLFM4^{high} colon stem cells were harvested, and subjected to lipotransfection of CRISPR/Cas9 plasmids using Lipofectamine-2000 in a similar method as described earlier for mouse ISCs. CRISPR/Cas9 gRNAs targeting the NICD binding motif of human Notch1 (cloned into a pGS-gRNA-Cas9-Puro vector backbone) were designed and ordered from GenScript with the following inserted gRNA sequences: (gRNA1: TGCTTTT GGGGGATCCGCGT, gRNA2: CACTGCGGGAATTTCCCACG). A pGS-CAS9-PURO only vector (no gRNA) was used as a control. Transfected human colon stem cells were selected in medium lacking WNT-3A, and R-spondin1 and containing Y-27632 and 300 ng/µl puromycin for 48 h.

Subsequently, transfected cells were suspended in Matrigel, and overlaid with human colon stem cell medium containing Advanced DMEM/F12 supplemented with Glutamax, HEPES, N-2, B-27 without vitamin A, 1 mM N-acetyl-L-cysteine, nicotinamide, PGE2, Y-27632, human Noggin (Peprotech), human EGF (Life Technologies), gastrin (Sigma), TGF-β type I receptor inhibitor A83-01 (Tocris), P38 inhibitor SB202190 (Sigma-Aldrich), WNT3A-conditioned media (generated in house), and R-spondin1-conditioned medium (generated in house) (Jung et al, 2011). For organoid culture, full medium was replaced every 2 days. Transfected organoids were monitored for 14 days and then harvested and analyzed by FACS to isolate EPHB2^{high}OLFM4^{high} colon stem cells for RT–PCR and protein analysis. Individual CRISPR/Cas9-mutated organoid clones were also harvested and lysed for DNA extraction using a QIAmp DNA Mini kit (Qiagen: 51304) according to the manufacturer's instructions. Subsequently, the NICD binding site on human Notch1 was amplified by PCR and analyzed by Sanger sequencing to visualize mutations.

Quantitative RT–PCR and protein analysis

Total RNA from mouse ISCs or human colon stem cells was extracted using a Qiagen RNeasy Plus kit. Subsequently, isolated RNA was reverse transcribed to cDNA using ABI Taqman Reverse Transcription kit (Applied Biosystems). ABI Taqman Master mix and ABI Prism HT7900 were used to run quantitative real-time PCR. Taqman primers (ABI) purchased from Life Technologies were used for the following mouse genes: Notch1 (Product ID: Mm00627185_m1), Notch2 (Product ID: Mm00803077_m1), Hes1 (Product ID: Mm01342805_m1), Hes5 (Product ID: Mm00439311_g1), Dll1 (Product ID: Mm01279269_m1), Dll4 (Product ID: Mm00444619_m1), Jag1 (Product ID: Mm00496902_m1), Atoh1 (Product ID: Mm00476035_s1), Lgr5 (Product ID: Mm00438890_m1). Human Taqman primers purchased from Life Technologies include: Notch1 (Product ID: Hs01062014_m1), Notch2 (Product ID: Hs01050702_m1), Hes1 (Product ID: Hs00172878_m1), Hes5 (Product ID: Hs01387463_g1), Lgr5 (Product ID: Hs00969422_m1), and Olfm4 (Product ID: Hs00197437_m1). RT–PCR analysis represents the average of three independent experiments normalized to GAPDH expression. Error bars designate SEM. Protein extraction from mouse ISCs or human

colon stem cells and Western blotting were performed as previously described. β-actin was used as a control for normalization (Pan et al, 2008). Antibodies used for Western blotting are listed in Table 1.

ChIP-PCR

Mouse intestinal and human colonic organoids were harvested, and ChIP-PCR was performed according to manufacturer's instructions (EMD Millipore: 17-408). Briefly, normal rabbit IgG was used as a negative control while rabbit anti-acetyl histone H3 was used as a positive control for immunoprecipitation (IP). Subsequently, primer pairs specific for human or mouse GAPDH sequences were for positive PCR controls. Following IP using anti-mouse NICD, PCR primers (forward: AGATGAAGGTGGAGCATGTG, reverse: TTTTCC CACGGCCTAGAAG) were used for amplification of Notch1. Similarly, for ChIP assays involving anti-human NICD, PCR primers (forward: ACTAGGTGTCACCAAAGTGC, reverse: CATGACCATCTTG GCCTCTC) were used to amplify Notch1. Sanger sequencing was used to validate NICD binding motif on Notch1 for PCR products. Subsequently, ChIP-qPCR analyses were performed according to the manufacturer's instructions (Active Motif: 53029). Antibodies used for ChIP are listed in Table 1.

Immunofluorescence

Intestinal tissues from tamoxifen-induced Notch1 (tdTomato) reporter mice were harvested at various time points, fixed with 4% PFA, snap-frozen in O.C.T., cryo-sectioned, and visualized on a Zeiss LSM 510 laser scanning confocal microscope. DAPI was used as a nuclear counterstain. For in vitro imaging, wild-type or CRISPR/Cas9-mutated intestinal organoids derived from LGR5-EGFP mice were embedded in Matrigel on glass chamber slides. Cells were fixed for 15 min at room temperature using 4% PFA and rinsed three times with PBS. 0.2% Triton X-100 was used for permeabilization of cell membranes. Next, cells were incubated in a serum-free blocking solution (Dako) for 30 min. For co-immunofluorescence staining, an antibody diluent solution (Dako) was used to prepare primary and secondary antibodies. Primary antibodies were added overnight at room temperature followed by application of Alexa-flour 488/555 secondary antibodies for 1 h. Organoids were visualized using lysozyme (LYZ) and LGR5 (detected by GFP) expression. DAPI (Life Technologies) was as a nuclear counterstain on a Zeiss LSM 510 laser scanning confocal microscope using an Apo 40× NA 1.40 oil objective. Antibodies used for immunofluorescence are listed in Table 1.

Luciferase assay

The wild-type (WT) enhancer sequence and three mutated sequences were PCR amplified (WT:CTGTCAACCTTGCTTCCTCCC CttcccacgCACCTTCCATGCATGTACACAC, Mut1: CTGTCAACCTTG CTTCCTCCCCttcccgactcaCTTCCATGCATGTACACAC, Mut2: CTGT CAACCTTGCTTCCTCCCCttcccaCACCTTCCATGCATGTACACAC, and Mut3: CTGTCAACCTTGCTTCCTCCCCcgtaatacCACCTTCCATGCAT GTACACAC) and cloned into a pGL4.24 firefly luciferase reporter plasmid (Promega). These luciferase reporter vectors and Renilla luciferase vector (pRL-SV40, Promega) were co-transfected into mouse intestine cells using Lipofectamine 3000 (Life Technologies) according to the manufacturer's instructions. Cell lysates were collected, and luciferase samples were prepared using the Luc-Pair Duo-Luciferase Assay kit (Genecopoeia) in 48 h after transfection. Firefly luciferase activities were measured using an FLUOstar optima plate reader (BMG Labtech), and firefly luciferase activity was normalized to *Renilla* luciferase activity.

Biotinylated nucleotide pull-down assay

Oligonucleotides of the wild-type and three mutated sequences (same as in the luciferase assay) were labeled using a biotin labeling kit (Pierce) and annealed for pull-down assay. Mouse intestine crypt cell lysates were freshly prepared using RIPA buffer (Millipore) with proteinase inhibitor (Roche). After precleared using Dynabeads M-270 streptavidin (Invitrogen), the cell lysates were diluted in binding buffer and incubated with the biotinylated DNA duplex for 2 h at 4°C. Dynabeads M-270 streptavidin were then added into the mixture and incubated for 1 h at 4°C. After washing, the DNA-binding protein complexes were released from the Dynabeads. The retrieved proteins were collected for Western blot validation.

Statistical analysis

The data are displayed as mean ± SEM. Statistical comparisons between two groups were made using Student's *t*-test. $P < 0.05$ was used to establish statistical significance.

ODEs models of Notch signaling circuits

The mathematical model of Notch signaling includes three types of regulations: (i) trans-activation of Notch receptor by external ligand (TA), (ii) lateral Inhibition (LI), and (iii) positive feedback (PF). Transcriptional activation is modeled by Hill function $\sigma(x, k, p) = (x^p/(k^p + x^p))$, and transcriptional suppression is modeled by Hill function $\delta(x, k, h) = (k^h/(k^h + x^h))$, where x refers to the regulator, k refers to the saturation coefficient, and h is the Hill coefficient. Below are the ODE equations:

$$\dot{Notch}_{mRNA\,i} = \beta_{n0} + \beta_n \cdot \sigma(k_t \cdot NOTCH_i \langle DLL_j \rangle, k_p, p) - \alpha_n \cdot Notch_{mRNA\,i}$$
$$\dot{NOTCH}_i = \beta_N \cdot Notch_{mRNA\,i} - \alpha_N \cdot NOTCH_i - k_t \cdot NOTCH_i \langle DLL_j \rangle$$
$$\dot{Dll}_{mRNA\,i} = \beta_{d0} + \beta_d \cdot \delta(k_t \cdot NOTCH_i \langle DLL_j \rangle, k_d, h) - \alpha_d \cdot Dll_{mRNA\,i}$$
$$\dot{DLL}_i = \beta_D \cdot Dll_{mRNA\,i} - \alpha_D DLL_i - k_t \cdot \langle NOTCH_j \rangle DLL_i$$
$$\dot{R}_i = k_t NOTCH_i \langle DLL_j \rangle - \alpha_R R_i$$

where $Notch_{mRNA}$, $NOTCH$, Dll_{mRNA}, DLL, and R refer to the expression level of Notch mRNA, NOTCH receptor, Dll mRNA, DLL ligand, and cleaved NICD (activated Notch signaling), respectively. The annotation i and annotation j refer to cell j adjacent to cell i. β_s denote the synthesis rates (transcription rates for mRNAs and translation rates for protein), while α_s denote the degradation rates. $\langle X_j \rangle_i$ is the average expression of X from the neighboring j cells of cell i. k_t is the reaction rate of trans-activation. β_{n0} and β_{d0} are the basal transcriptional rates of $Notch_{mRNA}$ and Dll_{mRNA}. By changing the ratios of $\beta_n/(\beta_{n0} + \beta_n)$ and $\beta_d/(\beta_{d0} + \beta_d)$, we can adjust the regulatory strength of LI and PF.

For simplicity, we transform the equations into dimensionless equations with dimensionless parameters:

$$\tau \equiv t_0 t, Nm \equiv \frac{Notch_{mRNA}}{Nm_0}, N \equiv \frac{NOTCH}{N_0}, Dm \equiv \frac{Dll_{mRNA}}{Dm_0},$$
$$D \equiv \frac{DLL}{D_0}, R \equiv \frac{R}{R_0}, N_0 = D_0 = R_0 \equiv \frac{t_0}{k_t}.$$

Model 1:

$$\dot{Nm}_i = \beta_{n0} + \beta_n \cdot \sigma(NOTCH_i \langle DLL_j \rangle, k_p, p) - \alpha_n \cdot Nm_i$$
$$\dot{N}_i = \beta_N \cdot Nm_i - \alpha_N \cdot N_i - N_i \langle D_j \rangle$$
$$\dot{Dm}_i = \beta_{d0} + \beta_d \cdot \delta(NOTCH_i \langle DLL \rangle_j, k_d, h) - \alpha_d \cdot Dm_i$$
$$\dot{D}_i = \beta_D \cdot Dm_i - \alpha_D \cdot D_i - \langle N_j \rangle D_i$$
$$\dot{R}_i = NOTCH_i \langle DLL \rangle_j - \alpha_R R_i$$

Where:

$$\beta_{n0} \equiv \frac{\beta_{n0}}{t_0 \cdot Nm_0}, \beta_n \equiv \frac{\beta_n}{t_0 \cdot Nm_0}, \beta_N \equiv \frac{\beta_N \cdot Nm_0}{t_0 \cdot N_0}, \beta_{d0} \equiv \frac{\beta_{d0}}{t_0 \cdot Dm_0},$$
$$\beta_d \equiv \frac{\beta_d}{t_0 \cdot Dm_0}, \beta_D \equiv \frac{\beta_D \cdot Dm_0}{t_0 \cdot D_0}, \alpha_n \equiv \frac{\alpha_n}{t_0}, \alpha_N \equiv \frac{\alpha_N}{t_0}, \alpha_d \equiv \frac{\alpha_d}{t_0},$$
$$\alpha_D \equiv \frac{\alpha_D}{t_0}, \alpha_R \equiv \frac{\alpha_R}{t_0}, k_p \equiv \frac{k_p}{k_t \cdot N_0 \cdot D_0}, k_d \equiv \frac{k_d}{k_t \cdot N_0 \cdot D_0}.$$

To modulate the relative strength of the transcriptional regulations (LI, PF), we rescaled the ratios of basal transcriptional rates and regulated transcriptional rates as $S_{PF} \equiv (\beta_n)/(\beta_{n0} + \beta_n)$, $S_{LI} \equiv (\beta_d/(\beta_{d0} + \beta_d))$, and maximum transcriptional rate as $\beta_{nm} = (\beta_{n0} + \beta_n), \beta_{dm} = (\beta_{d0} + \beta_d)$, where $0 \leq S_{PF}, S_{LI} \leq 1$. A new set of equations can be shown as

Model 2:

$$\dot{Nm}_i = \beta_{nm}[(1 - S_{PF}) + S_{PF} \cdot \sigma(NOTCH_i \langle DLL_j \rangle, k_p, p)] - \alpha_n \cdot Nm_i$$
$$\dot{N}_i = \beta_N \cdot Nm_i - \alpha_N \cdot N_i - N_i \langle D_j \rangle$$
$$\dot{Dm}_i = \beta_{dm}[(1 - S_{LI}) + S_{LI} \cdot \delta(NOTCH_i \langle DLL_j \rangle, k_d, h)] - \alpha_d \cdot Dm_i$$
$$\dot{D}_i = \beta_D \cdot Dm_i - \alpha_D \cdot D_i - \langle N_j \rangle D_i$$
$$\dot{R}_i = N_i \langle D_j \rangle - \alpha_R R_i$$

We assume the timescale for mRNA is much faster than proteins, so quasi-steady state method is applied to reduce the mRNA species in the equations:

Set $\dot{Nm}_i = 0, \dot{Dm}_i = 0$

$$Nm_i^* = \frac{\beta_{nm}}{\alpha_n}[(1 - S_{PF}) + S_{PF} \cdot \sigma(NOTCH_i \langle DLL_j \rangle, k_p, p)],$$
$$Dm_i^* = \frac{\beta_{dm}}{\alpha_d}[(1 - S_{LI}) + S_{LI} \cdot \delta(NOTCH_i \langle DLL_j \rangle, k_d, h)]$$

replace $[Nm_i, Dm_i]$ with $[Nm_i^*, Dm_i^*]$ in $\dot{N}_i$ and $\dot{D}_i$ respectively, and set $\beta_N = (\beta_N (\beta_{nm}/\alpha_n))$, $\beta_D = (\beta_D (\beta_{dm}/\alpha_d))$. A simple protein model can be shown as

Model 3:

$$\dot{N}_i = \beta_N \cdot [(1 - S_{PF}) + S_{PF} \cdot \sigma(NOTCH_i \langle DLL_j \rangle, k_p, p)] - \alpha_n \cdot N_i - N_i \langle D_j \rangle$$
$$\dot{D}_i = \beta_D \cdot [(1 - S_{LI}) + S_{LI} \cdot \delta(NOTCH_i \langle DLL_j \rangle, k_d, h)] - \alpha_D \cdot D_i - \langle N_j \rangle D_i$$
$$\dot{R}_i = N_i \langle D_j \rangle - \alpha_R R_i$$

Computational modeling

The deterministic model was constructed and simulated in MATLAB, and the systems dynamics analysis was solved by numerical optimization in MATLAB. The 3D stochastic crypt model was designed and simulated based on the Glazier–Graner–Hogeweg (GGH) computational model using Compucell3D (Swat *et al*, 2012). A supporting layer covered by a single-cell layer of epithelial cells was designed to mimic the finger-like shape of intestine crypts. All *in silico* epithelial cells inherited an effective energy function programmed in the CGH's cell-lattice configuration to have the desired cell prosperities, behaviors, and interactions. The epithelial cells were programmed to possess the essential cellular prosperities including: (i) cell growth, (ii) cell divisions, (iii) cell–cell adhesion, and (iv) anoikis (when epithelial cells detach from supporting layers). In addition, a module of SBML (Systems Biology Markup Language) solver was applied to integrate the Notch signaling models (LI, PFLI) programmed in SBML (Hucka *et al*, 2003) format to every epithelial cell at the bottom of the crypt. Notch signaling and stochastic cellular dynamics were simulated simultaneously in the combined model. A Notch signaling threshold was assigned to determine Notch^{high} (stem) cells and Notch^{low} (Paneth) cells at the crypt bottom. Notch^{high} cells were programmed to actively grow and divide when their cell volume reached division threshold, while Notch^{low} cells were programmed to neither grew nor divide to mimic differentiated Paneth cells. These cells naturally migrate upward to leave the bottom of the crypt with the force generated by the growing and dividing cells at the crypt base. R was used to analyze and plot the statistical results.

Acknowledgements

We thank Dr. Spyros Artavanis-Tsakonas for providing the Notch transgenic mice and the Cornell imaging facility. This work was supported by DOD (DARPA): HR0011-16-C-0138, NIH R01GM95990, NIH R01GM114254, NSF 1350659 career award, NSF 1137269, NYSTEM C029543, and CAS Pioneer Hundred Talents Program, the Thousand Young Talents Program of China.

Author contributions

K-YC, TS, PB, and XS conceived the concept and designed the experiments. TS and K-LT conducted all experiments with murine and human organoids, flow cytometry, and CRISPR editing experiments with assistance from LW, PKLM, SK, NR, MW, JC, and AKV. K-YC and JMB developed the mathematical model together. K-YC performed ChIP-Seq analysis. PB performed the pull-down assay, ChIP-PCR, and additional organoid assays. K-YC, TS, and XS wrote the manuscript with comments from SML, NN, and JAG. All authors approved the paper.

References

Alon U, Surette MG, Barkai N, Leibler S (1999) Robustness in bacterial chemotaxis. *Nature* 397: 168–171

Artavanis-Tsakonas S, Rand MD, Lake RJ (1999) Notch signaling: cell fate control and signal integration in development. *Science* 284: 770–776

Barkai N, Leibler S (1997) Robustness in simple biochemical networks. *Nature* 387: 913–917

Barker N, van Es JH, Kuipers J, Kujala P, van den Born M, Cozijnsen M, Haegebarth A, Korving J, Begthel H, Peters PJ, Clevers H (2007) Identification of stem cells in small intestine and colon by marker gene Lgr5. *Nature* 449: 1003–1007

Barker N (2014) Adult intestinal stem cells: critical drivers of epithelial homeostasis and regeneration. *Nat Rev Mol Cell Biol* 15: 19–33

Buczacki SJ, Zecchini HI, Nicholson AM, Russell R, Vermeulen L, Kemp R, Winton DJ (2013) Intestinal label-retaining cells are secretory precursors expressing Lgr5. *Nature* 495: 65–69

Buske P, Galle J, Barker N, Aust G, Clevers H, Loeffler M (2011) A comprehensive model of the spatio-temporal stem cell and tissue organisation in the intestinal crypt. *PLoS Comput Biol* 7: e1001045

Castel D, Mourikis P, Bartels SJ, Brinkman AB, Tajbakhsh S, Stunnenberg HG (2013) Dynamic binding of RBPJ is determined by Notch signaling status. *Genes Dev* 27: 1059–1071

Collier JR, Monk NA, Maini PK, Lewis JH (1996) Pattern formation by lateral inhibition with feedback: a mathematical model of delta-notch intercellular signalling. *J Theor Biol* 183: 429–446

Consortium EP (2012) An integrated encyclopedia of DNA elements in the human genome. *Nature* 489: 57–74

Farin HF, Jordens I, Mosa MH, Basak O, Korving J, Tauriello DV, de Punder K, Angers S, Peters PJ, Maurice MM, Clevers H (2016) Visualization of a short-range Wnt gradient in the intestinal stem-cell niche. *Nature* 530: 340–343

van der Flier LG, Clevers H (2009) Stem cells, self-renewal, and differentiation in the intestinal epithelium. *Annu Rev Physiol* 71: 241–260

Fre S, Huyghe M, Mourikis P, Robine S, Louvard D, Artavanis-Tsakonas S (2005) Notch signals control the fate of immature progenitor cells in the intestine. *Nature* 435: 964–968

Fre S, Hannezo E, Sale S, Huyghe M, Lafkas D, Kissel H, Louvi A, Greve J, Louvard D, Artavanis-Tsakonas S (2011) Notch lineages and activity in intestinal stem cells determined by a new set of knock-in mice. *PLoS One* 6: e25785

Gordon WR, Zimmerman B, He L, Miles LJ, Huang J, Tiyanont K, McArthur DG, Aster JC, Perrimon N, Loparo JJ, Blacklow SC (2015) Mechanical allostery: evidence for a force requirement in the proteolytic activation of notch. *Dev Cell* 33: 729–736

Hicks C, Ladi E, Lindsell C, Hsieh JJ, Hayward SD, Collazo A, Weinmaster G (2002) A secreted Delta1-Fc fusion protein functions both as an activator and inhibitor of Notch1 signaling. *J Neurosci Res* 68: 655–667

Hucka M, Finney A, Sauro HM, Bolouri H, Doyle JC, Kitano H, Arkin AP, Bornstein BJ, Bray D, Cornish-Bowden A, Cuellar AA, Dronov S, Gilles ED, Ginkel M, Gor V, Goryanin II, Hedley WJ, Hodgman TC, Hofmeyr JH, Hunter PJ *et al* (2003) The systems biology markup language (SBML): a medium for representation and exchange of biochemical network models. *Bioinformatics* 19: 524–531

Johnston MD, Edwards CM, Bodmer WF, Maini PK, Chapman SJ (2007) Mathematical modeling of cell population dynamics in the colonic crypt and in colorectal cancer. *Proc Natl Acad Sci USA* 104: 4008–4013

Jung P, Sato T, Merlos-Suarez A, Barriga FM, Iglesias M, Rossell D, Auer H, Gallardo M, Blasco MA, Sancho E, Clevers H, Batlle E (2011) Isolation and *in vitro* expansion of human colonic stem cells. *Nat Med* 17: 1225–1227

Kim TH, Li F, Ferreiro-Neira I, Ho LL, Luyten A, Nalapareddy K, Long H, Verzi M, Shivdasani RA (2014) Broadly permissive intestinal chromatin underlies lateral inhibition and cell plasticity. *Nature* 506: 511–515

Kitano H (2007) Towards a theory of biological robustness. *Mol Syst Biol* 3: 137

Kopan R, Ilagan MX (2009) The canonical Notch signaling pathway: unfolding the activation mechanism. *Cell* 137: 216–233

Lander AD, Kimble J, Clevers H, Fuchs E, Montarras D, Buckingham M, Calof AL, Trumpp A, Oskarsson T (2012) What does the concept of the stem cell niche really mean today? *BMC Biol* 10: 19

Lefort N, Benne C, Lelievre JD, Dorival C, Balbo M, Sakano S, Coulombel L, Levy Y (2006) Short exposure to Notch ligand Delta-4 is sufficient to induce T-cell differentiation program and to increase the T cell potential of primary human CD34+ cells. *Exp Hematol* 34: 1720–1729

Lopez-Garcia C, Klein AM, Simons BD, Winton DJ (2010) Intestinal stem cell replacement follows a pattern of neutral drift. *Science* 330: 822–825

Mahapatro M, Foersch S, Hefele M, He GW, Giner-Ventura E, McHedlidze T, Kindermann M, Vetrano S, Danese S, Gunther C, Neurath MF, Wirtz S, Becker C (2016) Programming of intestinal epithelial differentiation by IL-33 derived from pericryptal fibroblasts in response to systemic infection. *Cell Rep* 15: 1743–1756

Metcalfe C, Kljavin NM, Ybarra R, de Sauvage FJ (2014) Lgr5$^+$ stem cells are indispensable for radiation-induced intestinal regeneration. *Cell Stem Cell* 14: 149–159

Mohtashami M, Shah DK, Nakase H, Kianizad K, Petrie HT, Zuniga-Pflucker JC (2010) Direct comparison of Dll1- and Dll4-mediated Notch activation levels shows differential lymphomyeloid lineage commitment outcomes. *J Immunol* 185: 867–876

Oh P, Lobry C, Gao J, Tikhonova A, Loizou E, Manent J, van Handel B, Ibrahim S, Greve J, Mikkola H, Artavanis-Tsakonas S, Aifantis I (2013) *In vivo* mapping of notch pathway activity in normal and stress hematopoiesis. *Cell Stem Cell* 13: 190–204

Pan Z, Sikandar S, Witherspoon M, Dizon D, Nguyen T, Benirschke K, Wiley C, Vrana P, Lipkin SM (2008) Impaired placental trophoblast lineage differentiation in Alkbh1(-/-) mice. *Dev Dyn* 237: 316–327

Pellegrinet L, Rodilla V, Liu Z, Chen S, Koch U, Espinosa L, Kaestner KH, Kopan R, Lewis J, Radtke F (2011) Dll1- and dll4-mediated notch signaling are required for homeostasis of intestinal stem cells. *Gastroenterology* 140: 1230–1240.e7

Rand MD, Grimm LM, Artavanis-Tsakonas S, Patriub V, Blacklow SC, Sklar J, Aster JC (2000) Calcium depletion dissociates and activates heterodimeric notch receptors. *Mol Cell Biol* 20: 1825–1835

Riccio O, van Gijn ME, Bezdek AC, Pellegrinet L, van Es JH, Zimber-Strobl U, Strobl LJ, Honjo T, Clevers H, Radtke F (2008) Loss of intestinal crypt progenitor cells owing to inactivation of both Notch1 and Notch2 is accompanied by derepression of CDK inhibitors p27Kip1 and p57Kip2. *EMBO Rep* 9: 377–383

Sato T, Vries RG, Snippert HJ, van de Wetering M, Barker N, Stange DE, van Es JH, Abo A, Kujala P, Peters PJ, Clevers H (2009) Single Lgr5 stem cells build crypt-villus structures *in vitro* without a mesenchymal niche. *Nature* 459: 262–265

Sato T, van Es JH, Snippert HJ, Stange DE, Vries RG, van den Born M, Barker N, Shroyer NF, van de Wetering M, Clevers H (2011a) Paneth cells constitute the niche for Lgr5 stem cells in intestinal crypts. *Nature* 469: 415–418

Sato T, Stange DE, Ferrante M, Vries RG, Van Es JH, Van den Brink S, Van Houdt WJ, Pronk A, Van Gorp J, Siersema PD, Clevers H (2011b) Long-term expansion of epithelial organoids from human colon, adenoma, adenocarcinoma, and Barrett's epithelium. *Gastroenterology* 141: 1762–1772

Sato T, van Es JH, Snippert HJ, Stange DE, Vries RG, van den Born M, Barker N, Shroyer NF, van de Wetering M, Clevers H (2011c) Paneth cells constitute the niche for Lgr5 stem cells in intestinal crypts. *Nature* 469: 415–418

Savageau MA (1971) Parameter sensitivity as a criterion for evaluating and comparing the performance of biochemical systems. *Nature* 229: 542–544

Schuijers J, Junker JP, Mokry M, Hatzis P, Koo BK, Sasselli V, van der Flier LG, Cuppen E, van Oudenaarden A, Clevers H (2015) Ascl2 acts as an R-spondin/Wnt-responsive switch to control stemness in intestinal crypts. *Cell Stem Cell* 16: 158–170

Schwank G, Koo BK, Sasselli V, Dekkers JF, Heo I, Demircan T, Sasaki N, Boymans S, Cuppen E, van der Ent CK, Nieuwenhuis EE, Beekman JM, Clevers H (2013) Functional repair of CFTR by CRISPR/Cas9 in intestinal stem cell organoids of cystic fibrosis patients. *Cell Stem Cell* 13: 653–658

Shen X, Collier J, Dill D, Shapiro L, Horowitz M, McAdams HH (2008) Architecture and inherent robustness of a bacterial cell-cycle control system. *Proc Natl Acad Sci USA* 105: 11340–11345

Sikandar SS, Pate KT, Anderson S, Dizon D, Edwards RA, Waterman ML, Lipkin SM (2010) NOTCH signaling is required for formation and self-renewal of tumor-initiating cells and for repression of secretory cell differentiation in colon cancer. *Can Res* 70: 1469–1478

Snippert HJ, van der Flier LG, Sato T, van Es JH, van den Born M, Kroon-Veenboer C, Barker N, Klein AM, van Rheenen J, Simons BD, Clevers H (2010) Intestinal crypt homeostasis results from neutral competition between symmetrically dividing Lgr5 stem cells. *Cell* 143: 134–144

Sprinzak D, Lakhanpal A, Lebon L, Santat LA, Fontes ME, Anderson GA, Garcia-Ojalvo J, Elowitz MB (2010) Cis-interactions between Notch and Delta generate mutually exclusive signalling states. *Nature* 465: 86–90

Sprinzak D, Lakhanpal A, Lebon L, Garcia-Ojalvo J, Elowitz MB (2011) Mutual inactivation of notch receptors and ligands facilitates developmental patterning. *PLoS Comput Biol* 7: e1002069

Srinivasan T, Than EB, Bu P, Tung KL, Chen KY, Augenlicht L, Lipkin SM, Shen X (2016) Notch signalling regulates asymmetric division and inter-conversion between lgr5 and bmi1 expressing intestinal stem cells. *Sci Rep* 6: 26069

Stelling J, Sauer U, Szallasi Z, Doyle FJ III, Doyle J (2004) Robustness of cellular functions. *Cell* 118: 675–685

Swat MH, Thomas GL, Belmonte JM, Shirinifard A, Hmeljak D, Glazier JA (2012) Multi-scale modeling of tissues using CompuCell 3D. *Methods Cell Biol* 110: 325–366

Takeda N, Jain R, LeBoeuf MR, Wang Q, Lu MM, Epstein JA (2011) Interconversion between intestinal stem cell populations in distinct niches. *Science* 334: 1420–1424

Van Landeghem L, Santoro MA, Krebs AE, Mah AT, Dehmer JJ, Gracz AD, Scull BP, McNaughton K, Magness ST, Lund PK (2012) Activation of two distinct Sox9-EGFP-expressing intestinal stem cell populations during crypt regeneration after irradiation. *Am J Physiol Gastrointest Liver Physiol* 302: G1111–G1132

VanDussen KL, Carulli AJ, Keeley TM, Patel SR, Puthoff BJ, Magness ST, Tran IT, Maillard I, Siebel C, Kolterud A, Grosse AS, Gumucio DL, Ernst SA, Tsai YH, Dempsey PJ, Samuelson LC (2012) Notch signaling modulates proliferation and differentiation of intestinal crypt base columnar stem cells. *Development* 139: 488–497

Wang X, Ha T (2013) Defining single molecular forces required to activate integrin and notch signaling. *Science* 340: 991–994

Wang H, Zang C, Taing L, Arnett KL, Wong YJ, Pear WS, Blacklow SC, Liu XS, Aster JC (2014) NOTCH1-RBPJ complexes drive target gene expression through dynamic interactions with superenhancers. *Proc Natl Acad Sci USA* 111: 705–710

Wu Y, Cain-Hom C, Choy L, Hagenbeek TJ, de Leon GP, Chen Y, Finkle D, Venook R, Wu X, Ridgway J, Schahin-Reed D, Dow GJ, Shelton A, Stawicki S, Watts RJ, Zhang J, Choy R, Howard P, Kadyk L, Yan M *et al* (2010) Therapeutic antibody targeting of individual Notch receptors. *Nature* 464: 1052−1057

Xu D, Hu J, De Bruyne E, Menu E, Schots R, Vanderkerken K, Van Valckenborgh E (2012) Dll1/Notch activation contributes to bortezomib resistance by upregulating CYP1A1 in multiple myeloma. *Biochem Biophys Res Commun* 428: 518−524

Yamamura H, Yamamura A, Ko EA, Pohl NM, Smith KA, Zeifman A, Powell FL, Thistlethwaite PA, Yuan JX (2014) Activation of Notch signaling by short-term treatment with Jagged-1 enhances store-operated Ca(2+) entry in human pulmonary arterial smooth muscle cells. *Am J Physiol Cell Physiol* 306: C871−C878

Yang Q, Bermingham NA, Finegold MJ, Zoghbi HY (2001) Requirement of Math1 for secretory cell lineage commitment in the mouse intestine. *Science* 294: 2155−2158

12

Exploiting native forces to capture chromosome conformation in mammalian cell nuclei

Lilija Brant[1], Theodore Georgomanolis[1,†], Milos Nikolic[1,†], Chris A Brackley[2], Petros Kolovos[3], Wilfred van Ijcken[4] (iD), Frank G Grosveld[3], Davide Marenduzzo[2] & Argyris Papantonis[1,*] (iD)

Abstract

Mammalian interphase chromosomes fold into a multitude of loops to fit the confines of cell nuclei, and looping is tightly linked to regulated function. Chromosome conformation capture (3C) technology has significantly advanced our understanding of this structure-to-function relationship. However, all 3C-based methods rely on chemical cross-linking to stabilize spatial interactions. This step remains a "black box" as regards the biases it may introduce, and some discrepancies between microscopy and 3C studies have now been reported. To address these concerns, we developed "i3C", a novel approach for capturing spatial interactions without a need for cross-linking. We apply i3C to intact nuclei of living cells and exploit native forces that stabilize chromatin folding. Using different cell types and loci, computational modeling, and a methylation-based orthogonal validation method, "TALE-iD", we show that native interactions resemble cross-linked ones, but display improved signal-to-noise ratios and are more focal on regulatory elements and CTCF sites, while strictly abiding to topologically associating domain restrictions.

Keywords chromatin looping; chromosome conformation capture; cross-linking; nuclear compartments; nuclear organization

Subject Categories Chromatin, Epigenetics, Genomics & Functional Genomics; Genome-Scale & Integrative Biology; Methods & Resources

Introduction

The higher-order folding of mammalian chromosomes has long been linked to the regulation of their function. However, over the last decade, studies exploited 3C technology to significantly advance our understanding of this structure-to-function relationship (Dekker et al, 2013; Pombo & Dillon, 2015; Denker & de Laat, 2016) and allowed us to address diverse biological questions (e.g., Tolhuis et al, 2002; Papantonis et al, 2012; Zhang et al, 2012; Naumova et al, 2013; Rao et al, 2014). We now know that interphase chromosomes are partitioned into topologically associating domains (TADs; Dixon et al, 2012) ranging from 0.1 to few Mbp. TADs are rich in intradomain (versus interdomain) multi-loop interactions connecting genes and cis-regulatory elements, and their boundaries remain largely invariant between different cell types (Dixon et al, 2012; Rao et al, 2014) or upon cytokine signaling (Jin et al, 2013; Le Dily et al, 2014).

3C methods rely on chemical cross-linking for stabilizing and capturing spatial interactions. Although formaldehyde is widely used in molecular biology, and its chemistry is well understood, its in vivo effects remain obscure (Gavrilov et al, 2015). For instance, not all nuclear proteins/loci are equally efficiently cross-linked (Teytelman et al, 2013); cross-linking may trigger the DNA damage response to induce polyADP-ribosylation of the proteome and thus alter its susceptibility to fixation (Beneke et al, 2012), while fixation is sensitive to even slight changes in temperature, pH, and duration (Schmiedeberg et al, 2009). Then, it is conceivable that variations in fixation efficiency within the different nuclear microenvironments (e.g., in hetero- versus euchromatin) can skew experimental readouts. Discrepancies between DNA FISH and 5C studies in the HoxD locus were recently reported (Williamson et al, 2014). FISH requires harsher fixation than the 5C procedure and, in multiple cases, microscopy and 3C results do correlate well; nonetheless, such discrepancies may, at least in part, stem from differential fixation effects in the dense chromatin mesh within TADs. Moreover, any interactions that end up being detected via 3C assays must survive harsh detergent treatment, prolonged heating and shaking, plus unphysiological salt concentrations (Stadhouders et al, 2013). To address these concerns, we developed "intrinsic 3C" (i3C), a novel approach to capture chromatin conformation in living cells without a need for cross-linking. i3C exploits native forces that preserve the relative spatial positioning of chromatin fragments. We generated i3C profiles in a number of cell types and loci to investigate different features of chromatin looping in the absence of chemical fixation.

1 Center for Molecular Medicine, University of Cologne, Cologne, Germany
2 School of Physics and Astronomy, University of Edinburgh, Edinburgh, UK
3 Department of Cell Biology, Erasmus Medical Center, Rotterdam, The Netherlands
4 Biomics Department, Erasmus Medical Center, Rotterdam, The Netherlands
 *Corresponding author. E-mail: argyris.papantonis@uni-koeln.de
 †These authors contributed equally to this work

Results and Discussion

We apply i3C (overview in Fig 1A) to intact nuclei from living, uncross-linked, cells harvested in a buffer (PB) that closely approximates physiological salt concentrations and deters aggregation of nuclear components (Kimura *et al*, 1999; Melnik *et al*, 2011). Thus, > 95% transcriptional activity is retained (Appendix Fig S1A), suggesting that chromatin organization is also maintained. Nuclei are treated with a restriction endonuclease for ~30 min at a suboptimal temperature (not all enzymes work equally well in PB; Appendix Fig S1B), washed, and spun to remove any unattached chromatin. This removes > 40% of total chromatin (Fig 1B) and so reduces the fraction of "bystander/baseline" ligations (Dekker *et al*, 2013) to improve signal quality. Cohesive DNA ends are then ligated within intact nuclei, where native interactions are inherently preserved (Gavrilov *et al*, 2013; Rao *et al*, 2014), before the i3C template is purified (see Materials and Methods for details). The i3C workflow takes place in a single tube to minimize material losses, is faster than the conventional one (Stadhouders *et al*, 2013) and just as efficient (Appendix Fig S1C). In principle, i3C can also be applied to solid tissue (e.g., mouse liver) so long as single nuclei can be obtained.

To ensure that ligation occurs exclusively within single nuclei under native conditions, we mixed an equal number of human endothelial (HUVEC) and mouse embryonic stem cell (mESC) nuclei, performed i3C, and sequenced the resulting ligation products; < 0.7% of the reads pairs mapped one end to the human and the other end to the mouse genome. Then, as a substantial amount of DNA is lost in i3C, we asked whether any biases arise during cutting (which does not trigger the DNA damage response; Appendix Fig S1D). We treated nuclei with *Nla*III, isolated DNA at different stages along the procedure and sequenced it (steps 2–4, Fig 1A). Read profiles from "lost" (step 3) and "retained" fractions (step 4) overlap (~70% *Nla*III fragments have reads in both fractions) are equally enriched in active and inactive loci (e.g., at enhancers, CTCF–CTCF loops, lamin-associated domains; Fig 1C and Appendix Fig S1E–I), and display very similar fragment size contents (Appendix Fig S1J). Moreover, cutting chromatin in the presence or absence of cross-linking does not yield very different profiles, nor does it display a preference for "open" chromatin (despite the short incubation times used; Appendix Fig S2). Thus, the different chromatin regions are equally represented in i3C ligations.

Next, we produced an i3C template in HUVECs to query using qPCR (i3C-qPCR) interactions seen by conventional 4C for the *EDN1* housekeeping gene (Diermeier *et al*, 2014). We paired an "anchor" primer at the *EDN1* TSS with tandem primers at eight *Apo*I fragments within its locus and faithfully recapitulated all major interactions (with no interactions in the "lost" chromatin fraction; Appendix Fig S3). Some interacting fragments were separated by as many as 500 kbp, encouraging us to apply i4C-seq to our model *SAMD4A* locus (Diermeier *et al*, 2014). i4C templates were produced in HUVECs by cutting with *Apo*I, recutting and circularizing, amplifying fragments contacted by the *SAMD4A* TSS by inverse PCR, and amplimer sequencing. In parallel, the same viewpoint and primers were used to generate conventional 4C profiles. The resulting data were processed via "fourSig" (Williams *et al*, 2014) to correct for mapping biases and identify significant interactions. Comparison of i4C and conventional 4C *SAMD4A* cis-interactions

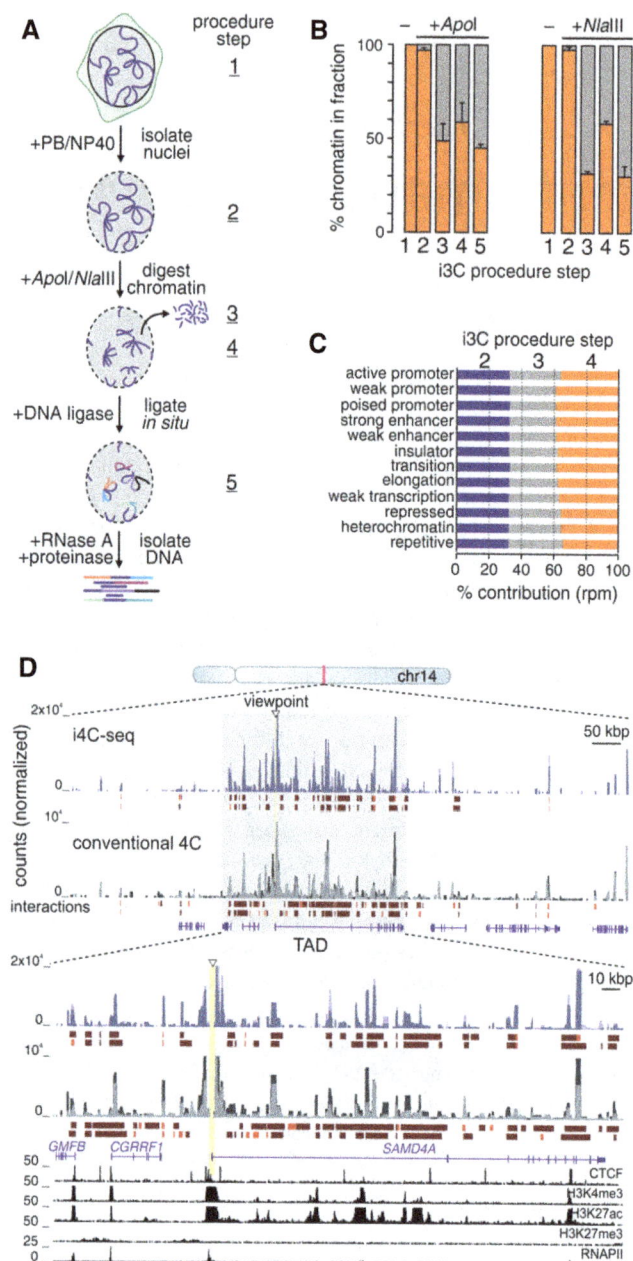

Figure 1. Features of i3C performed in HUVECs.

A Overview of the i3C protocol. Living cells are harvested in a close-to-physiological buffer (PB; step 1); intact nuclei isolated by mild NP-40 treatment (step 2); chromatin digested using *Apo*I or *Nla*III, nuclei spun to release unattached chromatin (step 3); and leave cut chromatin bound to the nuclear substructure (step 4). Then, ligation takes places *in situ*, and DNA is isolated (step 5).

B Percentage of total cell chromatin present at the different steps of the procedure (± SD; *n* = 2).

C Relative contribution of the different HUVEC ChromHMM features in each i3C fraction.

D i4C-seq (blue shades) and conventional 4C (gray shades) were performed side by side in HUVECs, using *Apo*I and the *SAMD4A* TSS as a viewpoint (triangle); profiles from two replicates are overlaid. The browser view shows interactions in the ~1 Mbp around *SAMD4A*. The zoom-in shows interactions in the *SAMD4A* TAD (gray rectangle). Strong (red) and intermediate (brown) interactions called by *fourSig*, RefSeq gene models, and ENCODE ChIP-seq data are shown below.

revealed extensive similarities, especially within the viewpoint's TAD (Fig 1D). Few contacts were seen only in the absence of cross-linking (which was also confirmed by a differential analysis via "FourCSeq"; Klein *et al*, 2015; Appendix Fig S4). Compared to high-resolution Hi-C (Rao *et al*, 2014) and ChIA-PET contact maps (Papantonis *et al*, 2012) from HUVECs, i4C showed matching inter-action profiles and aligned well within TAD boundaries, while also reproducibly detecting some longer-range contacts (Appendix Fig S5A and B). However, i4C profiles were more enriched in interactions with *cis*-regulatory regions (e.g., enhancers, CTCF sites; Appendix Fig S5C–E), and interactions unique to i4C (~65% of all *cis*-contacted fragments in i4C and conventional 4C overlap) are with genomic regions carrying the expected histone marks (e.g., H3K4me1/2, H3K36me3, H3K9ac; Appendix Fig S5F). Importantly, the signal at i4C contacts is more focal and allows deconvolution of single interactions (Fig 1D and Appendix Fig S6A and B).

Next, setting a "background" threshold at 100 rpm (as both positions and enrichments below this threshold showed the most variance in our replicates), we found 83% of i4C-seq reads over the threshold compared to < 60% of conventional 4C reads (Appendix Fig S7A). Moreover, i4C displayed significantly lower numbers of uncut and self-ligation reads for the *SAMD4A* viewpoint (a trend associated with milder fixation; van de Werken *et al*, 2012), as well as more reads mapping within its TAD (Appendix Fig S7B). This held also true for i4C from the *BMP4*, *CDKN3*, and *CNIH* genes that reside in TADs of different sizes; all displayed > 40% reads mapping within their respective TAD (Fig 2 and Appendix Fig S7C), suggesting TADs impose strong topological restrictions under native conditions.

Similar *SAMD4A* i4C profiles were also obtained in a different cell type (IMR-90) or when *Apo*I was replaced by *Nla*III (Appendix Fig S8). However, the *SAMD4A* locus is densely populated by genes and *cis*-elements. Hence, we also applied i4C to the gene-poor *EDN1* and the hetero-chromatinized *TBX5* loci. For the *EDN1* TSS viewpoint, we essentially only record i4C contacts to other active promoters and *cis*-elements, again markedly more focal and enriched for relevant chromatin marks compared to conventional data (Appendix Fig S9), while *TBX5* interacts with other H3K27me3-bound regions, including the neighboring, inactive, *TBX3* locus (Appendix Fig S10). In addition, we could reproduce previously recorded interactions at and between the *Nanog* and *Sox2* loci in mESCs (de Wit *et al*, 2013; Appendix Fig S11), confirming that i4C also captures *trans*-interactions. Finally, one tends to think that methods not using fixation require large numbers of primary material, but we typically use 5 million cells for i4C (below what is recommended for conventional 4C; Stadhouders *et al*, 2013). We also tested increasingly lower cell counts in i4C of the *SAMD4A* viewpoint; similar *cis*-profiles were obtained with < 10^6 cells, albeit at the expense of signal-to-noise ratios (Appendix Fig S12).

Recently, *in situ* Hi-C was performed in uncross-linked lympho-blasts (typically by embedding cells in agar); despite their relative sparsity, these profiles largely matched those obtained using cross-linking (Rao *et al*, 2014). To compare this and different experimental conditions to i3C, we generated interaction profiles for the *SAMD4A* TSS. Omitting formaldehyde fixation from the *in situ* protocol results in a markedly de-enriched interactome; for instance, the *SAMD4A* TSS is looped to a cluster of enhancers in its first intron—this interaction is significantly diminished when conventional 4C is performed without cross-linking, and essentially lost once cells are treated with RNase A (Appendix Fig S13A–C). Similarly, we previously used 3C-PCR to probe interactions between DNA fragments attached to isolated transcription factories (Melnik *et al*, 2011). We incorporated the factory isolation step (using Group-III caspases) in i3C and produced "factory 4C" data for

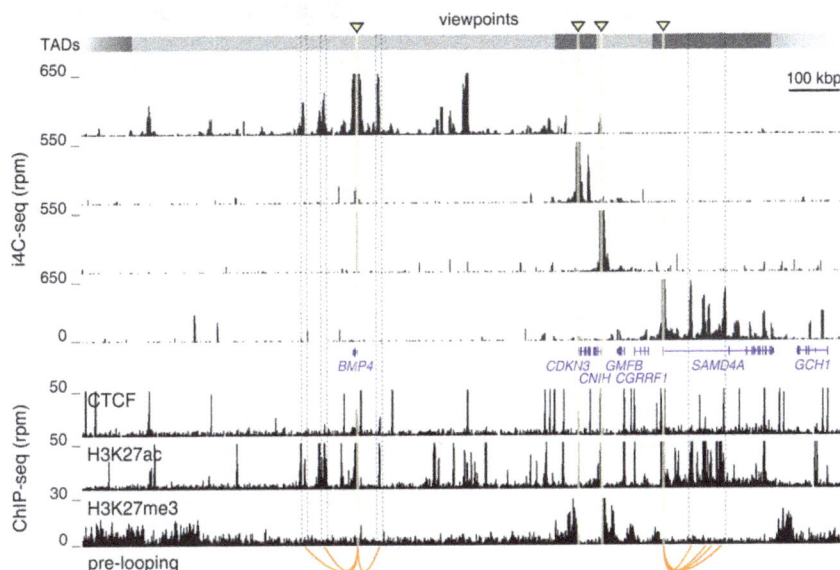

Figure 2. Native interactions are confined by TAD boundaries and describe prelooping.
i4C-seq was performed in HUVECs using *Nla*III and the TSSs of *BMP4*, *CDKN3*, *CNIH*, and *SAMD4A* as viewpoints (triangles). Interactions are shown aligned to TAD boundaries (gray rectangles; from Dixon *et al*, 2012) and HUVEC ENCODE ChIP-seq data (below). Prelooping of the *SAMD4A* and *BMP4* TNF-responsive TSSs to enhancers is indicated (orange lines).

SAMD4A; as expected, interactions are largely preserved, but such "factory 4C" can suffer from a bias for active gene interactions, as well as from the unknown effects of caspase digestion (Appendix Fig S13D). Then, native interactions seem best detected using i3C (which avoids SDS treatment, caspase digestion, or heating) and RNA, like that produced at the *SAMD4A* enhancer cluster, may stabilize particular interactions and reduce the release of cut fragments from the nuclear substructure (Appendix Fig S13E).

Next, we used a predictive polymer modeling approach that can faithfully reproduce spatial chromatin organization based on ENCODE ChIP-seq and ChromHMM data (Brackley *et al*, 2016a) to simulate the interactome of the ~2.8-Mbp locus shown in Fig 2. We generated 500 *in silico* conformations at 1-kbp resolution, from which average simulated interaction profiles were obtained and compared to experimental 4C/i4C data (see Appendix Supplementary Methods and Appendix Fig S14A and B). In agreement with all other comparisons, i4C and conventional 4C profiles closely resemble simulated ones (e.g., *BMP4* i4C shows a correlation of 0.697 to the simulations, and 4C one of 0.745; Appendix Fig S14C).

We also devised "TALE-iD", a new orthogonal method for validating i4C interactions, as we sought to avoid FISH approaches, which require cross-linking (Williamson *et al*, 2014). We fused a custom TAL-effector DNA-binding domain (that specifically binds an enhancer in the first intron of *ZFPM2*; Mendenhall *et al*, 2013) with a bacterial adenosine methylase (Dam; Vogel *et al*, 2006). Once the targeted enhancer is found in physical proximity to other genomic regions (due to looping), the Dam methylase will methylate adenine residues thereon (Fig 3A). This construct was introduced into K562 cells, where the target enhancer is active and i4C finds it looped to the *ZFPM2* TSS (Fig 3B). Genomic DNA from transfected K562 was then isolated and digested using *DpnI* (that cuts only methylated sites). Cutting efficiency at 12 different sites was quantified by qPCR and showed that the targeted enhancer contacts the TSS ("p1–p4") and an upstream enhancer ("m1"). However, another enhancer further downstream is not contacted, and no interactions are detected when a "ΔDam" construct is used (Fig 3B and C).

We now understand that the promoters of stimulus-inducible genes are often prelooped to cognate enhancers (Jin *et al*, 2013). This motivated us to examine whether prelooping might arise as a result of cross-linking within tightly packed TADs. First, we applied i3C-qPCR to the *IL1A* TNF-responsive locus in HUVECs and verified prelooping under native conditions (Appendix Fig S15). We next generated i4C data for the TSSs of four genes in the same locus following a 60-min TNF pulse. Of these, the TNF-responsive *BMP4* and *SAMD4A* are prelooped to H3K27ac-decorated enhancers (Fig 2 and Appendix Fig S16). We also reasoned that the focal i4C contacts can be used to track dynamic changes in interactions upon TNF stimulation. We compared i4C and 4C profiles before and after stimulation to find more changes in the absence of cross-linking (Appendix Figs S16 and S17). For *SAMD4A*, the loops between its TSS and the downstream enhancer cluster are partially remodeled

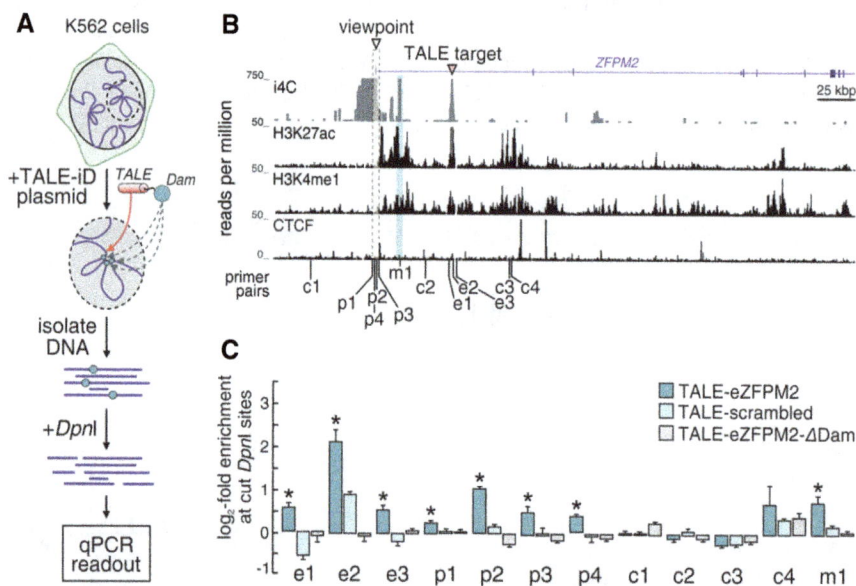

Figure 3. **TALE-iD verifies native looping at the human *ZFPM2* locus.**

A   An overview of TALE-iD. A construct encoding a TALE DNA-binding domain that targets an active enhancer in the *ZFPM2* first intron is fused to a bacterial Dam methylase and introduced into K562 cells. Cells are harvested 48 h after transfection; genomic DNA is isolated and digested with *DpnI* to reveal sites specifically methylated by the Dam activity. Finally, qPCR using primers flanking different *DpnI* sites is used as readout.

B   i4C performed in K562 cells using *ApoI* and the *ZFPM2* TSS as a viewpoint (triangle). i4C interaction in the 458-kbp *ZFPM2* locus is shown, and the enhancer targeted by the TALE-iD construct is indicated (red triangle). K562 ENCODE ChIP-seq data are also shown below.

C   qPCR readout at different *DpnI* sites. *DpnI* sites at the *ZFPM2* promoter (p1–p4) and enhancer (e1–e3; positions in panel B) were targeted in qPCRs after restriction digest. Bar plots show log₂-fold enrichment of cut sites (1/ΔΔC$_t$) over background *DpnI* cutting levels in untransfected K562 cells. Regions c1–c4 serve as controls; region m1 (an enhancer shown to interact with the TSS by i4C) is also methylated as part of a multi-loop structure. The same *DpnI* sites were also tested in transfections involving a construct encoding either a non-targeting ("scrambled") TALE domain or the targeting domain fused to an inactive Dam protein ("ΔDam"). *$P < 0.05$; two-tailed unpaired Student's *t*-test ($n = 3$). The bars and error bars denote mean ± SEM.

to follow NF-κB binding (seen using either *Apo*I or *Nla*III; Appendix Fig S18A), which was further verified by differential analysis (especially at two sites in the *SAMD4A* enhancer cluster, where changes are dampened in conventional 4C; Appendix Figs S19 and S20). TNF stimulation does not change the fraction of reads mapping within the *SAMD4A* TAD (Appendix Fig S18B), many of which overlap NF-κB binding sites (Appendix Fig S18C). Then, prelooping and interaction remodeling to follow NF-κB binding (predominantly within their respective TADs) is seen for responsive TSSs, indicating these are prevalent and dynamic features of native folding.

Finally, as i4C is a "one-to-all" approach, we sought to map interactions in a global fashion. We applied a capture-based 3C method, "T2C" (Kolovos *et al*, 2014), where probes targeting every *Apo*I

fragment in our 2.8-Mbp model locus on chromosome 14 are used to retrieve and sequence a subset of 3C/i3C ligations. T2C was reproducibly performed in HUVECs to yield 1.5-kbp-resolution interaction maps (Fig 4A and Appendix Fig S21A–C). In the presence of cross-linking, all known features of genomic organization are seen; when performed natively, the outline of sub-TADs (or contact domains) is mapped at lower resolutions (Fig 4A, top), while at higher resolutions individual chromatin loops are resolved against ultra-low background (Fig 4A; bottom). In fact, using the "directionality index" approach to call TADs at 10-kbp resolution (Dixon *et al*, 2012), we find very similar organization in the presence or absence of cross-linking, with some additional subdomains emerging in iT2C (Appendix Fig S21D). Moreover, all seven CTCF–CTCF loops

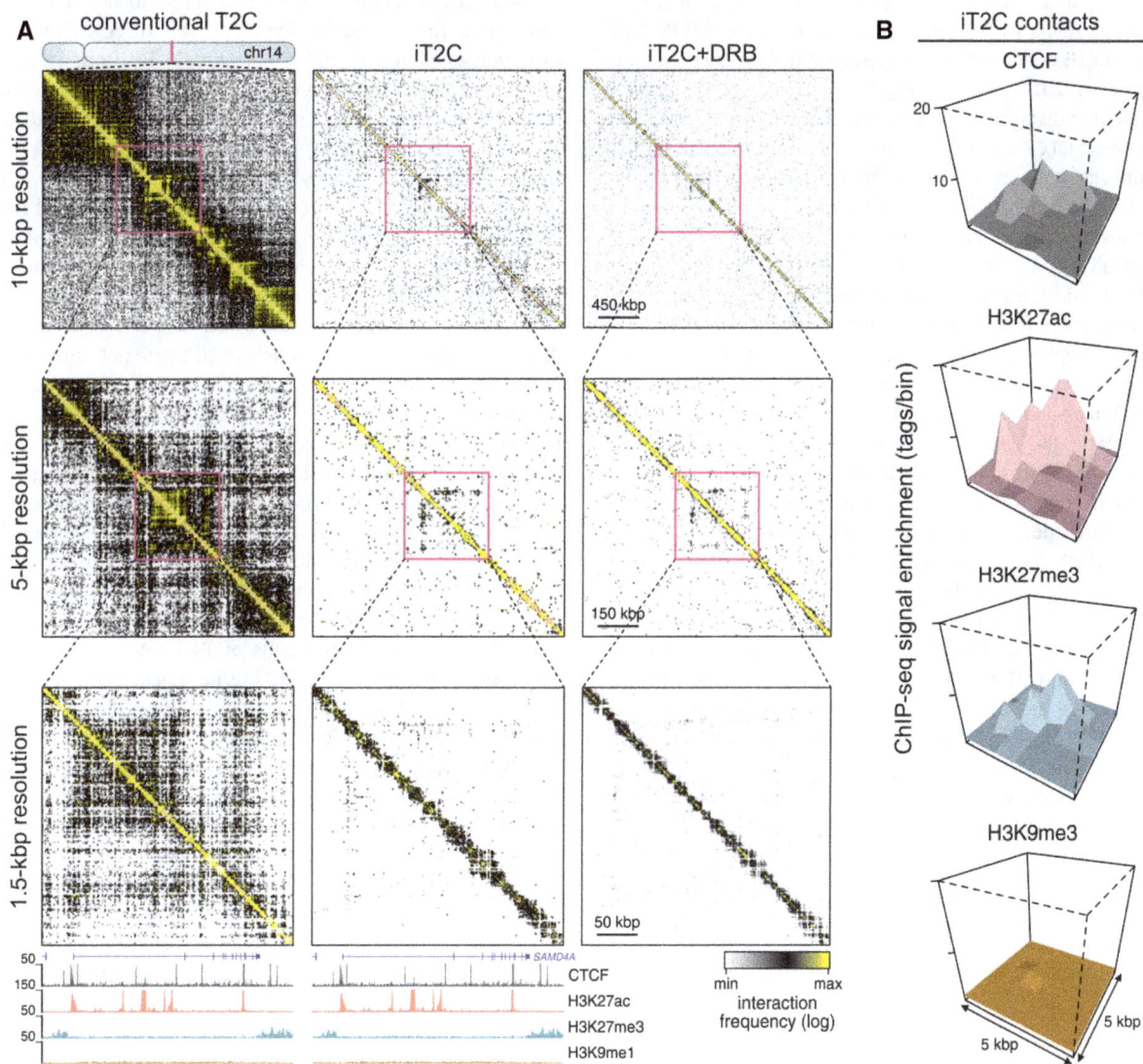

Figure 4. 3D organization of a 2.8-Mbp human locus analyzed by iT2C/conventional T2C.

A   Interaction maps from conventional T2C (left) and iT2C (middle, right) in the 2.8 Mbp around *SAMD4A* on chromosome 14. Magnifications show interactions at increasingly higher resolution. Bottom: HUVEC ENCODE ChIP-seq data are aligned to interactions mapped at 1.5-kbp resolution in the 250 kbp around *SAMD4A*.

B   PE-SCAN graphs (see de Wit *et al*, 2013) show the enrichment of iT2C interactions (± 5 kbp) for CTCF (gray), H3K27ac (pink), and H3K27me3 (blue), while H3K9me3 (brown) that is absent from this region serves as a control.

previously called in this 2.8-Mbp region (Rao *et al*, 2014) are picked up by iT2C (Appendix Fig S21E) and, overall, iT2C contacts are enriched for CTCF, H3K27ac, or H3K27me3 at the interacting sites (Fig 4B). Treatment of HUVECs with the transcriptional inhibitor DRB does not dramatically alter the iT2C map (e.g., prelooping in *SAMD4A* persists; Fig 4A). This supports the notion of an overarching organization that is in part independent of transcription, and was affirmed when iT2C for the same locus was applied to IMR90s (Appendix Fig S22). Importantly, iT2C is devoid of signal from "bystander/baseline" interactions, allowing us to detect contacts over ultra-low background (Appendix Fig S21F), while contact structure close to the diagonal is reminiscent of that seen by "micro-C" (Hsieh *et al*, 2015).

Finally, we provide proof-of-principle iHi-C ("all-to-all" i3C) under native conditions by using *Dpn*II, incorporating biotin-dATP in ligation junctions, and sequencing to ~3 × 10^8 reads. The resulting interaction maps were compared to conventional HUVEC Hi-C (and to *in situ* Hi-C from uncross-linked agar-encapsulated lymphoblasts; Rao *et al*, 2014; Appendix Fig S23A and B). Contact strength distribution over distance is not very different in these datasets, yet iHi-C can detect CTCF–CTCF loops more robustly, and resolves individual contacts with as few as 50 million reads (Appendix Fig S23C–E). Last, in order to assess whether any contacts not captured in i3C are lost in the fraction that diffuses from nuclei upon cutting, we performed side-by-side iHi-C on the "lost" (in dilution) and "retained" (*in situ*) fractions, sequenced to > 2 × 10^8 reads, and compared the resulting interactions. Our analysis, exemplified by a representative region on chromosome 18, showed that very few interpretable contacts can be retrieved from the "lost" fraction, and these are often also found in the "retained" one (Appendix Fig S24).

In summary, i3C contact profiles display great similarity to conventional ones, thus alleviating most concerns about discrepancies due to fixation. Critically, our data verify the importance of topological restrictions imposed by TAD formation under native conditions, which highlights their regulatory implications *in vivo*. Moreover, the unattached chromatin lost in i3C renders captured interactions more focal, which can be advantageous when studying regions densely populated by *cis*-elements (like "super-enhancers"). Similarly, iT2C and iHi-C offer the potential to call loops at high resolution against essentially zero background without a need for excessive sequencing depth. Thus, we suggest that i3C offers a rapid and robust means for interrogating native spatial interactions; it dampens "bystander/baseline" ligations to increase signal quality, and so complements the existing toolkit for investigating 3D genome architecture in eukaryotes.

Materials and Methods

Cell culture

HUVECs from pooled donors (Pan Biotech.) were grown in Endothelial Basal Medium with supplements and 3% fetal bovine serum (FBS; Pan Biotech.). IMR90s (Coriell Repository) were grown in MEM (Sigma-Aldrich) with 20% FBS (Gibco) and 1% non-essential amino acids (Sigma-Aldrich). K562 cells were grown in RPMI (Sigma-Aldrich) with 10% FBS, and mouse embryonic stem cells (E14 mESCs) in Knockout-DMEM medium (Life Technologies) containing 15% FBS and LIF (a gift by Alvaro Rada-Iglesias) on gelatin-coated plates. All cells were grown to ~90% confluency before harvesting for further processing or passaging. Where appropriate, cells were serum-starved overnight and treated with TNF (10 ng/ml; Peprotech) or with 50 μM DRB (Sigma-Aldrich) for 1 h at 37°C.

i3C

An adapted close-to-physiological, isotonic buffer (PB; 100 mM KCH$_3$COO, 30 mM KCl, 10 mM Na$_2$HPO$_4$, 1 mM MgCl$_2$, 1 mM Na$_2$ATP, 1 mM DTT, 10 mM β-glycerophosphate, 10 mM NaF, 0.2 mM Na$_3$VO$_4$, and pH is adjusted to 7.4 using 100 mM KH$_2$PO$_4$; Kimura *et al*, 1999) is prepared fresh every time in nuclease-free water (Millipore MilliQ), supplemented with 25 U/ml RiboLock (Thermo Scientific) and protease inhibitors (Roche), and kept on ice throughout the procedure. Typically, 5 × 10^6 cells are used per experiment, harvested in 4 ml of ice-cold PB from 15-cm culture plates using a soft rubber cell scraper (Roth) on ice. Harvested cells are spun at 600 *g* (4°C, 5 min), resuspended and incubated (ice, 10 min) in 10 ml of PB supplemented with 0.4% NP-40 to release nuclei. This step is usually repeated 1–2 times (ice, 5 min), and nuclei integrity is checked on a hemocytometer. Isolated nuclei are collected via centrifugation at 600 *g* (4°C, 5 min), gently resuspended in 500 μl of ice-cold PB/0.4% NP-40, and transferred to 2-ml round-bottom, low-retention tubes. Next, chromatin is digested with 500 units of *Apo*I or *Nla*III (New England Biolabs; 33°C, 30–45 min) without shaking. Aliquots of 10 μl are put aside right before and after digestion as "uncut" and "cut" chromatin controls. Treated nuclei are then spun at 600 *g* (4°C, 5 min) to separate cut, unattached chromatin fragments that are released into the supernatant, washed in 500 μl of ice-cold PB, and respun. Following resuspension in 1 ml of ice-cold PB, spatially proximal chromatin ends are ligated together in intact nuclei (an idea based on the original "proximity ligation" assay by Cullen *et al*, 1993) and supported by recent findings that, even under cross-linked conditions, ligations predominantly occur within the "chromatin cage" of intact nuclei; Gavrilov *et al*, 2013) by adding 100 units of T4 DNA ligase (5 U/μl stock; Invitrogen) and 10 μl BSA (10 mg/ml stock; Sigma-Aldrich), and incubating at 16°C for 6–12 h without shaking. Finally, 25 μl proteinase K (10 mg/ml stock; AppliChem) are added to the samples, which are kept at 42°C overnight. Next day, samples are treated with 25 μl RNase A (10 mg/ml stock; AppliChem; 37°C, 1 h) and purified by phenol/chloroform extraction (pH 8.0) and ethanol precipitation. To reduce co-precipitating DTT, the aqueous phase volume is increased to 1 ml using nuclease-free water, 200 μl 3 M sodium acetate, and 5 ml absolute ethanol are added, and tubes are placed at −80°C for 30 min. Following centrifugation at 4,500 *g* (4°C, 1.5 h), pellets are washed in 5 ml 70% ethanol, air-dried for ~20 min at room temperature, dissolved in 70–100 μl of TE (pH 8.0) at 37°C for 20 min, and the concentration of the i3C template determined using a Qubit 2.0 Fluorometer (Life technologies).

For i4C-seq, circularization and inverse PCR were as described previously (Stadhouders *et al*, 2013); ~25 μg of i3C template are digested with 25 units of *Dpn*II (New England Biolabs; 37°C, overnight). After heat inactivation (65°C, 25 min), DNA is diluted in ligation buffer to a volume of 7 ml, religated using 20 μl T4 DNA ligase (5 U/μl stock; Invitrogen; 16°C, 6–8 h), and purified. Then,

~150 ng of the circularized i3C template is used in inverse PCRs as follows: one cycle at 95°C for 2 min, followed by 34 cycles at 94°C for 15 s/56–58°C for 1 min/72°C for 3 min, before a final extension at 72°C for 7 min using 3.75 units of the Expand long template HF DNA Polymerase (Roche). Typically, eight such PCRs are pooled, purified using the DNA Clean & Concentrator kit (Zymo Research), and amplicons checked by electrophoresis on a 1.5% (wt/vol) agarose gel. The rest of the sample are directly sequenced on a HiSeq2500 platform (Illumina) as the primers used in inverse PCRs carry the P5/P7 Illumina adapters as overhangs. All primers used in this study are listed in Appendix Table S1.

For i3C-qPCR, ~100 ng of the i3C/3C template was used. Primers were designed using Primer3Plus (www.bioinformatics.nl/cgi-bin/primer3plus/primer3plus.cgi) to have a length of 18–23 nucleotides, a Tm of 58–62°C, and to yield amplimers of 70–150 bp. qPCRs (15 μl) were performed using the SYBR Green JumpStart Ready Mix (Sigma) on a Rotor-Gene Q cycler (Qiagen; one cycle of 95°C for 5 min, followed by 40 cycles at 95°C for 15 s, 61°C for 40 s, and 72°C for 20 s). i3C/3C amplimer levels were normalized to both a "loading" primer pair (for equiloading) and to templates prepared by cutting and ligating bacterial artificial chromosomes (BACs) spanning the studied loci of interest (controls for primer efficiency). All primers are available on request; all BAC's used in this study are listed in Appendix Table S2.

Data analysis

The analysis of high-throughput sequencing data from i4C-/4C-seq experiments was carried out using the fourSig (Williams *et al*, 2014) or FourCSeq (Klein *et al*, 2015) packages. In brief, 76-bp single-end reads from a HiSeq2000 platform (Illumina) were trimmed to remove the viewpoint primer sequence using homerTools (http://homer.salk.edu/homer/). Trimmed reads were then mapped to the reference genome (hg19) using the short read aligner BWA-MEM (Li & Durbin, 2010; exact parameters were as follows: BWA MEM -t 8 -k 15 -r 1 -B 1 –M) and processed via fourSig or FourCSeq. Data were then visualized by uploading. BedGraph files to the UCSC genome browser (https://genome.ucsc.edu/; hg19) and using its embedded smoothing option. i4C-seq replicates and mapping efficiencies are listed in Appendix Table S3, and quality control/correlation plots (as in van de Werken *et al*, 2012) shown in Appendix Fig S7D and E.

Acknowledgements

We thank Erez Lieberman Aiden, Alvaro Rada-Iglesias and Leo Kurian for discussions; Peter Cook and Svitlana Melnik for help during the early stages of this work; Elzo de Wit for the PE-SCAN code; Bradley Bernstein and Eric Mendenhall for the TALE plasmids; Michaela Bozukov for help with experiments; and the Cologne Center for Genomics for sequencing iHi-C libraries. Work in AP's laboratory is supported by the Fritz Thyssen Stiftung, the DFG Excellence Initiative via a "UoC Advanced Researcher Grant", and by CMMC Junior Research Group funding; DM is supported by an ERC consolidator grant.

Author contributions

LB and AP designed experiments and developed the method. LB performed experiments. WI sequenced i4C/iT2C libraries. PK and FGG helped with T2C experiments. TG and MN analyzed high-throughput data. CAB and DM performed simulations. LB and AP wrote the manuscript with input from all co-authors.

References

Beneke S, Meyer K, Holtz A, Hüttner K, Bürkle A (2012) Chromatin composition is changed by poly(ADP-ibosyl)ation during chromatin immunoprecipitation. *PLoS One* 7: e32914

Brackley CA, Brown JM, Waithe D, Babbs C, Davies J, Hughes JR, Buckle VJ, Marenduzzo D (2016a) Predicting the three-dimensional folding of cis-regulatory regions in mammalian genomes using bioinformatic data and polymer models. *Genome Biol* 17: 59

Cullen KE, Kladde MP, Seyfred MA (1993) Interaction between transcription regulatory regions of prolactin chromatin. *Science* 261: 203–206

Dekker J, Marti-Renom MA, Mirny LA (2013) Exploring the three-dimensional organization of genomes: interpreting chromatin interaction data. *Nat Rev Genet* 14: 390–403

Denker A, de Laat W (2016) The second decade of 3C technologies: detailed insights into nuclear organization. *Genes Dev* 30: 1357–1382

Diermeier S, Kolovos P, Heizinger L, Schwartz U, Georgomanolis T, Zirkel A, Wedemann G, Grosveld F, Knoch TA, Merkl R, Cook PR, Längst G, Papantonis A (2014) TNF signaling primes chromatin for NF-κB binding and induces rapid and widespread nucleosome repositioning. *Genome Biol* 15: 536

Dixon JR, Selvaraj S, Yue F, Kim A, Li Y, Shen Y, Hu M, Liu JS, Ren B (2012) Topological domains in mammalian genomes identified by analysis of chromatin interactions. *Nature* 485: 376–380

Gavrilov AA, Gushchanskaya ES, Strelkova O, Zhironkina O, Kireev II, Iarovaia OV, Razin SV (2013) Disclosure of a structural milieu for the proximity ligation reveals the elusive nature of an active chromatin hub. *Nucleic Acids Res* 41: 3563–3575

Gavrilov A, Razin SV, Cavalli G (2015) In vivo formaldehyde cross-linking: it is time for black box analysis. *Brief Funct Genomics* 14: 163–165

Hsieh TH, Weiner A, Lajoie B, Dekker J, Friedman N, Rando OJ (2015) Mapping nucleosome resolution chromosome folding in yeast by micro-C. *Cell* 162: 108–119

Jin F, Li Y, Dixon JR, Selvaraj S, Ye Z, Lee AY, Yen CA, Schmitt AD, Espinoza CA, Ren B (2013) A high-resolution map of the three-dimensional chromatin interactome in human cells. *Nature* 503: 290–294

Kimura H, Tao Y, Roeder RG, Cook PR (1999) Quantitation of RNA polymerase II and its transcription factors in an HeLa cell: little soluble holoenzyme but significant amounts of polymerases attached to the nuclear substructure. *Mol Cell Biol* 19: 5383–5392

Klein FA, Pakozdi T, Anders S, Ghavi-Helm Y, Furlong EE, Huber W (2015) FourCSeq: analysis of 4C sequencing data. *Bioinformatics* 31: 3085–3091

Kolovos P, van de Werken HJ, Kepper N, Zuin J, Brouwer RW, Kockx CE, Wendt KS, van IJcken WF, Grosveld F, Knoch TA (2014) Targeted Chromatin Capture (T2C): a novel high resolution high throughput method to detect genomic interactions and regulatory elements. *Epigenetics Chrom* 7: 10

Le Dily F, Baù D, Pohl A, Vicent GP, Serra F, Soronellas D, Castellano G, Wright RH, Ballare C, Filion G, Marti-Renom MA, Beato M (2014) Distinct structural transitions of chromatin topological domains correlate with coordinated hormone-induced gene regulation. *Genes Dev* 28: 2151–2162

Li H, Durbin R (2010) Fast and accurate long-read alignment with Burrows-Wheeler transform. *Bioinformatics* 26: 589–595

Melnik S, Deng B, Papantonis A, Baboo S, Carr IM, Cook PR (2011) The proteomes of transcription factories containing RNA polymerases I, II or III. *Nat Methods* 8: 963–968

Mendenhall EM, Williamson KE, Reyon D, Zou JY, Ram O, Joung JK, Bernstein BE (2013) Locus-specific editing of histone modifications at endogenous enhancers. *Nat Biotechnol* 31: 1133−1136

Naumova N, Imakaev M, Fudenberg G, Zhan Y, Lajoie BR, Mirny LA, Dekker J (2013) Organization of the mitotic chromosome. *Science* 342: 948−953

Papantonis A, Kohro T, Baboo S, Larkin JD, Deng B, Short P, Tsutsumi S, Taylor S, Kanki Y, Kobayashi M, Li G, Poh HM, Ruan X, Aburatani H, Ruan Y, Kodama T, Wada Y, Cook PR (2012) TNF signals through specialized factories where responsive coding and miRNA genes are transcribed. *EMBO J* 31: 4404−4414

Pombo A, Dillon N (2015) Three-dimensional genome architecture: players and mechanisms. *Nat Rev Mol Cell Biol* 16: 245−257

Rao SS, Huntley MH, Durand NC, Stamenova EK, Bochkov ID, Robinson JT, Sanborn AL, Machol I, Omer AD, Lander ES, Aiden EL (2014) A 3D map of the human genome at kilobase resolution reveals principles of chromatin looping. *Cell* 159: 1665−1680

Schmiedeberg L, Skene P, Deaton A, Bird A (2009) A temporal threshold for formaldehyde crosslinking and fixation. *PLoS One* 4: e4636

Stadhouders R, Kolovos P, Brouwer R, Zuin J, van den Heuvel A, Kockx C, Palstra RJ, Wendt KS, Grosveld F, van Ijcken W, Soler E (2013) Multiplexed chromosome conformation capture sequencing for rapid genome-scale high-resolution detection of long-range chromatin interactions. *Nat Protoc* 8: 509−524

Teytelman L, Thurtle DM, Rine J, van Oudenaarden A (2013) Highly expressed loci are vulnerable to misleading ChIP localization of multiple unrelated proteins. *Proc Natl Acad Sci USA* 110: 18602−18607

Tolhuis B, Palstra RJ, Splinter E, Grosveld F, de Laat W (2002) Looping and interaction between hypersensitive sites in the active beta-globin locus. *Mol Cell* 10: 1453−1465

Vogel MJ, Guelen L, de Wit E, Peric-Hupkes D, Lodén M, Talhout W, Feenstra M, Abbas B, Classen AK, van Steensel B (2006) Human heterochromatin proteins form large domains containing KRAB-ZNF genes. *Genome Res* 16: 1493−1504

van de Werken HJ, de Vree PJ, Splinter E, Holwerda SJ, Klous P, de Wit E, de Laat W (2012) 4C technology: protocols and data analysis. *Methods Enzymol* 513: 89−112

Williams RL Jr, Starmer J, Mugford JW, Calabrese JM, Mieczkowski P, Yee D, Magnuson T (2014) fourSig: a method for determining chromosomal interactions in 4C-Seq data. *Nucleic Acids Res* 42: e68

Williamson I, Berlivet S, Eskeland R, Boyle S, Illingworth RS, Paquette D, Dostie J, Bickmore WA (2014) Spatial genome organization: contrasting views from chromosome conformation capture and fluorescence in situ hybridization. *Genes Dev* 28: 2778−2791

de Wit E, Bouwman BA, Zhu Y, Klous P, Splinter E, Verstegen MJ, Krijger PH, Festuccia N, Nora EP, Welling M, Heard E, Geijsen N, Poot RA, Chambers I, de Laat W (2013) The pluripotent genome in three dimensions is shaped around pluripotency factors. *Nature* 501: 227−231

Zhang Y, McCord RP, Ho YJ, Lajoie BR, Hildebrand DG, Simon AC, Becker MS, Alt FW, Dekker J (2012) Spatial organization of the mouse genome and its role in recurrent chromosomal translocations. *Cell* 148: 908−921

13

Temporal fluxomics reveals oscillations in TCA cycle flux throughout the mammalian cell cycle

Eunyong Ahn[1,†], Praveen Kumar[2,†], Dzmitry Mukha[2], Amit Tzur[3,4] & Tomer Shlomi[1,2,5,*] (ID)

Abstract

Cellular metabolic demands change throughout the cell cycle. Nevertheless, a characterization of how metabolic fluxes adapt to the changing demands throughout the cell cycle is lacking. Here, we developed a temporal-fluxomics approach to derive a comprehensive and quantitative view of alterations in metabolic fluxes throughout the mammalian cell cycle. This is achieved by combining pulse-chase LC-MS-based isotope tracing in synchronized cell populations with computational deconvolution and metabolic flux modeling. We find that TCA cycle fluxes are rewired as cells progress through the cell cycle with complementary oscillations of glucose versus glutamine-derived fluxes: Oxidation of glucose-derived flux peaks in late G1 phase, while oxidative and reductive glutamine metabolism dominates S phase. These complementary flux oscillations maintain a constant production rate of reducing equivalents and oxidative phosphorylation flux throughout the cell cycle. The shift from glucose to glutamine oxidation in S phase plays an important role in cell cycle progression and cell proliferation.

Keywords cell cycle; cellular metabolism; isotope tracing; LC-MS; metabolic flux analysis

Subject Categories Cell Cycle; Genome-Scale & Integrative Biology; Metabolism

Introduction

Cell cycle progression is tightly interlinked with cellular metabolism (Kaplon *et al*, 2015). The availability of sufficient metabolic nutrients and intracellular energy status controls the ability of cells to enter and progress through cell cycle. The absence of glucose was first shown to arrest cells at the G1/S restriction point (Blagosklonny & Pardee, 2002). More recently, cellular energy status (ATP/AMP ratio) was found to regulate canonical cell cycle signaling pathways via AMP-activated protein kinase (AMPK; Banko *et al*, 2011). The mammalian target of rapamycin (mTOR) plays a central role in regulating cell cycle progression and growth, integrating stimuli of amino acid, energy, and oxygen availability (Fingar & Blenis, 2004; Cuyàs *et al*, 2014). Cell cycle progression is further controlled by intracellular metabolites affecting epigenetics: Nuclear acetyl-CoA levels, determined by nuclear ATP citrate lyase (ACL; Wellen *et al*, 2009) and pyruvate dehydrogenase (PDH; Sutendra *et al*, 2014), regulate the acetylation of histones and thus control cell cycle progression (Berger, 2007; Li *et al*, 2007). Additionally, several metabolic enzymes were shown to directly regulate the cell cycle machinery, including PFKFB3 and PKM2, controlling the activity of cyclins and cyclin-dependent kinase (CDK) inhibitors in the nucleus (Yalcin *et al*, 2009; Yang *et al*, 2011, 2012).

Signaling pathways that coordinate cell cycle progression further regulate metabolic activity to support the changing metabolic demands throughout the cell cycle. The ubiquitin proteasome system, which tightly controls the concentration of cyclins, regulates the activity of two key enzymes in glucose and glutamine metabolism (Almeida *et al*, 2010; Tudzarova *et al*, 2011; Estevez-Garcia *et al*, 2014); the ubiquitin ligase anaphase-promoting complex/cyclosome (APC/C) and ligase Skp1/cullin/F-box protein (SCF) complex control glycolytic flux via PFKFB3, restricting its expression to late G1 and early S; APC/C also regulates glutaminolysis via glutaminase 1 (GLS1), whose expression is induced in S and G2/M. Cyclins and cyclin-CDK complexes were further suggested to regulate central metabolic activities, including glycolysis, lipogenesis, and mitochondrial activity (Hsieh *et al*, 2008; Bienvenu *et al*, 2010). Furthermore, central oncogenes and tumor suppressors that control proliferation, growth, and cell cycle can stimulate the expression of enzymes that mediate glycolysis and glutaminolysis (Levine & Puzio-Kuter, 2010).

While the cell cycle machinery was found to regulate the concentration of key metabolic enzymes, an understanding of how the actual rate of metabolic reactions and pathway (i.e., metabolic flux) changes throughout the cell cycle is still fundamentally missing. Metabolic flux is a not a directly measurable quantity and is typically inferred using isotope tracing techniques coupled with computational metabolic flux analysis (MFA; Wiechert, 2002; Sauer, 2006).

---

1 Department of Computer Science, Technion, Haifa, Israel
2 Department of Biology, Technion, Haifa, Israel
3 Faculty of Life Sciences, Bar-Ilan University, Ramat Gan, Israel
4 The Institute of Nanotechnology and Advanced Materials, Bar-Ilan University, Ramat Gan, Israel
5 Lokey Center for Life Science and Engineering, Technion, Haifa, Israel
*Corresponding author. E-mail: tomersh@cs.technion.ac.il
†These authors contributed equally to this work

Isotope tracing coupled with MFA is commonly used to address problems in biotechnology and medicine and has recently become a central technique in studies of cancer cellular metabolism (Metallo *et al*, 2009; Duckwall *et al*, 2013). Applied to cell populations with cells at different phases of the cell cycle, this approach typically estimates the average flux throughout the cell cycle.

Here, we present a temporal-fluxomics approach for quantifying cell cycle-dependent oscillations in metabolic flux, combining isotope tracing in synchronized cell populations with computational deconvolution and metabolic network modeling. We applied this approach to derive a first comprehensive and quantitative view of flux dynamics in central metabolism of proliferating cancer cells. The analysis adds a temporal dimension to our understanding of TCA cycle metabolism, showing major oscillations in the oxidation/reduction of glucose versus glutamine-derived fluxes as cells progress through the cell cycle.

Results

Cellular concentration of central metabolic intermediates oscillate throughout the cell cycle

To study metabolic dynamics throughout cell cycle, we synchronized HeLa cells using double thymidine block and applied high-throughput LC-MS-based targeted metabolomics analysis to synchronized cell populations (> 10^6 cells per sample) in 3-h intervals for two complete cell cycles (see cell synchronization dynamics measured via propodium iodide staining/FACS analysis in Fig 1A; Appendix Fig S1; Materials and Methods). To obtain a reliable and accurate view of periodic metabolic oscillations and to overcome a potential perturbation of metabolism due to synchronization-induced growth arrest, we let synchronized cells complete one cell cycle before starting the LC-MS analysis (9 h after cells are released in G1/S). Measured metabolite abundances in the synchronized cells were normalized by total cell volume in each time point to determine metabolite concentrations.

As cell synchronization is gradually lost with time due to inherent non-genetic cell-to-cell variability (a phenomenon also known as "dispersion"), the distribution of cell cycle phases in the synchronized cell population becomes similar to that of non-synchronized cells after completing three rounds of replications (Fig 1A). To account for the loss of synchrony and to precisely quantify oscillations in metabolite levels, we employed "computational synchronization" (Bar-Joseph *et al*, 2008): We constructed a probabilistic model that describes the dynamics of the cell population loosing synchrony, assuming that each cell has its own "internal clock" which controls the cell cycle progression rate (see Synchronization loss model in Materials and Methods). The parameters of the model were estimated by fitting a simulation of how the synchronized cell population progresses through the different phases of the cell cycle with corresponding FACS measurements, finding that cell–cell variability in the rate of cell cycle progression through the cell cycle is 11% (Fig 1B, Materials and Methods). We used this model for *computational deconvolution* of measured metabolomics data, estimating metabolite concentration dynamics throughout the cell cycle, circumventing the impact of cell dispersion (Materials and Methods). Inferring the dynamics of metabolite concentrations throughout the cell cycle (rather than that of metabolite abundances) further required estimates of the dynamics of cell volume throughout the cell cycle. The latter was estimated based on deconvolution of total cell volume measurements performed in the synchronized cell population (Materials and Methods, Fig 1C). For example, the concentration of the nucleotide cytidine triphosphate (CTP) was found to oscillate throughout cell cycle, showing a ~50% increase in concentration in G1 phase versus G2/M phase (measured and deconvoluted concentrations shown in Fig 1D). As shown, the magnitude of the oscillation drops with time and converges to the steady-state concentration measured in non-synchronized cells.

Our analysis reveals 57 metabolites whose concentrations significantly oscillate throughout the cell cycle (Fig 2, Dataset EV1, FDR-corrected *P*-value < 0.05, Materials and Methods). Oscillations in nearly 44% of these metabolites could be detected only when *in silico* synchronization via computational deconvolution was applied (i.e., equation 7 in Materials and Methods), emphasizing the strength of our pipeline. The median size of the observed oscillations is ~60% (difference between maximal and minimal concentration throughout the cell cycle); roughly one-quarter of these metabolites show concentration changes larger than twofold throughout cell cycle. A significantly high fraction of the metabolites peaks either in late G1 (~50% in the second half of G1, *P*-value < 10^{-5}, compared with the expected fraction assuming that concentration peaks are uniformly distributed throughout the entire cell cycle) or early S (~35% in the first half of S, *P*-value < 0.002).

The oscillating metabolites include glycolytic and TCA cycle intermediates, nucleotides, amino acids, and energy and redox cofactors. Expectedly, the concentrations of the deoxynucleotides dCTP and dATP increase in S phase, when utilized for DNA replication. Intermediates in polyamine metabolism (5-methylthioadenosine, acetyl-putrescine, S-adenosyl-L-methionine, and S-adenosyl-L-homocysteine) show marked oscillations, in accordance with the known cell cycle-dependent activity of this pathway (Oredsson, 2003). Several glycolytic metabolites peak during G1/S transition, in accordance with reports of increased glycolytic flux at this cell cycle phase (Colombo *et al*, 2011; Tudzarova *et al*, 2011). Cellular ATP/ADP ratio and redox potential (NADH/NAD$^+$) further show a ~50% increase in the G1/S transition (Appendix Fig S2). Intracellular concentration of non-essential amino acids synthesized from consumed glutamine, including glutamate, ornithine, proline, and aspartate peak in S phase, in accordance with a reported increase in glutamine dependence in S phase (Gaglio *et al*, 2009; Colombo *et al*, 2011). Intriguingly, we find that different TCA cycle metabolites peak in distinct cell cycle phases: Acetyl-CoA and citrate peak in G1/S, while malic acid and α-ketoglutarate peak in late S, suggesting that the TCA cycle is rewired as cells progress through the cell cycle.

Time-resolved fluxomics reveals increased glycolytic flux into TCA cycle in G1/S transition

To observe metabolic flux dynamics in TCA cycle and in branching pathways throughout the cell cycle, we performed pulse-chase isotopic tracing experiments in synchronized HeLa cells with [U-^{13}C]-glucose and [U-^{13}C]-glutamine (1-h feeding), every 3 h for two cell cycles (Fig 3A and B, Materials and Methods). Here, LC-MS was utilized to measure the mass-isotopomer distribution of metabolites (i.e., the fraction of each metabolite pool having zero,

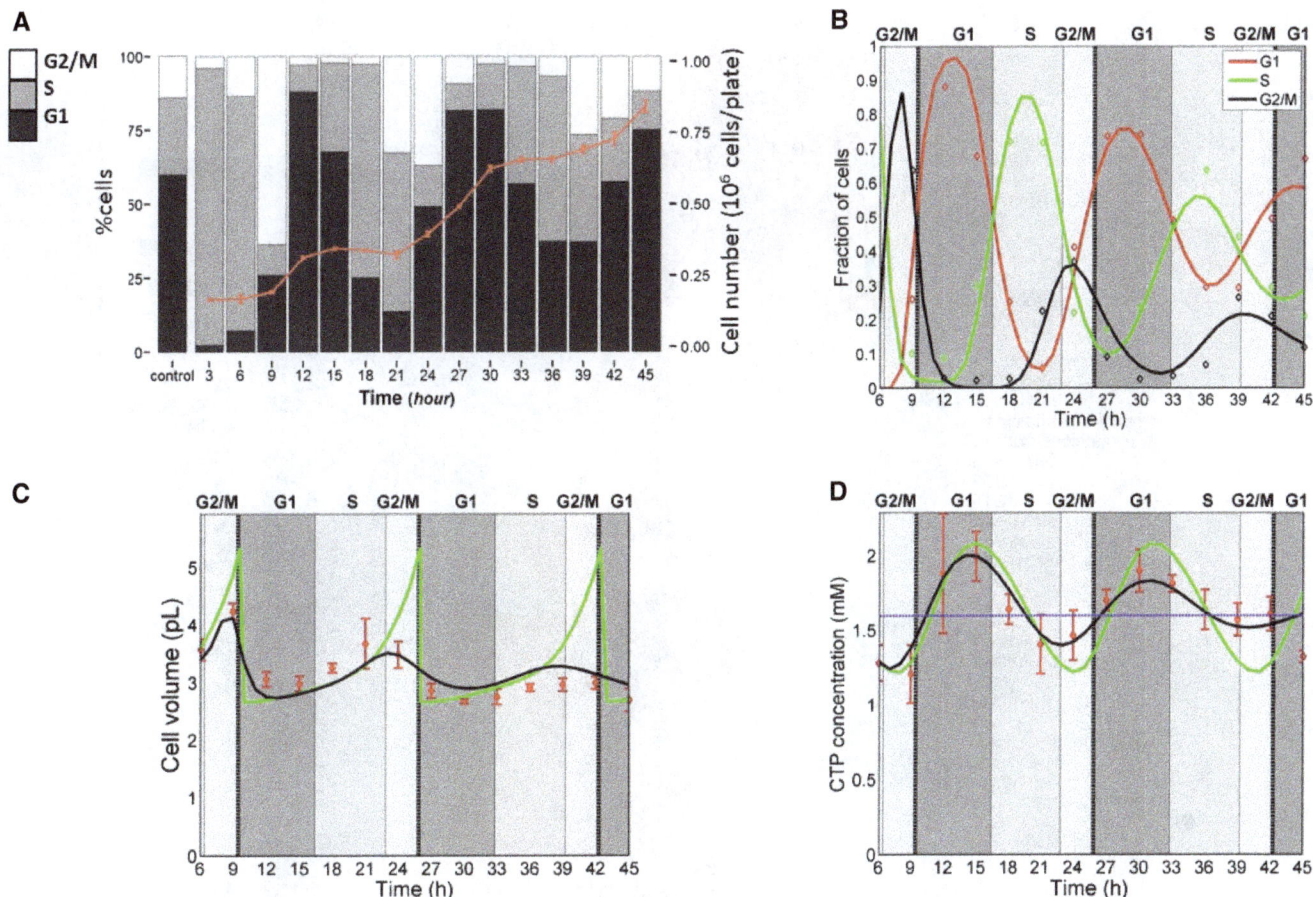

Figure 1. Detection of oscillations in metabolite concentrations throughout the cell cycle in HeLa cells revealed by LC-MS-based metabolomics of synchronized HeLa cells and computational deconvolution.

A Synchronization dynamics of a population of HeLa cells within almost three complete cell cycles measured via PI staining followed by FACS analysis. The increase in cell number following each mitosis is shown by an overlaid curve (in orange, mean and s.d. of $n = 3$).

B Computational modeling of the synchronization loss, considering 11% cell–cell variation in doubling time, shows that the simulated fraction of the cells in G1, S, and G2/M in the synchronized cells throughout the cell cycle (straight lines) match experimental measurements (marked with asterisk).

C The measured average cell volume in the synchronized cell population (red, mean and s.d. of $n = 3$, $v(t)$ in equation 6), the deconvoluted signal (in case of no synchronization loss, green, $v'(x)$ in equation 6), and the simulated average cell volume considering the loss in synchronization (black, matching the measured concentration data, equation 6).

D The measured concentration of CTP in synchronized cells shown in red (mean and s.d. of $n = 5$, $u_i(t)$ in equation 7), the deconvoluted concentration dynamics, in case of no synchronization loss (green, $u'(x)$ in equation 7), and the expected concentration dynamics based on the deconvoluted concentrations and considering the loss in synchronization, matching the measured concentrations (black, equation 7). The measured concentrations converge toward the steady-state concentrations measured in non-synchronized cells (horizontal blue line).

one, and two labeled carbon atoms) after 1 h of feeding with the isotopic tracers. Computational deconvolution was employed to analyze oscillations in metabolite isotopic labeling patterns while considering cell dispersion. The deconvolution approach is based on the observation that the measured fractional isotopic labeling of a metabolite in the synchronized cell population represents the average labeling in cells with distinct intrinsic times, weighted by the metabolite pool size in these cells (i.e., the measured isotopic labeling pattern is biased toward that of cells with an intrinsic time in which the metabolite pool size is larger than in others, Materials and Methods). Overall, we detected statistically significant oscillations in the isotopic labeling pattern of 21 metabolites when feeding isotopic glucose, and 16 metabolites when feeding isotopic glutamine (FDR-corrected P-value < 0.05, Figs 3 and 4, Dataset EV2).

The inferred oscillations in metabolite isotopic labeling and concentrations were used to computationally analyze metabolic flux dynamics throughout the cell cycle, utilizing a variant of kinetic flux profiling (KFP; Yuan *et al*, 2008; Fig 5, in units of nmole/µl-cells/h, i.e., mM/h, Materials and Methods). Specifically, given a metabolite whose isotopic labeling dynamics throughout the cell cycle was inferred as explained above, we search for the most likely transient production and consumption fluxes in each 1-h interval through the cell cycle, such that the simulated labeling kinetics of this metabolite (within the 1-h interval) would optimally match the experimental measurements (Materials and Methods, Appendix Figs S3–S8). The simulation of the isotopic labeling kinetics of a metabolite of interest within a 1-h time interval is performed via an ordinary differential equations (ODE) model,

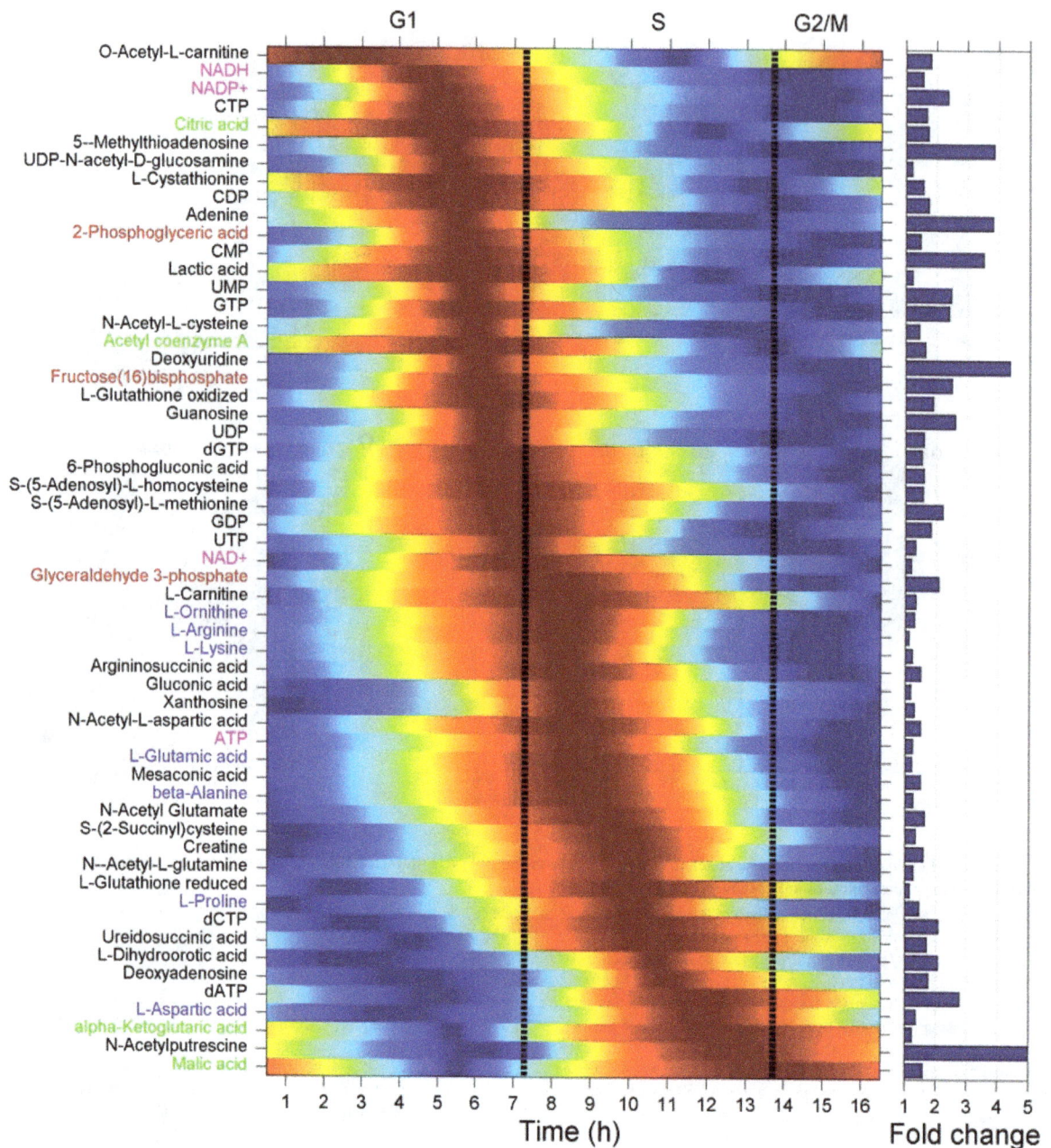

Figure 2. Oscillation in metabolite concentrations throughout the cell cycle in HeLa cells.

The figure shows metabolites found to significantly oscillate throughout the cell cycle (concentrations normalized per metabolite, maximal concentration in red, minimal concentration in blue). The amplitude of the oscillations is shown on the right. Metabolites are color-coded according to metabolic pathways: energy/redox cofactors (purple), amino acids (blue), glycolytic metabolites (red), and TCA cycle metabolites (green).

relying on the inferred concentration of the metabolite within this time interval (considering that a metabolite with a larger pool size would take more time to label, per unit of flux), as well as the isotopic labeling kinetics of intermediates that produce this metabolite. While KFP is typically applied to estimate fluxes under metabolic steady state (in which fluxes satisfy a stoichiometric mass-balance constraint), here, we constrain the difference between transient fluxes that produce and consume a certain metabolite according to the measured momentary change in the concentration of that metabolite.

Oscillations in the isotopic labeling pattern of TCA cycle intermediates when feeding isotopic glucose suggest that glucose-derived flux into TCA cycle increases in G1 phase and then drops in S phase. The fractional labeling of the m + 2 form of the TCA cycle intermediates citrate, α-ketoglutarate, and malate drops in S phase (Fig 3C–E). Feeding cells with isotopic glutamine, we further observed a drop in citrate m + 4 produced from oxaloacetate via citrate synthase in S phase (Fig 3F). Combined with the drop in citrate concentration during S phase (Fig 3G), metabolic modeling reveals a ~2-fold decrease in glycolytic flux into TCA cycle as cells progress through S phase;

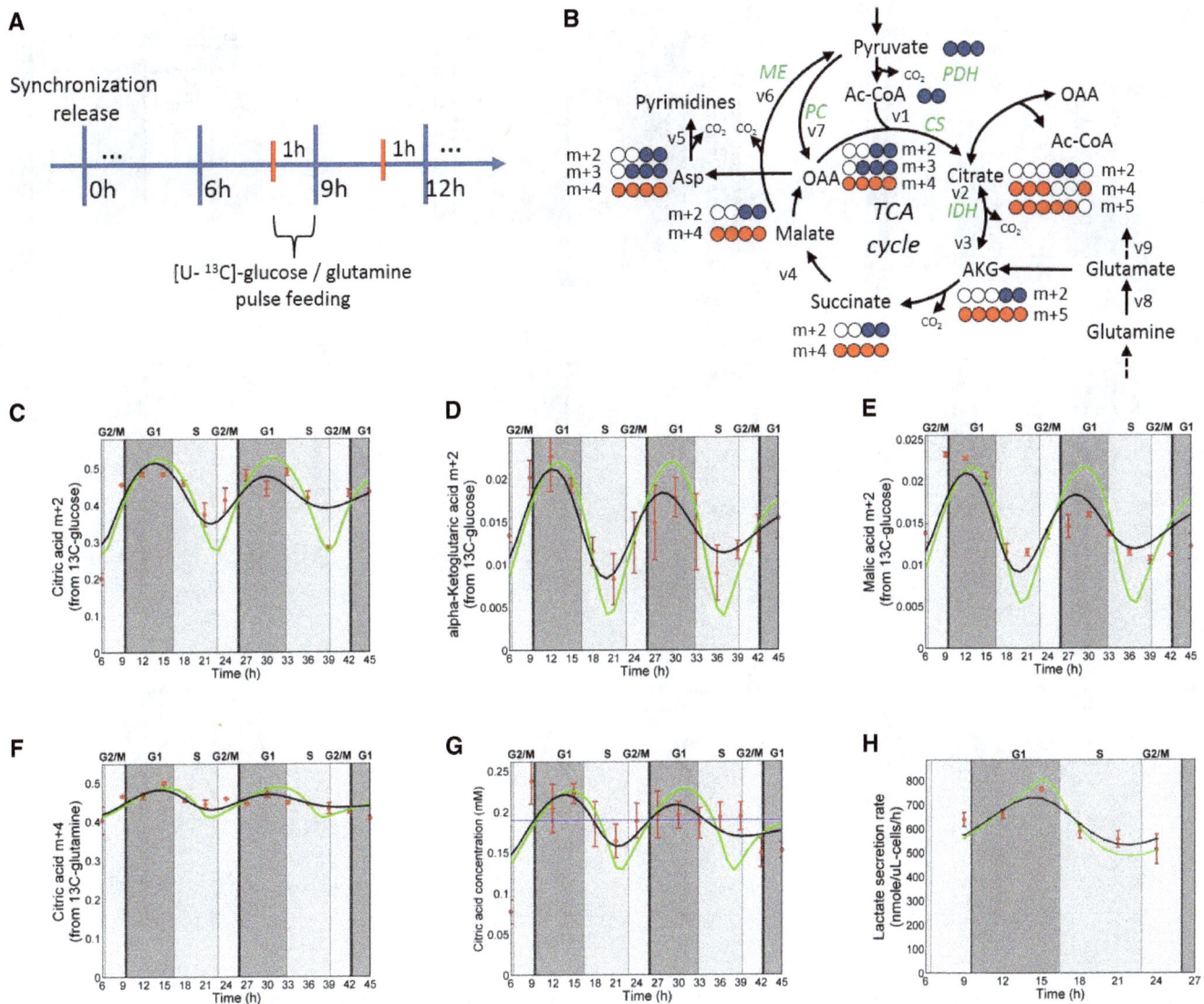

Figure 3. Oscillations in isotopic labeling of TCA cycle metabolites throughout the cell cycle from [U-¹³C]-glucose show induced glycolytic flux into TCA cycle in G1/S.

A Experimental scheme for a series of pulse-chase isotope tracing experiments in synchronized cells.

B Atom tracing of TCA cycle metabolites from [U-¹³C]-glucose (blue) and [U-¹³C]-glutamine (red).

C–E Measured relative fraction of the m + 2 labeling of TCA cycle intermediates after feeding [U-¹³C]-glucose (red, mean and s.d. of $n = 3$), the deconvoluted signal (green), and the expected labeling dynamics considering the loss in synchronization (black, representing TCA cycle oxidation of glucose-derived acetyl-CoA).

F Oscillations in citrate m + 4 labeling after feeding [U-¹³C]-glutamine (mean and s.d. of $n = 3$; experimentally measured labeling dynamics shown in red, the deconvoluted signal in green, and the expected labeling dynamics considering the synchronization loss in black).

G Oscillations in the total citrate concentration throughout the cell cycle (mean and s.d. of $n = 5$; experimentally measured concentration dynamics shown in red, the deconvoluted signal in green, the expected concentration dynamics considering the synchronization loss in black, and the measured concentration in non-synchronized cells in blue).

H The measured lactate secretion flux in synchronized cells shown in red (mean and s.d. of $n = 5$, $f_i(t)$ in equation 9), the deconvoluted secretion flux dynamics, in case of no synchronization loss (green, $f_i'(x)$ in equation 10), and the expected secretion flux based on the deconvoluted fluxes and considering the loss in synchronization, matching the measured fluxes (black, equation 10).

citrate synthase flux drops from ~6 mM/h in G1/S phase to ~3 mM/h in late S phase (Fig 5A, Appendix Fig S3). TCA cycle oxidation of citrate via isocitrate dehydrogenase (IDH) shows similar flux dynamics, with ~2-fold drop in S phase (Fig 5C, Appendix Fig S4).

To examine whether the increase in glucose-derived flux into TCA cycle in G1/S is associated with increased in glycolytic flux,

we measured lactate concentrations in the culture media in the synchronized cell population followed by computational deconvolution (Fig 3H, Materials and Methods). We find a ~65% increase in lactate secretion rate in G1/S transition. Considering that the average lactate secretion rate throughout the cell cycle is two orders of magnitude higher than that of pathways that

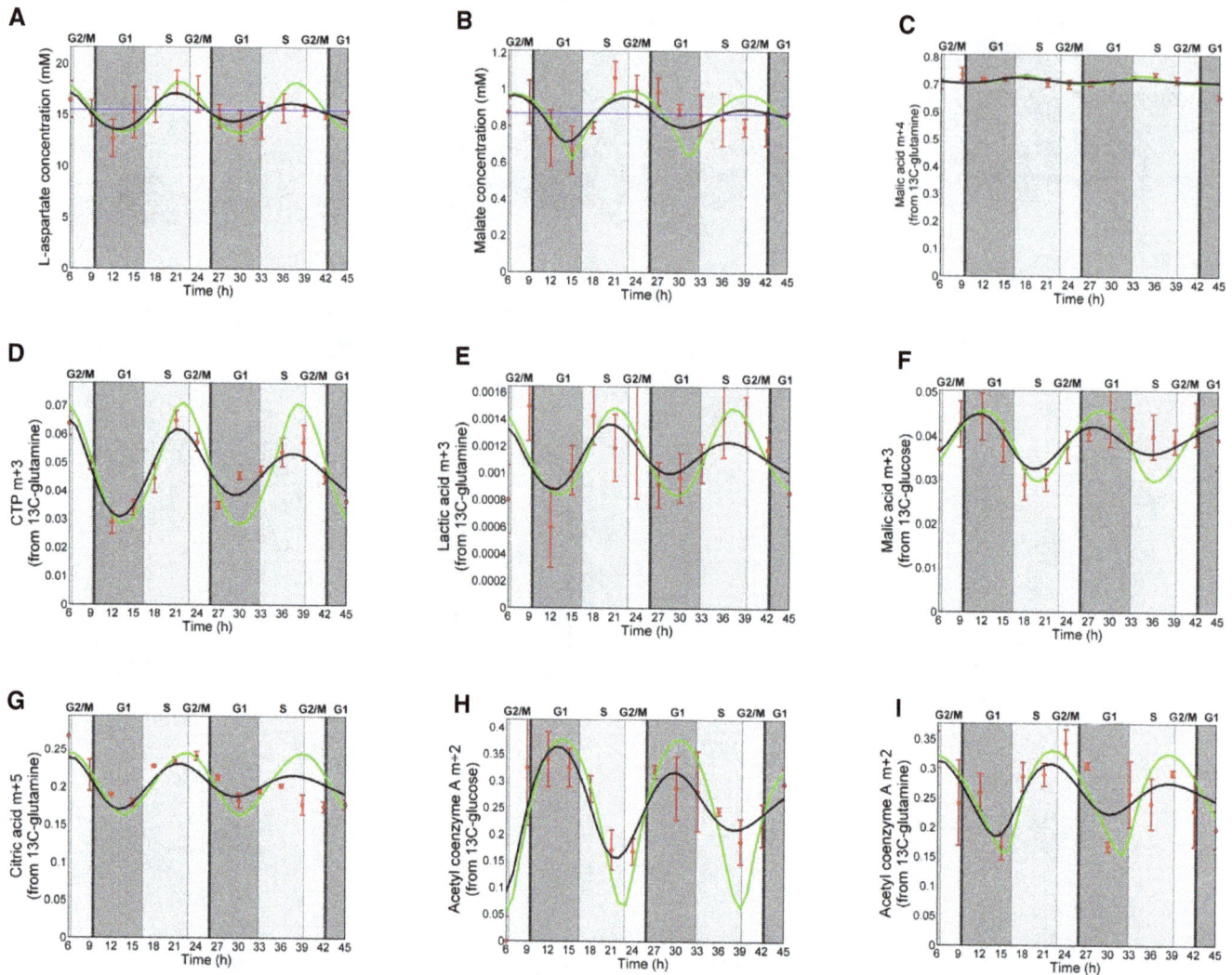

Figure 4. Oscillations in isotopic labeling of TCA cycle metabolites throughout the cell cycle from [U-¹³C]-glutamine show induced oxidative and reductive glutamine metabolism in S phase.

A, B Oscillations in aspartate (A) and malate (B) concentrations throughout the cell cycle when feeding [U-¹³C]-glutamine (representing oxidative TCA cycle activity).

C Uniform malate m + 4 labeling throughout the cell cycle (combined with the increase in malate concentration in S phase representing increased oxidative TCA cycle flux in S phase).

D Oscillations in pyrimidines m + 3 labeling throughout the cell cycle when feeding [U-¹³C]-glutamine (representing *de novo* pyrimidine biosynthesis).

E Oscillations in lactate m + 3 labeling throughout the cell cycle when feeding [U-¹³C]-glutamine (representing malic enzyme activity).

F Oscillations in malate m + 3 throughout the cell cycle when feeding [U-¹³C]-glucose (representing pyruvate carboxylase activity).

G Oscillations in citrate m + 5 when feeding [U-¹³C]-glutamine throughout the cell cycle (representing reductive IDH flux).

H Oscillations in acetyl-CoA m + 2 when feeding [U-¹³C]-glucose, representing oxidative glucose metabolism.

I Oscillations in acetyl-CoA m + 2 when feeding [U-¹³C]-glutamine, representing reductive glutamine metabolism.

Data information: (A, B) Measured metabolite concentrations are shown in red (showing mean and s.d. of *n* = 5), the deconvoluted signal in green, the expected concentration dynamics considering the synchronization loss in black, and the measured concentrations in non-synchronized cells in blue. (C–I) Measured fractional isotopic labeling are shown in red (showing mean and s.d. of *n* = 3), the deconvoluted signal in green, and the expected isotopic labeling dynamics considering the synchronization loss in black.

branch out from glycolysis, the observed oscillation in lactate secretion represents cell cycle-dependent changes in glycolytic flux: The average lactate secretion throughout the cell cycle is ~600 mM/h, while oxidative pentose-phosphate pathway (PPP) is ~3 mM/h, reductive PPP is ~3 mM/h, glycogenesis is ~0.3 mM/h, and serine biosynthesis is below 1 mM/h (Appendix Fig S8). Overall, our data show that the increase in glycolytic flux in G1/S phase co-occurs with the increased glucose-driven flux entering

the TCA cycle. Notably, analyzing oscillations in glycolytic flux based on direct measurement of changes in glucose consumption throughout the cell cycle (rather than based on lactate secretion) was not possible due to technical difficulty in accurately quantifying glucose consumption by synchronized cells within 3-h time intervals (considering that the synchronized cell population consumes ~1% of the glucose in media within this short time period).

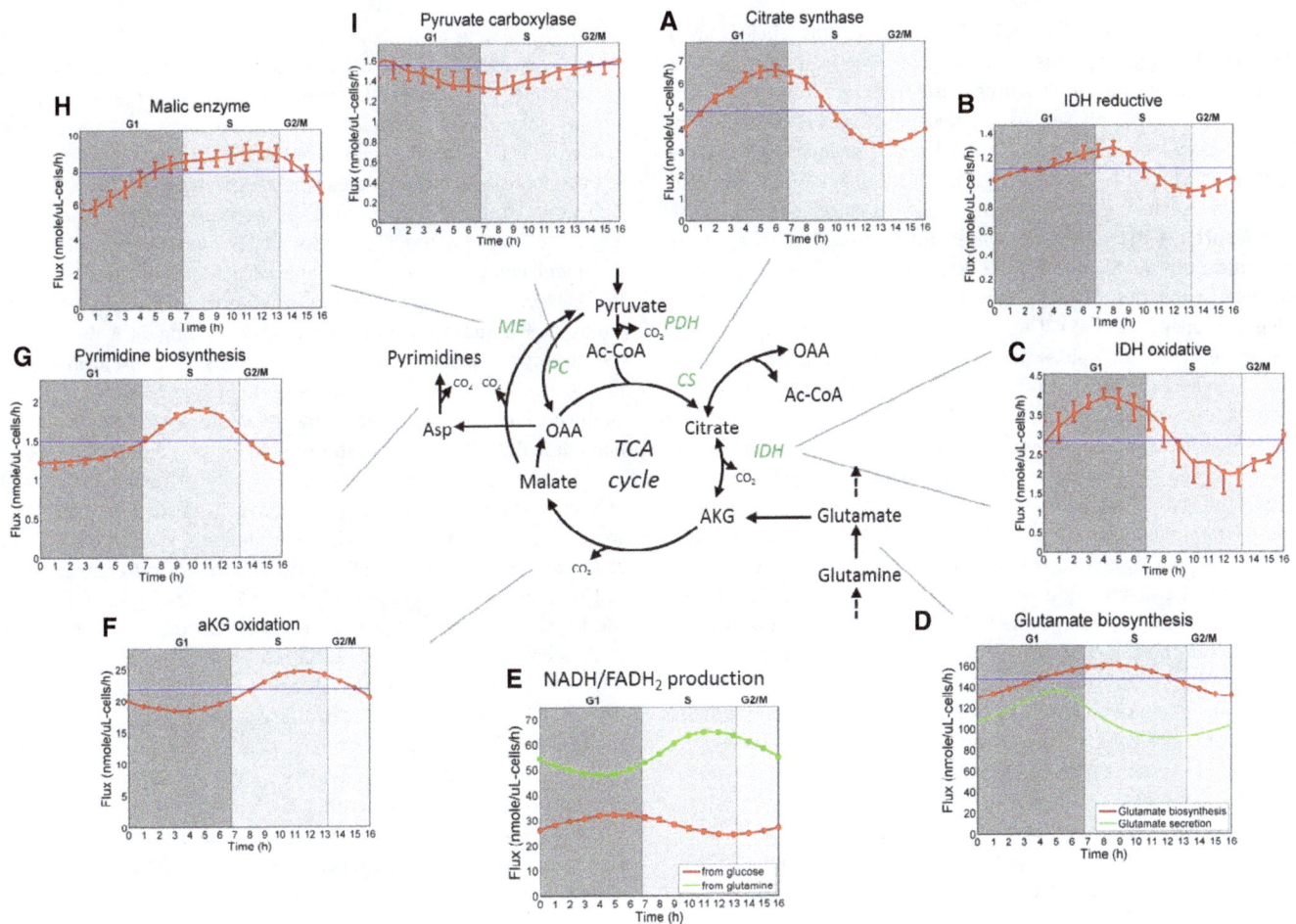

Figure 5. Complementary oscillations of glucose versus glutamine-derived fluxes in TCA cycle.

A–I Oscillations in metabolic flux throughout the cell cycle (in mM/h), computed based on metabolic modeling of measured oscillations in metabolite concentrations and isotopic labeling (red and green marks represent optimal estimates of transient flux with 95% CI). Blue lines represent average fluxes inferred in a non-synchronized cell population. As shown, glucose-derived flux into TCA cycle peaks in late G1 phase, while oxidative and reductive glutamine metabolism dominates S phase.

Induced oxidative and reductive glutamine metabolism compensates for the decreased glycolytic flux into TCA cycle in S phase

Glutamine feeds TCA cycle flux by producing glutamate, which is converted to α-ketoglutarate either via transamination or by glutamate dehydrogenase. As a first estimation of the cell cycle dynamics of glutamine-derived flux into the TCA cycle, we quantified cell cycle-dependent glutamate production from glutamine versus glutamate secretion to the media. Tracing the m + 5 labeling dynamics of glutamate when feeding [U-^{13}C]-glutamine and glutamate concentration throughout the cell cycle suggests that glutamate production flux increases by 25% in S phase compared to G1 (Fig 5D). This is evident by a marked increase in glutamate concentration in S phase and similar m + 5 glutamate labeling kinetics throughout the cell cycle (Appendix Fig S4). Glutamate secretion rate to the culture medium shows a marked drop in S phase, suggesting increased availability of glutamate for feeding the TCA cycle flux in S phase (Fig 5D). The increased entry of glutamine-derived flux into the

TCA cycle in S phase is followed by a ~40% increase in α-ketoglutarate oxidation (Fig 5F, Appendix Fig S5). This is evident by the marked increase in malate and aspartate concentration in S phase (Fig 4A and B) and barely altered m + 4 and m + 3 labeling kinetics of these metabolites throughout the cell cycle, respectively (Fig 4C and Appendix Fig S9).

The increased glutamine-derived anaplerotic flux into the TCA cycle in S phase (via net production of the TCA cycle intermediate α-ketoglutarate) is balanced by oscillations in cataplerotic fluxes, consuming TCA cycle intermediates for biosynthetic and bioenergetics purposes: We find a ~70% increase in pyrimidine biosynthesis flux in S phase, consuming oxaloacetate from TCA cycle (transaminated to produce aspartate, Fig 5G). This is evident by a marked increase in the m + 3 labeling of pyrimidines in S phase (Fig 4D and Appendix Fig S10), while considering the oscillations in the labeling kinetics of carbamoyl-aspartate in pyrimidine biosynthesis and pyrimidine concentrations (Appendix Fig S6). Oscillations in the biosynthetic flux of pyrimidines as well as purines is further supported by an increased m + 5 labeling of pyrimidines and purines in S phase (i.e., having all five ribose carbons labeled) upon feeding with isotopic glucose (Dataset EV2). The malic enzyme flux (decarboxylating malate dehydrogenases) further shows a marked ~65% increase in S phase (Fig 5H, Appendix Fig S7), as evident by the increased lactate m + 3 labeling in S phase when feeding isotopic glutamine (Fig 4E). Consistently, an increased concentration of lactate m + 3 in the culture media is further observed in S phase (Appendix Fig S11). Notably, while glutamine-derived anaplerotic flux increases in S phase, there is no major change in glucose-derived anaplerotic flux through pyruvate carboxylase in S phase (Fig 5I, Appendix Fig S5). This is evident by the drop in malate and aspartate m + 3 in S phase when feeding isotopic glucose (Fig 4F and Appendix Fig S12) occurring while the concentration of malate and aspartate increases (Fig 4A and B).

While the increased glutamine-derived flux into the TCA cycle in S phase supports an increase in α-ketoglutarate oxidation, we further detect a surprisingly high ~55% increase in the rate of α-ketoglutarate reduction in early S phase (Fig 5B, Appendix Fig S3). This is evident by the marked increase in m + 5 citrate in S phase when feeding isotopic glutamine (Fig 4G). Considering the major drop in glycolytic flux into the TCA cycle in S phase, the relative contribution of reductive IDH to citrate production increases from ~15% in G1 to ~24% in S phase. Cell cycle oscillations in the relative contribution of glucose versus glutamine to citrate biosynthesis are further observed in the labeling of acetyl-CoA (produced in cytosol from citrate via ATP citrate lyase), where the fractional labeling of acetyl-CoA m + 2 from [U-^{13}C]-glucose peaks in G1 phase (Fig 4H) and acetyl-CoA m + 2 from [U-^{13}C]-glutamine in S phase (Fig 4I). Accordingly, acetylated amino acids show increased m + 2 labeling from glucose in G1 and from glutamine in S phase (Appendix Fig S13).

The complementary oscillations in glucose versus glutamine-derived flux into the TCA cycle result in an overall uniform production rate of reducing equivalents (NADH/FADH$_2$) ~85 ± 5 mM/h throughout the cell cycle (Fig 5E, Materials and Methods): Glucose-derived production of reducing equivalents peaks in G1/S while glutamine-derived production of reducing equivalents peaks in late S phase. While the relative contribution of glutamine to NADH/FADH$_2$ production oscillates (between ~60% in G1/S and ~75% in late S phase), it remains the prime source of reducing power for driving oxidative phosphorylation all throughout the cell cycle, in accordance with previous measurements in non-synchronized cells (Fan *et al*, 2013). Consistent with the total production rate of NADH/FADH$_2$ remaining constant throughout the cell cycle, we find that oxygen consumption rate does not change throughout the cell cycle (Appendix Fig S14). Hence, the complementary oscillations in glucose versus glutamine oxidation in TCA cycle result in a constant rate of reducing equivalent production, sustaining a constant rate of mitochondrial oxidative phosphorylation flux throughout the cell cycle.

Suppression of glycolytic flux into TCA cycle in S phase is important for cellular progression through the cell cycle

The drop in glycolytic flux into TCA cycle in S phase (Fig 5A) involves a major twofold decrease in flux through pyruvate dehydrogenase (PDH), the prime source for acetyl groups for TCA cycle oxidation. PDH is negatively regulated by pyruvate dehydrogenase

kinase (PDK), and treatment with the PDK inhibitor dichloroacetate (DCA) was previously shown to enhance glycolytic flux into TCA cycle while decreasing reductive glutamine metabolism toward citrate biosynthesis (Fendt *et al*, 2013). Accordingly, treating the synchronized HeLa cells with 4 mM of DCA for 1 h leads to a marked increase in the fractional labeling of citrate m + 2 and acetyl-CoA m + 2 from isotopic glucose, representing a major increase in glycolytic flux into the TCA cycle (Fig 6A and B). Notably, DCA treatment completely eliminates the oscillations in glycolytic flux into the TCA cycle, as evident by a uniform fractional labeling of citrate m + 2 and acetyl-CoA m + 2 when feeding DCA throughout the cell cycle. DCA treatment further leads to a uniform citrate m + 5/m + 4 ratio and fractional labeling of acetyl-CoA m + 2 from isotopic glutamine, eliminating the oscillations in glucose versus glutamine flux toward acetyl-CoA production (Fig 6C and D).

To test whether the suppression of glycolytic flux into TCA cycle in S phase is important for progression of cells through the cell cycle, we measured the cell cycle phase distribution in synchronized HeLa cells after 3-h treatment with DCA. We find that DCA treatment leads to 16% increase in the fraction of cells in S phase (Fig 6E, two-tailed *t*-test P-value = 0.006). Treating of non-synchronized HeLa cells as well as colon carcinoma cells (HCT116) with DCA for 24 h further shows a significant increase in the fraction of cells in S phase (Fig 6F, 30% increase in HeLa, two-tailed *t*-test P-value < 10^{-3}, 17% increase in HCT116, P-value < 10^{-5}). Notably, the observed increase in the fraction of cells in S phase represents slower progression rate of cells through S phase rather than cell cycle arrest at that phase, as almost all HeLa cells (~97%) complete at least one cell cycle after a 72-h treatment with DCA (Appendix Fig S15). Conversely, treating of cells with a mitochondrial pyruvate carrier inhibitor (UK5099), which slows glycolytic flux into TCA cycle, shows the opposite effect of lowering the fraction of cells in S phase (Appendix Fig S16). Overall, our results indicate that the shift from glucose to glutamine-derived flux into TCA cycle plays an important role in cellular progression through S phase.

Discussion

We described a temporal-fluxomics approach for analyzing the dynamics of intracellular metabolic flux throughout the cell cycle in proliferating human cells. Inferring cell cycle-dependent changes in flux is technically challenging due to several factors including the perturbative nature of synchronization-induced growth arrest, the gradual loss of population synchrony, and the requirement for accurate measurements of oscillations in metabolite pool sizes that in many cases vary by less than twofold at maximum. Addressing these challenges, we tracked synchronized cells for three complete cell cycles, performed LC-MS-based metabolomics and pulse-chase isotope tracing in the synchronized cells, and employed computational deconvolution techniques to reliably detect oscillations in metabolite concentrations and isotopic labeling dynamics. Inferring transient fluxes within each 1-h interval throughout the cell cycle was complicated by the fact that the labeling of TCA cycle intermediates in the synchronized cells does not reach isotopic steady state within 1-h feeding with the isotopic nutrients. This was addressed by modeling the isotopic labeling kinetics of metabolites within each 1-h interval throughout the cell cycle, for each one time interval

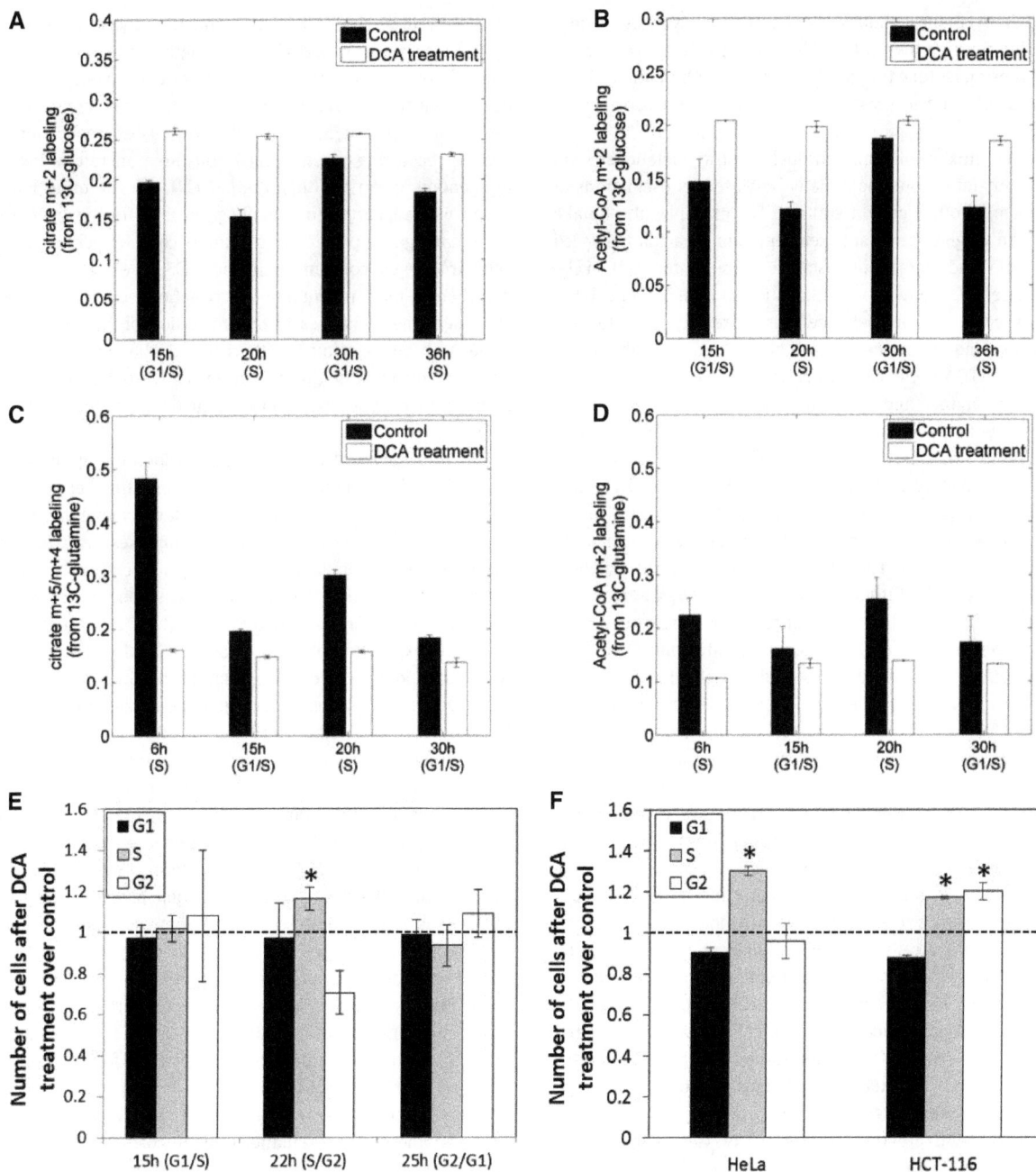

Figure 6. PDK inhibition via DCA treatment eliminates the oscillation of glycolytic flux into TCA cycle and inhibits cellular progression through S phase.

A–D One-hour treatment of synchronized cells with DCA inhibits the oscillations in citrate m + 2 (A) and acetyl-CoA m + 2 (B) from isotopic glucose (representing glycolytic flux into TCA cycle). It further inhibits oscillations in citrate m + 5/m + 4 ratio (C) and acetyl-CoA m + 2 (D) from isotopic glutamine (representing oxidative versus reductive TCA cycle flux).

E The fraction of cells in G1, S, and G2/M phases in synchronized HeLa cells after 3-h treatment with DCA (normalized by measurements in untreated control cells).

F The fraction of cells in G1, S, and G2/M phases in non-synchronized HeLa and HCT-116 cells after 24-h treatment with DCA (normalized by measurements in untreated control cells).

Data information: As shown, DCA significantly increases the fraction of cells in S phase, inhibiting cellular progression into G2 phase, showing mean and s.d. of $n = 3$ for all isotopic labeling forms and FACS measurements (A–F). Statistical significance of changes in the fraction of cells in each cell cycle phase following DCA treatment is calculated based on two-tailed, unequal variance t-test (E, F; *P-value < 0.01).

using an approach conceptually similar to non-stationary metabolic flux analysis (MFA; Noh *et al*, 2006; Noack *et al*, 2011). Applied to HeLa cells, we derived a first comprehensive and quantitative view

of metabolic flux oscillations at a high temporal resolution in central metabolism throughout the cell cycle of human cells, showing complementary oscillations between glucose and glutamine-derived flux in the TCA cycle throughout the cell cycle. Cell cycle-dependent changes in flux through the pentose-phosphate pathway were previously studied using isotope tracing in synchronized human cell lines (Vizan et al, 2009) and in yeast by also utilizing MFA (Costenoble et al, 2007).

The inferred flux oscillations through central metabolism via glucose and glutamine could potentially be biased by oscillations in the metabolism of other carbon sources. For example, the uptake and catabolism of glucogenic and ketogenic amino acids may feed into TCA cycle and potentially oscillate through the cell cycle. Though, apparently, the relative contribution of amino acid catabolism to TCA cycle flux in HeLa cells is extremely low; feeding isotopic glucose and glutamine for 24 h shows that more than 97% of the carbons in TCA cycle metabolites are derived exclusively from glucose and glutamine (and from atmospheric CO_2, Appendix Fig S17). Hence, a potential change in amino acid metabolism throughout the cell cycle would have little or no effect on the reported flux oscillations from glucose and glutamine. Not accounting for potential oscillations through other reactions implicated in central metabolism of glucose and glutamine could in principle also bias the presented flux estimations. Though, notably, the flux analysis here is performed separately through groups of converging reactions producing different metabolites (e.g., for reactions producing citrate, malate/oxaloacetate, and lactate); hence, a potential bias in some of these independent flux estimations would not change the overall emerging view of glucose and glutamine oscillations; for example, the increased TCA cycle metabolism of glutamine in S phase is supported by several independent flux estimations showing increased glutamine-derived flux into TCA cycle in S phase, oxidation of α-ketoglutarate, reduction of α-ketoglutarate, malic enzyme flux, and increased nucleotide biosynthesis. Another simplifying assumption that facilitated the estimation of flux dynamics throughout the cell cycle is of rapid mixing of mitochondrial and cytosolic metabolite pools, giving rise to estimates of whole-cell level fluxes. This assumption is typically made when analyzing flux in eukaryotic cells due to experimental complications in measuring metabolite concentrations and labeling dynamics in distinct subcellular compartments. Methodological advancements in subcellular level metabolomics are required for further studies on cell cycle oscillations in metabolic flux in mitochondria versus cytosol.

Our result of an induced glycolytic flux in G1/S phase is qualitatively consistent with previous reports ~3- to 10-fold increase in glycolytic flux in G1/S in HeLa cells (Colombo et al, 2011; Tudzarova et al, 2011). However, here, analyzing the metabolism of synchronized cells after resuming exponential growth (i.e., completing an entire cell cycle after released from synchronization growth arrest) suggests a moderate increase in glycolytic flux of only ~65% in G1/S phase. The smaller magnitude of the oscillations in glycolytic flux inferred here is in agreement with the moderate changes in concentration of glycolytic intermediates, which increase only ~2-fold in G1/S phase. Our finding of increased glutamine-derived flux into the TCA cycle and support of pyrimidine biosynthesis in S phase agree with reports of the essentiality of glutamine (and not glucose) for entering and progressing through S phase, which can be rescued by nucleotide feeding (Gaglio et al, 2009; Colombo et al, 2011).

We showed that complementary oscillations in the rate of glucose versus glutamine oxidation in the TCA cycle result in a constant rate of $NADH/FADH_2$ production and reduction of oxygen by the electron transport chain throughout the cell cycle. Notably, while here we observe a constant rate of oxygen consumption throughout the cell cycle, fluctuations in oxygen consumption were previously reported in yeast cells, with DNA replication and cell division occurring when oxygen consumption rate is low, protecting genome integrity (Chen et al, 2007). On the other hand, respiration was suggested to actually protect against oxygen-associated DNA damage in proliferating human cells by reducing the intracellular oxygen concentration and ROS levels (Sung et al, 2010) and hence may potentially be beneficial during S phase. Cell cycle-dependent changes in the utilization of glucose versus glutamine may be associated with reported changes in mitochondrial structure throughout the cell cycle, converted from isolated fragments into a hyperfused network in G1/S transition (Mitra et al, 2009).

While the current study focuses on identifying and quantifying oscillations in metabolic flux throughout the cell cycle, further research is required to decipher how these metabolic changes are regulated. The ubiquitin ligase complexes APC/C and SCF complex were claimed to control glycolytic flux by limiting the expression of PFKFB3 to the G1/S transition, in accordance with the identified increase in glycolytic flux. SCF complex further limits the expression of GLS1 to S and G2/M phases, in agreement with our finding of induced glutamine to glutamate conversion in these cell cycle phases. More generally, metabolic enzymes are typically not regulated at the level of mRNA or protein throughout the cell cycle based on cell cycle transcriptomics and proteomics studies (Olsen et al, 2010). However, post-translational modification of central metabolic enzymes is highly abundant and oscillations in phosphorylation levels of metabolic enzymes have been described (Olsen et al, 2010). Considering our finding of oscillations in the concentration of numerous metabolic intermediates, as well as energy and redox cofactors, suggests that metabolic regulation via changes in enzyme-binding site occupancy and allosteric regulation may also play a key role in regulating cell cycle flux dynamics. This is further supported by the fact that metabolic substrate and inhibitor levels in mammalian cells are typically in the same range as the K_m values for the corresponding enzymes (Park et al, 2016).

We showed that treatment of HeLa cells with the PDK inhibitor DCA eliminates the oscillations in glucose flux into TCA cycle, suggesting that cell cycle-specific regulation of PDH activity may be involved in regulating these flux oscillations. Consistently, PDK4 was reported to be induced by the E2F-pRB pathway which controls cell entry to S phase (Hsieh et al, 2008; Kaplon et al, 2015). Furthermore, analyzing published phosphoproteomics data for HeLa cells measured throughout the cell cycle (Olsen et al, 2010) shows more than twofold increase in the phosphorylation of the PDH E1 component in early and late S phase versus in G1 (Appendix Fig S18). Validating this observation here, we find a significant twofold increase in the phosphorylation of PDH- S232 specifically in S phase (Appendix Fig S19). A drop in glycolytic flux into TCA cycle in S phase is further supported by a recent report of a drop in the abundance of the pyruvate dehydrogenase complex in mitochondria in S phase, following translocation of the complex components to the nucleus (Sutendra et al, 2014). Further research is required to determine the precise regulatory mechanism that underlies cell cycle oscillations in glycolytic flux into TCA cycle, which may potentially spread quantitatively among several enzymes [in accordance with the view of metabolic control analysis (Fell, 1997)].

In the conventional view of mammalian metabolism, acetyl-CoA (a major anabolic precursor for fatty acid biosynthesis) is primarily produced by the oxidation of glucose-derived pyruvate in mitochondria. Previous studies have employed isotope tracers to show that in cancer cells grown under hypoxia (Metallo et al, 2012), in cells with defective mitochondria (Mullen et al, 2012), and in anchorage-independent growth (Jiang et al, 2016), a major fraction of acetyl-CoA is produced via another route, reductive carboxylation of glutamine-derived α-ketoglutarate (catalyzed by reverse flux through isocitrate dehydrogenase, IDH). Under these conditions, feeding cells with isotopic glutamine leads to a marked increase in the fractional labeling of citrate m + 5 and consequently in the isotopic labeling of synthesized fatty acids. Here, we showed that the fractional labeling of citrate m + 5 significantly oscillates throughout the cell cycle under standard normoxic conditions. This reflects major oscillations in the relative contribution of oxidative TCA cycle flux (peaking in G1) and in the reductive metabolism of glutamine-derived α-ketoglutarate (peaking in S) to the production of acetyl-CoA throughout the cell cycle. Notably, though, the oxidative IDH flux remains several-fold higher than the reductive flux all throughout the cell cycle, reflecting an overall net flux in the oxidative direction.

Understanding the metabolic adaptation of cells to tumorigenic mutations is a central goal of cancer metabolic research. Considering that tumorigenic mutations typically alter cell cycle progression, flux alterations observed at a cell population level may merely reflect a change in the distribution of cell cycle phases in the population (due to cells in different phases having different metabolic fluxes). Hence, the presented temporal-fluxomics approach will enable to revisit our understanding of oncogene-induced metabolic alterations, disentangling population level artifacts from directly regulated flux alterations with important tumorigenic role and revealing potential targets for therapy. Combined targeting of cell cycle-specific flux alterations with drugs that block progression through the same cell cycle phases is expected to have important therapeutic applications (Diaz-Moralli et al, 2013; Saqcena et al, 2015).

Materials and Methods

Materials

Most of the chemical reagents were purchased from Sigma-Aldrich unless otherwise specified. Stable isotopes [U-^{13}C]-glucose and [U-^{13}C]-glutamine were obtained from Cambridge Isotope Laboratories, Inc. Cell culture media and reagents were purchased from Biological Industries, unless otherwise specified. HeLa and HCT-116 cells were purchased from ATCC. For oxygen consumption measurements, chemicals (inhibitors and culture medium) were obtained from Sigma and other materials from Agilent. Antibodies used were as follows: phospho S232 PDH E1 alpha protein (PDHA1) Profiling ELISA Kit, (ABCam-ab115343); human PDH E1 alpha protein ELISA kit (PDHA1), (Abcam-ab181415).

Cell culture and synchronization

HeLa cells were cultured in Dulbecco's modified Eagle's medium (high glucose, Biological Industries, 01-055-1A) supplemented with 10% (v/v) heat-inactivated dialyzed fetal bovine serum (Biological Industries), 3 mM L-glutamine, 100 U/ml penicillin, and 100 µg/ml streptomycin with 5% CO_2 in a humidified incubator at 37°C. Culture medium was additionally supplemented with 84 mg/ml L-serine, and 48 mg/ml L-cystine to maintain sufficient amount of nutrients for three cell doublings. HeLa cells have been validated by the vendors, and we tested for mycoplasma using EZ-PCR mycoplasma detection kit (Biological Industries). Cell number and volume analysis were performed using a Z2 Beckman Coulter Counter (100 µm aperture); cells were trypsinized and resuspended in IsoFlow Sheath Fluid (Beckman Coulter) immediately before counting.

Cell synchronization was achieved using double thymidine block. Briefly, 2 mM thymidine (Sigma-T1895) was added to 10-cm culture plates at 25–30% confluence for 17 h. Cells were released from the first block by washing twice with phosphate buffer saline (PBS) and replacing with fresh culture medium. After 9 h, cells were incubated with 2 mM thymidine for a second block time for 17 h. Cells were replated in smaller plates (60 or 35 mm) for further analysis. Cell cycle analysis of synchronized cells was performed by quantitation of DNA content using propodium iodide (PI) staining followed by flow cytometry. For PI staining, cells were fixed using 75% ethanol/PBS and then resuspended in 0.5 ml of PI staining solution (3.8 mM sodium citrate, 40 µg/ml PI, 50 ng/ml RNase A) for 40 min at room temperature in dark. Flow cytometric analysis was performed using LSRII (BD Biosciences, with at least 50,000 cells per FACS run). Cell cycle stages from raw FACS data were quantified using Modfit (Verity House Software).

To check the effect of dichloroacetate treatment on cell cycle progression, synchronized cells were treated with 4 mM pyruvate dehydrogenase kinase inhibitor dichloroacetate (Sigma-Aldrich) for 3 h. Non-synchronized cells were treated with 16 mM DCA for 24 h, followed by cytometric analysis after DNA staining with PI.

LC-MS-based metabolomics and isotope tracing

To measure intracellular metabolite pools, cells were washed with 2 ml of ice-cold PBS for three times and metabolites extracted with 200 µl of 50:30:20 (v/v/v) methanol:acetonitrile:water solution at −20°C. The cells were quickly scraped on dry ice. For the extraction of metabolites from the culture medium, 50 µl of media was mixed with 200 µl of 50:30 (v/v) methanol:acetonitrile solution at −20°C. All metabolite extractions were stored at −80°C for at least 1 h, followed by centrifugation, twice at 20,000 g for 20 min to obtain protein-free metabolite extraction.

Metabolite pool sizes are expressed per total cell volume, measured using a Coulter counter in cells grown in parallel in different plates. Absolute metabolite concentrations for specific metabolites of interest were determined based on isotope ratio using chemical standards (Bennett et al, 2008; Dataset EV3). Pulse isotopic labeling was performed by feeding synchronized cells at each time point with either [U-^{13}C]-glucose or [U-^{13}C]-glutamine for 1 h. To minimize the perturbation to cells due to the replacement with fresh media, we used conditioned medium obtained from a previous culture of HeLa cells. Specifically, conditioned medium with either isotopic glucose or isotopic glutamine was incubated with fully attached HeLa cells at ~30% confluence for 4 h and then stored in 4°C until used for pulse-chase labeling experiments.

Chromatographic separation was achieved on a SeQuant ZIC-pHILIC column (2.1 × 150 mm, 5 μm bead size, Merck Millipore). Flow rate was set to 0.2 ml/min, column compartment was set to 30°C, and autosampler tray was maintained at 4°C. Mobile phase A consisted of 20 mM ammonium carbonate with 0.01% (v/v) ammonium hydroxide. Mobile Phase B was 100% acetonitrile. The mobile phase linear gradient (%B) was as follows: 0 min 80%, 15 min 20%, 15.1 min 80%, and 23 min 80%. A mobile phase was introduced to Thermo Q-Exactive mass spectrometer with an electrospray ionization source working in polarity switching mode. Metabolites were analyzed using full-scan method in the range 70–1,000 m/z and with a resolution of 70,000. Positions of metabolites in the chromatogram were identified by corresponding pure chemical standards. Data were analyzed with MAVEN (Clasquin et al, 2012).

Measurement of oscillations in oxygen consumption

Measurement of oxygen consumption was done using the XFp Extracellular Flux Analyzer (Agilent). Briefly, HeLa cells after synchronization were plated (20,000 cells/well) into XFp culture mini plates and grown at 37°C with 5% CO_2 in a humidified incubator for various times to enrich the cell population with cells at distinct cell cycle phases. Cells were washed and incubated with prewarmed XF assay medium (Sigma D5030, pH 7.4) supplemented with 25 mM glucose and 3 mM glutamine for 1 h in a non-CO_2 incubator at 37°C. Appropriate dilutions of the inhibitors (final well concentrations: oligomycin 1 μM, FCCP 1 μM, rotenone/antimycin A 1 μM) were prepared in the assay medium as per the instructions in the manual. Hydrated sensor cartridges were calibrated prior to the measurement on SeaHorse XFp Extracellular Flux Analyzer (Agilent). Data acquisition consisted of a baseline measurement followed by oligomycin, FCCP, and rotenone/antimycin A injections, respectively. OCR data were normalized against cell volume obtained from Coulter counter measurements of the cells from a parallel plate without any treatment and expressed in pmoles/min/μl.

Synchronization loss model

We construct a probabilistic model that describes the loss of synchronization following release from double thymidine block, due to cell–cell variability in the rate of progression through the cell cycle. Each cell is assumed to have its own "internal clock", which controls the speed at which it progresses through the cell cycle (denoted by γ). The relative progression rates of cells released from synchronization arrest are assumed to be normally distributed: $\gamma \sim N(1, \sigma^2)$. We estimate the variance of the distribution (σ^2) as well as the duration of G1, S, and G2/M (denoted by d_G, d_S, and d_M, respectively, in hours), given the FACS measurements of the fraction of cells in each cell cycle phase in the synchronized cell population (as described below). We denote the cell doubling time by d_{CYC} ($= d_G + d_S + d_M$, in hours). For a cell whose rate of progression through the cell cycle is γ, the cell-intrinsic time x (in hours) within the cell cycle ($0 \le x \le d_{CYC}$) at time t after the release from synchronization-induced growth arrest is

$$x = \gamma \cdot t + d_G - \left\lfloor \frac{(\gamma \cdot t + d_G)}{d_{CYC}} \right\rfloor d_{CYC}, \tag{1}$$

considering that cells resume growth in G1/S transition after released from double thymidine block. For example, for a cell with a relative progression rate through the cell cycle of $\gamma = 1$ released from growth arrest, it will take $d_S + d_M$ hours to complete one cell cycle (and then have an intrinsic time of $x = 0$), while for a cell with double the rate ($\gamma = 2$), it will take half the time (i.e., $(d_S + d_M)/2$). At time t after the release from synchronization arrest, a cell having an intrinsic time of x has a relative progression rate through the cell cycle γ equal to $\frac{1}{t}(x + k \cdot d_{CYC} - d_G)$, with k representing the number of completed cell cycles since the release from growth arrest (based on equation 1). Hence, considering that γ is normally distributed, we can compute the number of cells in the synchronized population at time t whose cell-intrinsic time is x (denoted $g'(x,t)$) as:

$$g'(x,t) = \sum_{k=0}^{2} 2^k \cdot e^{-\frac{\left(\frac{x + k \cdot d_{CYC} - d_G}{t} - 1\right)^2}{2\sigma^2}}, \tag{2}$$

considering values of k between zero and two, representing three complete cell cycle. We denote by $g(x,t)$ the probability density function of the number of cells at time t whose cell-intrinsic time is x, with $g(x,t) = \frac{1}{C} g'(x,t)$, where C is a normalization factor. The expected fraction of the cells in S, G1, and G2/M phases at time t (denoted by $m_s(t)$, $m_G(t)$ and $m_M(t)$, respectively) are calculated as:

$$m_s(t) = \int_0^{d_S} g(x,t)\,dx, \tag{3}$$

$$m_G(t) = \int_{d_S}^{d_S + d_G} g(x,t)\,dx, \tag{4}$$

$$m_M(t) = \int_{d_S + d_G}^{d_{CYC}} g(x,t)\,dx, \tag{5}$$

We perform a maximum log-likelihood estimation of the four parameters of the model (σ, d_G, d_S, and d_M), minimizing the variance-weighted sum of squared residuals between the simulated fraction of cells in the different cell cycle phases ($m_s(t)$, $m_G(t)$, and $m_M(t)$) and the FACS measurements (assuming Gaussian noise in FACS measurements with an empirically estimated standard deviation of ~10%). This minimization was performed via an implementation of sequential quadratic optimization (SQP) available in MATLAB. Confidence intervals were computed by the likelihood ratio test, comparing the maximum log-likelihood estimates with that obtained when constraining each of the four parameters to increasing and then decreasing value (considering the 95% quantile of chi-squared distribution with one degree of freedom). The optimal parameters found were $\sigma = 11\% \pm 1\%$, $d_G = 6.8$ h ± 0.8 h, $d_S = 6.4$ h ± 0.5 h, and $d_M = 3.2$ h ± 0.4 h. Overall, the good fit between the model prediction and experimental data shown in Fig 1B supports the underlying assumptions of this model. Computing the duration of each cell cycle phase based on PI staining/FACS measurements in a population of non-synchronized HeLa cells and considering a decreasing exponential cell age distribution show 8.2 h for G1, 4.9 h for S, and 2.9 h for G2/M. The small under estimation of the duration of G1 and over estimation of the duration of S by the analysis of the synchronized cell (both not more than 1 h off the measurements in the non-synchronized cells) may be due to a slight perturbation to cell cycle dynamics due to synchronization-induced growth arrest.

Computational deconvolution of cell volume measurements in the synchronized cell population

Computational deconvolution was used to estimate cell volume dynamics throughout the cell cycle, correcting for cell dispersion that bias Coulter counter measurements of cell volume performed in the synchronized cells. Specifically, denoting the average cell volume in the synchronized cell population at time t by $v(t)$ ($9 \leq t \leq 45$), we estimate the average cell volume in the cell-intrinsic time x within the cell cycle ($0 \leq x \leq d_{CYC}$), denoted $v'(x)$, as following:

$$v(t) = \int_0^{d_{CYC}} v'(x)g(x,t)\,\mathrm{d}x + \varepsilon(t), \tag{6}$$

where $\varepsilon(t)$ represents experimental noise in the volume measurement performed at time t. We represent $v'(x)$ using a cubic spline, which is a commonly used approach for fitting biological time-series data (considering splines with four segments, defined based on five knots). Non-convex optimization was used to find the optimal position of the knots and corresponding value of $v'(t)$ minimizing the sum-of-square of the error terms. Non-convex optimizations were solved using MATLAB's implementation of SQP.

Computational deconvolution of metabolite concentration, isotope labeling, and uptake and secretion rate measurements

Given a metabolite i whose measured concentration in the synchronized cell population at time t ($9 \leq t \leq 45$) is denoted by $u_i(t)$ (measured metabolite pool size normalized by the measured average cell volume at time t, $v(t)$), we estimate the concentration in cell-intrinsic time x within the cell cycle ($0 \leq x \leq d_{CYC}$), denoted $u_i'(x)$ as:

$$u_i(t) = \frac{1}{\int_0^{d_{CYC}} v'(x)g(x,t)\,\mathrm{d}x} \int_0^{d_{CYC}} u_i'(x)v'(x)g(x,t)\,\mathrm{d}x + \varepsilon(t), \tag{7}$$

considering that the measured concentration of metabolite i at time t represents the average concentration in cells with cell-intrinsic time x, weighted by the total volume of cells with intrinsic time x at time t (i.e., $v'(x)g(x,t)$). We represent $u_i'(x)$ using a cubic spline and estimate its coefficients as described above.

We denote the measured relative abundance of the kth mass isotopomer of metabolite i (i.e., the fraction of the metabolite pool having k labeled carbons) after 1-h feeding with an isotopic substrate of synchronized cells at time t by $u_{i,k}(t)$. We estimate the relative abundance of the kth mass isotopomer of metabolite i in cell-intrinsic time x within the cell cycle denoted $u_{i,k}'(x)$ as:

$$u_{i,k}(t) = \frac{1}{\int_0^\infty u_i'(x)v'(x)g(x,t)\,\mathrm{d}x} \int_0^\infty u_{i,k}'(x)u_i'(x)v'(x)g(x,t)\,\mathrm{d}x + \varepsilon(t), \tag{8}$$

considering that the measured fractional isotopic labeling of a metabolite i at time t represents the average labeling in cells with intrinsic cell cycle time x, weighted by the metabolite pool size in cells with cell-intrinsic time x (i.e., with $u_i'(x)v'(x)g(x,t)$).

We denote the measured change in pool size of metabolite i in the culture media between time $t-\Delta t$ and time t by $\Delta e_i(t)$. We estimate the transport flux of metabolite i by the synchronized cell population at time t, denoted $f_i(t)$ (in molar amount per unit of cell volume per hour, with positive and negative flux representing secretion and uptake, respectively) by dividing the change in pool size of metabolite i by the accumulated volume of cells in the culture metabolite between time $t-\Delta t$ and time t:

$$f_i(t) = \frac{1}{\int_{t-\Delta t}^{t} \int_0^{d_{CYC}} v'(x)g(x,t')\,\mathrm{d}x\mathrm{d}t'} \Delta e_i(t), \tag{9}$$

The transport flux of metabolite i at cell-intrinsic time x, denoted $f_i'(x)$ is estimated as:

$$\Delta e_i(t) = \int_{t-\Delta t}^{t} \int_0^{d_{CYC}} f_i'(x)v'(x)g(x,t')\,\mathrm{d}x\mathrm{d}t' + \varepsilon(t), \tag{10}$$

considering that the measured change in pool size of metabolite i at time t represents the cumulative transport within the Δt time interval by cells with different intrinsic cell cycle time x.

Statistical significance of oscillations in metabolite concentrations and isotopic labeling patterns

To assess the statistical significance of observed oscillations in the deconvoluted concentration of a certain metabolite, we compared the observed amplitude of the oscillation to the amplitude expected by chance (considering the noise in LC-MS measurements). Specifically, for each metabolite i, we define the amplitude of its oscillation by a_i as:

$$a_i = \max_{0 \leq x \leq d_{CYC}} u_i'(x) - \min_{0 \leq x \leq d_{CYC}} u_i'(x). \tag{11}$$

Next, we compute the distribution of amplitudes expected by chance by repeating the following steps 10,000 times: For each time t for which LC-MS measurements were performed on the synchronized cell population ($9 \leq t \leq 45$), we generate a random metabolite concentration [denoted $r(t)$] by sampling from a normal distribution whose mean is the average concentration of metabolite i measured throughout all time points in the synchronized cells, and with the standard deviation of the experimental measurement of metabolite i at time t, denoted $\sigma_i(t)$:

$$r(t) \sim N\left(\frac{1}{k}\sum_{t=1}^{k} u_i(t), \sigma_i^2(t)\right). \tag{12}$$

Computational deconvolution (equation 7) is applied on the randomly generated metabolite concentration data [i.e., $r(t)$] and an empirical P-value computed based on the fraction of randomly sampled concentration vectors for which the derived amplitude is equal or larger than that computed for the concentration measurements of metabolite i. FDR correction for multiple testing is computed using the approach of Benjamini–Hochberg. A conceptually similar method was employed to assess the statistical significance of oscillations in the relative abundance of metabolite isotopic labeling [applying computational deconvolution (equation 8) to randomly generated isotopic labeling data].

Computational inference of metabolic flux dynamics throughout the cell cycle

For every 1-h interval j throughout the cell cycle $j \in \{0 \ldots \lfloor d_{CYC} \rfloor\}$ (referred to as the jth cell cycle interval), we computed the most likely momentary fluxes through nine reactions in TCA cycle and in branching pathways (Fig 3B). Toward this end, we employed a variant of kinetic flux profiling (KFP) to separately infer fluxes producing each metabolite i in cell cycle interval j, for which the simulated isotopic labeling kinetics optimally match the experimental measurements.

The relative abundance of the kth mass isotopomer of metabolite i after 1-h feeding of cells within cell cycle interval j was inferred based on the pulse-chase labeling experiments in the synchronized cells followed by deconvolution as described above (denoted by $u'_{i,k}(j)$). To estimate the dynamics of the isotopic labeling form of this metabolite within the 1-h cell cycle interval (i.e., within time periods shorter than 1 h), we performed pulse-chase labeling experiments with isotopic glucose and glutamine in non-synchronized cells, measuring the relative abundance of the kth mass isotopomer of metabolite i at different times $t \in T = \{10,20,30,60\}$ (in minutes), denoted by $X_{i,k}(t)$. These measurements were used to estimate the relative abundance of the kth mass isotopomer of metabolite i, t minutes after the beginning of the jth 1-h time interval within the cell cycle (denoted by $X^j_{i,k}(t)$) by scaling the measurements performed on the non-synchronized cells:

$$X^j_{i,k}(t) = \frac{u'_{i,k}(j+1)}{X_{i,k}(60)} X_{i,k}(t). \tag{13}$$

To validate these estimated labeling kinetics, we performed rapid pulse-chase labeling experiments (10 and 30 min) in synchronized cells grown for 15 h (G1/S) and 20 h (S), finding a good match between the estimated and measured labeling dynamics (Appendix Fig S20).

We describe the inference of metabolic flux through reactions producing citrate while other fluxes are obtained similarly (see below). The analysis accounts for citrate synthase (v_1) and reductive isocitrate dehydrogenase (IDH, v_2) producing citrate (Fig 3B). We denote the total citrate consumption flux by v_{out}, which may be lower or higher than the sum of $v1$ and $v2$ in case citrate is accumulated or depleted within a cell cycle interval, respectively (as the synchronized cells are not in metabolic steady state). The expected mass-isotopomer distribution of citrate after t minutes into the jth cell cycle interval, considering the fluxes v_1, v_2, and v_{out}, is denoted $Y^j_{cit}(t, v_1, v_2, v_{out})$. Assuming that the error in the measured isotope labeling data is normally distributed, maximum likelihood estimate of fluxes is obtained by minimizing the variance-weighted sum of squared residuals between measured and computed mass-isotopomer distributions, where $\sigma^j_{cit,k}$ is the standard deviation in the measurement of the relative abundance of the kth mass isotopomer of citrate in the jth time interval:

$$\min_{v_1, v_2, v_{out}} \sum_{t \in T} \sum_{k \in \{4,5\}} \left(\frac{X^j_{cit,k}(t) - Y^j_{cit,k}(t, v_1, v_2, v_{out})}{\sigma^j_{cit,k}} \right)^2 \tag{14}$$

s.t.

$$v_{out} = v_1 + v_2 + \left(u'_{cit}(j+1) - u'_{cit}(j) \right), \tag{14.1}$$

$$v_1, v_2, v_{out} \geq 0, \tag{14.2}$$

where $u'_{cit}(j)$ represents the deconvoluted concentration of citrate (in mM) in the jth cell cycle interval and is used to constrain the difference between the total citrate producing and consuming flux within the jth 1-h cell cycle interval (equation 14.1). We accounted for the two major mass isotopomers of citrate, m + 4 and m + 5 (Appendix Fig S3). To simulate the labeling kinetics of the kth mass isotopomer of citrate within the jth cell cycle interval, denoted by $Y^j_{cit,k}(t, v_1, v_2, v_{out})$, we utilized the following system of ordinary differential equations:

$$\frac{dY^j_{cit,k}(t, v_1, v_2, v_{out})}{dt}$$
$$= \frac{1}{u'_{cit}(j)} \left(v_1 X^j_{mal,k}(t) + v_2 X^j_{aKG,k}(t) - v_{out} Y^j_{cit,k}(t, v_1, v_2, v_{out}) \right),$$

where $X^j_{mal,k}(t)$ and $X^j_{aKG,k}(t)$ represent the relative abundance of the kth mass isotopomer of malate and α-ketoglutarate, t minutes into the jth cell cycle interval, and $v_1 X^j_{mal,k}(t)$ and $v_2 X^j_{aKG,k}(t)$ represent the momentary production of the kth mass isotopomer of citrate from of malate and from α-ketoglutarate. The term $v_{out} Y^j_{cit,k}(t, v_1, v_2, v_{out})$ represents the total momentary consumption of the kth mass isotopomer of citrate. The difference between the momentary production and consumption rate of the different mass isotopomers of citrate (term in parenthesis on right hand side of the equation) is normalized by the concentration of citrate to give the momentary change in fractional labeling.

A similar approach was employed to infer fluxes through reactions producing the following metabolites: (i) α-ketoglutarate—rapid isotopic exchange with glutamate results in essentially a single intracellular pool of α-ketoglutarate and glutamate (as reflected by similar labeling kinetics of the two metabolites). We considered α-ketoglutarate/glutamate m + 5 production by oxidative IDH from citrate m + 5 when feeding isotopic glutamine (reaction $v3$ in Fig 3B) and from glutamine m + 5 when feeding isotopic glutamine ($v8$), as well α-ketoglutarate/glutamate m + 3 production by oxidative IDH from citrate m + 4 when feeding isotopic glutamine ($v3$) (Appendix Fig S4), (ii) Malate—considering the rapid isotopic exchange with aspartate (> 100 mM/h based KFP analysis of malate and aspartate labeling kinetics, we account for a single malate/aspartate pool. We consider for malate/aspartate m + 4 production from α-ketoglutarate m + 5 ($v4$), when feeding isotopic glutamine, and malate/aspartate m + 4 production from pyruvate m + 3 ($v7$), when feeding isotopic glucose (Appendix Fig S5), (iii) UTP—considering UTP m + 3 production from carbamoyl-aspartate m + 3 when feeding isotopic glutamine ($v5$) (Appendix Fig S6), (iv) Lactate—considering lactate m + 3 production by malic enzyme when feeding isotopic glutamine ($v6$) and the production of non-labeled lactate by glycolysis (Appendix Fig S7). We consider an average malic enzyme flux of 7.9 mM/h throughout the cell cycle (considering a fractional labeling of 1.3% m + 3 lactate under isotopic steady state, when feeding isotopic glutamine, and lactate secretion rate of 610 mM/h).

Non-convex optimizations were solved using MATLAB's implementation of sequential quadratic optimization (SQP). All optimizations were run 10 times, starting from different sets of random fluxes, to overcome potential local minima. To compute confidence intervals for estimated fluxes, we iteratively ran the SQP optimization to compute the maximum log-likelihood estimation while constraining the flux to increasing (and then decreasing) values (with a step size equal to 5% of the flux predicted in the initial maximum log-likelihood estimation; Antoniewicz *et al*, 2006; Fan *et al*, 2013). The confidence interval bounds were determined based on the 95% quantile of chi-squared distribution with one degree of freedom. Notably, while all flux estimates are given in mM/h (i.e., fmole/pl-cells/h), multiplying a flux estimate with intrinsic time x with the estimated cell volume at that time (i.e., $v'(x)$ in pl, see Fig 1C) gives a flux value per cell (fmole/cell/h).

The rate of production of reducing equivalents for driving oxidative phosphorylation generated by glucose oxidation was calculated by summing the flux through the following NADH producing reactions: PDH (according to reaction $v1$ in Fig 3B, considering that ~99% of pyruvate is produced by glucose oxidation throughout the cell cycle, with the fractional labeling of pyruvate from isotopic glutamine under isotopic steady state being < 0.01), oxidative IDH (reaction $v3$), and the rate of shuttling of NADH produced in glycolysis for oxidation in mitochondria (estimated based on pyruvate secretion, a uniform flux of ~17 mM/h measured throughout the cell cycle). The rate of reducing equivalents production from glutamine oxidation was calculated based on the rate of α-ketoglutarate oxidation in TCA cycle (reaction $v4$ in Fig 3B, considering that > 95% of α-ketoglutarate is produced from glutamine all throughout the cell cycle; the fractional labeling of α-ketoglutarate from isotopic glutamine under isotopic steady state is > 0.95), producing NADH by α-ketoglutarate dehydrogenase and FADH$_2$ by succinate dehydrogenase (SDH). Malate dehydrogenase ($v4$ in Fig 3B) further produces another NADH.

Acknowledgements

We would like to thank Jing Fan and Won Dong Lee for providing valuable comments on this manuscript. The research leading to these results has received funding from the European Research Council/ERC Grant Agreement No. 714738, and the Israel Science Foundation (ISF), Grant No. 1717/16. AT is funded by the Israel Cancer Research Fund (ICRF), Grant No. RCDA00102, the Israeli Centers of Research Excellence (I-CORE), Gene Regulation in Complex Human Disease, Center no. 41/11, and the Israel Science Foundation (ISF), Grant No. 659/16.

Author contributions

EA and PK performed the biological experiments. DM assisted with LC-MS work. EA, PK, and TS designed the research, analyzed data, and wrote the article. AT participated in the research design.

References

Almeida A, Bolanos JP, Moncada S (2010) E3 ubiquitin ligase APC/C-Cdh1 accounts for the Warburg effect by linking glycolysis to cell proliferation. *Proc Natl Acad Sci USA* 107: 738–741

Antoniewicz MR, Kelleher JK, Stephanopoulos G (2006) Determination of confidence intervals of metabolic fluxes estimated from stable isotope measurements. *Metab Eng* 8: 324–337

Banko MR, Allen JJ, Schaffer BE, Wilker EW, Tsou P, White JL, Villen J, Wang B, Kim SR, Sakamoto K, Gygi SP, Cantley LC, Yaffe MB, Shokat KM, Brunet A (2011) Chemical genetic screen for AMPKalpha2 substrates uncovers a network of proteins involved in mitosis. *Mol Cell* 44: 878–892

Bar-Joseph Z, Siegfried Z, Brandeis M, Brors B, Lu Y, Eils R, Dynlacht BD, Simon I (2008) Genome-wide transcriptional analysis of the human cell cycle identifies genes differentially regulated in normal and cancer cells. *Proc Natl Acad Sci USA* 105: 955–960

Bennett BD, Yuan J, Kimball EH, Rabinowitz JD (2008) Absolute quantitation of intracellular metabolite concentrations by an isotope ratio-based approach. *Nat Protoc* 3: 1299–1311

Berger SL (2007) The complex language of chromatin regulation during transcription. *Nature* 447: 407–412

Bienvenu F, Jirawatnotai S, Elias JE, Meyer CA, Mizeracka K, Marson A, Frampton GM, Cole MF, Odom DT, Odajima J, Geng Y, Zagozdzon A, Jecrois M, Young RA, Liu XS, Cepko CL, Gygi SP, Sicinski P (2010) Transcriptional role of cyclin D1 in development revealed by a genetic-proteomic screen. *Nature* 463: 374–378

Blagosklonny MV, Pardee AB (2002) The restriction point of the cell cycle. *Cell Cycle* 1: 103–110

Chen Z, Odstrcil EA, Tu BP, McKnight SL (2007) Restriction of DNA replication to the reductive phase of the metabolic cycle protects genome integrity. *Science* 316: 1916–1919

Clasquin MF, Melamud E, Rabinowitz JD (2012) LC-MS data processing with MAVEN: a metabolomic analysis and visualization engine. *Curr Protoc Bioinformatics.* 37: 14.11.1–14.11.23

Colombo SL, Palacios-Callender M, Frakich N, Carcamo S, Kovacs I, Tudzarova S, Moncada S (2011) Molecular basis for the differential use of glucose and glutamine in cell proliferation as revealed by synchronized HeLa cells. *Proc Natl Acad Sci USA* 108: 21069–21074

Costenoble R, Müller D, Barl T, Van Gulik WM, Van Winden WA, Reuss M, Heijnen JJ (2007) 13C-Labeled metabolic flux analysis of a fed-batch culture of elutriated *Saccharomyces cerevisiae*. *FEMS Yeast Res* 7: 511–526

Cuyàs E, Corominas-Faja B, Joven J, Menendez JA (2014) Cell cycle regulation by the nutrient-sensing mammalian target of rapamycin (mTOR) pathway. *Methods Mol Biol* 1170: 113–144

Diaz-Moralli S, Tarrado-Castellarnau M, Miranda A, Cascante M (2013) Targeting cell cycle regulation in cancer therapy. *Pharmacol Ther* 138: 255–271

Duckwall CS, Murphy TA, Young JD (2013) Mapping cancer cell metabolism with(13)C flux analysis: recent progress and future challenges. *J Carcinog* 12: 13

Estevez-Garcia IO, Cordoba-Gonzalez V, Lara-Padilla E, Fuentes-Toledo A, Falfan-Valencia R, Campos-Rodriguez R, Abarca-Rojano E (2014) Glucose and glutamine metabolism control by APC and SCF during the G1-to-S phase transition of the cell cycle. *J Physiol Biochem* 70: 569–581

Fan J, Kamphorst JJ, Mathew R, Chung MK, White E, Shlomi T, Rabinowitz JD (2013) Glutamine-driven oxidative phosphorylation is a major ATP source in transformed mammalian cells in both normoxia and hypoxia. *Mol Syst Biol* 9: 712

Fell D (1997) Understanding the control of metabolism. *Front Metab* 2: 300

Fendt S-M, Bell EL, Keibler MA, Olenchock BA, Mayers JR, Wasylenko TM, Vokes NI, Guarente L, Vander Heiden MG, Stephanopoulos G (2013)

Reductive glutamine metabolism is a function of the α-ketoglutarate to citrate ratio in cells. *Nat Commun* 4: 2236

Fingar DC, Blenis J (2004) Target of rapamycin (TOR): an integrator of nutrient and growth factor signals and coordinator of cell growth and cell cycle progression. *Oncogene* 23: 3151–3171

Gaglio D, Soldati C, Vanoni M, Alberghina L, Chiaradonna F (2009) Glutamine deprivation induces abortive s-phase rescued by deoxyribonucleotides in k-ras transformed fibroblasts. *PLoS ONE* 4: e4715

Hsieh MCF, Das D, Sambandam N, Zhang MQ, Nahlé Z (2008) Regulation of the PDK4 isozyme by the Rb-E2F1 complex. *J Biol Chem* 283: 27410–27417

Jiang L, Shestov AA, Swain P, Yang C, Parker SJ, Wang QA, Terada LS, Adams ND, McCabe MT, Pietrak B, Schmidt S, Metallo CM, Dranka BP, Schwartz B, DeBerardinis RJ (2016) Reductive carboxylation supports redox homeostasis during anchorage-independent growth. *Nature* 532: 255–258

Kaplon J, van Dam L, Peeper D (2015) Two-way communication between the metabolic and cell cycle machineries: the molecular basis. *Cell Cycle* 14: 2022–2032

Levine AJ, Puzio-Kuter AM (2010) The control of the metabolic switch in cancers by oncogenes and tumor suppressor genes. *Science* 330: 1340–1344

Li B, Carey M, Workman JL (2007) The role of chromatin during transcription. *Cell* 128: 707–719

Metallo CM, Walther JL, Stephanopoulos G (2009) Evaluation of 13C isotopic tracers for metabolic flux analysis in mammalian cells. *J Biotechnol* 144: 167–174

Metallo CM, Gameiro PA, Bell EL, Mattaini KR, Yang J, Hiller K, Jewell CM, Johnson ZR, Irvine DJ, Guarente L, Kelleher JK, Vander Heiden MG, Iliopoulos O, Stephanopoulos G (2012) Reductive glutamine metabolism by IDH1 mediates lipogenesis under hypoxia. *Nature* 481: 380–384

Mitra K, Wunder C, Roysam B, Lin G, Lippincott-Schwartz J (2009) A hyperfused mitochondrial state achieved at G1-S regulates cyclin E buildup and entry into S phase. *Proc Natl Acad Sci USA* 106: 11960–11965

Mullen AR, Wheaton WW, Jin ES, Chen PH, Sullivan LB, Cheng T, Yang Y, Linehan WM, Chandel NS, DeBerardinis RJ (2012) Reductive carboxylation supports growth in tumour cells with defective mitochondria. *Nature* 481: 385–388

Noack S, Noh K, Moch M, Oldiges M, Wiechert W (2011) Stationary versus non-stationary (13)C-MFA: a comparison using a consistent dataset. *J Biotechnol* 154: 179–190

Noh K, Wahl A, Wiechert W (2006) Computational tools for isotopically instationary 13C labeling experiments under metabolic steady state conditions. *Metab Eng* 8: 554–577

Olsen JV, Vermeulen M, Santamaria A, Kumar C, Miller ML, Jensen LJ, Gnad F, Cox J, Jensen TS, Nigg EA, Brunak S, Mann M (2010) Quantitative phosphoproteomics reveals widespread full phosphorylation site occupancy during mitosis. *Sci Signal* 3: ra3

Oredsson SM (2003) Polyamine dependence of normal cell-cycle progression. *Biochem Soc Trans* 31: 366–370

Park JO, Rubin SA, Xu Y-F, Amador-Noguez D, Fan J, Shlomi T, Rabinowitz JD (2016) Metabolite concentrations, fluxes and free energies imply efficient enzyme usage. *Nat Chem Biol* 12: 482–489

Saqcena M, Mukhopadhyay S, Hosny C, Alhamed A, Chatterjee A, Foster DA (2015) Blocking anaplerotic entry of glutamine into the TCA cycle sensitizes K-Ras mutant cancer cells to cytotoxic drugs. *Oncogene* 34: 2672–2680

Sauer U (2006) Metabolic networks in motion: 13C-based flux analysis. *Mol Syst Biol* 2: 62

Sung HJ, Ma W, Wang P, Hynes J, O'Riordan TC, Combs CA, McCoy JP, Bunz F, Kang J, Hwang PM (2010) Mitochondrial respiration protects against oxygen-associated DNA damage. *Nat Commun* 1: 5

Sutendra G, Kinnaird A, Dromparis P, Paulin R, Stenson TH, Haromy A, Hashimoto K, Zhang N, Flaim E, Michelakis ED (2014) A nuclear pyruvate dehydrogenase complex is important for the generation of Acetyl-CoA and histone acetylation. *Cell* 158: 84–97

Tudzarova S, Colombo SL, Stoeber K, Carcamo S, Williams GH, Moncada S (2011) Two ubiquitin ligases, APC/C-Cdh1 and SKP1-CUL1-F (SCF)-beta-TrCP, sequentially regulate glycolysis during the cell cycle. *Proc Natl Acad Sci USA* 108: 5278–5283

Vizan P, Alcarraz-Vizan G, Diaz-Moralli S, Solovjeva ON, Frederiks WM, Cascante M (2009) Modulation of pentose phosphate pathway during cell cycle progression in human colon adenocarcinoma cell line HT29. *Int J Cancer* 124: 2789–2796

Wellen KE, Hatzivassiliou G, Sachdeva UM, Bui TV, Cross JR, Thompson CB (2009) ATP-citrate lyase links cellular metabolism to histone acetylation. *Science* 324: 1076–1080

Wiechert W (2002) An introduction to 13C metabolic flux analysis. *Genet Eng (N Y)* 24: 215–238

Yalcin A, Clem BF, Simmons A, Lane A, Nelson K, Clem AL, Brock E, Siow D, Wattenberg B, Telang S, Chesney J (2009) Nuclear targeting of 6-phosphofructo-2-kinase (PFKFB3) increases proliferation via cyclin-dependent kinases. *J Biol Chem* 284: 24223–24232

Yang W, Xia Y, Ji H, Zheng Y, Liang J, Huang W, Gao X, Aldape K, Lu Z (2011) Nuclear PKM2 regulates β-catenin transactivation upon EGFR activation. *Nature* 478: 118–122

Yang W, Zheng Y, Xia Y, Ji H, Chen X, Guo F, Lyssiotis CA, Aldape K, Cantley LC, Lu Z (2012) ERK1/2-dependent phosphorylation and nuclear translocation of PKM2 promotes the Warburg effect. *Nat Cell Biol* 14: 1295–1304

Yuan J, Bennett BD, Rabinowitz JD (2008) Kinetic flux profiling for quantitation of cellular metabolic fluxes. *Nat Protoc* 3: 1328–1340

Permissions

List of Contributors

Lorenz Adlung, Marie-Christine Wagner, Bin She, Susen Lattermann, Marcel Schilling and Sajib Chakraborty
Division of Systems Biology of Signal Transduction, German Cancer Research Center (DKFZ), Heidelberg, Germany

Thomas Höfer
Division of Theoretical Systems Biology, German Cancer Research Center (DKFZ), Heidelberg, Germany
BioQuant Center, University of Heidelberg, Heidelberg, Germany

Sandip Kar
Division of Theoretical Systems Biology, German Cancer Research Center (DKFZ), Heidelberg, Germany
BioQuant Center, University of Heidelberg, Heidelberg, Germany Department of Chemistry, Indian Institute of Technology, Mumbai, India

Jie Bao
Systems Biology of the Cellular Microenvironment Group, IMMZ, ALU, Freiburg, Germany

Melanie Boerries and Hauke Busch
Systems Biology of the Cellular Microenvironment Group, IMMZ, ALU, Freiburg, Germany
German Cancer Consortium (DKTK), Freiburg, Germany
German Cancer Research Center (DKFZ), Heidelberg, Germany

Anthony D Ho
Department of Medicine V, University of Heidelberg, Heidelberg, Germany

Patrick Wuchter
Department of Medicine V, University of Heidelberg, Heidelberg, Germany
Institute for Transfusion Medicine and Immunology, University of Heidelberg, Mannheim, Germany

Jens Timmer
Center for Biological Signaling Studies (BIOSS), Institute of Physics, University of Freiburg, Freiburg, Germany

Ursula Klingmüller
Division of Systems Biology of Signal Transduction, German Cancer Research Center (DKFZ), Heidelberg, Germany
Translational Lung Research Center (TLRC), Member of the German Center for Lung Research (DZL), Heidelberg, Germany

Omer Karin and Uri Alon
Department of Molecular Cell Biology, Weizmann Institute of Science, Rehovot, Israel

Elena Torlai Triglia, Tiago Rito and Alexander Kukalev
Epigenetic Regulation and Chromatin Architecture, Max Delbrück Center for Molecular Medicine, Berlin, Germany

Carmelo Ferrai
Epigenetic Regulation and Chromatin Architecture, Max Delbrück Center for Molecular Medicine, Berlin, Germany
Genome Function, MRC London Institute of Medical Sciences (previously MRC Clinical Sciences Centre), London, UK
Institute of Clinical Sciences (ICS), Faculty of Medicine, Imperial College London, London, UK

Inês de Santiago
Genome Function, MRC London Institute of Medical Sciences (previously MRC Clinical Sciences Centre), London, UK
Institute of Clinical Sciences (ICS), Faculty of Medicine, Imperial College London, London, UK

Jessica R Risner-Janiczek
Institute of Clinical Sciences (ICS), Faculty of Medicine, Imperial College London, London, UK
Stem Cell Neurogenesis, MRC London Institute of Medical Sciences (previously MRC Clinical Sciences Centre), London, UK
Neurophysiology Group, MRC London Institute of Medical Sciences (previously MRC Clinical Sciences Centre), London, UK

Meng Li
Institute of Clinical Sciences (ICS), Faculty of Medicine, Imperial College London, London, UK
Stem Cell Neurogenesis, MRC London Institute of Medical Sciences (previously MRC Clinical Sciences Centre), London, UK

Mark A Ungless
Institute of Clinical Sciences (ICS), Faculty of Medicine, Imperial College London, London, UK

Neurophysiology Group, MRC London Institute of Medical Sciences (previously MRC Clinical Sciences Centre), London, UK

Owen JL Rackham
Duke-NUS Medical School, Singapore, Singapore

Mario Nicodemi
Dipartimento di Fisica, Università di Napoli Federico II and INFN Napoli, Complesso Universitario di Monte Sant'Angelo, Naples, Italy

Altuna Akalin
Scientific Bioinformatics Platform, Berlin Institute for Medical Systems Biology, Max Delbrück Center for Molecular Medicine, Berlin, Germany

Ana Pombo
Epigenetic Regulation and Chromatin Architecture, Max Delbrück Center for Molecular Medicine, Berlin, Germany
Genome Function, MRC London Institute of Medical Sciences (previously MRC Clinical Sciences Centre), London, UK
Institute of Clinical Sciences (ICS), Faculty of Medicine, Imperial College London, London, UK
Institute for Biology, Humboldt-Universität zu Berlin, Berlin, Germany

Katharina Kramer
Plant Proteomics, Max Planck Institute for Plant Breeding Research, Cologne, Germany

Markus Hartl
Plant Proteomics, Max Planck Institute for Plant Breeding Research, Cologne, Germany
Plant Molecular Biology, Department Biology I, Ludwig-Maximilians-University Munich, Martinsried, Germany
Mass Spectrometry Facility, Max F. Perutz Laboratories (MFPL), Vienna Biocenter (VBC), University of Vienna, Vienna, Austria

Magdalena Plöchinger and Dario Leister
Plant Molecular Biology, Department Biology I, Ludwig-Maximilians-University Munich, Martinsried, Germany

Magdalena Füßl and Iris Finkemeier
Plant Proteomics, Max Planck Institute for Plant Breeding Research, Cologne, Germany
Plant Molecular Biology, Department Biology I, Ludwig-Maximilians-University Munich, Martinsried, Germany
Plant Physiology, Institute of Plant Biology and Biotechnology, University of Muenster, Muenster, Germany

Ahmet Bakirbas
Plant Proteomics, Max Planck Institute for Plant Breeding Research, Cologne, Germany
Plant Physiology, Institute of Plant Biology and Biotechnology, University of Muenster, Muenster, Germany

Paul J Boersema, Jürgen Cox and Matthias Mann
Proteomics and Signal Transduction, Max-Planck Institute of Biochemistry, Martinsried, Germany

Jan-Oliver Jost, Julia Sindlinger and Dirk Schwarzer
Interfaculty Institute of Biochemistry, University of Tübingen, Tübingen, Germany

Glen Uhrig and Greg BG Moorhead
Department of Biological Sciences, University of Calgary, Calgary, AB, Canada

Michael E Salvucci
US Department of Agriculture, Agricultural Research Service, Arid-Land Agricultural Research Center, Maricopa, AZ, USA

Tobias Fuhrer, Mattia Zampieri, Daniel C Sévin, Uwe Sauer and Nicola Zamboni
Institute of Molecular Systems Biology, ETH Zürich, Zürich, Switzerland

Elaine Johnstone
Department of Oncology, University of Oxford, Oxford, UK

Mathias Wilhelm, Chen Meng, Karl Kramer, Runsheng Zheng and Anna Jarzab
Chair of Proteomics and Bioanalytics, Technical University of Munich, Freising, Germany

Martin Frejno
Department of Oncology, University of Oxford, Oxford, UK
Chair of Proteomics and Bioanalytics, Technical University of Munich, Freising, Germany

Riccardo Zenezini Chiozzi
Chair of Proteomics and Bioanalytics, Technical University of Munich, Freising, Germany
Department of Chemistry, Sapienza – Università di Roma, Rome, Italy

Heiner Koch, Susan Klaeger and Stephanie Heinzlmeir
Chair of Proteomics and Bioanalytics, Technical University of Munich, Freising, Germany
German Cancer Consortium (DKTK), Munich, Germany
German Cancer Research Center (DKFZ), Heidelberg, Germany

Benjamin Ruprecht
Chair of Proteomics and Bioanalytics, Technical University of Munich, Freising, Germany
Center for Integrated Protein Science (CIPSM), Munich, Germany

Enric Domingo
Department of Oncology, University of Oxford, Oxford, UK
Wellcome Trust Centre for Human Genetics (WTCHG), University of Oxford, Oxford, UK

David Kerr
Nuffield Division of Clinical Laboratory Sciences (NDCLS), University of Oxford, Oxford, UK

Moritz Jesinghaus, Julia Slotta-Huspenina and Wilko Weichert
Institute of Pathology, Technical University of Munich, Munich, Germany

Stefan Knapp
Institute of Pharmaceutical Chemistry, Goethe University, Frankfurt am Main, Germany

Stephan M Feller
Weatherall Institute of Molecular Medicine, University of Oxford, Oxford, UK
Institute of Molecular Medicine, Martin-Luther-University, Halle, Germany

Bernhard Kuster
Chair of Proteomics and Bioanalytics, Technical University of Munich, Freising, Germany
German Cancer Consortium (DKTK), Munich, Germany
Center for Integrated Protein Science (CIPSM), Munich, Germany
Bavarian Biomolecular Mass Spectrometry Center (BayBioMS), Freising, Germany

Georg Kustatscher
Wellcome Trust Centre for Cell Biology, University of Edinburgh, Edinburgh, UK

Piotr Grabowski
Chair of Bioanalytics, Institute of Biotechnology, Technische Universität Berlin, Berlin, Germany

Juri Rappsilber
Wellcome Trust Centre for Cell Biology, University of Edinburgh, Edinburgh, UK
Chair of Bioanalytics, Institute of Biotechnology, Technische Universität Berlin, Berlin, Germany

Clement Gallay, Jelle Slager, Robin A Sorg, Arnau Domenech and Sebastiaan Pvan Kessel
Molecular Genetics Group, Groningen Biomolecular Sciences and Biotechnology Institute, Centre for Synthetic Biology, University of Groningen, Groningen, The Netherlands
Department of Chemistry, Biotechnology and Food Science, Norwegian University of Life Sciences, Ås, Norway

Jing-Ren Zhang
Center for Infectious Disease Research, School of Medicine, Tsinghua University, Beijing, China

Xue Liu
Molecular Genetics Group, Groningen Biomolecular Sciences and Biotechnology Institute, Centre for Synthetic Biology, University of Groningen, Groningen, The Netherlands
Center for Infectious Disease Research, School of Medicine, Tsinghua University, Beijing, China

Morten Kjos
Molecular Genetics Group, Groningen Biomolecular Sciences and Biotechnology Institute, Centre for Synthetic Biology, University of Groningen, Groningen, The Netherlands
Department of Chemistry, Biotechnology and Food Science, Norwegian University of Life Sciences, Ås, Norway

Kèvin Knoops
Molecular Cell Biology, Groningen Biomolecular Sciences and Biotechnology Institute, University of Groningen, Groningen, The Netherlands

Jan-Willem Veening
Molecular Genetics Group, Groningen Biomolecular Sciences and Biotechnology Institute, Centre for Synthetic Biology, University of Groningen, Groningen, The Netherlands
Department of Fundamental Microbiology, Faculty of Biology and Medicine, University of Lausanne, Lausanne, Switzerland

Luke S Tain, Chirag Jain, Paul Essers, Mark Rassner, Sebastian Grönke and Jenny Froelich
Max-Planck Institute for Biology of Ageing, Cologne, Germany

Manopriya Chokkalingam
CECAD Cologne Excellence Cluster on Cellular Stress Responses in Aging Associated Diseases, Cologne, Germany

Matthias Mann and Nagarjuna Nagaraj
Department of Proteomics and Signal Transduction, Max-Planck-Institute of Biochemistry, Martinsried, Germany

Robert Sehlke
Max-Planck Institute for Biology of Ageing, Cologne, Germany
CECAD Cologne Excellence Cluster on Cellular Stress Responses in Aging Associated Diseases, Cologne, Germany

Christoph Dieterich
Section of Bioinformatics and Systems Cardiology, Department of Internal Medicine III and Klaus Tschira Institute for Integrative Computational Cardiology, University of Heidelberg, Heidelberg, Germany
DZHK (German Centre for Cardiovascular Research), Partner site Heidelberg/Mannheim, Heidelberg, Germany

Nazif Alic
Institute of Healthy Ageing, and GEE, UCL, London, UK

Linda Partridge
Max-Planck Institute for Biology of Ageing, Cologne, Germany
Institute of Healthy Ageing, and GEE, UCL, London, UK

Andreas Beyer
CECAD Cologne Excellence Cluster on Cellular Stress Responses in Aging Associated Diseases, Cologne, Germany
Center for Molecular Medicine Cologne (CMMC), University of Cologne, Cologne, Germany

Mohammad Fallahi-Sichani, Verena Becker, Gregory J Baker and Sarah A Boswell
Department of Systems Biology, Program in Therapeutic Sciences, Harvard Medical School, Boston, MA, USA

Parin Shah and Asaf Rotem
Department of Medical Oncology, Dana–Farber Cancer Institute, Boston, MA, USA

Benjamin Izar
Department of Medical Oncology, Dana–Farber Cancer Institute, Boston, MA, USA
Broad Institute of Harvard and MIT, Cambridge, MA, USA

Jia-Ren Lin
HMS LINCS Center and Laboratory of Systems Pharmacology, Harvard Medical School, Boston, MA, USA

Levi A Garraway
Department of Medical Oncology, Dana–Farber Cancer Institute, Boston, MA, USA
Broad Institute of Harvard and MIT, Cambridge, MA, USA
Ludwig Center at Harvard, Harvard Medical School, Boston, MA, USA

Peter K Sorger
Department of Systems Biology, Program in Therapeutic Sciences, Harvard Medical School, Boston, MA, USA
HMS LINCS Center and Laboratory of Systems Pharmacology, Harvard Medical School, Boston, MA, USA
Ludwig Center at Harvard, Harvard Medical School, Boston, MA, USA

Kai-Yuan Chen and Nikolai Rakhilin
School of Electrical and Computer Engineering, Cornell University, Ithaca, NY, USA
Department of Biomedical Engineering, Duke University, Durham, NC, USA

Jiahn Choi
Department of Biomedical Engineering, Duke University, Durham, NC, USA

Sarah King, Tara Srinivasan and Nozomi Nishimura
Department of Biomedical Engineering, Cornell University, Ithaca, NY, USA

Kuei-Ling Tung and Anastasia Kristine Varanko
Department of Biological and Environmental Engineering, Cornell University, Ithaca, NY, USA

Lihua Wang
Department of Biomedical Engineering, Duke University, Durham, NC, USA
Department of Biological and Environmental Engineering, Cornell University, Ithaca, NY, USA

Julio M Belmonte and James A Glazier
Biocomplexity Institute and Department of Physics, Indiana University, Bloomington, IN, USA

Preetish Kadur Lakshminarasimha Murthy and Mavee Witherspoon
School of Mechanical Aerospace Engineering, Cornell University, Ithaca, NY, USA

Steven M Lipkin
Departments of Medicine, Genetic Medicine and Surgery, Weill Cornell Medical College, New York, NY, USA

Pengcheng Bu
Department of Biomedical Engineering, Duke University, Durham, NC, USA
Key Laboratory of RNA Biology, Key Laboratory of Protein and Peptide Pharmaceutical, Institute of Biophysics, Chinese Academy of Sciences, Beijing, China

Xiling Shen
School of Electrical and Computer Engineering, Cornell University, Ithaca, NY, USA
Department of Biomedical Engineering, Duke University, Durham, NC, USA
Department of Biomedical Engineering, Cornell University, Ithaca, NY, USA

Lilija Brant, Theodore Georgomanolis, Milos Nikolic and Argyris Papantonis
Center for Molecular Medicine, University of Cologne, Cologne, Germany

Davide Marenduzzo and Chris A Brackley
School of Physics and Astronomy, University of Edinburgh, Edinburgh, UK

Petros Kolovos and Frank G Grosveld
Department of Cell Biology, Erasmus Medical Center, Rotterdam, The Netherlands

Wilfred van Ijcken
Biomics Department, Erasmus Medical Center, Rotterdam, The Netherlands

Eunyong Ahn
Department of Computer Science, Technion, Haifa, Israel

Praveen Kumar and Dzmitry Mukha
Department of Biology, Technion, Haifa, Israel

Amit Tzur
Faculty of Life Sciences, Bar-Ilan University, Ramat Gan, Israel
The Institute of Nanotechnology and Advanced Materials, Bar-Ilan University, Ramat Gan, Israel

Tomer Shlomi
Department of Computer Science, Technion, Haifa, Israel
Department of Biology, Technion, Haifa, Israel
Lokey Center for Life Science and Engineering, Technion, Haifa, Israel

Index

www.ingramcontent.com/pod-product-compliance
Lightning Source LLC
Chambersburg PA
CBHW080529200326

41458CB00012B/4384